Categorical Data Analysis Using SAS®

Third Edition

Maura E. Stokes
Charles S. Davis
Gary G. Koch

§.sas.

The correct bibliographic citation for this manual is as follows: Stokes, Maura E., Charles S. Davis, and Gary G. Koch. 2012. *Categorical Data Analysis Using SAS®, Third Edition*. Cary, NC: SAS Institute Inc.

Categorical Data Analysis Using SAS®, Third Edition

Copyright © 2012, SAS Institute Inc., Cary, NC, USA

ISBN 978-1-61290-090-2 (electronic book)
ISBN 978-1-60764-664-8

All rights reserved. Produced in the United States of America.

For a hard-copy book: No part of this publication may be reproduced, stored in a retrieval system, or transmitted, in any form or by any means, electronic, mechanical, photocopying, or otherwise, without the prior written permission of the publisher, SAS Institute Inc.

For a Web download or e-book: Your use of this publication shall be governed by the terms established by the vendor at the time you acquire this publication.

The scanning, uploading, and distribution of this book via the Internet or any other means without the permission of the publisher is illegal and punishable by law. Please purchase only authorized electronic editions and do not participate in or encourage electronic piracy of copyrighted materials. Your support of others' rights is appreciated.

U.S. Government Restricted Rights Notice: Use, duplication, or disclosure of this software and related documentation by the U.S. government is subject to the Agreement with SAS Institute and the restrictions set forth in FAR 52.227-19, Commercial Computer Software-Restricted Rights (June 1987).

SAS Institute Inc., SAS Campus Drive, Cary, North Carolina 27513-2414

1st printing, July 2012

SAS Institute Inc. provides a complete selection of books and electronic products to help customers use SAS software to its fullest potential. For more information about our e-books, e-learning products, CDs, and hard-copy books, visit the SAS Books Web site at **support.sas.com/bookstore** or call 1-800-727-3228.

SAS® and all other SAS Institute Inc. product or service names are registered trademarks or trademarks of SAS Institute Inc. in the USA and other countries. ® indicates USA registration.

Other brand and product names are registered trademarks or trademarks of their respective companies.

Contents

Chapter 1.	Introduction	1
Chapter 2.	The 2×2 Table	15
Chapter 3.	Sets of 2×2 Tables	47
Chapter 4.	$2 \times r$ and $s \times 2$ Tables	73
Chapter 5.	The $s \times r$ Table	107
Chapter 6.	Sets of $s \times r$ Tables	141
Chapter 7.	Nonparametric Methods	175
Chapter 8.	Logistic Regression I: Dichotomous Response	189
Chapter 9.	Logistic Regression II: Polytomous Response	259
Chapter 10.	Conditional Logistic Regression	297
Chapter 11.	Quantal Response Data Analysis	345
Chapter 12.	Poisson Regression and Related Loglinear Models	373
Chapter 13.	Categorized Time-to-Event Data	409
Chapter 14.	Weighted Least Squares	427
Chapter 15.	Generalized Estimating Equations	487
References		557
Index		573

Preface to the Third Edition

The third edition accomplishes several purposes. First, it updates the use of SAS® software to current practices. Since the last edition was published more than 10 years ago, numerous sets of example statements have been modified to reflect best applications of SAS/STAT® software.

Second, the material has been expanded to take advantage of the many graphs now provided by SAS/STAT software through ODS Graphics. Beginning with SAS/STAT 9.3, these graphs are available with SAS/STAT—no other product license is required (a SAS/GRAPH® license was required for previous releases). Graphs displayed in this edition include:

- mosaic plots
- effect plots
- odds ratio plots
- predicted cumulative proportions plot
- regression diagnostic plots
- agreement plots

Third, the book has been updated and reorganized to reflect the evolution of categorical data analysis strategies. The previous Chapter 14, "Repeated Measurements Using Weighted Least Squares," has been combined with the previous Chapter 13, "Weighted Least Squares," to create the current Chapter 14, "Weighted Least Squares." The material previously in Chapter 16, "Loglinear Models," is found in the current Chapter 12, "Poisson Regression and Related Loglinear Models." The material in Chapter 10, "Conditional Logistic Regression," has been rewritten, and Chapter 8, "Logistic Regression I: Dichotomous Response," and Chapter 9, "Logistic Regression II: Polytomous Response," have been expanded. In addition, the previous Chapter 16, "Categorized Time-to-Event Data" is the current Chapter 13.

Numerous additional techniques are covered in this edition, including:

- incidence density ratios and their confidence intervals
- additional confidence intervals for difference of proportions
- exact Poisson regression
- difference measures to reflect direction of association in sets of tables
- partial proportional odds model
- use of the QIC statistic in GEE analysis

- odds ratios in the presence of interactions
- Firth penalized likelihood approach for logistic regression

In addition, miscellaneous revisions and additions have been incorporated throughout the book. However, the scope of the book remains the same as described in Chapter 1, "Introduction."

Computing Details

The examples in this third edition were executed with SAS/STAT 12.1, although the revision was largely based on SAS/STAT 9.3. The features specific to SAS/STAT 12.1 are:

- mosaic plots in the FREQ procedure
- partial proportional odds model in the LOGISTIC procedure
- Miettinen-Nurminen confidence limits for proportion differences in PROC FREQ
- headings for the estimates from the FIRTH option in PROC LOGISTIC

Because of limited space, not all of the output that is produced with the example SAS code is shown. Generally, only the output pertinent to the discussion is displayed. An ODS SELECT statement is sometimes used in the example code to limit the tables produced. The ODS GRAPHICS ON and ODS GRAPHICS OFF statements are used when graphs are produced. However, these statements are not needed when graphs are produced as part of the SAS windowing environment beginning with SAS 9.3. Also, the graphs produced for this book were generated with the STYLE=JOURNAL option of ODS because the book does not feature color.

For More Information

The website `http://www.sas.com/catbook` contains further information that pertains to topics in the book, including data (where possible) and errata.

Acknowledgments

We are grateful to the many people who have contributed to this revision. Bob Derr, Amy Herring, Michael Hussey, Diana Lam, Siying Li, Michela Osborn, Ashley Lauren Paynter, Margaret Polinkovsky, John Preisser, David Schlotzhauer, Todd Schwartz, Valerie Smith, Daniela Soltres-Alvarez, Donna Watts, Catherine Wiener, Laura Elizabeth Weiner, and Laura Zhou provided reviews, suggestions, proofing, and numerous other contributions that are greatly appreciated.

And, of course, we remain thankful to those persons who contributed to the earlier editions. They include Diane Catellier, Sonia Davis, Bob Derr, William Duckworth II, Suzanne Edwards, Stuart Gansky, Greg Goodwin, Wendy Greene, Duane Hayes, Allison Kinkead, Gordon Johnston, Lisa LaVange, Antonio Pedroso-de-Lima, Annette Sanders, John Preisser, David Schlotzhauer, Todd Schwartz, Dan Spitzner, Catherine Tangen, Lisa Tomasko, Donna Watts, Greg Weier, and Ozkan Zengin.

Anne Baxter and Ed Huddleston edited this book.

Tim Arnold provided documentation programming support.

Chapter 1

Introduction

Contents

1.1	Overview	1
1.2	Scale of Measurement	2
1.3	Sampling Frameworks	4
1.4	Overview of Analysis Strategies	5
	1.4.1 Randomization Methods	6
	1.4.2 Modeling Strategies	6
1.5	Working with Tables in SAS Software	8
1.6	Using This Book	13

1.1 Overview

Data analysts often encounter response measures that are categorical in nature; their outcomes reflect categories of information rather than the usual interval scale. Frequently, categorical data are presented in tabular form, known as contingency tables. Categorical data analysis is concerned with the analysis of categorical response measures, regardless of whether any accompanying explanatory variables are also categorical or are continuous. This book discusses hypothesis testing strategies for the assessment of association in contingency tables and sets of contingency tables. It also discusses various modeling strategies available for describing the nature of the association between a categorical response measure and a set of explanatory variables.

An important consideration in determining the appropriate analysis of categorical variables is their scale of measurement. Section 1.2 describes the various scales and illustrates them with data sets used in later chapters. Another important consideration is the sampling framework that produced the data; it determines the possible analyses and the possible inferences. Section 1.3 describes the typical sampling frameworks and their ramifications. Section 1.4 introduces the various analysis strategies discussed in this book and describes how they relate to one another. It also discusses the target populations generally assumed for each type of analysis and what types of inferences you are able to make to them. Section 1.5 reviews how SAS software handles contingency tables and other forms of categorical data. Finally, Section 1.6 provides a guide to the material in the book for various types of readers, including indications of the difficulty level of the chapters.

1.2 Scale of Measurement

The scale of measurement of a categorical response variable is a key element in choosing an appropriate analysis strategy. By taking advantage of the methodologies available for the particular scale of measurement, you can choose a well-targeted strategy. If you do not take the scale of measurement into account, you may choose an inappropriate strategy that could lead to erroneous conclusions. Recognizing the scale of measurement and using it properly are very important in categorical data analysis.

Categorical response variables can be

- dichotomous
- ordinal
- nominal
- discrete counts
- grouped survival times

Dichotomous responses are those that have two possible outcomes—most often they are yes and no. Did the subject develop the disease? Did the voter cast a ballot for the Democratic or Republican candidate? Did the student pass the exam? For example, the objective of a clinical trial for a new medication for colds is whether patients obtained relief from their pain-producing ailment. Consider Table 1.1, which is analyzed in Chapter 2, "The 2 × 2 Table."

Table 1.1 Respiratory Outcomes

Treatment	Favorable	Unfavorable	Total
Placebo	16	48	64
Test	40	20	60

The placebo group contains 64 patients, and the test medication group contains 60 patients. The columns contain the information concerning the categorical response measure: 40 patients in the Test group had a favorable response to the medication, and 20 subjects did not. The outcome in this example is thus dichotomous, and the analysis investigates the relationship between the response and the treatment.

Frequently, categorical data responses represent more than two possible outcomes, and often these possible outcomes take on some inherent ordering. Such response variables have an *ordinal* scale of measurement. Did the new school curriculum produce little, some, or high enthusiasm among the students? Does the water exhibit low, medium, or high hardness? In the former case, the order of the response levels is clear, but there is no clue as to the relative distances between the levels. In the latter case, there is a possible distance between the levels: medium might have twice the hardness of low, and high might have three times the hardness of low. Sometimes the distance is even clearer: a 50% potency dose versus a 100% potency dose versus a 200% potency dose. All three cases are examples of ordinal data.

An example of an ordinal measure occurs in data displayed in Table 1.2, which is analyzed in Chapter 9, "Logistic Regression II: Polytomous Response." A clinical trial investigated a treatment for rheumatoid arthritis. Male and female patients were given either the active treatment or a placebo; the outcome measured was whether they showed marked, some, or no improvement at the end of the clinical trial. The analysis uses the proportional odds model to assess the relationship between the response variable and gender and treatment.

Table 1.2 Arthritis Data

Sex	Treatment	Improvement			Total
		Marked	Some	None	
Female	Active	16	5	6	27
Female	Placebo	6	7	19	32
Male	Active	5	2	7	14
Male	Placebo	1	0	10	11

Note that categorical response variables can often be managed in different ways. You could combine the Marked and Some columns in Table 1.2 to produce a dichotomous outcome: No Improvement versus Improvement. Grouping categories is often done during an analysis if the resulting dichotomous response is also of interest.

If you have more than two outcome categories, and there is no inherent ordering to the categories, you have a *nominal* measurement scale. Which of four candidates did you vote for in the town council election? Do you prefer the beach, mountains, or lake for a vacation? There is no underlying scale for such outcomes and no apparent way in which to order them.

Consider Table 1.3, which is analyzed in Chapter 5, "The $s \times r$ Table." Residents in one town were asked their political party affiliation and their neighborhood. Researchers were interested in the association between political affiliation and neighborhood. Unlike ordinal response levels, the classifications Bayside, Highland, Longview, and Sheffeld lie on no conceivable underlying scale. However, you can still assess whether there is association in the table, which is done in Chapter 5.

Table 1.3 Distribution of Parties in Neighborhoods

Party	Neighborhood			
	Bayside	Highland	Longview	Sheffeld
Democrat	221	160	360	140
Independent	200	291	160	311
Republican	208	106	316	97

Categorical response variables sometimes contain *discrete counts*. Instead of falling into categories that are labeled (yes, no) or (low, medium, high), the outcomes are numbers themselves. Was the litter size 1, 2, 3, 4, or 5 members? Did the house contain 1, 2, 3, or 4 air conditioners? While the usual strategy would be to analyze the mean count, the assumptions required for the standard linear model for continuous data are often not met with discrete counts that have small range; the counts are not distributed normally and may not have homogeneous variance.

For example, researchers examining respiratory disease in children visited children in different regions two times and determined whether they showed symptoms of respiratory illness. The response measure was whether the children exhibited symptoms in 0, 1, or 2 periods. Table 1.4 contains these data, which are analyzed in Chapter 14, "Weighted Least Squares."

Table 1.4 Colds in Children

Sex	Residence	Periods with Colds			Total
		0	1	2	
Female	Rural	45	64	71	180
Female	Urban	80	104	116	300
Male	Rural	84	124	82	290
Male	Urban	106	117	87	310

The table represents a cross-classification of gender, residence, and number of periods with colds. The analysis is concerned with modeling mean colds as a function of gender and residence.

Finally, another type of response variable in categorical data analysis is one that represents *survival times*. With survival data, you are tracking the number of patients with certain outcomes (possibly death) over time. Often, the times of the condition are grouped together so that the response variable represents the number of patients who fail during a specific time interval. Such data are called *grouped survival times*. For example, the data displayed in Table 1.5 are from Chapter 13, "Categorized Time-to-Event Data." A clinical condition is treated with an active drug for some patients and with a placebo for others. The response categories are whether there are recurrences, no recurrences, or whether the patients withdrew from the study. The entries correspond to the time intervals 0–1 years, 1–2 years, and 2–3 years, which make up the rows of the table.

Table 1.5 Life Table Format for Clinical Condition Data

Controls				
Interval	No Recurrences	Recurrences	Withdrawals	At Risk
0–1 Years	50	15	9	74
1–2 Years	30	13	7	50
2–3 Years	17	7	6	30
Active				
Interval	No Recurrences	Recurrences	Withdrawals	At Risk
0–1 Years	69	12	9	90
1–2 Years	59	7	3	69
2–3 Years	45	10	4	59

1.3 Sampling Frameworks

Categorical data arise from different sampling frameworks. The nature of the sampling framework determines the assumptions that can be made for the statistical analyses and in turn influences the type of analysis that can be applied. The sampling framework also determines the type of inference that is possible. Study populations are limited to target populations, those populations to which inferences can be made, by assumptions justified by the sampling framework.

Generally, data fall into one of three sampling frameworks: historical data, experimental data, and sample survey data. *Historical data* are observational data, which means that the study population has a geographic or circumstantial definition. These may include all the occurrences of an infectious disease in a multicounty area, the children attending a particular elementary school, or those persons appearing in court during a specified time period. Highway safety data concerning injuries in motor vehicles is another example of historical data.

Experimental data are drawn from studies that involve the random allocation of subjects to different treatments of one sort or another. Examples include studies where types of fertilizer are applied to agricultural plots and studies where subjects are administered different dosages of drug therapies. In the health sciences, experimental data may include patients randomly administered a placebo or treatment for their medical condition.

In *sample survey studies*, subjects are randomly chosen from a larger study population. Investigators may randomly choose students from their school IDs and survey them about social behavior; national health care studies may randomly sample Medicare users and investigate physician utilization patterns. In addition, some sampling designs may be a combination of sample survey and experimental data processes. Researchers may randomly select a study population and then randomly assign treatments to the resulting study subjects.

The major difference in the three sampling frameworks described in this section is the use of randomization to obtain them. Historical data involve no randomization, and so it is often difficult to assume that they are representative of a convenient population. Experimental data have good coverage of the possibilities of alternative treatments for the restricted protocol population, and sample survey data have very good coverage of the larger population from which they were selected.

Note that the unit of randomization can be a single subject or a cluster of subjects. In addition, randomization may be applied within subsets, called strata or blocks, with equal or unequal probabilities. In sample surveys, all of this can lead to more complicated designs, such as stratified random samples, or even multistage cluster random samples. In experimental design studies, such considerations lead to repeated measurements (or split-plot) studies.

1.4 Overview of Analysis Strategies

Categorical data analysis strategies can be classified into those that are concerned with hypothesis testing and those that are concerned with modeling. Many questions about a categorical data set can be answered by addressing a specific hypothesis concerning association. Such hypotheses are often investigated with randomization methods. In addition to making statements about association, you may also want to describe the nature of the association in the data set. Statistical modeling techniques using maximum likelihood estimation or weighted least squares estimation are employed to describe patterns of association or variation in terms of a parsimonious statistical model. Imrey (2011) includes a historical perspective on numerous methods described in this book.

Most often the hypothesis of interest is whether association exists between the rows of a contingency table and its columns. The only assumption that is required is randomized allocation of subjects, either through the study design (experimental design) or through the hypothesis itself (necessary for historical data). In addition, particularly for the use of historical data, you often want to control for other explanatory variables that may have influenced the observed outcomes.

1.4.1 Randomization Methods

Table 1.1, the respiratory outcomes data, contains information obtained as part of a randomized allocation process. The hypothesis of interest is whether there is an association between treatment and outcome. For these data, the randomization is accomplished by the study design.

Table 1.6 contains data from a similar study. The main difference is that the study was conducted in two medical centers. The hypothesis of association is whether there is an association between treatment and outcome, controlling for any effect of center.

Table 1.6 Respiratory Improvement

Center	Treatment	Yes	No	Total
1	Test	29	16	45
1	Placebo	14	31	45
Total		43	47	90
2	Test	37	8	45
2	Placebo	24	21	45
Total		61	29	90

Chapter 2, "The 2 × 2 Table," is primarily concerned with the association in 2 × 2 tables; in addition, it discusses measures of association, that is, statistics designed to evaluate the strength of the association. Chapter 3, "Sets of 2 × 2 Tables," discusses the investigation of association in sets of 2 × 2 tables. When the table of interest has more than two rows and two columns, the analysis is further complicated by the consideration of scale of measurement. Chapter 4, "Sets of 2 × r and s × 2 Tables," considers the assessment of association in sets of tables where the rows (columns) have more than two levels.

Chapter 5 describes the assessment of association in the general $s \times r$ table, and Chapter 6, "Sets of $s \times r$ Tables," describes the assessment of association in sets of $s \times r$ tables. The investigation of association in tables and sets of tables is further discussed in Chapter 7, "Nonparametric Methods," which discusses traditional nonparametric tests that have counterparts among the strategies for analyzing contingency tables.

Another consideration in data analysis is whether you have enough data to support the asymptotic theory required for many tests. Often, you may have an overall table sample size that is too small or a number of zero or small cell counts that make the asymptotic assumptions questionable. Recently, exact methods have been developed for a number of association statistics that permit you to address the same hypotheses for these types of data. The above-mentioned chapters illustrate the use of exact methods for many situations.

1.4.2 Modeling Strategies

Often, you are interested in describing the variation of your response variable in your data with a statistical model. In the continuous data setting, you frequently fit a model to the expected mean response. However, with categorical outcomes, there are a variety of response functions that you can model. Depending on the response function that you choose, you can use weighted least squares or maximum likelihood methods to estimate the model parameters.

Perhaps the most common response function modeled for categorical data is the logit. If you have a dichotomous response and represent the proportion of those subjects with an event (versus no event) outcome as p, then the logit can be written

$$\log\left(\frac{p}{1-p}\right)$$

Logistic regression is a modeling strategy that relates the logit to a set of explanatory variables with a linear model. One of its benefits is that estimates of odds ratios, important measures of association, can be obtained from the parameter estimates. Maximum likelihood estimation is used to provide those estimates.

Chapter 8, "Logistic Regression I: Dichotomous Response," discusses logistic regression for a dichotomous outcome variable. Chapter 9, "Logistic Regression II: Polytomous Response," discusses logistic regression for the situation where there are more than two outcomes for the response variable. Logits called *generalized logits* can be analyzed when the outcomes are nominal. And logits called *cumulative logits* can be analyzed when the outcomes are ordinal. Chapter 10, "Conditional Logistic Regression," describes a specialized form of logistic regression that is appropriate when the data are highly stratified or arise from matched case-control studies. These chapters describe the use of exact conditional logistic regression for those situations where you have limited or sparse data, and the asymptotic requirements for the usual maximum likelihood approach are not met.

Poisson regression is a modeling strategy that is suitable for discrete counts, and it is discussed in Chapter 12, "Poisson Regression and Related Loglinear Models." Most often the log of the count is used as the response function.

Some application areas have features that led to the development of special statistical techniques. One of these areas for categorical data is bioassay analysis. Bioassay is the process of determining the potency or strength of a reagent or stimuli based on the response it elicits in biological organisms. Logistic regression is a technique often applied in bioassay analysis, where its parameters take on specific meaning. Chapter 11, "Quantal Bioassay Analysis," discusses the use of categorical data methods for quantal bioassay. Another special application area for categorical data analysis is the analysis of grouped survival data. Chapter 13, "Categorized Time-to-Event Data," discusses some features of survival analysis that are pertinent to grouped survival data, including how to model them with the piecewise exponential model.

In logistic regression, the objective is to predict a response outcome from a set of explanatory variables. However, sometimes you simply want to describe the structure of association in a set of variables for which there are no obvious outcome or predictor variables. This occurs frequently for sociological studies. The loglinear model is a traditional modeling strategy for categorical data and is appropriate for describing the association in such a set of variables. It is closely related to logistic regression, and the parameters in a loglinear model are also estimated with maximum likelihood estimation. Chapter 12, "Poisson Regression and Related Loglinear Models," includes a discussion of the loglinear model, including a typical application.

Besides the logit and log counts, other useful response functions that can be modeled include proportions, means, and measures of association. Weighted least squares estimation is a method of analyzing such response functions, based on large sample theory. These methods are appropriate when you have sufficient sample size and when you have a randomly selected sample, either directly through study design or indirectly via assumptions concerning the representativeness of the data. Not only can you model a variety of useful functions, but weighted least squares estimation also provides a useful framework for the analysis of repeated categorical measurements, particularly those limited to a small number of repeated values. Chapter 14, "Weighted Least Squares," addresses modeling categorical data with weighted least squares methods, including the analysis of repeated measurements data.

Generalized estimating equations (GEE) is a widely used method for the analysis of correlated responses, particularly for the analysis of categorical repeated measurements. The GEE method applies to a broad range of repeated measurements situations, such as those including time-dependent covariates and continuous explanatory variables, that weighted least squares doesn't handle. Chapter 15, "Generalized Estimating Equations," discusses the GEE approach and illustrates its application with a number of examples.

1.5 Working with Tables in SAS Software

This section discusses some considerations of managing tables with SAS. If you are already familiar with the FREQ procedure, you may want to skip this section.

Many times, categorical data are presented to the researcher in the form of tables, and other times, they are presented in the form of case record data. SAS procedures can handle either type of data. In addition, many categorical data have ordered categories, so that the order of the levels of the rows and columns takes on special meaning. There are numerous ways that you can specify a particular order to SAS procedures.

Consider the following SAS DATA step that inputs the data displayed in Table 1.1.

```
data respire;
   input treat $ outcome $ count;
   datalines;
placebo  f 16
placebo  u 48
test     f 40
test     u 20
;

proc freq;
   weight count;
   tables treat*outcome;
run;
```

The data set RESPIRE contains three variables: TREAT is a character variable containing values for treatment, OUTCOME is a character variable containing values for the outcome (f for favorable and u for unfavorable), and COUNT contains the number of observations that have the respective TREAT and OUTCOME values. Thus, COUNT effectively takes values corresponding to the cells of Table 1.1. The PROC FREQ statements request that a table be constructed using TREAT as the row variable and OUTCOME as the column variable. By default, PROC FREQ orders the values of the rows (columns) in alphanumeric order. The WEIGHT statement is necessary to tell the procedure that the data are count data, or frequency data; the variable listed in the WEIGHT statement contains the values of the count variable.

Output 1.1 contains the resulting frequency table.

Output 1.1 Frequency Table

Frequency Percent Row Pct Col Pct	Table of treat by outcome		
		outcome	
treat	f	u	Total
placebo	16 12.90 25.00 28.57	48 38.71 75.00 70.59	64 51.61
test	40 32.26 66.67 71.43	20 16.13 33.33 29.41	60 48.39
Total	56 45.16	68 54.84	124 100.00

Suppose that a different sample produced the numbers displayed in Table 1.7.

Table 1.7 Respiratory Outcomes

Treatment	Favorable	Unfavorable	Total
Placebo	5	10	15
Test	8	20	28

These data may be stored in case record form, which means that each individual is represented by a single observation. You can also use this type of input with the FREQ procedure. The only difference is that the WEIGHT statement is not required.

The following statements create a SAS data set for these data and invoke PROC FREQ for case record data. The @@ symbol in the INPUT statement means that the data lines contain multiple observations.

```
data respire;
   input treat $ outcome $ @@;
   datalines;
placebo f placebo f  placebo f
placebo f placebo f
placebo u placebo u  placebo u
placebo u placebo u  placebo u
placebo u placebo u  placebo u
placebo u
test    f test    f test    f
test    f test    f test    f
test    f test    f
test    u test    u test    u
```

```
test     u  test    u  test    u
test     u  test    u  test    u
test     u  test    u  test    u
test     u  test    u  test    u
test     u  test    u  test    u
test     u  test    u
;

proc freq;
   tables treat*outcome;
run;
```

Output 1.2 displays the resulting frequency table.

Output 1.2 Frequency Table

Frequency Percent Row Pct Col Pct	Table of treat by outcome		
		outcome	
treat	f	u	Total
placebo	5 11.63 33.33 38.46	10 23.26 66.67 33.33	15 34.88
test	8 18.60 28.57 61.54	20 46.51 71.43 66.67	28 65.12
Total	13 30.23	30 69.77	43 100.00

In this book, the data are generally presented in count form.

When ordinal data are considered, it becomes quite important to ensure that the levels of the rows and columns are sorted correctly. By default, the data are going to be sorted alphanumerically. If this isn't suitable, then you need to alter the default behavior.

Consider the data displayed in Table 1.2. Variable IMPROVE is the outcome, and the values marked, some, and none are listed in decreasing order. Suppose that the data set ARTHRITIS is created with the following statements.

```
data arthritis;
   length treatment $7. sex $6. ;
   input sex $ treatment $ improve $ count @@;
   datalines;
female active   marked 16 female active   some 5 female active   none  6
female placebo marked  6 female placebo some 7 female placebo none 19
male   active   marked  5 male   active   some 2 male   active   none  7
male   placebo marked  1 male   placebo some 0 male   placebo none 10
;
```

If you invoked PROC FREQ for this data set and used the default sort order, the levels of the columns would be ordered marked, none, and some, which would be incorrect. One way to change this default sort order is to use the ORDER=DATA option in the PROC FREQ statement. This specifies that the sort order is the same order in which the values are encountered in the data set. Thus, since 'marked' comes first, it is first in the sort order. Since 'some' is the second value for IMPROVE encountered in the data set, then it is second in the sort order. And 'none' would be third in the sort order. This is the desired sort order. The following PROC FREQ statements produce a table displaying the sort order resulting from the ORDER=DATA option.

```
proc freq order=data;
    weight count;
    tables treatment*improve;
run;
```

Output 1.3 displays the frequency table for the cross-classification of treatment and improvement for these data; the values for IMPROVE are in the correct order.

Output 1.3 Frequency Table from ORDER=DATA Option

Frequency Percent Row Pct Col Pct	Table of treatment by improve				
		improve			
	treatment	marked	some	none	Total
	active	21 25.00 51.22 75.00	7 8.33 17.07 50.00	13 15.48 31.71 30.95	41 48.81
	placebo	7 8.33 16.28 25.00	7 8.33 16.28 50.00	29 34.52 67.44 69.05	43 51.19
	Total	28 33.33	14 16.67	42 50.00	84 100.00

Other possible values for the ORDER= option include FORMATTED, which means sort by the formatted values. The ORDER= option is also available with the CATMOD, LOGISTIC, and GENMOD procedures. For information on the ORDER= option for the FREQ procedure, refer to the *SAS/STAT User's Guide*. This option is used frequently in this book.

Often, you want to analyze sets of tables. For example, you may want to analyze the cross-classification of treatment and improvement for both males and females. You do this in PROC FREQ by using a three-way crossing of the variables SEX, TREAT, and IMPROVE.

```
proc freq order=data;
    weight count;
    tables sex*treatment*improve / nocol nopct;
run;
```

The two rightmost variables in the TABLES statement determine the rows and columns of the table, respectively. Separate tables are produced for the unique combination of values of the other variables in the crossing. Since SEX has two levels, one table is produced for males and one table is produced for females. If there were four variables in this crossing, with the two variables on the left having two levels each, then four tables would be produced, one for each unique combination of the two leftmost variables in the TABLES statement.

Note also that the options NOCOL and NOPCT are included. These options suppress the printing of column percentages and cell percentages, respectively. Since generally you are interested in row percentages, these options are often specified in the code displayed in this book.

Output 1.4 contains the two tables produced with the preceding statements.

Output 1.4 Producing Sets of Tables

Frequency Row Pct	Table 1 of treatment by improve					
	Controlling for sex=female					
		improve			
	treatment	marked	some	none	Total
	active	16 59.26	5 18.52	6 22.22	27
	placebo	6 18.75	7 21.88	19 59.38	32
	Total	22	12	25	59

Frequency Row Pct	Table 2 of treatment by improve					
	Controlling for sex=male					
		improve			
	treatment	marked	some	none	Total
	active	5 35.71	2 14.29	7 50.00	14
	placebo	1 9.09	0 0.00	10 90.91	11
	Total	6	2	17	25

This section reviewed some of the basic table management necessary for using the FREQ procedure. Other related options are discussed in the appropriate chapters.

1.6 Using This Book

This book is intended for a variety of audiences, including novice readers with some statistical background (solid understanding of regression analysis), those readers with substantial statistical background, and those readers with a background in categorical data analysis. Therefore, not all of this material will have the same importance to all readers. Some chapters include a good deal of tutorial material, while others have a good deal of advanced material. This book is not intended to be a comprehensive treatment of categorical data analysis, so some topics are mentioned briefly for completeness and some other topics are emphasized because they are not well documented.

The data used in this book come from a variety of sources and represent a wide breadth of application. However, due to the biostatistical background of all three authors, there is a certain inevitable weighting of biostatistical examples. Most of the data come from practice, and the original sources are cited when this is true; however, due to confidentiality concerns and pedagogical requirements, some of the data are altered or created. However, they still represent realistic situations.

Chapters 2–4 are intended to be accessible to all readers, as is most of Chapter 5. Chapter 6 is an integration of Mantel-Haenszel methods at a more advanced level, but scanning it is probably a good idea for any reader interested in the topic. In particular, the discussion about the analysis of repeated measurements data with extended Mantel-Haenszel methods is useful material for all readers comfortable with the Mantel-Haenszel technique.

Chapter 7 is a special interest chapter relating Mantel-Haenszel procedures to traditional nonparametric methods used for continuous data outcomes.

Chapters 8 and 9 on logistic regression are intended to be accessible to all readers, particularly Chapter 8. The last section of Chapter 8 describes the statistical methodology more completely for the advanced reader. Most of the material in Chapter 9 should be accessible to most readers. Chapter 10 is a specialized chapter that discusses conditional logistic regression and requires somewhat more statistical expertise. Chapter 11 discusses the use of logistic regression in analyzing bioassay data.

Parts of the subsequent chapters discuss more advanced topics and are necessarily written at a higher statistical level. Chapter 12 describes Poisson regression and loglinear models; much of the Poisson regression should be fairly accessible but the loglinear discussion is somewhat advanced. Chapter 13 discusses the analysis of categorized time-to-event data and most of it should be fairly accessible.

Chapter 14 discusses weighted least squares and is written at a somewhat higher statistical level than Chapters 8 and 9, but most readers should find this material useful, particularly the examples. Chapter 15 describes the use of generalized estimating equations. The opening section includes a basic example that is intended to be accessible to a wide range of readers.

All of the examples were executed with SAS/STAT 12.1, and the few exceptions where options and results are only available with SAS/STAT 12.1 are noted in the "Preface to the Third Edition."

Chapter 2
The 2×2 Table

Contents

2.1	Introduction	**15**
2.2	Chi-Square Statistics	**17**
2.3	Exact Tests	**20**
	2.3.1 Exact p-values for Chi-Square Statistics	23
2.4	Difference in Proportions	**25**
2.5	Odds Ratio and Relative Risk	**31**
	2.5.1 Exact Confidence Limits for the Odds Ratio	38
2.6	Sensitivity and Specificity	**39**
2.7	McNemar's Test	**41**
2.8	Incidence Densities	**43**
2.9	Sample Size and Power Computations	**46**

2.1 Introduction

The 2×2 contingency table is one of the most common ways to summarize categorical data. Categorizing patients by their favorable or unfavorable response to two different drugs, asking health survey participants whether they have regular physicians and regular dentists, and asking residents of two cities whether they desire more environmental regulations all result in data that can be summarized in a 2×2 table.

Generally, interest lies in whether there is an association between the row variable and the column variable that produce the table; sometimes there is further interest in describing the strength of that association. The data can arise from several different sampling frameworks, and the interpretation of the hypothesis of no association depends on the framework. Data in a 2×2 table can represent the following:

- simple random samples from two groups that yield two independent binomial distributions for a binary response

    Asking residents from two cities whether they desire more environmental regulations is an example of this framework. This is a stratified random sampling setting, since the subjects from each city represent two independent random samples. Because interest lies in whether the proportion favoring regulation is the same for the two cities, the hypothesis of interest is the hypothesis of homogeneity. Is the distribution of the response the same in both groups?

- a simple random sample from one group that yields a single multinomial distribution for the cross-classification of two binary responses

  Taking a random sample of subjects and asking whether they see both a regular physician and a regular dentist is an example of this framework. The hypothesis of interest is one of independence. Are having a regular dentist and having a regular physician independent of each other?

- randomized assignment of patients to two equivalent treatments, resulting in the hypergeometric distribution

  This framework occurs when patients are randomly allocated to one of two drug treatments, regardless of how they are selected, and their response to that treatment is the binary outcome. Under the null hypothesis that the effects of the two treatments are the same for each patient, a hypergeometric distribution applies to the response distributions for the two treatments.

- incidence densities for counts of subjects who responded with some event versus the extent of exposure for the event

  These counts represent independent Poisson processes. This framework occurs less frequently than the others but is still important.

Table 2.1 summarizes the information from a randomized clinical trial that compared two treatments (test and placebo) for a respiratory disorder.

Table 2.1 Respiratory Outcomes

Treatment	Favorable	Unfavorable	Total
Placebo	16	48	64
Test	40	20	60

The question of interest is whether the rates of favorable response for test (67%) and placebo (25%) are the same. You can address this question by investigating whether there is a statistical association between treatment and outcome. The null hypothesis is stated

H_0: There is no association between treatment and outcome.

There are several ways of testing this hypothesis; many of the tests are based on the chi-square statistic. Section 2.2 discusses these methods. However, sometimes the counts in the table cells are too small to meet the sample size requirements necessary for the chi-square distribution to apply, and exact methods based on the hypergeometric distribution are used to test the hypothesis of no association. Exact methods are discussed in Section 2.3.

In addition to testing the hypothesis concerning the presence of association, you may be interested in describing the association or gauging its strength. Section 2.4 discusses the estimation of the difference in proportions from 2×2 tables. Section 2.5 discusses measures of association, which assess strength of association, and Section 2.6 discusses measures called sensitivity and specificity, which are useful when the two responses correspond to two different methods for determining whether a particular disorder is present. And 2×2 tables often display data for matched pairs; Section 2.7 discusses McNemar's test for assessing association for matched pairs data. Finally, Section 2.8 discusses computing incidence density ratios when the 2×2 table represents counts from Poisson processes.

2.2 Chi-Square Statistics

Table 2.2 displays the generic 2 × 2 table, including row and column marginal totals.

Table 2.2 2 × 2 Contingency Table

Row Levels	Column 1	Column 2	Total
1	n_{11}	n_{12}	n_{1+}
2	n_{21}	n_{22}	n_{2+}
Total	n_{+1}	n_{+2}	n

Under the randomization framework that produced Table 2.1, the row marginal totals n_{1+} and n_{2+} are fixed since 60 patients were randomly allocated to one of the treatment groups and 64 to the other. The column marginal totals can be regarded as fixed under the null hypothesis of no treatment difference for each patient (since each patient would have the same response regardless of the assigned treatment, under this null hypothesis). Then, given that all of the marginal totals n_{1+}, n_{2+}, n_{+1}, and n_{+2} are fixed under the null hypothesis, the probability distribution from the randomized allocation of patients to treatment can be written

$$\Pr\{n_{ij}\} = \frac{n_{1+}! n_{2+}! n_{+1}! n_{+2}!}{n! n_{11}! n_{12}! n_{21}! n_{22}!}$$

which is the hypergeometric distribution. The expected value of n_{ij} is

$$E\{n_{ij}|H_0\} = \frac{n_{i+} n_{+j}}{n} = m_{ij}$$

and the variance is

$$V\{n_{ij}|H_0\} = \frac{n_{1+} n_{2+} n_{+1} n_{+2}}{n^2(n-1)} = v_{ij}$$

For a sufficiently large sample, n_{11} approximately has a normal distribution, which implies that

$$Q = \frac{(n_{11} - m_{11})^2}{v_{11}}$$

approximately has a chi-square distribution with one degree of freedom. It is the ratio of a squared difference from the expected value versus its variance, and such quantities follow the chi-square distribution when the variable is distributed normally. Q is often called the randomization (or Mantel-Haenszel) chi-square. It doesn't matter how the rows and columns are arranged; Q takes the same value since

$$|n_{11} - m_{11}| = |n_{ij} - m_{ij}| = \frac{|n_{11} n_{22} - n_{12} n_{21}|}{n} = \frac{n_{1+} n_{2+}}{n} |p_1 - p_2|$$

where $p_i = (n_{i1}/n_{i+})$ is the observed proportion in column 1 for the ith row.

A related statistic is the Pearson chi-square statistic. This statistic is written

$$Q_P = \sum_{i=1}^{2}\sum_{j=1}^{2} \frac{(n_{ij} - m_{ij})^2}{m_{ij}} = \frac{n}{(n-1)}Q = \frac{(p_1 - p_2)^2}{\{(1/n_{1+} + 1/n_{2+})p_+(1 - p_+)\}}$$

where $p_+ = (n_{+1}/n)$ is the proportion in column 1 for the pooled rows.

If the cell counts are sufficiently large, Q_P is distributed as chi-square with one degree of freedom. As n grows large, Q_P and Q converge. A useful rule for determining adequate sample size for both Q and Q_P is that the expected value m_{ij} should exceed 5 (and preferable 10) for all of the cells. While Q is discussed here in the framework of a randomized allocation of patients to two groups, Q and Q_P are also appropriate for investigating the hypothesis of no association for all of the sampling frameworks described previously.

The following PROC FREQ statements produce a frequency table and the chi-square statistics for the data in Table 2.1. The data are supplied in frequency (count) form. An observation is supplied for each configuration of the values of the variables TREAT and OUTCOME. The variable COUNT holds the total number of observations that have that particular configuration. The WEIGHT statement tells the FREQ procedure that the data are in frequency form and names the variable that contains the frequencies. Alternatively, the data could be provided as case records for the individual patients; with this data structure, there would be 124 data lines corresponding to the 124 patients, and neither the variable COUNT nor the WEIGHT statement would be required.

The CHISQ option in the TABLES statement produces chi-square statistics.

```
    data respire;
       input treat $ outcome $ count;
       datalines;
    placebo f 16
    placebo u 48
    test    f 40
    test    u 20
    ;

    proc freq;
       weight count;
       tables treat*outcome / chisq;
    run;
```

Output 2.1 displays the data in a 2 × 2 table. With an overall sample size of 124, and all expected cell counts greater than 10, the sampling assumptions for the chi-square statistics are met. PROC FREQ prints out a warning message when more than 20% of the cells in a table have expected counts less than 5. (You can specify the EXPECTED option in the TABLE statement to produce the expected cell counts along with the cell percentages.)

2.2. Chi-Square Statistics

Output 2.1 Frequency Table

Frequency Percent Row Pct Col Pct	Table of treat by outcome		
		outcome	
treat	f	u	Total
placebo	16 12.90 25.00 28.57	48 38.71 75.00 70.59	64 51.61
test	40 32.26 66.67 71.43	20 16.13 33.33 29.41	60 48.39
Total	56 45.16	68 54.84	124 100.00

Output 2.2 contains the table with the chi-square statistics.

Output 2.2 Chi-Square Statistics

Statistic	DF	Value	Prob
Chi-Square	1	21.7087	<.0001
Likelihood Ratio Chi-Square	1	22.3768	<.0001
Continuity Adj. Chi-Square	1	20.0589	<.0001
Mantel-Haenszel Chi-Square	1	21.5336	<.0001
Phi Coefficient		-0.4184	
Contingency Coefficient		0.3860	
Cramer's V		-0.4184	

Fisher's Exact Test	
Cell (1,1) Frequency (F)	16
Left-sided Pr <= F	2.838E-06
Right-sided Pr >= F	1.0000
Table Probability (P)	2.397E-06
Two-sided Pr <= P	4.754E-06

Sample Size = 124

The randomization statistic Q is labeled "Mantel-Haenszel Chi-Square," and the Pearson chi-square Q_P is labeled "Chi-Square." Q has a value of 21.5336 and $p < 0.0001$; Q_P has a value of 21.7087 and $p < 0.0001$. Both of these statistics are clearly significant. There is a strong association between treatment and outcome such that the test treatment results in a more favorable response outcome than the placebo. The row percentages in Output 2.1 show that the test treatment resulted in 67% favorable response and the placebo treatment resulted in 25% favorable response.

The output also includes a statistic labeled "Likelihood Ratio Chi-Square." This statistic, often written Q_L, is asymptotically equivalent to Q and Q_P. The statistic Q_L is described in Chapter 8 in the context of hypotheses for the odds ratio, for which there is some consideration in Section 2.5. Q_L is not often used in the analysis of 2 × 2 tables. Some of the other statistics are discussed in the next section.

2.3 Exact Tests

Sometimes your data include small and zero cell counts. For example, consider the data in Table 2.3 from a study on treatments for healing severe infections. Randomly assigned test treatment and control are compared to determine whether the rates of favorable response are the same.

Table 2.3 Severe Infection Treatment Outcomes

Treatment	Favorable	Unfavorable	Total
Test	10	2	12
Control	2	4	6
Total	12	6	18

Obviously, the sample size requirements for the chi-square tests described in Section 2.2 are not met by these data. However, if you can consider the margins (12, 6, 12, 6) to be fixed, then the random assignment and the null hypothesis of no association imply the hypergeometric distribution

$$\Pr\{n_{ij}\} = \frac{n_{1+}!n_{2+}!n_{+1}!n_{+2}!}{n!n_{11}!n_{12}!n_{21}!n_{22}!}$$

The row margins may be fixed by the treatment allocation process; that is, subjects are randomly assigned to Test and Control. The column totals can be regarded as fixed by the null hypothesis; there are 12 patients with favorable response and 6 patients with unfavorable response, regardless of treatment. If the data are the result of a sample of convenience, you can still condition on marginal totals being fixed by addressing the null hypothesis that the patients are interchangeable; that is, the observed distributions of outcome for the two treatments are compatible with what would be expected from random assignment. That is, all possible assignments of the outcomes for 12 of the patients to Test and for 6 to Control are equally likely.

Recall that a p-value is the probability of the observed data or more extreme data occurring under the null hypothesis. With Fisher's exact test, you determine the p-value for this table by summing the probabilities of the tables that are as likely or less likely, given the fixed margins. Table 2.4 includes all possible table configurations and their associated probabilities.

Table 2.4 Table Probabilities

\multicolumn{4}{c}{Table Cell}	
(1,1)	(1,2)	(2,1)	(2,2)	Probabilities
12	0	0	6	0.0001
11	1	1	5	0.0039
10	2	2	4	0.0533
9	3	3	3	0.2370
8	4	4	2	0.4000
7	5	5	1	0.2560
6	6	6	0	0.0498

To find the one-sided *p*-value, you sum the probabilities that are as small or smaller than those computed for the table observed, in the direction specified by the one-sided alternative. In this case, it would be those tables in which the Test treatment had the more favorable response:

$$p = 0.0533 + 0.0039 + 0.0001 = 0.0573$$

To find the two-sided *p*-value, you sum all of the probabilities that are as small or smaller than that observed, or

$$p = 0.0533 + 0.0039 + 0.0001 + 0.0498 = 0.1071$$

Generally, you are interested in the two-sided *p*-value. Note that when the row (or column) totals are nearly equal, the *p*-value for the two-sided Fisher's exact test is approximately twice the *p*-value for the one-sided Fisher's exact test for the better treatment. When the row (or column) totals are equal, the *p*-value for the two-sided Fisher's exact test is exactly twice the value of the *p*-value for the one-sided Fisher's exact test.

The following SAS statements produce the 2 × 2 frequency table for Table 2.3. Specifying the CHISQ option also produces Fisher's exact test for a 2 × 2 table. In addition, the ORDER=DATA option specifies that PROC FREQ order the levels of the rows (columns) in the same order in which the values are encountered in the data set.

```
data severe;
   input treat $ outcome $ count;
   datalines;
Test     f 10
Test     u 2
Control  f 2
Control  u 4
;

proc freq order=data;
   weight count;
   tables treat*outcome / chisq nocol;
run;
```

The NOCOL option suppresses the column percentages, as seen in Output 2.3.

Output 2.3 Frequency Table

Frequency Percent Row Pct	Table of treat by outcome		
		outcome	
treat	f	u	Total
Test	10 55.56 83.33	2 11.11 16.67	12 66.67
Control	2 11.11 33.33	4 22.22 66.67	6 33.33
Total	12 66.67	6 33.33	18 100.00

Output 2.4 contains the chi-square statistics, including the exact test. Note that the sample size assumptions are not met for the chi-square tests: the warning beneath the table asserts that this is the case.

Output 2.4 Table Statistics

Statistic	DF	Value	Prob
Chi-Square	1	4.5000	0.0339
Likelihood Ratio Chi-Square	1	4.4629	0.0346
Continuity Adj. Chi-Square	1	2.5313	0.1116
Mantel-Haenszel Chi-Square	1	4.2500	0.0393
Phi Coefficient		0.5000	
Contingency Coefficient		0.4472	
Cramer's V		0.5000	
WARNING: 75% of the cells have expected counts less than 5. Chi-Square may not be a valid test.			

Fisher's Exact Test	
Cell (1,1) Frequency (F)	10
Left-sided Pr <= F	0.9961
Right-sided Pr >= F	0.0573
Table Probability (P)	0.0533
Two-sided Pr <= P	0.1070

Sample Size = 18

SAS produces both a left-tail and right-tail p-value for Fisher's exact test. The left-tail probability is the probability of all tables such that the (1,1) cell value is less than or equal to the one observed. The right-tail probability is the probability of all tables such that the (1,1) cell value is greater than or equal to the one observed. Thus, the one-sided p-value is the same as the right-tailed p-value in this case, since large values for the (1,1) cell correspond to better outcomes for Test treatment.

Both the two-sided p-value of 0.1070 and the one-sided p-value of 0.0573 are larger than the p-values associated with Q_P ($p = 0.0339$) and Q ($p = 0.0393$). Depending on your significance criterion, you might reach very different conclusions with these three test statistics. The sample size requirements for the chi-square distribution are not met with these data; hence the p-values from these test statistics with this approximation are questionable. This example illustrates the usefulness of Fisher's exact test when the sample size requirements for the usual chi-square tests are not met.

The output also includes a statistic labeled the "Continuity Adj. Chi-Square"; this is the continuity-adjusted chi-square statistic suggested by Yates (1934), which is intended to correct the Pearson chi-square statistic so that it more closely approximates Fisher's exact test. In this case, the correction produces a chi-square value of 2.5313 with $p = 0.1116$, which is certainly close to the two-sided Fisher's exact test value. And using half of the continuity-corrected chi-square approximates the one-sided Fisher's exact test well. However, many statisticians recommend that you should simply apply Fisher's exact test when the sample size requires it rather than try to approximate it. In particular, the continuity-corrected chi-square may be overly conservative for two-sided tests when the corresponding hypergeometric distribution is asymmetric; that is, the two row totals and the two column totals are very different, and the sample sizes are small.

Fisher's exact test is always appropriate, even when the sample size is large.

2.3.1 Exact p-values for Chi-Square Statistics

For many years, the only practical way to assess association in 2×2 tables that had small or zero counts was with Fisher's exact test. This test is computationally quite easy for the 2×2 case. However, you can also obtain exact p-values for the statistics discussed in Section 2.2. This is possible due to the development of fast and efficient network algorithms that provide a distinct advantage over direct enumeration. Although such enumeration is reasonable for Fisher's exact test, it can prove prohibitive in other instances. See Mehta, Patel, and Tsiatis (1984) for a description of these algorithms; Agresti (1992) provides a useful overview of the various algorithms for the computation of exact p-values.

In the case of Q, Q_P, and a closely related statistic, Q_L (likelihood ratio statistic), large values of the statistic imply a departure from the null hypothesis. The exact p-values for these statistics are the sum of the probabilities for the tables that have a test statistic greater than or equal to the value of the observed test statistic.

The EXACT statement enables you to request exact p-values or confidence limits for many of the statistics produced by the FREQ procedure. See the *SAS/STAT User's Guide* for details about specification and the options that control computation time. Exact computations might take a considerable amount of memory and time for large problems.

For the Table 2.3 data, the following SAS statements produce the exact p-values for the chi-square tests of association. You include the keyword(s) for the statistics for which to compute exact p-values, CHISQ in this case.

```
proc freq order=data;
   weight count;
   tables treat*outcome / chisq nocol;
   exact chisq;
run;
```

First, the usual table for the CHISQ statistics is displayed (not re-displayed here), and then individual tables for Q_P, Q_L, and Q are presented, including test values and both asymptotic and exact p-values, as shown in Output 2.5. Output 2.6, and Output 2.7.

Output 2.5 Pearson Chi-Square Test

Pearson Chi-Square Test	
Chi-Square	4.5000
DF	1
Asymptotic Pr > ChiSq	0.0339
Exact Pr >= ChiSq	0.1070

Output 2.6 Likelihood Ratio Chi-Square Test

Likelihood Ratio Chi-Square Test	
Chi-Square	4.4629
DF	1
Asymptotic Pr > ChiSq	0.0346
Exact Pr >= ChiSq	0.1070

Output 2.7 Mantel-Haenszel Chi-Square Test

Mantel-Haenszel Chi-Square Test	
Chi-Square	4.2500
DF	1
Asymptotic Pr > ChiSq	0.0393
Exact Pr >= ChiSq	0.1070

$Q_P = 4.5$ with an exact p-value of 0.1070 (asymptotic $p = 0.0339$). $Q = 4.25$ with an exact p-value of 0.1070 (asymptotic $p = 0.0393$). Q_L is similar, with a value of 4.4629 and an exact p-value 0.1070 (asymptotic $p = 0.0346$). Thus, a researcher using the asymptotic p-values in this case may have found an inappropriate significance that is not there when exact *p*-values are considered. Note that Fisher's exact test provides an identical *p*-value of 0.1070, but this is not always the case.

Using the exact *p*-values for the association chi-square versus applying the Fisher's exact test is a matter of preference. However, there might be some interpretation advantage in using the Fisher's exact test since the comparison is to your actual table rather than to a test statistic based on the table.

2.4 Difference in Proportions

The previous sections have addressed the question of whether there is an association between the rows and columns of a 2 × 2 table. In addition, you may be interested in describing the association in the table. For example, once you have established that the proportions computed from a table are different, you may want to estimate their difference.

Consider Table 2.5, which displays data from two independent groups.

Table 2.5 2 × 2 Contingency Table

	Yes	No	Total	Proportion Yes
Group 1	n_{11}	n_{12}	n_{1+}	$p_1 = n_{11}/n_{1+}$
Group 2	n_{21}	n_{22}	n_{2+}	$p_2 = n_{21}/n_{2+}$
Total	n_{+1}	n_{+2}	n	

If the two groups are arguably comparable to simple random samples from populations with corresponding population fractions for Yes as π_1 and π_2, respectively, you might be interested in estimating the difference between the proportions p_1 and p_2 with $d = p_1 - p_2$. You can show that the expected value with respect to the samples from the two groups having independent binomial distributions is

$$E\{p_1 - p_2\} = \pi_1 - \pi_2$$

and the variance is

$$V\{p_1 - p_2\} = \frac{\pi_1(1-\pi_1)}{n_{1+}} + \frac{\pi_2(1-\pi_2)}{n_{2+}}$$

for which a consistent estimator is

$$v_d = \frac{p_1(1-p_1)}{n_{1+}} + \frac{p_2(1-p_2)}{n_{2+}}$$

A $100(1-\alpha)\%$ confidence interval for $(\pi_1 - \pi_2)$ is written

$$d \pm \left\{ z_{\alpha/2}\sqrt{v_d} + \frac{1}{2}\left\{ \frac{1}{n_{1+}} + \frac{1}{n_{2+}} \right\} \right\}$$

where $z_{\alpha/2}$ is the $100(1 - (\alpha/2))$ percentile of the standard normal distribution; this confidence interval is based on Fleiss, Levin, and Paik (2003). These confidence limits include a continuity adjustment to the Wald asymptotic confidence limits that adjust for the difference between the normal approximation and the discrete binomial distribution. They are appropriate for moderate sample sizes—say cell counts of at least 8.

For example, consider Table 2.6, which reproduces the data analyzed in Section 2.2. In addition to determining that there is a statistical association between treatment and response, you may be interested in estimating the difference between the proportions of favorable response for the test and placebo treatments, including a 95% confidence interval.

Table 2.6 Respiratory Outcomes

Treatment	Favorable	Unfavorable	Total	Favorable Proportion
Placebo	16	48	64	0.250
Test	40	20	60	0.667
Total	56	68	124	0.452

The difference is $d = 0.667 - 0.25 = 0.417$, and the confidence interval is written

$$0.417 \pm \left\{ (1.96)\left[\frac{0.667(1-0.667)}{60} + \frac{0.25(1-0.25)}{64} \right]^{1/2} + \frac{1}{2}\left(\frac{1}{60} + \frac{1}{64} \right) \right\}$$

$$= 0.417 \pm 0.177$$
$$= (0.241, 0.592)$$

A related measure of association is the Pearson correlation coefficient. This statistic is proportional to the difference of proportions. Since Q_P is also proportional to the squared difference in proportions, the Pearson correlation coefficient is also proportional to $\sqrt{Q_P}$.

The Pearson correlation coefficient can be written

$$r = \left\{ (n_{11} - \frac{n_{1+}n_{+1}}{n})/\left[(n_{1+} - \frac{n_{1+}^2}{n})(n_{+1} - \frac{n_{+1}^2}{n}) \right]^{1/2} \right\}$$
$$= \left\{ (n_{11}n_{22} - n_{12}n_{21})/[(n_{1+}n_{2+}n_{+1}n_{+2})]^{1/2} \right\}$$
$$= [n_{1+}n_{2+}/n_{+1}n_{+2}]^{1/2} d$$
$$= (Q_P/n)^{1/2}$$

For the data in Table 2.6, r is computed as

$$r = [(60)(64)/(56)(68)]^{1/2}(0.417) = 0.418$$

2.4. Difference in Proportions

The FREQ procedure produces the difference in proportions and the continuity-corrected Wald interval. PROC FREQ also provides the uncorrected Wald confidence limits, but the Wald-based interval is known to have poor coverage, among other issues, especially when the proportions grow close to 0 or 1. See Newcombe (1998) and Agresti and Caffo (2000) for further discussion.

You can request the difference of proportions and the continuity-corrected Wald confidence limits with the RISKDIFF (CORRECT) option in the TABLES statement. The following statements produce the difference along with the Pearson correlation coefficient, which is requested with the MEASURES option.

The ODS SELECT statement restricts the output produced to the RiskDiffCol1 table and the Measures table. The RiskDiffCol1 table produces the difference for column 1 of the frequency table. There is also a table for the column 2 difference called RiskDiffCol2, which is not produced in this example.

```
data respire2;
   input treat $ outcome $ count @@;
   datalines;
test    f 40 test    u 20
placebo f 16 placebo u 48
;

ods select RiskDiffCol1 Measures;
proc freq order=data;
   weight count;
   tables treat*outcome / riskdiff (correct) measures;
run;
```

Output 2.8 displays the value for the Pearson correlation coefficient as 0.4184.

Output 2.8 Pearson Correlation Coefficient

Statistics for Table of treat by outcome

Statistic	Value	ASE
Gamma	0.7143	0.0974
Kendall's Tau-b	0.4184	0.0816
Stuart's Tau-c	0.4162	0.0814
Somers' D C|R	0.4167	0.0814
Somers' D R|C	0.4202	0.0818
Pearson Correlation	0.4184	0.0816
Spearman Correlation	0.4184	0.0816
Lambda Asymmetric C|R	0.3571	0.1109
Lambda Asymmetric R|C	0.4000	0.0966
Lambda Symmetric	0.3793	0.0983
Uncertainty Coefficient C|R	0.1311	0.0528
Uncertainty Coefficient R|C	0.1303	0.0525
Uncertainty Coefficient Symmetric	0.1307	0.0526

Output 2.9 contains the value for the difference of proportions for Test versus Placebo for the Favorable response, which is 0.4167 with confidence limits (0.2409, 0.5924). Note that this table also includes the proportions of column 1 response in both rows, along with the continuity-corrected asymptotic confidence limits and exact (Clopper-Pearson) confidence limits for the row proportions, which are based on inverting two equal-tailed binomial tests to identify the π_i that would not be contradicted by the observed p_i at the $(\alpha/2)$ significance level. See Clopper and Pearson (1934) for more information.

Output 2.9 Difference in Proportions

	Column 1 Risk Estimates					
	Risk	ASE	(Asymptotic) 95% Confidence Limits		(Exact) 95% Confidence Limits	
Row 1	0.6667	0.0609	0.5391	0.7943	0.5331	0.7831
Row 2	0.2500	0.0541	0.1361	0.3639	0.1502	0.3740
Total	0.4516	0.0447	0.3600	0.5432	0.3621	0.5435
Difference	0.4167	0.0814	0.2409	0.5924		
Difference is (Row 1 - Row 2)						
The asymptotic confidence limits include a continuity correction.						

2.4. Difference in Proportions

Another way to generate a confidence interval for the difference of proportions is to invert a score test. For testing goodness of fit for a specified difference Δ, Q_P is the score test. Consider that $E\{p_1\} = \pi$ and $E\{p_2\} = \pi + \Delta$. Then you can write

$$Q_P = \frac{(n_{11} - n_{1+}\hat{\pi})^2}{n_{1+}\hat{\pi}} + \frac{(n_{1+} - n_{11} - n_{1+}(1 - \hat{\pi}))^2}{n_{1+}(1 - \hat{\pi})} +$$
$$\frac{(n_{21} - n_{2+}(\hat{\pi} + \Delta))^2}{n_{2+}(\hat{\pi} + \Delta)} + \frac{(n_{2+} - n_{21} - n_{2+}(1 - \hat{\pi} - \Delta))^2}{n_{2+}(1 - \hat{\pi} - \Delta)}$$
$$= Z_\alpha^2$$
$$= 3.84$$

for $\alpha = 0.05$. You then identify the Δ so that $Q_P \leq 3.84$ for a 0.95 confidence interval, which requires iterative methods. This Miettinen-Nurminen interval (1985) has mean coverage somewhat above the nominal value (Newcombe 1998) and is also appealing theoretically (Newcombe and Nurminen 2011). The score interval is available with the FREQ procedure, which produces a bias-corrected interval by default (as specified in Miettinen and Nurminen 1985).

The following statements request the Miettinen-Nurminen interval, along with a corrected Wald interval. You specify these additional confidence intervals with the CL=(WALD MN) suboption of the RISKDIFF option. Adding the CORRECT option means that the Wald interval will be the corrected one.

```
proc freq order=data;
   weight count;
   tables treat*outcome / riskdiff(cl=(wald mn) correct) measures;
run;
```

Output 2.10 contains both the Miettinen-Nurminen and corrected Wald confidence intervals.

Output 2.10 Miettinen and Nurminen Confidence Interval

Confidence Limits for the Proportion (Risk) Difference		
Column 1 (outcome = f)		
Proportion Difference = 0.4167		
Type	95% Confidence Limits	
Miettinen-Nurminen	0.2460	0.5627
Wald (Corrected)	0.2409	0.5924

The Miettinen-Nurminen confidence interval is a bit narrower than the corrected Wald interval. In general, it might be preferred when the cell count size is marginal.

But what if the cell counts are smaller than 8? Consider the data in Table 2.3 again. One asymptotic method that does well for small sample sizes is the Newcombe hybrid score interval (Newcombe 1998), which uses Wilson score confidence limits for the binomial proportion (Wilson 1927) in its construction. You compute these limits by inverting the normal test that uses the null proportion for the variance (score test) and solving the resulting quadratic equation:

$$\frac{(p-P)^2}{P(1-P)} = \frac{z_{\alpha/2}^2}{n}$$

The solutions (limits) are

$$\left(p + z_{\alpha/2}^2/2n \pm z_{\alpha/2}\sqrt{\left(p(1-p) + z_{\alpha/2}^2/4n\right)/n} \right) / \left(1 + z_{\alpha/2}^2/n\right)$$

You can produce Wilson score confidence limits for the binomial proportion in PROC FREQ by specifying the BINOMIAL (WILSON) option for a one-way table.

You then compute the Newcombe confidence interval for the difference of proportions by plugging in the Wilson score confidence limits P_{U1}, P_{L1} and P_{U2}, P_{L2}, which correspond to the row 1 and row 2 proportions, respectively, to obtain the lower (L) and upper (U) bounds for the confidence interval for the proportion difference:

$$L = (p_1 - p_2) - \sqrt{(p_1 - P_{L1})^2 + (P_{U2} - p_2)^2}$$

and

$$U = (p_1 - p_2) + \sqrt{(P_{U1} - p_1)^2 + (p_2 - P_{L2})^2}$$

The Newcombe confidence interval for the difference of proportions has been shown to have good coverage properties and avoids overshoot (Newcombe 1998); it's the choice of many practitioners regardless of sample size. In general, it attains near nominal coverage when the proportions are away from 0 and 1, and it can have higher than nominal coverage when the proportions are both close to 0 or 1 (Agresti and Caffo 2000). A continuity-corrected Newcombe's method also exists, and it should be considered if a row count is less than 10. You obtain a continuity-corrected confidence interval for the difference of proportions by plugging in the continuity-corrected Wilson score confidence limits.

There are also exact methods for computing the confidence intervals for the difference of proportions; they are unconditional exact methods which contend with a nuisance parameter by maximizing the p-value over all possible values of the parameter (versus, say, Fisher's exact test, which is a conditional exact test that conditions on the margins). The unconditional exact intervals do have the property that the nominal coverage is the lower bound of the actual coverage. One type of these intervals is computed by inverting two separate one-sided tests where the size of each test is $\alpha/2$ at most; the actual coverage is bounded by the nominal coverage. This is called the tail method. However, these intervals have excessively higher than nominal coverage, especially when the proportions are near 0 or 1, in which case the lower bound of the coverage is $1 - \alpha/2$ instead of $1 - \alpha$ (Agresti 2002).

The following PROC FREQ statements request the Wald, Newcombe, and unconditional exact confidence intervals for the difference of the favorable proportion for Test and Placebo. The CORRECT option specifies that the continuity correction be applied where possible, and the NORISK option suppresses the rest of the relative risk difference results.

```
proc freq order=data data=severe;
   weight count;
   tables treat*outcome / riskdiff(cl=(wald newcombe exact) correct );
   exact riskdiff;
run;
```

Output 2.11 displays the confidence intervals.

Output 2.11 Confidence Intervals for Difference of Proportions

Statistics for Table of treat by outcome

Confidence Limits for the Proportion (Risk) Difference		
Column 1 (outcome = f)		
Proportion Difference = 0.5000		
Type	95% Confidence Limits	
Exact	-0.0296	0.8813
Newcombe Score (Corrected)	-0.0352	0.8059
Wald (Corrected)	-0.0571	1.0000

The continuity-corrected Wald-based confidence interval is the widest interval at $(-0.0571, 1.000)$, and it might have boundary issues with the upper limit of 1. The exact unconditional confidence interval at $(-0.0296, 0.8813)$ also includes zero. The corrected Newcombe interval is the narrowest at $(-0.0352, 0.8059)$. All of these confidence intervals are in harmony with the Fisher's exact test result (two-sided $p = 0.1071$), but the corrected Newcombe interval might be the most suitable for these data.

2.5 Odds Ratio and Relative Risk

Measures of association are used to assess the strength of an association. Numerous measures of association are available for the contingency table, some of which are described in Chapter 5, "The $s \times r$ Table." For the 2×2 table, one measure of association is the *odds ratio*, and a related measure of association is the *relative risk*.

Consider Table 2.5. The *odds ratio* (OR) compares the odds of the Yes proportion for Group 1 to the odds of the Yes proportion for Group 2. It is computed as

$$\text{OR} = \frac{p_1/(1-p_1)}{p_2/(1-p_2)} = \frac{n_{11}n_{22}}{n_{12}n_{21}}$$

The odds ratio ranges from 0 to infinity. When OR is 1, there is no association between the row variable and the column variable. When OR is greater than 1, Group 1 is more likely than Group 2 to have the Yes response; when OR is less than 1, Group 1 is less likely than Group 2 to have the Yes response.

Define the *logit* for general p as

$$\text{logit}(p) = \log\left\{\frac{p}{1-p}\right\}$$

with log as the natural logarithm. If you take the log of the odds ratio,

$$f = \log\{OR\} = \log\left\{\frac{p_1(1-p_2)}{p_2(1-p_1)}\right\}$$
$$= \log\{p_1/(1-p_1)\} - \log\{p_2/(1-p_2)\}$$

you see that the odds ratio can be written in terms of the difference between two logits. The logit is the function that is modeled in logistic regression. As you will see in Chapter 8, "Logistic Regression I: Dichotomous Response," the odds ratio and logistic regression are closely connected.

Since

$$f = \log\{n_{11}\} - \log\{n_{12}\} - \log\{n_{21}\} + \log\{n_{22}\}$$

a consistent estimate of its variance with usefulness when all $n_{ij} \geq 5$ (preferably ≥ 10) is

$$v_f = \left\{\frac{1}{n_{11}} + \frac{1}{n_{12}} + \frac{1}{n_{21}} + \frac{1}{n_{22}}\right\}$$

so a $100(1-\alpha)\%$ confidence interval for OR can be written as

$$\exp(f \pm z_{\alpha/2}\sqrt{v_f})$$

The odds ratio is a useful measure of association regardless of how the data are collected. However, it has special meaning for retrospective studies because it can be used to estimate a quantity called *relative risk*, which is commonly used in epidemiological work. The relative risk (RR) is the risk of developing a particular condition (often a disease) for one group compared to another group. For data collected prospectively, the relative risk is written

$$RR = \frac{p_1}{p_2}$$

You can show that

$$RR = OR \times \frac{\{1 + (n_{21}/n_{22})\}}{\{1 + (n_{11}/n_{12})\}}$$

or that OR approximates RR when n_{11} and n_{21} are small relative to n_{12} and n_{22}, respectively. This is called the *rare outcome paradigm*. Usually, the outcome of interest needs to occur less than 10% of the time for OR and RR to be similar. However, many times when the event under investigation is a relatively common occurrence, you are more interested in looking at the difference in proportions rather than at the odds ratio or the relative risk.

For a retrospective study, estimates for p_1, p_2, and RR are not available because they involve the unknown risk of the disease, but the OR estimator for $p_1(1 - p_2)/p_2(1 - p_1)$ is still valid.

For cross-sectional data, the quantity p_1/p_2 is called the *prevalence ratio*; it does not indicate risk because the disease and risk factor are assessed at the same time, but it does give you an idea of the prevalence of a condition in one group compared to another.

It is important to realize that the odds ratio can always be used as a measure of association, and that relative risk and the odds ratio as an estimator of relative risk have meaning for certain types of studies and require certain assumptions.

Table 2.7 contains data from a study about how general daily stress affects one's opinion on a proposed new health policy. Since information about stress level and opinion were collected at the same time, the data are cross-sectional.

Table 2.7 Opinions on New Health Policy

Stress	Favorable	Unfavorable	Total
Low	48	12	60
High	96	94	190

To produce the odds ratio and other measures of association from PROC FREQ, you specify the MEASURES option in the TABLES statement. The ORDER=DATA option is used in the PROC FREQ statement to produce a table that looks the same as that displayed in Table 2.7. Without this option, the row that corresponds to high stress would come first, and the row that corresponds to low stress would come last.

```
data stress;
   input stress $ outcome $ count;
   datalines;
low    f  48
low    u  12
high   f  96
high   u  94
;

proc freq order=data;
   weight count;
   tables stress*outcome / chisq measures nocol nopct;
run;
```

Output 2.12 contains the resulting frequency table. Since the NOCOL and NOPCT options are specified, only the row percentages are printed. You can see that 80% of the low stress group were favorable, while the high-stress group was nearly evenly split between favorable and unfavorable.

Output 2.12 Frequency Table

Frequency Row Pct	Table of stress by outcome		
		outcome	
stress	f	u	Total
low	48 80.00	12 20.00	60
high	96 50.53	94 49.47	190
Total	144	106	250

Output 2.13 displays the chi-square statistics. The statistics Q and Q_P indicate a strong association, with values of 16.1549 and 16.2198, respectively. Note how close the values for these statistics are for a sample size of 250.

Output 2.13 Chi-Square Statistics

Statistic	DF	Value	Prob
Chi-Square	1	16.2198	<.0001
Likelihood Ratio Chi-Square	1	17.3520	<.0001
Continuity Adj. Chi-Square	1	15.0354	0.0001
Mantel-Haenszel Chi-Square	1	16.1549	<.0001
Phi Coefficient		0.2547	
Contingency Coefficient		0.2468	
Cramer's V		0.2547	

Output 2.14 contains the measures of association. Chapter 5 contains more information about these measures, which include Somers' $DC|R = (p_1 - p_2)$ and gamma = $(OR-1)/(OR+1)$ for the 2 × 2 table.

Output 2.14 Measures of Association

Statistic	Value	ASE
Gamma	0.5932	0.1147
Kendall's Tau-b	0.2547	0.0551
Stuart's Tau-c	0.2150	0.0489
Somers' D C|R	0.2947	0.0631
Somers' D R|C	0.2201	0.0499
Pearson Correlation	0.2547	0.0551
Spearman Correlation	0.2547	0.0551
Lambda Asymmetric C|R	0.0000	0.0000
Lambda Asymmetric R|C	0.0000	0.0000
Lambda Symmetric	0.0000	0.0000
Uncertainty Coefficient C|R	0.0509	0.0231
Uncertainty Coefficient R|C	0.0630	0.0282
Uncertainty Coefficient Symmetric	0.0563	0.0253

Output 2.15 displays the odds ratio information.

Output 2.15 Odds Ratio

Estimates of the Relative Risk (Row1/Row2)			
Type of Study	Value	95% Confidence Limits	
Case-Control (Odds Ratio)	3.9167	1.9575	7.8366
Cohort (Col1 Risk)	1.5833	1.3104	1.9131
Cohort (Col2 Risk)	0.4043	0.2389	0.6841

Sample Size = 250

The odds ratio value is listed beside "Case-Control" in the section labeled "Estimates of the Relative Risk (Row1/Row2)." The estimated OR is 3.9167, which means that the odds of a favorable response are roughly four times higher for those persons with low stress than for those persons with high stress. The confidence intervals are labeled "Confidence Limits" and are 95% confidence intervals by default. To change them, use the ALPHA= option in the TABLES statement.

The values listed for "Cohort (Col1 Risk)" and "Cohort (Col2 Risk)" are the estimates of relative risk for a cohort (prospective) study. Since these data are cross-sectional, (p_1/p_2) does not represent relative risk. However, the value 1.5833 is the ratio of the prevalence of favorable opinions for the low stress group compared to the high stress group. (The value 0.4043 is the prevalence ratio of the unfavorable opinions of the low stress group compared to the high stress group.)

Table 2.8 contains data that concern respiratory illness. Two groups having the same symptoms of respiratory illness were selected via simple random sampling: one group was treated with a test treatment, and one group was treated with a placebo. This is an example of a cohort study since the comparison groups were chosen before the responses were measured. They are considered to come from independent binomial distributions.

Table 2.8 Respiratory Improvement

Treatment	Yes	No	Total
Test	29	16	45
Placebo	14	31	45

The following statements are submitted to produce chi-square statistics, odds ratios, and relative risk measures for these data. The ALL option has the same action as specifying both the CHISQ and the MEASURES options (and the CMH option, discussed in Chapter 3).

```
data respire;
   input treat $ outcome $ count;
   datalines;
test      yes  29
test      no   16
placebo   yes  14
placebo   no   31
;

proc freq order=data;
   weight count;
   tables treat*outcome / all nocol nopct;
run;
```

Output 2.16 displays $Q = 9.9085$ and $Q_P = 10.0198$, as well as the two-sided $p = 0.003$ for Fisher's exact test. Clearly, there is a strong association between treatment and improvement.

Output 2.16 Table Statistics

Frequency Row Pct	Table of treat by outcome		
		outcome	
treat	yes	no	Total
test	29 64.44	16 35.56	45
placebo	14 31.11	31 68.89	45
Total	43	47	90

Output 2.16 *continued*

Statistic	DF	Value	Prob
Chi-Square	1	10.0198	0.0015
Likelihood Ratio Chi-Square	1	10.2162	0.0014
Continuity Adj. Chi-Square	1	8.7284	0.0031
Mantel-Haenszel Chi-Square	1	9.9085	0.0016
Phi Coefficient		0.3337	
Contingency Coefficient		0.3165	
Cramer's V		0.3337	

Fisher's Exact Test	
Cell (1,1) Frequency (F)	29
Left-sided Pr <= F	0.9997
Right-sided Pr >= F	0.0015
Table Probability (P)	0.0011
Two-sided Pr <= P	0.0029

Output 2.17 displays the estimates of relative risk and the odds ratio (other measures of association produced by the ALL option are not displayed here). Two versions of the relative risk are supplied: one is the relative risk of the attribute corresponding to the first column, or the risk of improvement. The column 2 risk is the risk of no improvement. The relative risk for improvement is 2.0714, with a 95% confidence interval of (1.2742, 3.3675).

Note that if these data had been obtained retrospectively, the odds ratio would be used to represent the association between group and outcome, but it would not represent the relative risk since the proportions with improvement are 0.64 and 0.31 for test and control, and so the rare outcome paradigm does not apply. Note that, in this case where $0.25 < p_1, p_2 < 0.75$, $\log OR/4 \approx (p_1 - p_2)$, where $\log OR/4 = 0.35$ and $(p_1 - p_2) = 0.33$.

Output 2.17 Odds Ratio and Relative Risk

Estimates of the Relative Risk (Row1/Row2)			
Type of Study	Value	95% Confidence Limits	
Case-Control (Odds Ratio)	4.0134	1.6680	9.6564
Cohort (Col1 Risk)	2.0714	1.2742	3.3675
Cohort (Col2 Risk)	0.5161	0.3325	0.8011

Sample Size = 90

2.5.1 Exact Confidence Limits for the Odds Ratio

Section 2.3 discussed Fisher's exact test for assessing association in 2×2 tables that were too sparse for the usual asymptotic chi-square tests to apply. You may want to compute the odds ratio as a measure of association for these data, but the usual asymptotic confidence limits would not be appropriate because, again, the sparseness of the data violates the asymptotic assumptions.

You can obtain exact confidence limits for the odds ratio by using the FREQ procedure. The computation is based on work presented by Thomas (1971) and Gart (1971). These confidence limits are conservative by having higher than nominal coverage; the confidence coefficient is not exactly $1 - \alpha$, but it is at least $1 - \alpha$.

Consider the severe infection data in Table 2.3. To compute an odds ratio estimate for the odds of having a favorable outcome for the treatment group compared to the control group, you submit the following statements, including the EXACT statement with the OR keyword.

```
data severe;
   input treat $ outcome $ count;
   datalines;
Test     f 10
Test     u 2
Control  f 2
Control  u 4
;

proc freq order=data;
   weight count;
   tables treat*outcome / nocol;
   exact or;
run;
```

Output 2.18 displays the estimate of the odds ratio, which is 10. Test subjects have 10 times higher odds for the favorable response than the control subjects.

Output 2.18 Odds Ratio (Case-Control Study)

Statistics for Table of treat by outcome

Odds Ratio (Case-Control Study)	
Odds Ratio	10.0000
Asymptotic Conf Limits	
95% Lower Conf Limit	1.0256
95% Upper Conf Limit	97.5005
Exact Conf Limits	
95% Lower Conf Limit	0.6896
95% Upper Conf Limit	166.3562

The exact confidence limits for the odds ratio are (0.6896, 166.3562), indicating low precision. Note that the exact confidence bounds are much wider than the asymptotic ones, and they are in harmony with the Fisher's exact test $p = 0.1071$ by including the value 1.00 for no association.

2.6 Sensitivity and Specificity

Some other measures frequently calculated for 2 × 2 tables are sensitivity and specificity. These measures are of particular interest when you are determining the efficacy of screening tests for various disease outcomes. *Sensitivity* is the true proportion of positive results that a test elicits when performed on subjects known to have the disease; *specificity* is the true proportion of negative results that a test elicits when performed on subjects known to be disease free.

Often, a standard screening method is used to determine whether disease is present and compared to a new test method. Table 2.9 contains the results of a study investigating a new screening device for a skin disease. The distributions for positive and negative results for the test method are assumed to result from simple random samples from the corresponding populations of persons with disease present and those with disease absent.

Table 2.9 Skin Disease Screening Test Results

Status	Test +	Test −	Total
Disease Present	52	8	60
Disease Absent	20	100	120

Sensitivity and specificity for these data are estimated by

$$\text{sensitivity} = (n_{11}/n_{1+}) \doteq \Pr(\text{Test} + | \text{disease present})$$

and

$$\text{specificity} = (n_{22}/n_{2+}) \doteq \Pr(\text{Test} - | \text{disease absent})$$

For these data, sensitivity = 52/60 = 0.867 and specificity = 100/120 = 0.833.

You can generate these estimates plus their exact confidence intervals by using PROC FREQ with the RISKDIFF option.

```
data screening;
   input disease $ outcome $ count @@;
   datalines;
present  + 52 present  -   8
absent   + 20 absent   - 100
;

proc freq data=screening order=data;
   weight count;
```

```
    tables disease*outcome / riskdiff;
run;
```

Output 2.19 displays the results. The sensitivity is the Column 1 Estimate for Row 1, 0.8667, with exact confidence limits (0.7541, 0.9406).

Output 2.19 Sensitivity Estimate and Confidence Interval

Statistics for Table of disease by outcome

Column 1 Risk Estimates						
	Risk	ASE	(Asymptotic) 95% Confidence Limits		(Exact) 95% Confidence Limits	
Row 1	0.8667	0.0439	0.7807	0.9527	0.7541	0.9406
Row 2	0.1667	0.0340	0.1000	0.2333	0.1049	0.2456
Total	0.4000	0.0365	0.3284	0.4716	0.3278	0.4755
Difference	0.7000	0.0555	0.5912	0.8088		
Difference is (Row 1 - Row 2)						

The specificity is the Column 2 Estimate for Row 2, 0.8333, with exact confidence limits (0.7544, 0.8951), as displayed in Output 2.20.

Output 2.20 Specificity Estimate and Confidence Interval

Column 2 Risk Estimates						
	Risk	ASE	(Asymptotic) 95% Confidence Limits		(Exact) 95% Confidence Limits	
Row 1	0.1333	0.0439	0.0473	0.2193	0.0594	0.2459
Row 2	0.8333	0.0340	0.7667	0.9000	0.7544	0.8951
Total	0.6000	0.0365	0.5284	0.6716	0.5245	0.6722
Difference	-0.7000	0.0555	-0.8088	-0.5912		
Difference is (Row 1 - Row 2)						

You may know the underlying percentage of those subjects with and without the disease in a population of interest. You may want to estimate the proportion of subjects with the disease among those who have a positive test. You can determine these proportions with the use of Bayes' theorem.

Suppose that the underlying prevalence of disease for an appropriate target population for these data is 15%. That is, 15% of the population have the disease and 85% do not. You can compute joint probabilities by multiplying the conditional probabilities by the marginal probabilities:

$$\Pr(T, D) = \Pr(T|D) \times \Pr(D)$$

Table 2.10 contains these results.

Table 2.10 How Test Should Perform in General Population

Status	Test +	Test −	Total
Disease Present	0.867(.15) = 0.130	0.133(.15) = 0.020	0.15
Disease Absent	0.167(.85) = 0.142	0.833(.85) = 0.708	0.85
Total	0.130 + 0.142 = 0.272	0.020 + 0.708 = 0.728	

The values in the row titled "Total" are Pr(Test +) and Pr(Test −), respectively. You can now determine the probability of those subjects with the disease among those with a positive test:

$$\Pr(D|T) = \frac{\Pr(T, D)}{\Pr(T)}$$

Thus, Pr (disease|Test +) = 0.130/0.272 = 0.478 and Pr(no disease|Test −) = 0.708/0.728 = 0.972. See Fleiss, Levin, and Paik (2003) for more detail, including the calculation of false negative and false positive rates.

2.7 McNemar's Test

The 2 × 2 table often contains information collected from *matched pairs*, experimental units for which two related responses are made. The sampling unit is no longer one individual but a pair of related individuals, which could be two locations on the same individual such as left and right eyes or two occasions for the same individual, such as before and after measurements. For example, in case-control studies, cases are often matched to controls on the basis of demographic, environmental, or genetic characteristics; interest lies in determining whether there is a difference between control exposure to a risk factor and case exposure to the same risk factor. Another example of matched pairs is husband and wife voting preference. Measurements at two different time points can also be considered a matched pair, such as before and after measurements.

Data from a study on matched pairs are represented in Table 2.11. The n_{11} in the (1,1) cell means that n_{11} pairs responded Yes for both Response 1 and Response 2; the n_{21} in the (2,1) cell means that n_{21} pairs responded Yes for Response 1 and No for Response 2.

Table 2.11 Matched Pairs Data

	Response 1		
Response 2	Yes	No	Total
Yes	n_{11}	n_{12}	n_{1+}
No	n_{21}	n_{22}	n_{2+}
Total	n_{+1}	n_{+2}	n

The question of interest for such data is whether the proportion of pairs responding Yes for Response 1 is the same as the proportion of pairs responding Yes for Response 2. This question cannot be addressed with the chi-square tests of association of previous sections, since the relevant counts are marginal totals rather than cell counts.

The question is whether

$$p_1 = \frac{n_{1+}}{n}$$

and

$$p_2 = \frac{n_{+1}}{n}$$

are the same. Recognizing that $(p_1 - p_2) = (n_{12} - n_{21})/n$, McNemar (1947) developed a chi-square test based on the conditional binomial distribution of (n_{12}, n_{21}) to address this situation, and so only the off-diagonal elements are important in determining the test statistic.

$$Q_M = \frac{(n_{12} - n_{21})^2}{(n_{12} + n_{21})}$$

and is approximately chi-square with one degree of freedom.

Table 2.12 displays data collected by political science researchers who polled husbands and wives on whether they approved of one of their U.S. senators. The cell counts represent the number of pairs of husbands and wives who fit the configurations indicated by the row and column levels.

Table 2.12 State Senator Approval Ratings

Husband Approval	Wife Approval		Total
	Yes	No	
Yes	20	5	25
No	10	10	20
Total	30	15	45

McNemar's test is easy to compute by hand.

$$Q_M = \frac{(5-10)^2}{(5+10)} = 1.67$$

Compared to a chi-square distribution with 1 df, this statistic is clearly nonsignificant.

The FREQ procedure computes McNemar's test with the AGREE option in the TABLE statement (see Chapter 5 for other analyses available with the AGREE option for tables of other dimensions). The following SAS statements request McNemar's test. The ODS SELECT statement is used to restrict the output to that test.

```
data approval;
    input hus_resp $ wif_resp $ count;
    datalines;
```

```
    yes yes  20
    yes no    5
    no  yes  10
    no  no   10
    ;

    ods select McNemarsTest;
    proc freq order=data;
       weight count;
       tables hus_resp*wif_resp / agree;
    run;
```

Output 2.21 displays the output that is produced. $Q_M = 1.67$, the same value as computed previously.

Note that McNemar's test is identical to the Mantel-Haenszel test in Chapter 3 and Chapter 6 when matched pairs are treated as strata. Also, (n_{12}/n_{21}) is the adjusted odds ratio for the within matched pairs association between exposure (for what is compared within matched pairs) and response for a conditional logistic regression model as described in Chapter 10.

Output 2.21 McNemar's Test

Statistics for Table of hus_resp by wif_resp

McNemar's Test	
Statistic (S)	1.6667
DF	1
Pr > S	0.1967

Exact p-values are also available for McNemar's test. You would include the statement

```
    exact mcnem;
```

in your PROC FREQ invocation. The computations work in a similar fashion to those for the chi-square tests of association; the exact p-value is the sum of the probabilities of those outcomes with a Q_M greater than or equal to the actual one (or more extreme than (n_{12}, n_{21}) for the binomial distribution with $(n_{12} + n_{21}) = n_M$ fixed and $\pi = 0.5$).

2.8 Incidence Densities

The 2×2 table can also represent incidence densities, in which you have counts of subjects who responded with an event versus extent of exposure for that event. These counts often follow the Poisson distribution. Some examples are:

- colony counts for bacteria or viruses

- accidents or equipment failure
- incidences for disease

Table 2.13 is based on the U.S. participants in an international study of the effect of a rotavirus vaccine in nearly 70,000 infants on health care events (hospitalizations and emergency room visits). Subjects received three doses of the vaccine or a placebo at 4- to 12- week intervals. They were followed for one year (Vesikari et al. 2007). Table 2.13 displays the failures (hospitalizations or emergency room visits) and the person-years of exposure since not all subjects were in the study for an entire year.

Table 2.13 Vaccine Study Results for U.S.

Treatment	Events	Person Years
Vaccine	3	7500
Placebo	58	7250

It is of interest to compute the ratio of incidence densities (IDR) of events for vaccine compared to placebo. Consider Table 2.14.

Table 2.14 Vaccine Study

Treatment	Events	Person Years
Vaccine	n_v	N_v
Placebo	n_p	N_p

You can assume that n_p and n_v have independent Poisson distributions with expected values $\lambda_p N_p$ and $\lambda_v N_v$, respectively, particularly if the time to event has an exponential distribution; see Chapter 13. You are interested in whether the incidence densities λ_v and λ_p are the same, which can be addressed by seeing whether the incidence density ratio (IDR) is equal to 1; you determine this by evaluating its confidence limits. When the counts are small, such as the count of 3 in Table 2.13, then the exact confidence limits are appropriate.

Because n_v given $(n_v + n_p)$ has the conditional binomial distribution (under the assumption that n_v and n_p are independent Poisson),

$$\text{Bin}\left(n = n_v + n_p, P = \frac{\lambda_v N_v}{\lambda_p N_p + \lambda_v N_v}\right)$$

then

$$P = \frac{\frac{\lambda_v}{\lambda_p}\left(\frac{N_v}{N_p}\right)}{\frac{\lambda_v}{\lambda_p}\left(\frac{N_v}{N_p}\right) + 1} = \frac{RC}{RC + 1}$$

2.8. Incidence Densities

You then compute $p = n_v/(n_v + n_p)$ to produce a $100(1-\alpha)\%$ confidence interval (P_L, P_U) for P. If you want the exact confidence interval for the incidence density ratios, you compute the exact confidence interval for the binomial proportion P. Use

$$\frac{N_p p}{(1-p)N_v}$$

as an estimator for R, and then

$$\left\{ \frac{P_L}{(1-P_L)C}, \frac{P_U}{(1-P_U)C} \right\}$$

serves as a $100(1-\alpha)\%$ confidence interval for the IDR.

The following SAS statements compute the exact binomial confidence intervals for the binomial proportion P in the table and output them to a SAS data set.

```
data vaccine2;
   input Outcome $ Count @@;
datalines;
fail 3 success 58
;

ods select BinomialCLs;
proc freq;
   weight count;
   tables Outcome / binomial (exact);
ods output BinomialCLs=BinomialCLs;
run;
```

The following SAS/IML® statements then produce the exact confidence interval for the IDR.

```
proc iml ;
   Use BinomialCLs var{LowerCL UpperCL};
   read all into CL;
   print CL;
   q = { 3, 7500, 58 , 7250 };
   C= q[2]/q[4];
   P= q[1]/ (q[1] + q[3]);
   R= P/ ((1-P) *C) ;
   CI= CL[1]/((1-CL[1])*C) || CL[2]/((1-CL[2])*C) ;
   print r CI;
quit;
```

The IDR that compares vaccine failure to placebo is 0.05000 with an exact confidence interval of (0.01002, 0.15355). Obviously the vaccine is much more effective than the placebo. The percent rate reduction in failures, a quantity often used in the assessment of vaccines, is computed as 100(1–IDR)=95% with (84.645%, 98.998%) as the exact 0.95 confidence interval.

You can also use the GENMOD procedure to compute the exact confidence interval for the IDR, as illustrated on page 392.

2.9 Sample Size and Power Computations

Sample size and power computations are often an integral part of the analysis of 2×2 tables, for example, when a certain power is required for the comparison of a test treatment and an active control in a clinical trial setting. You might need to address a question such as the following:

> What sample size per treatment is needed to have 0.80 power for the comparison between the test treatment and the active control at the two-sided $\alpha = 0.05$ significance level (given that the sample sizes are equal for these two treatments)?

However, the topic of sample size and power computations is beyond the scope of this book. See Fleiss, Levin, and Paik (2003) for a discussion of the methodology involved, and see the documentation for the POWER procedure in SAS/STAT software for how to implement with SAS. In particular, the TWOSAMPLEFREQ statement of PROC POWER enables you to compute sample sizes and power for the Pearson chi-square test, likelihood ratio chi-square test, and Fisher's exact test.

Chapter 3
Sets of 2×2 Tables

Contents

3.1	Introduction		**47**
3.2	Mantel-Haenszel Test		**48**
	3.2.1	Respiratory Data Example	49
	3.2.2	Health Policy Data	53
	3.2.3	Soft Drink Example	57
3.3	Measures of Association		**62**
	3.3.1	Homogeneity of Odds Ratios	63
	3.3.2	Coronary Artery Disease Data Example	64
3.4	Exact Confidence Intervals for the Common Odds Ratio		**69**

3.1 Introduction

The respiratory data displayed in Table 2.8 in the previous chapter are only a subset of the data collected in the clinical trial. The study included patients at two medical centers and produced the complete data shown in Table 3.1. These data comprise a set of two 2×2 tables.

Table 3.1 Respiratory Improvement

Center	Treatment	Yes	No	Total
1	Test	29	16	45
1	Placebo	14	31	45
Total		43	47	90
2	Test	37	8	45
2	Placebo	24	21	45
Total		61	29	90

Investigators were interested in whether there were overall differences in proportions with improvement; however, they were concerned that the patient populations at the two centers were sufficiently different that center needed to be accounted for in the analysis. One strategy for examining the association between two variables while adjusting for the effects of others is *stratified analysis*.

In general, the strata may represent explanatory variables, or they may represent research sites or hospitals in a multicenter study. Each table corresponds to one stratum; the strata are determined by the levels of the explanatory variables (one for each unique combination of the levels of the explanatory variables). The idea is to evaluate the association between the row variable and the response variable, while *adjusting*, or *controlling*, for the effects of the stratification variables. In some cases, the stratification results from the study design, such as in the case of a multicenter clinical trial; in other cases, it may arise from a prespecified poststudy stratification performed to control for the effects of certain explanatory variables that are thought to be related to the response variable.

The analysis of sets of tables addresses the same questions as the analysis of a single table: is there an association between the row and column variables in the tables and what is the strength of that association? These questions are investigated with similar strategies involving chi-square statistics and measures of association such as the odds ratios; the key difference is that you are investigating overall association instead of the association in just one table.

3.2 Mantel-Haenszel Test

For the data in Table 3.1, interest lies in determining whether there is a difference in the favorable rates between Test and Placebo. Patients in both centers were randomized into two treatment groups, which induces independent hypergeometric distributions for the within-center frequencies under the hypothesis that treatments have equal effects for all patients. Thus, the distribution for the two tables is the product of these two hypergeometric distributions. You can induce the hypergeometric distribution via conditional distribution arguments when you have postrandomization stratification or when you have independent binomial distributions from simple random sampling.

Consider the following table as representative of q 2×2 tables, $h = 1, 2, \ldots, q$.

Table 3.2 hth 2×2 Contingency Table

	Yes	No	Total
Group 1	n_{h11}	n_{h12}	n_{h1+}
Group 2	n_{h21}	n_{h22}	n_{h2+}
Total	n_{h+1}	n_{h+2}	n_h

Under the null hypothesis of no treatment difference, the expected value of n_{h11} is

$$E\{n_{h11}|H_0\} = \frac{n_{h1+}n_{h+1}}{n_h} = m_{h11}$$

and its variance is

$$V\{n_{h11}|H_0\} = \frac{n_{h1+}n_{h2+}n_{h+1}n_{h+2}}{n_h^2(n_h - 1)} = v_{h11}$$

One method for assessing the overall association of group and response, adjusting for the stratification factor, is the Mantel-Haenszel (1959) statistic.

$$Q_{MH} = \frac{\{\sum_{h=1}^{q} n_{h11} - \sum_{h=1}^{q} m_{h11}\}^2}{\sum_{h=1}^{q} v_{h11}}$$

$$= \frac{\{\sum_{h=1}^{q} (n_{h1+}n_{h2+}/n_h)(p_{h11} - p_{h21})\}^2}{\sum_{h=1}^{q} v_{h11}}$$

where $p_{hi1} = n_{hi1}/n_{hi+}$ is the proportion of subjects from the hth stratum and the ith group who have a favorable response. Q_{MH} approximately has the chi-square distribution with one degree of freedom when the combined row sample sizes ($\sum_{h=1}^{q} n_{hi+} = n_{+i+}$) are large, for example, greater than 30. This means that individual cell counts and table sample sizes may be small, so long as the overall row sample sizes are large. For the case of two tables, such as for Table 3.1, $q = 2$.

The Mantel-Haenszel strategy potentially removes the confounding influence of the explanatory variables that comprise the stratification. It can provide increased power for detecting association in a randomized study by comparing like subjects with like subjects. It can also remove the bias that can occur in an observational study from inbalances in confounding factors, but possibly at the cost of decreased power. In some sense, the strategy is similar to adjustment for blocks in a two-way analysis of variance for randomized blocks; it is also like covariance adjustment for a categorical explanatory variable.

Q_{MH} is effective for detecting patterns of association across q strata when there is a consistent tendency to expect the predominant majority of differences $\{p_{h11} - p_{h21}\}$ to have the same sign. For this reason, Q_{MH} is often called an *average partial association statistic*. Q_{MH} may fail to detect association when the differences are in opposite directions and are of similar magnitude. Q_{MH} as formulated here is directed at the n_{h11} cell; however, it is invariant to whatever cell is chosen. For an overview of Mantel-Haenszel methods, see to Landis et al. (1998).

Mantel and Fleiss (1980) proposed a criterion for determining whether the chi-square approximation is appropriate for the distribution of the Mantel-Haenszel statistic for q strata:

$$\min\left\{\left[\sum_{h=1}^{q} m_{h11} - \sum_{h=1}^{q} (n_{h11})_L\right], \left[\sum_{h=1}^{q} (n_{h11})_U - \sum_{h=1}^{q} m_{h11}\right]\right\} > 5$$

where $(n_{h11})_L = \max(0, n_{h1+} - n_{h+2})$ and $(n_{h11})_U = \min(n_{h+1}, n_{h1+})$. The criterion specifies that the across-strata sum of expected values for a particular cell has a difference of at least 5 from both the minimum possible sum and the maximum possible sum of the observed values.

3.2.1 Respiratory Data Example

For the data in Table 3.1, there is interest in the association between treatment and respiratory outcome, after adjusting for the effects of the centers. The following DATA step puts all the respiratory data into the SAS data set RESPIRE. (The indicator variable N_RESPONSE is created for use in future computations described below.)

```
data respire;
   input center treatment $ response $ count @@;
   n_response=(response='y');
   datalines;
1 test      y 29 1 test       n 16
1 placebo y 14 1 placebo n 31
2 test      y 37 2 test       n  8
2 placebo y 24 2 placebo n 21
;
```

Producing a Mantel-Haenszel analysis from PROC FREQ requires the specification of multiway tables. The triple crossing CENTER*TREATMENT*RESPONSE specifies that the data consists of sets of two-way tables. The two rightmost variables TREATMENT and RESPONSE determine the rows and columns of the tables, respectively, and the variables to the left (CENTER) determine the stratification scheme. There will be one table for each value of CENTER. If there are more variables to the left of the variables determining the rows and columns of the tables, there will be strata for each unique combination of values for those variables.

```
proc freq order=data;
   weight count;
   tables center*treatment*response /
       nocol nopct chisq cmh(mf);
run;
```

The ORDER=DATA option specifies that PROC FREQ order the rows and columns according to the order in which the variable values are encountered in the input data. The CHISQ option specifies that chi-square statistics be printed for each table. The CMH option requests the Mantel-Haenszel statistics for the stratified analysis; these are also called summary statistics. The MF option in the TABLES statement requests the Mantel-Fleiss criterion.

Output 3.1 and Output 3.2 display the frequency tables and chi-square statistics for each center. For Center 1, the favorable rate for test treatment is 64%, versus 31% for placebo. For Center 2, the favorable rate for test treatment is 82%, versus 53% for placebo. Q (the randomization statistic discussed in Chapter 2) for Center 1 is 9.908; Q for Center 2 is 8.503. With 1 df, both of these statistics are strongly significant.

Output 3.1 Table 1 Results

Frequency Row Pct	Table 1 of treatment by response		
	Controlling for center=1		
	response		
treatment	y	n	Total
test	29 64.44	16 35.56	45
placebo	14 31.11	31 68.89	45
Total	43	47	90

Output 3.1 *continued*

Statistic	DF	Value	Prob
Chi-Square	1	10.0198	0.0015
Likelihood Ratio Chi-Square	1	10.2162	0.0014
Continuity Adj. Chi-Square	1	8.7284	0.0031
Mantel-Haenszel Chi-Square	1	9.9085	0.0016
Phi Coefficient		0.3337	
Contingency Coefficient		0.3165	
Cramer's V		0.3337	

Output 3.2 Table 2 Results

Frequency Row Pct	Table 2 of treatment by response			
	Controlling for center=2			
		response		
	treatment	y	n	Total
	test	37 82.22	8 17.78	45
	placebo	24 53.33	21 46.67	45
	Total	61	29	90

Statistic	DF	Value	Prob
Chi-Square	1	8.5981	0.0034
Likelihood Ratio Chi-Square	1	8.8322	0.0030
Continuity Adj. Chi-Square	1	7.3262	0.0068
Mantel-Haenszel Chi-Square	1	8.5025	0.0035
Phi Coefficient		0.3091	
Contingency Coefficient		0.2953	
Cramer's V		0.3091	

Following the information for the individual tables, PROC FREQ prints out a section titled "Summary Statistics for treatment by response Controlling for center." This includes tables containing Mantel-Haenszel (MH) statistics, estimates of the common relative risk, and the Breslow-Day test for homogeneity of the odds ratio.

Output 3.3 Summary Statistics

Cochran-Mantel-Haenszel Statistics (Based on Table Scores)				
Statistic	Alternative Hypothesis	DF	Value	Prob
1	Nonzero Correlation	1	18.4106	<.0001
2	Row Mean Scores Differ	1	18.4106	<.0001
3	General Association	1	18.4106	<.0001

Mantel-Fleiss Criterion	36.0000

To find the value of Q_{MH}, read the value for "General Association." The other statistics pertain to the situation where you have sets of tables with two or more rows or columns; they are discussed in Chapter 4, "$2 \times r$ and $s \times 2$ Tables" and Chapter 6, "Sets of $s \times r$ Tables." However, they all reduce to the MH statistic when you have 2×2 tables and use the CMH option in its default mode (that is, no SCORE= option specified). Note that the General Association statistic is always appropriate for sets of 2×2 tables regardless of the scores used.

Q_{MH} for these data is $Q_{MH} = 18.4106$, with 1 df. This is clearly significant. The associations in the individual tables reinforce each other so that the overall association is stronger than that seen in the individual tables. There is a strong association between treatment and response, adjusting for center. The test treatment had a significantly higher favorable response rate than placebo.

The information in the rest of the summary statistics output is discussed later in this chapter. Note that for these data, the Mantel-Fleiss criterion is satisfied:

$$\sum_{h=1}^{2} m_{h11} = 21.5 + 30.5 = 52$$

$$\sum_{h=1}^{2} (n_{h11})_L = 0 + 16 = 16$$

$$\sum_{h=1}^{2} (n_{h11})_U = 43 + 45 = 88$$

so that $(52 - 16) \geq 5$ and $(88 - 52) \geq 5$.

Sometimes the direction of the effect being assessed with the MH statistic is not apparent, especially when some of the cell counts are small. The following measure serves to quantify the effect and its direction; it is a weighted average of the difference of proportions for the first and second rows of the tables, averaged across the set of tables.

$$d = \frac{\sum_h w_h (p_{h11} - p_{h21})}{\sum_h w_h}$$

where $w_h = (n_{h1+} n_{h2+}/n_h)$.

One way to produce d is to use the GLM procedure to fit a model to the responses represented by a (0, 1) response variable with group and strata as the explanatory variables and then form d by computing a group difference with the ESTIMATE statement. Here, you fit a linear model based on the previously created variable N_RESPONSE, which takes the value 1 for Yes and 0 otherwise, and include CENTER and TREATMENT as the explanatory variables. The following ESTIMATE statement produces an estimate of d. Since the GLM procedure orders the levels of the effects alphanumerically, the coefficients [–1 1] form the difference of placebo from test, averaged across center.

```
proc glm;
   class center treatment;
   freq count;
   model n_response=center treatment;
   estimate 'direction' treatment -1 1;
run;
```

Output 3.4 displays the results from the ESTIMATE statement, with $d = 0.3111$. The test treatment has an average proportion of Yes outcomes that is 0.3111 higher than the comparable proportion for placebo across centers. However, note that the associated standard error is inappropriate since the GLM procedure assumes homogeneous variance. This computing device is suitable only for determining the distance measure, not its standard error.

Output 3.4 Direction Estimate

Dependent Variable: n_response

Parameter	Estimate	Standard Error	t Value	Pr > |t|
direction	0.31111111	0.06884899	4.52	<.0001

You can construct an appropriate confidence interval $d \pm z_{\alpha/2}\sqrt{v_d}$ for this measure based on the binomial distribution where

$$v_d = \frac{\sum_{h=1}^{q} w_h^2 \left\{ \frac{p_{h11}^*(1-p_{h11}^*)}{n_{h1+}} + \frac{p_{h21}^*(1-p_{h21}^*)}{n_{h2+}} \right\}}{(\sum_h w_h)^2}$$

and

$$p_{hi}^* = \frac{(n_{hi1} + 0.5)}{(n_{hi+} + 1)}$$

Note that this computation includes a continuity correction for the variance. For these data, the difference measure and confidence interval is 0.3111 ± 0.1343.

3.2.2 Health Policy Data

Another data set discussed in Chapter 2 was also a subset of the complete data. The health policy data displayed in Table 2.7 comes from a study that included interviews with subjects from both rural and urban geographic regions. Table 2.7 displays the information from the rural region, and Table 3.3 includes the complete data.

Table 3.3 Health Policy Opinion Data

Residence	Stress	Favorable	Unfavorable	Total
Urban	Low	48	12	60
Urban	High	96	94	190
	Total	144	106	250
Rural	Low	55	135	190
Rural	High	7	53	60
	Total	62	188	250

If you ignored region and pooled these two tables, you would obtain Table 3.4.

Table 3.4 Pooled Health Policy Opinion Data

Stress	Favorable	Unfavorable	Total
Low	103	147	250
High	103	147	250
Total	206	294	500

There is clearly no association in this table; the proportions for favorable opinion are the same for low stress and high stress. For this table, Q_P and Q take the value 0, and the odds ratio is exactly 1. These data illustrate the need to consider the sampling framework in any data analysis. If you note the row totals in Table 3.3, you see that high stress subjects were more prevalent for the urban region, and the low stress subjects were more prevalent for the rural region. This oversampling causes the pooled table to take its form, even though favorable response is more likely for low stress persons in both regions.

The fact that a marginal table (pooled over residence) may exhibit an association completely different from the partial tables (individual tables for urban and rural) is known as *Simpson's Paradox* (Simpson 1951, Yule 1903).

The following statements request a Mantel-Haenszel analysis for the health policy data.

```
data stress;
   input region $ stress $ outcome $ count @@;
   n_outcome=(outcome='f');
   datalines;
urban low  f 48 urban low  u  12
urban high f 96 urban high u  94
rural low  f 55 rural low  u 135
rural high f  7 rural high u  53
;

proc freq order=data;
   weight count;
   tables region*stress*outcome / chisq cmh nocol nopct;
run;
```

Output 3.5 and Output 3.6 display the results for the individual tables. The urban region has a Q of 16.1549 for the association of stress level and health policy opinion; the Q for the rural region is 7.2724. The rate of favorable response is higher for the low stress group than for the high stress group in each region.

Output 3.5 Table 1 Results

Frequency Row Pct	Table 1 of stress by outcome			
	Controlling for region=urban			
		outcome		
	stress	f	u	Total
	low	48 80.00	12 20.00	60
	high	96 50.53	94 49.47	190
	Total	144	106	250

Statistic	DF	Value	Prob
Chi-Square	1	16.2198	<.0001
Likelihood Ratio Chi-Square	1	17.3520	<.0001
Continuity Adj. Chi-Square	1	15.0354	0.0001
Mantel-Haenszel Chi-Square	1	16.1549	<.0001
Phi Coefficient		0.2547	
Contingency Coefficient		0.2468	
Cramer's V		0.2547	

Fisher's Exact Test	
Cell (1,1) Frequency (F)	48
Left-sided Pr <= F	1.0000
Right-sided Pr >= F	3.247E-05
Table Probability (P)	2.472E-05
Two-sided Pr <= P	4.546E-05

Output 3.6 Table 2 Results

Frequency Row Pct	Table 2 of stress by outcome		
	Controlling for region=rural		
	outcome		
stress	f	u	Total
low	55 28.95	135 71.05	190
high	7 11.67	53 88.33	60
Total	62	188	250

Statistic	DF	Value	Prob
Chi-Square	1	7.3016	0.0069
Likelihood Ratio Chi-Square	1	8.1976	0.0042
Continuity Adj. Chi-Square	1	6.4044	0.0114
Mantel-Haenszel Chi-Square	1	7.2724	0.0070
Phi Coefficient		0.1709	
Contingency Coefficient		0.1685	
Cramer's V		0.1709	

Fisher's Exact Test	
Cell (1,1) Frequency (F)	55
Left-sided Pr <= F	0.9988
Right-sided Pr >= F	0.0041
Table Probability (P)	0.0029
Two-sided Pr <= P	0.0061

From Output 3.7 you can see that Q_{MH} has the value 23.050, which is strongly significant. Stress is highly associated with health policy opinion, adjusting for regional effects.

Output 3.7 Summary Statistics

Cochran-Mantel-Haenszel Statistics (Based on Table Scores)				
Statistic	Alternative Hypothesis	DF	Value	Prob
1	Nonzero Correlation	1	23.0502	<.0001
2	Row Mean Scores Differ	1	23.0502	<.0001
3	General Association	1	23.0502	<.0001

Note that, for these tables, $d = 0.2335$. The favorable proportion is 0.2335 higher on the average for the low stress groups than for the high stress groups across the regions.

3.2.3 Soft Drink Example

The following data come from a study on soft drink tastes by a company interested in reactions to a new soft drink that was being targeted for both the United States and Great Britain. Investigators poststratified on gender because they thought it was potentially related to the response. After receiving a supply of the new soft drink and being given a week in which to try it, subjects were asked whether they would want to switch from their current soft drinks to this new soft drink.

Table 3.5 Soft Drink Data

Gender	Country	Switch? Yes	No	Total
Male	American	29	6	35
Male	British	19	15	34
Total		48	21	69
Female	American	7	23	30
Female	British	24	29	53
Total		31	52	83

The following statements produce a Mantel-Haenszel analysis.

```
data soft;
   input gender $ country $ question $ count @@;
   datalines;
male    American  y 29 male    American  n  6
male    British   y 19 male    British   n 15
female  American  y  7 female  American  n 23
female  British   y 24 female  British   n 29
;

proc freq order=data;
   weight count;
   tables gender*country*question /
       chisq cmh nocol nopct;
run;
```

58 Chapter 3: Sets of 2 × 2 Tables

Output 3.8 and Output 3.9 display the table results for males and females.

Output 3.8 Summary Statistics for Males

Frequency Row Pct	Table 1 of country by question		
	Controlling for gender=male		
	question		
country	y	n	Total
American	29 82.86	6 17.14	35
British	19 55.88	15 44.12	34
Total	48	21	69

Statistic	DF	Value	Prob
Chi-Square	1	5.9272	0.0149
Likelihood Ratio Chi-Square	1	6.0690	0.0138
Continuity Adj. Chi-Square	1	4.7216	0.0298
Mantel-Haenszel Chi-Square	1	5.8413	0.0157
Phi Coefficient		0.2931	
Contingency Coefficient		0.2813	
Cramer's V		0.2931	

Fisher's Exact Test	
Cell (1,1) Frequency (F)	29
Left-sided Pr <= F	0.9968
Right-sided Pr >= F	0.0143
Table Probability (P)	0.0112
Two-sided Pr <= P	0.0194

Output 3.9 Summary Statistics for Females

Frequency Row Pct	Table 2 of country by question		
	Controlling for gender=female		
	question		
country	y	n	Total
American	7 23.33	23 76.67	30
British	24 45.28	29 54.72	53
Total	31	52	83

Statistic	DF	Value	Prob
Chi-Square	1	3.9443	0.0470
Likelihood Ratio Chi-Square	1	4.0934	0.0431
Continuity Adj. Chi-Square	1	3.0620	0.0801
Mantel-Haenszel Chi-Square	1	3.8968	0.0484
Phi Coefficient		-0.2180	
Contingency Coefficient		0.2130	
Cramer's V		-0.2180	

Fisher's Exact Test	
Cell (1,1) Frequency (F)	7
Left-sided Pr <= F	0.0385
Right-sided Pr >= F	0.9881
Table Probability (P)	0.0267
Two-sided Pr <= P	0.0602

As indicated by Q for males (5.8413) and Q for females (3.8968), there is significant association in both tables between country and willingness to switch. However, look at Q_{MH} in Output 3.10.

Output 3.10 Summary Statistics

Cochran-Mantel-Haenszel Statistics (Based on Table Scores)				
Statistic	Alternative Hypothesis	DF	Value	Prob
1	Nonzero Correlation	1	0.0243	0.8762
2	Row Mean Scores Differ	1	0.0243	0.8762
3	General Association	1	0.0243	0.8762

Q_{MH} takes the value 0.024, thus not detecting any association between country and willingness to switch, after adjusting for gender. However, if you examine the individual tables more closely, you see that the association is manifested in opposite directions. For males, Americans are overwhelmingly favorable, and the British are a little more favorable than unfavorable. For females, Americans are very opposed, while the British are mildly opposed.

One way to determine whether the association is in the same direction is through graphical means. Plotting the differences of the binomial proportions across tables tells you what is going on visually. You can generate this plot with the PLOTS=RISKDIFFPLOT option in the TABLES statement of PROC FREQ. Additionally, you can examine the odds ratios plot, requested with the ODDSRATIOPLOT option. The suboption LOGBASE=2 specifies that the horizonal axis be presented on the log scale.

```
ods graphics on;
proc freq order=data;
   weight count;
   tables gender*country*question / riskdiff(cl=(wald) correct) measures
          plots=(riskdiffplot oddsratioplot(logbase=2));
run;
ods graphics off;
```

Output 3.11 displays the proportion difference graph, which clearly shows that the nature of the association is completely different for males and females.

Output 3.11 Direction of Association for Soft Drink Data – Proportion Differences

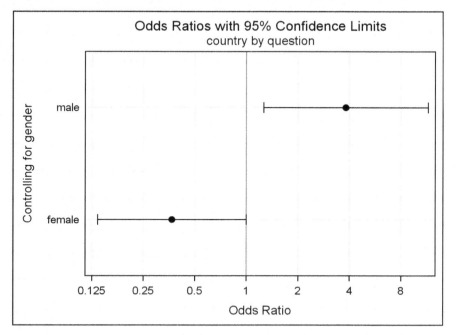

Output 3.12 displays the odds ratios, which also illustrate that the nature of the association is the opposite for males and females.

Output 3.12 Direction of Association for Soft Drink Data – Odds Ratios

Thus, for these data, Q_{MH} fails to detect an association because the association is of opposite directions with roughly the same magnitude. As discussed previously, Q_{MH} has power against the alternative hypothesis of consistent patterns of association; it has low power for detecting association in opposite directions. However, regardless of these matters of power, the method always performs at the specified significance level (or less) under the null hypothesis of no association in all q strata, so it is always valid.

Generally, the power of Q_{MH} isn't a problem because if there is association, it is usually in the same direction across a set of tables, although often to varying degrees. However, you should always examine the individual tables, especially if your results are puzzling, to determine whether you have a situation in which the association is inconsistent and the Q_{MH} statistic is not very powerful. A graph can point this out easily.

3.3 Measures of Association

Section 2.5 discusses the odds ratio as a measure of association for the 2×2 table. You can compute average odds ratios for sets of 2×2 tables. For the hth stratum,

$$\text{OR}_h = \frac{p_{h1}/(1-p_{h1})}{p_{h2}/(1-p_{h2})} = \frac{n_{h11}n_{h22}}{n_{h12}n_{h21}}$$

so OR_h estimates ψ_h, the population odds ratio for the hth stratum. If the ψ_h are homogeneous, then you can compute the Mantel-Haenszel estimator for the common odds ratio ψ.

$$\hat{\psi}_{MH} = \sum_{h=1}^{q} \frac{n_{h11}n_{h22}}{n_h} \bigg/ \sum_{h=1}^{q} \frac{n_{h12}n_{h21}}{n_h}$$

The standard error for this estimator is based on work by Robins, Breslow, and Greenland (1986) in which they provide an estimated variance for $\log \hat{\psi}_{MH}$. The $100\%(1-\alpha)$ confidence interval for $\hat{\psi}_{MH}$ is

$$\left(\hat{\psi}_{MH} \cdot \exp(-z_{\alpha/2}\hat{\sigma}),\ \hat{\psi}_{MH} \cdot \exp(z_{\alpha/2}\hat{\sigma}) \right)$$

where

$$\hat{\sigma}^2 = \hat{\text{var}}[\log(\hat{\psi}_{MH})]$$
$$= \frac{\sum_h (n_{h11} + n_{h22})(n_{h11}\,n_{h22})/n_h^2}{2\left(\sum_h n_{h11}\,n_{h22}/n_h\right)^2}$$
$$+ \frac{\sum_h [(n_{h11} + n_{h22})(n_{h12}\,n_{h21}) + (n_{h12} + n_{h21})(n_{h11}\,n_{h22})]/n_h^2}{2\left(\sum_h n_{h11}\,n_{h22}/n_h\right)\left(\sum_h n_{h12}\,n_{h21}/n_h\right)}$$
$$+ \frac{\sum_h (n_{h12} + n_{h21})(n_{h12}\,n_{h21})/n_h^2}{2\left(\sum_h n_{h12}\,n_{h21}/n_h\right)^2}$$

Another estimator of ψ is the logit estimator. This is a weighted regression estimate with the form

$$\hat{\psi}_L = \exp\left\{ \sum_{h=1}^{q} w_h f_h \bigg/ \sum_{h=1}^{q} w_h \right\} = \exp\{\bar{f}\}$$

where $f_h = \log OR_h$ and

$$w_h = \left\{ \frac{1}{n_{h11}} + \frac{1}{n_{h12}} + \frac{1}{n_{h21}} + \frac{1}{n_{h22}} \right\}^{-1}$$

You can write a $100(1-\alpha)\%$ confidence interval for $\hat{\psi}_L$ as

$$\exp\left\{ \bar{f} \pm z_{\alpha/2} \left[\sum_{h=1}^{q} w_h \right]^{-1/2} \right\}$$

The logit estimator is also reasonable but requires adequate sample sizes (all $n_{hij} \geq 5$); it has problems with zero cells for the n_{hij}, in which case you should proceed cautiously. The Mantel-Haenszel estimator is not as sensitive to sample size.

Logistic regression provides a better strategy for estimating the common odds ratio and produces a confidence interval based on maximum likelihood methods. This topic is discussed in Chapter 8, "Logistic Regression I: Dichotomous Response" and Chapter 10, "Conditional Logistic Regression."

3.3.1 Homogeneity of Odds Ratios

You are generally interested in whether the odds ratios in a set of tables are homogeneous. There are several test statistics that address the hypothesis of homogeneity, one of which is the Breslow-Day statistic.

Consider Table 3.6. The top table shows the expected counts m_{ij} for a 2 × 2 table, and the bottom table shows how you can write the expected counts for the rest of the cells if you know the (1,1) expected count m_{11}.

Table 3.6 Odds Ratios

	Yes	No	Total
Group 1	m_{11}	m_{12}	n_{1+}
Group 2	m_{21}	m_{22}	n_{2+}
Total	n_{+1}	n_{+2}	n
Group 1	m_{11}	$n_{1+} - m_{11}$	n_{1+}
Group 2	$n_{+1} - m_{11}$	$n - n_{1+} - n_{+1} + m_{11}$	$n - n_{1+}$
Total	n_{+1}	$n - n_{+1}$	n

If you assume that the odds ratio takes a certain value, $\psi = \psi_0$, then

$$\frac{m_{11}(n - n_{1+} - n_{+1} + m_{11})}{(n_{+1} - m_{11})(n_{1+} - m_{11})} = \psi_0$$

You can put this expression into the form of a quadratic equation and then solve for m_{11}; once you have m_{11}, you can solve for the other expected counts.

To compute the Breslow-Day statistic, you use ψ_{MH} as ψ_0 for each stratum and perform the preceding computations for the expected counts for each table; that is, you compute the m_{hij}. Then,

$$Q_{BD} = \sum_{h=1}^{q} \sum_{i=1}^{2} \sum_{j=1}^{2} \frac{(n_{hij} - m_{hij})^2}{m_{hij}}$$

Under the null hypothesis of homogeneity, Q_{BD} approximately has a chi-square distribution with $(q-1)$ degrees of freedom. In addition, the cells in all of the tables must have expected cell counts greater than 5 (or at least 80% of them should). Note that a chi-square approximation for Q_{MH} requires only the total sample size to be large, but the chi-square approximation for Q_{BD} requires each table to have a large sample size. If the odds ratios are not homogeneous, then the overall odds ratio should be viewed cautiously; the within-strata odds ratios should be emphasized.

The Mantel-Haenszel statistics do not require homogeneous odds ratios, so the Breslow-Day test should not be interpreted as an indicator of their validity. See Breslow and Day (1980, p. 182) for more information. Also, Chapter 8, "Logistic Regression I: Dichotomous Response," provides better test statistics than Q_{BD} to address homogeneity through goodness of fit statistics for logistic regression models which invoke homogeneous odds ratios in their structure.

3.3.2 Coronary Artery Disease Data Example

Table 3.7 contains data that are based on a study on coronary artery disease (Koch, Imrey et al. 1985). The sample is one of convenience since the patients studied were people who came to a clinic and requested an evaluation.

Table 3.7 Coronary Artery Disease Data

Sex	ECG	No Disease	Disease	Total
Female	< 0.1 ST segment depression	11	4	15
Female	≥ 0.1 ST segment depression	10	8	18
Male	< 0.1 ST segment depression	9	9	18
Male	≥ 0.1 ST segment depression	6	21	27

Investigators were interested in whether electrocardiogram (ECG) measurement was associated with disease status. Gender was thought to be associated with disease status, so investigators poststratified the data into male and female groups. In addition, there was interest in examining the odds ratios.

The following statements produce the SAS data set CA and request a stratified analysis. The first TABLES statement requests chi-square tests for the association of gender and disease status. The second TABLES statement requests the stratified analysis, including the generation of odds ratios with the MEASURES option.

```
data ca;
   input gender $ ECG $ disease $ count;
   datalines;
female <0.1   yes    4
```

```
female  <0.1   no    11
female  >=0.1  yes    8
female  >=0.1  no    10
male    <0.1   yes    9
male    <0.1   no     9
male    >=0.1  yes   21
male    >=0.1  no     6
;

proc freq;
   weight count;
   tables gender*disease / nocol nopct chisq;
   tables gender*ECG*disease / nocol nopct cmh chisq measures;
run;
```

Output 3.13 contains the table of GENDER by DISEASE. Q takes the value 6.9444, and Q_P takes the value 7.0346. Obviously there is a strong association between gender and disease status. Males with symptoms are much more likely to have a confirmed diagnosis of coronary artery disease than females. The idea to control for gender in a stratified analysis is a good one.

Note that you are controlling for confounding in this example. Confounding variables are those related to both the response and the factor under investigation. In the respiratory data analysis, the stratification variable (center) was part of the study design, and in the soft drink analysis, the stratification variable (gender) was thought to be related to the response. Adjusting for confounding is often required in epidemiological studies.

Output 3.13 GENDER × DISEASE

Frequency Row Pct	Table of gender by disease		
		disease	
gender	no	yes	Total
female	21 63.64	12 36.36	33
male	15 33.33	30 66.67	45
Total	36	42	78

Statistic	DF	Value	Prob
Chi-Square	1	7.0346	0.0080
Likelihood Ratio Chi-Square	1	7.1209	0.0076
Continuity Adj. Chi-Square	1	5.8681	0.0154
Mantel-Haenszel Chi-Square	1	6.9444	0.0084
Phi Coefficient		0.3003	
Contingency Coefficient		0.2876	
Cramer's V		0.3003	

Output 3.13 *continued*

Fisher's Exact Test	
Cell (1,1) Frequency (F)	21
Left-sided Pr <= F	0.9981
Right-sided Pr >= F	0.0075
Table Probability (P)	0.0056
Two-sided Pr <= P	0.0114

Output 3.14 and Output 3.15 display the individual tables results for ECG × disease status; included are the table of chi-square statistics generated by the CHISQ option and only the "Estimates of the Relative Risk" table part of the output generated by the MEASURES option.

Output 3.14 Results for Females

Statistic	DF	Value	Prob
Chi-Square	1	1.1175	0.2905
Likelihood Ratio Chi-Square	1	1.1337	0.2870
Continuity Adj. Chi-Square	1	0.4813	0.4879
Mantel-Haenszel Chi-Square	1	1.0836	0.2979
Phi Coefficient		0.1840	
Contingency Coefficient		0.1810	
Cramer's V		0.1840	

Fisher's Exact Test	
Cell (1,1) Frequency (F)	11
Left-sided Pr <= F	0.9233
Right-sided Pr >= F	0.2450
Table Probability (P)	0.1683
Two-sided Pr <= P	0.4688

Estimates of the Relative Risk (Row1/Row2)			
Type of Study	Value	95% Confidence Limits	
Case-Control (Odds Ratio)	2.2000	0.5036	9.6107
Cohort (Col1 Risk)	1.3200	0.7897	2.2063
Cohort (Col2 Risk)	0.6000	0.2240	1.6073

Q_{MH} is 1.084 for females, with a p-value of 0.2979. The odds ratio for the females is OR = 2.2, with a 95% confidence interval that includes 1. Those females with higher ST segment depression levels had 2.2 times the odds of CA disease than those with lower levels.

Output 3.15 Results for Males

Statistic	DF	Value	Prob
Chi-Square	1	3.7500	0.0528
Likelihood Ratio Chi-Square	1	3.7288	0.0535
Continuity Adj. Chi-Square	1	2.6042	0.1066
Mantel-Haenszel Chi-Square	1	3.6667	0.0555
Phi Coefficient		0.2887	
Contingency Coefficient		0.2774	
Cramer's V		0.2887	

Fisher's Exact Test	
Cell (1,1) Frequency (F)	9
Left-sided Pr <= F	0.9880
Right-sided Pr >= F	0.0538
Table Probability (P)	0.0417
Two-sided Pr <= P	0.1049

Estimates of the Relative Risk (Row1/Row2)			
Type of Study	Value	95% Confidence Limits	
Case-Control (Odds Ratio)	3.5000	0.9587	12.7775
Cohort (Col1 Risk)	2.2500	0.9680	5.2298
Cohort (Col2 Risk)	0.6429	0.3883	1.0642

Q_{MH} takes the value 3.667 for males, with a p-value of 0.056. The odds ratio for the males is OR = 3.5, with a 95% confidence interval that barely contains the value 1. Men with higher ST segment depression levels had 3.5 times the odds of CA disease than those with lower levels.

Output 3.16 contains the Q_{MH} statistic, which takes the value 4.503 with a p-value of 0.0338. By combining the genders, the power has been increased so that the association detected by Q_{MH} is significant at the $\alpha = 0.05$ level of significance.

Output 3.16 Stratified Analysis

Cochran-Mantel-Haenszel Statistics (Based on Table Scores)				
Statistic	Alternative Hypothesis	DF	Value	Prob
1	Nonzero Correlation	1	4.5026	0.0338
2	Row Mean Scores Differ	1	4.5026	0.0338
3	General Association	1	4.5026	0.0338

Output 3.17 contains the estimates of the common odds ratios. $\hat{\psi}_{MH} = 2.847$ and $\hat{\psi}_L = 2.859$. The confidence intervals do not contain the value 1. For the combined genders, persons with higher ST segment depression levels had nearly three times the odds of CA disease than those with lower levels.

Output 3.17 Odds Ratios

Estimates of the Common Relative Risk (Row1/Row2)				
Type of Study	Method	Value	95% Confidence Limits	
Case-Control	Mantel-Haenszel	2.8467	1.0765	7.5279
(Odds Ratio)	Logit	2.8593	1.0807	7.5650
Cohort	Mantel-Haenszel	1.6414	1.0410	2.5879
(Col1 Risk)	Logit	1.5249	0.9833	2.3647
Cohort	Mantel-Haenszel	0.6299	0.3980	0.9969
(Col2 Risk)	Logit	0.6337	0.4046	0.9926

Common measures of relative risk are also printed by the FREQ procedure. However, since these data do not come from a prospective study, these statistics are not relevant and should be ignored.

Finally, Output 3.18 displays the results of the Breslow-Day test. It does not contradict homogeneity of the odds ratios for these data, with $Q_{BD} = 0.215$ and $p = 0.6425$.

Output 3.18 Breslow-Day Test

Breslow-Day Test for Homogeneity of the Odds Ratios	
Chi-Square	0.2155
DF	1
Pr > ChiSq	0.6425

3.4 Exact Confidence Intervals for the Common Odds Ratio

Consider the data in Table 3.8. A small company initiated exercise programs at both of its locations, downtown and a satellite office in a nearby suburb. The office program consisted of directed aerobic activities such as running, walking, and bicycling, conducted under the guidance of an exercise counselor. The home program consisted of a range of activities that were self-monitored. Each employee signed an agreement to participate in a program and to check in monthly to ensure continual effort. After a year, participants and non-participants underwent a cardiovascular stress test to assess their fitness, and their result was recorded as good or not good depending on age-adjusted criteria. The exercise counselor was interested in whether type of program was associated with good test results.

Table 3.8 Cardiovascular Test Outcomes

Location	Program	Good	Not Good	Total
Downtown	Office	12	5	17
Downtown	Home	3	5	8
	Total	15	10	25
Satellite	Office	6	1	7
Satellite	Home	1	3	4
	Total	7	4	11

Interest lies in computing an odds ratio comparing good results for the office program compared to the home program. However, the counts in these tables are too small to be able to justify the asymptotic confidence limits for the odds ratio.

Exact counterparts are available. The exact confidence limits for the odds ratios are constructed from the distribution of $S = \sum_h n_{h11}$, conditional on the marginal totals of the 2 × 2 tables. This makes the assumption that the odds ratio is constant over the tables. Note that the confidence coefficient for the exact limits is at least $(1-\alpha)$. See Agresti (1992) for a comprehensive discussion.

These exact confidence limits can also be constructed by performing exact logistic regression, which is discussed in Chapter 8. While the exact confidence limits will be identical, that method also produces a regression-adjusted estimate of the odds ratio instead of the MH estimator produced by PROC FREQ.

Zelen's test is an exact analogue to the Breslow-Day test for equal odds ratios for sets of 2 × 2 tables. It is constructed in a similar fashion to Fisher's exact test. The *p*-value is the sum of all the table probabilities that are less than or equal to the observed table probability, where that sum is computed for all tables with the same two-way fixed margins as the observed set of tables. See Zelen (1971) and Hirji et al. (1996) for more information. Note that the test of the interaction of location and program with exact logistic regression would serve the same purpose as Zelen's test.

The following DATA step inputs the cardiovascular test data into the SAS data set EXERCISE.

```
data exercise;
   input location $ program $ outcome $ count @@;
   datalines;
```

```
downtown office   good 12 downtown office not 5
downtown home     good  3 downtown home   not 5
satellite office  good  6 satellite office not 1
satellite home    good  1 satellite home   not 3
;
```

The following PROC FREQ statements request the exact analysis. The CMH(MF) option requests that the Mantel-Fleiss criterion be produced for the CMH results. The COMOR option in the EXACT statement requests the exact common odds ratio confidence limits, and the EQOR option requests Zelen's test.

```
proc freq order=data;
   weight count;
   tables location*program*outcome /cmh(mf);
   exact comor eqor;
run;
```

The Mantel-Fleiss criterion has the value 4.6545, which is less than the proscribed limit of 5. This result isn't unexpected given the small numbers in the tables, and you would proceed cautiously in reporting the CMH results.

Output 3.19 Mantel-Fleiss Criterion

Summary Statistics for program by outcome
Controlling for location

Cochran-Mantel-Haenszel Statistics (Based on Table Scores)				
Statistic	Alternative Hypothesis	DF	Value	Prob
1	Nonzero Correlation	1	5.5739	0.0182
2	Row Mean Scores Differ	1	5.5739	0.0182
3	General Association	1	5.5739	0.0182

Mantel-Fleiss Criterion	4.6545	Warning: Criterion < 5

The odds ratio is a valid measure of association for these data. Output 3.20 provides the MH estimate of the common odds ratio as 5.8421 and both the asymptotic and exact confidence limits. The exact limits (1.0486, 33.3124) are the appropriate confidence limits for the common odds ratio for these data. In this case, these exact limits are wider than the asymptotic limits and have the lower limit closer to 1 for no association.

Output 3.20 Exact Confidence Limits

Common Odds Ratio	
Mantel-Haenszel Estimate	5.8421
Asymptotic Conf Limits	
95% Lower Conf Limit	1.3012
95% Upper Conf Limit	26.2296
Exact Conf Limits	
95% Lower Conf Limit	1.0486
95% Upper Conf Limit	33.3124

Output 3.21 reports the Zelen test for homogeneity of the odds ratio; with a p-value of 1, there is no evidence to the contrary.

Output 3.21 Zelen's Test Results

Tests for Homogeneity of Odds Ratios	
Breslow-Day Chi-Square	0.6970
DF	1
Pr > ChiSq	0.4038
Zelen's Exact Test (P)	0.4917
Exact Pr <= P	1.0000

Chapter 4

$2 \times r$ and $s \times 2$ Tables

Contents

4.1	Introduction		**73**
4.2	The $2 \times r$ Table		**74**
4.3	Sets of $2 \times r$ Tables		**76**
	4.3.1	Choosing Scores	78
	4.3.2	Analyzing the Arthritis Data	80
	4.3.3	Rank Statistics for Ordered Data	82
	4.3.4	Colds Example	85
4.4	The $s \times 2$ Table		**89**
4.5	Sets of $s \times 2$ Tables		**93**
	4.5.1	Correlation Statistic	93
	4.5.2	Analysis of Smokeless Tobacco Data	94
	4.5.3	Pain Data Analysis	95
4.6	Relationships between Sets of Tables		**100**
4.7	Exact Analysis of Association for the $s \times 2$ Table		**102**

4.1 Introduction

While 2×2 and sets of 2×2 tables are very common, many tables and sets of tables have other dimensions. This chapter focuses on tables and sets of tables that also occur frequently: $2 \times r$ tables in which the column variable is ordinally scaled and $s \times 2$ tables in which the row variable is ordinally scaled. For $2 \times r$ tables, there is interest in investigating a response variable with multiple ordered outcomes for a single table or for a combined set of strata. For example, you may be comparing a new treatment and a placebo on the extent of patient improvement that is rated as none, some, or marked. For $s \times 2$ tables, there is interest in the trend of proportions across ordered groups for a single table or for a combined set of strata. For example, you may be comparing the proportion of successful outcomes for different dosage levels of a new drug.

Section 4.2 addresses the $2 \times r$ table, and Section 4.4 addresses the $s \times 2$ table. Extensions of the Mantel-Haenszel strategy address association in sets of tables with these characteristics. Sections 4.3 and 4.5 discuss these strategies for sets of $2 \times r$ and sets of $s \times 2$ tables, respectively.

4.2 The 2 × r Table

Consider the data from Koch and Edwards (1988) displayed in Table 4.1. The information comes from a randomized, double-blind clinical trial investigating a new treatment for rheumatoid arthritis. Investigators compared the new treatment with a placebo; the response measured was whether there was no, some, or marked improvement in the symptoms of rheumatoid arthritis when the active treatment was administered.

Table 4.1 Combined Rheumatoid Arthritis Data

Treatment	Improvement			Total
	None	Some	Marked	
Active	13	7	21	41
Placebo	29	7	7	43
Total	42	14	28	84

As discussed in Chapter 1, "Introduction," you want to use the information in the ordinal column variable in forming a test statistic. This involves assigning scores to the response levels, forming means, and then examining location shifts of the means across the levels of the row variable.

Define the mean for the Active drug group as

$$\bar{f}_1 = \sum_{j=1}^{3} \frac{a_j n_{1j}}{n_{1+}}$$

where $\mathbf{a} = \{a_j\} = (a_1, a_2, a_3)$ is a set of scores reflecting the response levels with $a_1 \leq a_2 \leq a_3$ and at least one '$\leq$' is a '$<$'. Then, if the null hypothesis H_0 is no association whereby patients would have the same responses regardless of the treatment assignment

$$E\{\bar{f}_1|H_0\} = \sum_{j=1}^{3} \left(a_j \frac{n_{1+} n_{+j}}{n_{1+} n} \right) = \sum_{j=1}^{3} a_j \frac{n_{+j}}{n} = \mu_\mathbf{a}$$

It can be shown that

$$V\{\bar{f}_1|H_0\} = \frac{n - n_{1+}}{n_{1+}(n-1)} \sum_{j=1}^{3} (a_j - \mu_\mathbf{a})^2 \left(\frac{n_{+j}}{n} \right)$$

$$= \frac{(n - n_{1+}) v_\mathbf{a}}{n_{1+}(n-1)}$$

where $\mu_\mathbf{a}$ and $v_\mathbf{a}$ are the finite population mean and variance of scores $\mathbf{a}$ for the patients in the study. The quantity $\bar{f}_1$ approximately has a normal distribution by randomization central limit theory, so the quantity

$$Q_S = \frac{(\bar{f}_1 - \mu_a)^2}{\{(n - n_{1+})/[n_{1+}(n-1)]\}v_a}$$

$$= \frac{(\bar{f}_1 - \bar{f}_2)^2}{\{n^2/[n_{1+}n_{2+}(n-1)]\}v_a}$$

$$= \frac{(\bar{f}_1 - \bar{f}_2)^2}{\{\frac{1}{n_{1+}} + \frac{1}{n_{2+}}\}\{\frac{nv_a}{(n-1)}\}}$$

approximately has the chi-square distribution with one degree of freedom. Q_S is called the mean score statistic. By taking advantage of the ordinality of the response variable, Q_S can target the alternative hypothesis of location shifts to the hypothesis of no association with fewer degrees of freedom. While counterparts to Q and Q_P are useful for detecting general types of association (as discussed in Chapter 5, "The s × r Table"), they are not as effective as Q_S in detecting location shifts. Q_S is also a trend statistic for the tendency for the patients in one treatment group to have better scores than the patients in the other treatment group.

A very conservative sample size guideline is the guideline used for the Pearson chi-square statistic (that is, all expected values $n_{i+}n_{+j}/n = m_{ij}$ being greater than or equal to 5). However, one of the advantages of the mean score statistic is that it has less stringent sample size requirements. A more realistic but still conservative sample size guideline is to choose one or more cutpoints $j = (2, \ldots, (r-1))$, add the first through jth columns together and add the $(j+1)$th through rth columns together. If both of these sums are 5 or greater for at least one cutpoint in each row, then the sample size is adequate.

For example, for Table 4.1, choose $j = 2$. Adding the first and second columns together yields the sums 20 for the first row and 36 for the second; the remaining sums are just the third column cells (21 and 7, respectively). Thus, according to this criterion, the sample size is adequate; also, the added columns could differ for the two rows in order to satisfy this criterion.

The following PROC FREQ statements generate Q_S. Note the use of the ORDER=DATA option to ensure that the values for the variable RESPONSE are put in the correct order. If they are not, the resulting statistics do not account for the intended ordering. Ensuring the correct sort order is critical when you use statistics that assume ordered values.

```
data arth;
   input treat $ response $ count @@;
   datalines;
active    none 13   active    some 7   active   marked 21
placebo   none 29   placebo   some 7   placebo  marked  7
;

proc freq data=arth order=data;
   weight count;
   tables treat*response / chisq nocol nopct;
run;
```

The results are contained in Output 4.1.

Output 4.1 Mean Score Statistic

Frequency Row Pct	Table of treat by response				
		response			
	treat	none	some	marked	Total
	active	13 31.71	7 17.07	21 51.22	41
	placebo	29 67.44	7 16.28	7 16.28	43
	Total	42	14	28	84

Statistic	DF	Value	Prob
Chi-Square	2	13.0550	0.0015
Likelihood Ratio Chi-Square	2	13.5298	0.0012
Mantel-Haenszel Chi-Square	1	12.8590	0.0003
Phi Coefficient		0.3942	
Contingency Coefficient		0.3668	
Cramer's V		0.3942	

For a $2 \times r$ table, the statistic labeled "Mantel-Haenszel Chi-Square" is Q_S. The scores (1, 2, 3) are used for the response levels none, some, and marked in Table 4.1. Q_S takes the value 12.8590, which is strongly significant. The active treatment performs better than the placebo treatment.

You can also produce Q_S by specifying the CMH option and generating the summary statistics, which will be for just one stratum. Q_S is the statistic labeled "Row Mean Scores Differ" in the resulting summary statistics table.

4.3 Sets of $2 \times r$ Tables

Table 4.2 displays the data corresponding to Table 4.1 that have been broken out by gender.

Table 4.2 Rheumatoid Arthritis Data

Gender	Treatment	Improvement			Total
		None	Some	Marked	
Female	Active	6	5	16	27
Female	Placebo	19	7	6	32
Total		25	12	22	59
Male	Active	7	2	5	14
Male	Placebo	10	0	1	11
Total		17	2	6	25

These data comprise a set of two 2 × 3 tables. Interest lies in the association between treatment and degree of improvement, adjusting for gender effects. Degree of improvement is an ordinal response, since None, Some, and Marked are gradations of improvement.

Mantel (1963) proposed an extension of the Mantel-Haenszel strategy for the analysis of 2 × r tables when the response variable is ordinal. The extension involves computing mean scores for the responses and using the mean score differences across tables in the computation of a suitable test statistic, much as the difference in proportions across tables was the basis of the Mantel-Haenszel statistic.

Consider the following table as representative of q 2 × r tables, $h = 1, 2, \ldots, q$.

Table 4.3 *h*th Contingency Table

	Level of Column Variable				
	1	2	...	r	Total
Group 1	n_{h11}	n_{h12}	...	n_{h1r}	n_{h1+}
Group 2	n_{h21}	n_{h22}	...	n_{h2r}	n_{h2+}
Total	n_{h+1}	n_{h+2}	...	n_{h+r}	n_h

For the rheumatoid arthritis data in Table 4.2, $r = 3$ and $q = 2$. Under the null hypothesis of no difference in treatment effects for each patient, the appropriate probability model is

$$\Pr\{n_{hij}\} = \prod_{h=1}^{2} \frac{\prod_{i=1}^{2} n_{hi+}! \prod_{j=1}^{3} n_{h+j}!}{n_h! \prod_{i=1}^{2} \prod_{j=1}^{3} n_{hij}!}$$

Here, n_{hij} represents the number of patients in the *h*th stratum who received the *i*th treatment and had the *j*th response.

Suppose $\{a_{hj}\}$ is a set of scores for the response levels in the *h*th stratum. Then you can compute the sum of strata scores for the first treatment, active, as

$$f_{+1+} = \sum_{h=1}^{2} \sum_{j=1}^{3} a_{hj} n_{h1j} = \sum_{h=1}^{2} n_{h1+} \bar{f}_{h1}$$

where

$$\bar{f}_{h1} = \sum_{j=1}^{3} (a_{hj} n_{h1j} / n_{h1+})$$

is the mean score for Group 1 in the *h*th stratum. Under the null hypothesis of no association, f_{+1+} has the expected value

$$E\{f_{+1+} | H_0\} = \sum_{h=1}^{2} n_{h1+} \mu_h = \mu_*$$

and variance

$$V\{f_{+1+}|H_0\} = \sum_{h=1}^{2} \frac{n_{h1+}(n_h - n_{h1+})}{(n_h - 1)} v_h = v_*$$

where $\mu_h = \sum_{j=1}^{3}(a_{hj}n_{h+j}/n_h)$ is the finite subpopulation mean and

$$v_h = \sum_{j=1}^{3}(a_{hj} - \mu_h)^2 (n_{h+j}/n_h)$$

is the variance of scores for the hth stratum.

If the across-strata sample sizes $n_{+i+} = \sum_{h=1}^{q} \sum_{j=1}^{r} n_{hij}$ are sufficiently large, then f_{+1+} approximately has a normal distribution, and so the quantity

$$Q_{SMH} = \frac{(f_{+1+} - \mu_*)^2}{v_*}$$

approximately has a chi-square distribution with one degree of freedom. Q_{SMH} is known as the extended Mantel-Haenszel mean score statistic; it is sometimes called the ANOVA statistic. You can show that Q_{SMH} is a function of the weighted average of the differences in the mean scores of the two treatments for the q strata.

$$Q_{SMH} = \frac{\{\sum_{h=1}^{q} n_{h1+}(\bar{f}_{h1} - \mu_h)\}^2}{\sum_{h=1}^{q} n_{h1+}n_{h2+}v_h/(n_h - 1)}$$
$$= \frac{\{\sum_{h=1}^{q} (n_{h1+}n_{h2+}/n_h)(\bar{f}_{h1} - \bar{f}_{h2})\}^2}{\sum_{h=1}^{q} (n_{h1+}n_{h2+}/n_h)^2 \bar{v}_h}$$

where the

$$\bar{v}_h = \left\{\frac{1}{n_{h1+}} + \frac{1}{n_{h2+}}\right\} \frac{n_h v_h}{n_h - 1}$$

are the variances of the mean score differences $\{\bar{f}_{h1} - \bar{f}_{h2}\}$ for the respective strata.

Q_{SMH} is effective for detecting consistent patterns of differences across the strata when the $(\bar{f}_{h1} - \bar{f}_{h2})$ predominantly have the same sign.

Besides the guideline that the across strata row totals (n_{+i+}) be sufficiently large, another guideline for sample size requirements for Q_{SMH} is to choose cutpoints and add columns together so that each stratum table is collapsed to a 2×2 table, similar to what is described in Section 4.2; the cutpoints don't have to be the same for each table. Then, you apply the Mantel-Fleiss criterion to these 2×2 tables (see Section 3.2).

4.3.1 Choosing Scores

Ordinal data analysis strategies do involve some choice on the part of the analyst, and that is the choice of scores to apply to the response levels. There are a variety of scoring systems to consider; the following are often used.

- *integer scores*

  Integer scores are defined as $a_j = j$ for $j = 1, 2, \ldots, r$. They are useful when the response levels are ordered categories that can be viewed as equally spaced and when the response levels correspond to discrete counts. They are also useful if you have equal interest in detecting group differences for any binary partition $\leq j$ versus $> j$ of outcomes for $j = 1, 2, \ldots, r$. Note that if you add the same number to a set of scores, or multiply a set of scores by the same number, both sets of scores produce the same test statistic because multiplication is cancelled by division by the same factor in the variance and addition is cancelled by subtraction of the same factor in the expected value. Thus, the integer scores (1, 2, 3, ...) and (0, 1, 2, ...) produce the same results. Integer scores were used in the analysis of the arthritis data table in Section 4.2.

- *standardized midranks*

  These scores are defined as

  $$a_j = \frac{2[\sum_{k=1}^{j} n_{+k}] - n_{+j} + 1}{2(n+1)}$$

  The $\{a_j\}$ are constrained to lie between 0 and 1. Their advantage over integer scores is that they require no scaling of the response levels other than that implied by their relative ordering. For sets of $2 \times r$ tables, they provide somewhat more power than actual midranks since they produce the van Elteren (1960) extension of the Wilcoxon rank sum test (refer to Lehmann 1975 for a discussion). Standardized midranks are also known as *modified ridit scores*.

- *logrank scores*

  $$a_j = 1 - \sum_{k=1}^{j} \left(\frac{n_{+k}}{\sum_{m=k}^{r} n_{+m}} \right)$$

  Logrank scores are useful when the distribution is thought to be L-shaped, and there is greater interest in treatment differences for response levels with higher values than with lower values.

Other scores that are sometimes used are ridit and rank scores. For a single stratum, rank, ridit, and modified ridit scores produce the same result, which is the categorical counterpart of the Wilcoxon rank sum test. For stratified analyses, modified ridit scores produce van Elteren's extension of the Wilcoxon rank sum test, a property that makes them the preferred of these three types of scores. A possible shortcoming of rank scores, relative to ridit or modified ridit scores, is that their use tends to make the large strata overly influence the test statistic. See page 167 for additional discussion on choosing scores.

You specify the choice of scores in the FREQ procedure by using the SCORES= option in the TABLES statement. If you don't specify SCORES=, then the default table scores are applied. The column (row) numbers are the table scores for character data and the actual variable values are used as scores for numeric variables. Other SCORES= values are RANK, MODRIDIT, and RIDIT. If you are interested in using logrank scores, then you need to compute them in a DATA step and make them the values of the row and column variables you list in the TABLES statement.

4.3.2 Analyzing the Arthritis Data

The FREQ procedure also produces the extended Mantel-Haenzsel statistics. You specify the CMH option in the TABLES statement of PROC FREQ. Notice that the ORDER=DATA option is specified in the PROC statement to ensure that the levels of RESPONSE are sorted correctly. The columns will be ordered none, some, and marked; and the rows will be ordered active and placebo.

```
data arth;
   input gender $ treat $ response $ count @@;
   datalines;
female  active   none  6  female  active   some  5  female  active   marked 16
female  placebo  none 19  female  placebo  some  7  female  placebo  marked  6
male    active   none  7  male    active   some  2  male    active   marked  5
male    placebo  none 10  male    placebo  some  0  male    placebo  marked  1
;

proc freq data=arth order=data;
   weight count;
   tables gender*treat*response / cmh nocol nopct;
run;
```

Output 4.2 displays the frequency tables for females and males.

Output 4.2 Tables by Gender

Frequency Row Pct	Table 1 of treat by response				
	Controlling for gender=female				
		response			
	treat	none	some	marked	Total
	active	6 22.22	5 18.52	16 59.26	27
	placebo	19 59.38	7 21.88	6 18.75	32
	Total	25	12	22	59

Frequency Row Pct	Table 2 of treat by response				
	Controlling for gender=male				
		response			
	treat	none	some	marked	Total
	active	7 50.00	2 14.29	5 35.71	14
	placebo	10 90.91	0 0.00	1 9.09	11
	Total	17	2	6	25

Output 4.3 displays the table of Mantel-Haenszel statistics.

Note that the table heading includes "Based on Table Scores" in parentheses. Q_{SMH} is the "Row Mean Scores Differ" statistic. It has the value 14.6319, with 1 df, and it is clearly significant.

Note the small cell counts for several cells in the table for males. This is not a problem for Q_{SMH} since the adequacy of the sample sizes is determined by the across strata sample sizes n_{+i+}, which are $n_{+1+} = 41$ and $n_{+2+} = 43$ for these data.

Output 4.3 Mantel-Haenszel Results

Summary Statistics for treat by response
Controlling for gender

	Cochran-Mantel-Haenszel Statistics (Based on Table Scores)			
Statistic	Alternative Hypothesis	DF	Value	Prob
1	Nonzero Correlation	1	14.6319	0.0001
2	Row Mean Scores Differ	1	14.6319	0.0001
3	General Association	2	14.6323	0.0007

Total Sample Size = 84

If you can't make the case that the response levels for degree of improvement are equally spaced, then modified ridit scores are an alternative strategy. The following PROC FREQ invocation requests that modified ridit scores be used in the computation of Q_{SMH} through the use of the SCORES=MODRIDIT option in the TABLES statement.

```
proc freq data=arth order=data;
   weight count;
   tables gender*treat*response/cmh scores=modridit nocol nopct;
run;
```

Output 4.4 contains the table of CMH statistics based on modified ridit scores. Q_{SMH} takes the value 15.004 with 1 df, which is clearly significant. Note that the different scoring strategies produced similar results. This is often the case.

Output 4.4 Mantel-Haenszel Results for Modified Ridit Scores

Summary Statistics for treat by response
Controlling for gender

	Cochran-Mantel-Haenszel Statistics (Modified Ridit Scores)			
Statistic	Alternative Hypothesis	DF	Value	Prob
1	Nonzero Correlation	1	14.9918	0.0001
2	Row Mean Scores Differ	1	15.0041	0.0001
3	General Association	2	14.6323	0.0007

Total Sample Size = 84

4.3.3 Rank Statistics for Ordered Data

The Mann-Whitney rank measure of association statistics are useful for assessing the association between an ordinal outcome and a dichotomous explanatory variable. You may be interested in computing the Mann-Whitney rank measure of association as a way of assessing the extent to which patients with active treatment are more likely to have better response status than those with placebo.

The Mann-Whitney estimator for the hth table is computed as

$$g_h = \sum_{j=1}^{3} p_{hAj} \left\{ \left(\sum_{k=1}^{j} p_{hPk} \right) - 0.5 p_{hPj} \right\}$$
$$= \Pr(A > P) + 0.5 \Pr(A = P)$$
$$= \sum_{j} \Pr(A = j) \{ \Pr(P \le j) - 0.5 \Pr(P = j) \}$$

where A indicates active treatment and P indicates placebo.

Somers' D is another measure of association that is appropriate for ordinally scaled data; it is computed as

$$\text{Somers' } D = \frac{\Pr(A > P) - \Pr(A < P)}{\Pr(A > P) + \Pr(A < P) + \Pr(A = P)}$$

You can compute the Mann-Whitney statistic from Somers' D as $g_h = (\text{Somers' } D + 1)/2$, and the standard error of g_h is the standard error of Somers' D divided by 2. You can then write a test of homogeneity of the Mann-Whitney statistics for the tables as

$$Q_H = \frac{(g_F - g_M)^2}{(v_F + v_M)}$$

A general form of this statistic for q tables is

$$Q_H = \sum_{h=1}^{q} (g_h - \bar{g})^2 / v_h$$

where

$$\bar{g} = \sum_{h=1}^{q} (g_h/v_h) \bigg/ \sum_{h=1}^{q} (1/v_h)$$

If homogeneity is not contradicted, then you can proceed with computing an estimator of the common Mann-Whitney statistic. Two forms are available: $\bar{g}$ and $\tilde{g}$.

$$\bar{g} = \left[\sum_{h=F}^{M}(g_h/v_h)\right] \bigg/ \left[\sum_{h=F}^{M}(1/v_h)\right]$$

The estimate of the variance of $\bar{g}$ is

$$v_{\bar{g}} = \left[\sum_{h=F}^{M}(1/v_h)\right]^{-1}$$

You construct a hypothesis test that the common Mann-Whitney estimator is equal to 0.5 with

$$Q_{\bar{g}} = (\bar{g} - 0.5)^2/v_{\bar{g}}$$

Meanwhile,

$$\tilde{g} = \sum_{h=F}^{M} w_h g_h \bigg/ \sum_{h=F}^{M} w_h$$

with $w_h = n_{h1}n_{h2}/(n_{h1} + n_{h2})$.

The estimate of the variance of $\tilde{g}$ is

$$v_{\tilde{g}} = \sum_{h=F}^{M} w_h^2 (\text{s.e.}(g_h))^2 \bigg/ \left[\sum_{h=F}^{M} w_h^2\right]$$

The hypothesis test for the common Mann-Whitney estimator is performed as indicated for $\tilde{g}$. The choice of $\bar{g}$ or $\tilde{g}$ depends on sample size. In this instance, $\tilde{g}$ would be a better choice since the sample sizes are not large within strata. Both forms are appropriate when the same sizes are adequate, and $\bar{g}$ can be more powerful. Note that $\tilde{g}$ is similar to van Elteren's test.

To generate these quantities, you use the FREQ procedure to compute the Somers' D statistics and output them to a SAS data set, and then use the IML procedure to generate the Mann-Whitney statistics. First, the data are put into SAS data set ARTH2. Note that the placebo and active rows have been switched so that placebo is now the first row of the tables.

```
data arth2;
   input gender $ treat $ response $ count @@;
   datalines;
female placebo none 19 female placebo some 7  female placebo marked 6
female active  none 6  female active  some 5  female active  marked 16
male   placebo none 10 male   placebo some 0  male   placebo marked 1
male   active  none 7  male   active  some 2  male   active  marked 5
;
```

Chapter 4: $2 \times r$ and $s \times 2$ Tables

The following PROC FREQ statements request that Somers' D be computed for each of the two tables and that the two estimates and their standard errors be placed into SAS data set SOMEROUT. PROC FREQ produces two forms of Somers' D: Somers' $D(C|R)$ and Somers' $D(R|C)$. The former treats the row variable as the independent variable and the column variable as the dependent variable, and the reverse is true for Somers' $D(R|C)$. In this case, Somers' $D(C|R)$ is desired, and that statistic is output with the SMDCR option in the OUTPUT statement.

```
proc freq data=arth2 order=data;
   weight count;
   tables gender*treat*response /  measures scores=modridit;
   output out=somerout smdcr;
   ods output CrosstabFreqs=myFreqs;
   run;
data ns;
   set myFreqs; where _type_="110";
run;
```

Output 4.5 displays the contents of SOMEROUT.

Output 4.5 Data Set SOMEROUT

Obs	gender	_SMDCR_	E_SMDCR
1	female	0.46644	0.12352
2	male	0.39610	0.16375

The following SAS/IML statements generate the desired estimates.

```
proc iml;
   use somerout var{_SMDCR_ E_SMDCR};
   read all into combined;
   use ns var{Frequency};
   read all into total;
   print total;
   WH=(total[1,]#total[2,]) / (total[1,]+total[2,]) //
      (total[3,]#total[4,]) / (total[3,]+total[4,]);
   print wh;
   GH=(combined[,1]+1)/2;
   SEGH=(combined[,2])/2;
   print wh GH SEGH;
   QH= (GH[1] - GH[2])**2 / SSQ(SEGH);
   pvalue=1-probchi(QH,1);
   print GH SEGH QH pvalue;
   gbar1=sum(GH / SEGH##2);
   gbar2= sum (1/SEGH##2);
   gtilde1=sum(GH # WH);
   gtilde2=sum(WH);
   gbar=gbar1/gbar2;
   gtilde=gtilde1/gtilde2;
   gtildevar=(sum(WH##2 # SEGH##2)) /((sum(WH))**2);
   segtilde=sqrt(gtildevar);
   qgtilde=(gtilde-.5)**2/gtildevar;
```

```
        gbarvar=inv (sum(1/SEGH##2));
        qgbar=(gbar-.5)**2/gbarvar;
        qgtilde=(gtilde-.5)**2/gtildevar;
        print gbar gbarvar qgbar;
        print gtilde gtildevar qgtilde;
    quit;
```

Output 4.6 and Output 4.7 display the results.

The estimate of g_F is 0.7332 with a standard error of 0.0618, and the estimate of g_M is 0.6981 with a standard error of 0.0819. Q_H is 0.1176 ($p = 0.73$), so homogeneity is not contradicted.

Output 4.6 Computations for Mann-Whitney Statistics

GH	SEGH	QH	pvalue
0.7332176	0.0617604	0.1175713	0.7316837
0.6980519	0.0818762		

Output 4.7 displays the estimate of $\tilde{g}$, the common Mann-Whitney estimator.

Output 4.7 Hypothesis Test Results

gtilde	gtildevar	qgtilde
0.7228052	0.0024777	20.035832

Here, $\tilde{g}$ is 0.7228, with variance 0.002477, and $Q_{\tilde{g}} = 20.035$, which is clearly significant with 1 df. The common estimator is not equal to 1/2. You can form a 95% confidence interval for $\tilde{g}$ as $\tilde{g} \pm 1.96 \, \text{se}(\tilde{g})$, which is $0.7228 \pm 1.96(0.0497) = 0.7228 \pm 0.0974$.

4.3.4 Colds Example

The data in Table 4.4 come from a study on the presence of colds in children in two regions (Stokes 1986). Researchers visited children several times and noted whether they had any symptoms of colds. The outcome measure is the number of periods in which a child exhibited cold symptoms.

Table 4.4 Number of Periods with Colds by Gender and Residence

| Gender | Residence | Periods With Colds | | | Total |
		0	1	2	
Female	Urban	45	64	71	180
Female	Rural	80	104	116	300
Total		125	168	187	480
Male	Urban	84	124	82	290
Male	Rural	106	117	87	310
Total		190	141	169	600

These data consist of two 2 × 3 tables; interest lies in determining whether there is association between residence (urban or rural) and number of periods with colds (0, 1, or 2) while controlling for gender. The response levels for these data consist of small discrete counts, so number of periods with colds can be considered an ordinal variable in which the levels are equally spaced. The usual ANOVA strategy for interval-scaled response variables is not appropriate since there is no reason to think that the number of periods with colds is normally distributed with homogeneous variance.

The following statements produce an extended Mantel-Haenszel analysis. The default table scores are used, which will be the actual scores of the variable PER_COLD (0, 1, 2).

```
data colds;
   input gender $ residence $ per_cold count @@;
   datalines;
female urban 0   45  female urban 1  64  female urban 2  71
female rural 0   80  female rural 1 104  female rural 2 116
male   urban 0   84  male   urban 1 124  male   urban 2  82
male   rural 0  106  male   rural 1 117  male   rural 2  87
;

proc freq data=colds order=data;
   weight count;
   tables gender*residence*per_cold / all nocol nopct;
run;
```

Output 4.8 and Output 4.9 contain the frequency tables for females and males and their associated chi-square statistics. There is no significant association between residence and number of periods with colds for females or males; $Q = 0.1059$ ($p = 0.7448$) for females and $Q = 0.7412$ ($p = 0.3893$) for males.

Output 4.8 Results for Females

Frequency Row Pct	Table 1 of residence by per_cold			
	Controlling for gender=female			
		per_cold		
residence	0	1	2	Total
urban	45 25.00	64 35.56	71 39.44	180
rural	80 26.67	104 34.67	116 38.67	300
Total	125	168	187	480

Output 4.8 *continued*

Statistic	DF	Value	Prob
Chi-Square	2	0.1629	0.9218
Likelihood Ratio Chi-Square	2	0.1634	0.9215
Mantel-Haenszel Chi-Square	1	0.1059	0.7448
Phi Coefficient		0.0184	
Contingency Coefficient		0.0184	
Cramer's V		0.0184	

Output 4.9 Results for Males

Frequency Row Pct

Table 2 of residence by per_cold

Controlling for gender=male

residence	per_cold 0	per_cold 1	per_cold 2	Total
urban	84 28.97	124 42.76	82 28.28	290
rural	106 34.19	117 37.74	87 28.06	310
Total	190	241	169	600

Statistic	DF	Value	Prob
Chi-Square	2	2.2344	0.3272
Likelihood Ratio Chi-Square	2	2.2376	0.3267
Mantel-Haenszel Chi-Square	1	0.7412	0.3893
Phi Coefficient		0.0610	
Contingency Coefficient		0.0609	
Cramer's V		0.0610	

Output 4.10 contains the Mantel-Haenszel statistics. Q_{SMH} has the value 0.7379, with a *p*-value of 0.3903. Even controlling for gender, there appears to be no association between residence and number of periods with colds for these data.

Output 4.10 Q_{SMH} Statistic

Summary Statistics for residence by per_cold
Controlling for gender

Cochran-Mantel-Haenszel Statistics (Based on Table Scores)				
Statistic	Alternative Hypothesis	DF	Value	Prob
1	Nonzero Correlation	1	0.7379	0.3903
2	Row Mean Scores Differ	1	0.7379	0.3903
3	General Association	2	1.9707	0.3733

Total Sample Size = 1080

You can also compute a weighted difference of means for this analysis which serves as a distance measure (effect size), similar to what was done in Section 3.2.1 for sets of 2×2 tables.

$$d = \frac{\sum_h w_h (\bar{f}_{h1} - \bar{h}_2)}{\sum_h w_h}$$

and

$$v_d = \sum_h w_h^2 \left\{ \frac{v_{h1}}{n_{h1+}} + \frac{v_{h2}}{n_{h2+}} \right\} / \left(\sum_h w_h \right)^2$$

where

$$v_{hi} = \sum_{j=1}^{r} (a_{hj} - \bar{f}_{hi})^2 n_{hij} / (n_{hi+} - 1)$$

and

$$w_h = (n_{h1+} n_{h2+} / n_h)$$

You can use the GLM procedure for the purpose of computing d when you use the default table scores (actual values for the response variable). You specify PER_COLD as the response variable, with GENDER and RESIDENCE specified as the explanatory variables (for stratification and group). Then you use the ESTIMATE statement to produce the difference in effect for RESIDENCE.

```
proc glm;
   class gender residence;
   freq count;
   model per_cold = gender residence ;
   estimate 'd' residence -1 1;
run;
```

Here, $d = 0.0416$. v_d is calculated to be 0.0023.

Output 4.11 Distance Measure

Dependent Variable: per_cold

Parameter	Estimate	Standard Error	t Value	Pr > |t|
d	0.04155019	0.04839829	0.86	0.3908

4.4 The $s \times 2$ Table

Table 4.5 displays data from a study on adolescent usage of smokeless tobacco (Bauman, Koch, and Lentz 1989). Interest focused on factors that affected usage, such as perception of risk, father's usage of smokeless tobacco, and educational background. Table 4.5 contains two $s \times 2$ tables of risk perception (minimal, moderate, and substantial) and adolescent usage by father's usage. This time, the row variable is ordinally scaled. The question of interest is whether there is a discernible trend in the proportions of adolescent usage over the levels of risk perception. Does usage decline with higher risk perception?

Table 4.5 Adolescent Smokeless Tobacco Usage

Father's Usage	Risk Perception	Adolescent Usage		Total
		No	Yes	
No	Minimal	59	25	84
No	Moderate	169	29	198
No	Substantial	196	9	205
Yes	Minimal	11	8	19
Yes	Moderate	33	11	44
Yes	Substantial	22	2	24

Since the response variable is dichotomous, both risk perception and adolescent usage can be considered ordinal variables. The strategy for assessing association when both row and column variables are ordinal involves assigning scores to the levels of both variables and evaluating their correlation. First, consider the individual 3×2 tables.

Form the linear function

$$\bar{f} = \sum_{i=1}^{3} c_i \bar{f}_i \left(\frac{n_{i+}}{n}\right) = \sum_{i=1}^{3} \sum_{j=1}^{2} \frac{c_i a_j n_{ij}}{n}$$

where $\mathbf{c} = (c_1, c_2, c_3)$ represents scores for the groups and $\mathbf{a} = (a_1, a_2)$ represents scores for the columns (effectively 0, 1), with $a_1 < a_2$ and $c_1 \leq c_2 \leq c_3$ and at least one '$\leq$' is a '$<$'.

Under H_0,

$$E\{\bar{f}|H_0\} = \sum_{i=1}^{3} c_i \left(\frac{n_{i+}}{n}\right) \sum_{j=1}^{2} a_j \left(\frac{n_{+j}}{n}\right) = \mu_{\mathbf{c}} \mu_{\mathbf{a}}$$

and

$$V\{\bar{f}|H_0\} = \left\{\sum_{i=1}^{3}(c_i - \mu_c)^2\left(\frac{n_{i+}}{n}\right)\sum_{j=1}^{2}\frac{(a_j - \mu_a)^2(n_{+j}/n)}{(n-1)}\right\} = \frac{v_c v_a}{(n-1)}$$

The quantity $\bar{f}$ has an approximate normal distribution for large samples, so for these situations

$$\begin{aligned}Q_{CS} &= \frac{(\bar{f} - E\{\bar{f}|H_0\})^2}{\text{Var}\{\bar{f}|H_0\}} \\ &= \frac{(n-1)[\sum_{i=1}^{3}\sum_{j=1}^{2}(c_i - \mu_c)(a_j - \mu_a)n_{ij})]^2}{[\sum_{i=1}^{2}(c_i - \mu_c)^2 n_{i+}][\sum_{j=1}^{2}(a_i - \mu_a)^2 n_{+j}]} \\ &= (n-1)r_{ac}^2\end{aligned}$$

where r_{ac} is the Pearson correlation coefficient. Thus, Q_{CS} is known as the correlation statistic. It is approximately chi-square with one degree of freedom. This test is comparable to the Cochran-Armitage trend test (Cochran 1954, Armitage 1955), which tests for trends in binomial proportions across the levels of an ordinal covariate. In fact, multiplying Q_{CS} by $n/(n-1)$ yields the same value as the z^2 of the Cochran-Armitage test.

The following SAS statements request that association statistics be computed for the smokeless tobacco data. The Cochran-Armitage trend test is also requested directly with the TREND option in the TABLES statement. You can include as many TABLES statements in a PROC FREQ invocation as you like.

```
data tobacco;
   length risk $11. ;
   input f_usage $ risk $ usage $ count @@;
   datalines;
no  minimal       no   59 no  minimal      yes 25
no  moderate      no  169 no  moderate     yes 29
no  substantial   no  196 no  substantial  yes  9
yes minimal       no   11 yes minimal      yes  8
yes moderate      no   33 yes moderate     yes 11
yes substantial   no   22 yes substantial  yes  2
;

proc freq;
   weight count;
   tables f_usage*risk*usage /cmh chisq measures trend;
   tables f_usage*risk*usage /cmh scores=modridit;
run;
```

Output 4.12 contains the statistics for the table of risk perception by adolescent usage when there is no father's usage. Note that $Q_{CS} = 34.2843$, with 1 df, signifying a strong correlation between risk perception and smokeless tobacco usage.

Output 4.12 Results for No Father's Usage

Statistics for Table 1 of risk by usage
Controlling for f_usage=no

Statistic	DF	Value	Prob
Chi-Square	2	34.9217	<.0001
Likelihood Ratio Chi-Square	2	34.0684	<.0001
Mantel-Haenszel Chi-Square	1	34.2843	<.0001
Phi Coefficient		0.2678	
Contingency Coefficient		0.2587	
Cramer's V		0.2678	

Output 4.13 contains the Cochran-Armitage trend test table. The test statistic, Z, is 5.8613 and is highly significant. There is an increasing trend in binomial proportions as you go from minimal to substantial risk perception for the table with no father usage.

Output 4.13 Cochran-Armitage Trend Test

Cochran-Armitage Trend Test	
Statistic (Z)	5.8613
One-sided Pr > Z	<.0001
Two-sided Pr > |Z|	<.0001

Output 4.14 contains the measures of association. Somers' $D(C|R)$ and the Pearson correlation coefficient are of particular interest since they account for ordinality.

Output 4.14 Measures of Association

Statistic	Value	ASE
Gamma	-0.5948	0.0772
Kendall's Tau-b	-0.2477	0.0395
Stuart's Tau-c	-0.1863	0.0339
Somers' D C|R	-0.1484	0.0267
Somers' D R|C	-0.4135	0.0628
Pearson Correlation	-0.2656	0.0439
Spearman Correlation	-0.2602	0.0415
Lambda Asymmetric C|R	0.0000	0.0000
Lambda Asymmetric R|C	0.0709	0.0211
Lambda Symmetric	0.0580	0.0169
Uncertainty Coefficient C|R	0.0908	0.0290
Uncertainty Coefficient R|C	0.0339	0.0112
Uncertainty Coefficient Symmetric	0.0493	0.0161

Output 4.15 contains the same association test results for those whose fathers used smokeless tobacco, Output 4.16 contains the corresponding Cochran-Armitage trend test, and Output 4.17 displays the corresponding measures of association.

Output 4.15 Results for Father's Usage

Statistic	DF	Value	Prob
Chi-Square	2	6.6413	0.0361
Likelihood Ratio Chi-Square	2	7.0461	0.0295
Mantel-Haenszel Chi-Square	1	6.5644	0.0104
Phi Coefficient		0.2763	
Contingency Coefficient		0.2663	
Cramer's V		0.2763	

Output 4.16 Cochran-Armitage Trend Test

Cochran-Armitage Trend Test	
Statistic (Z)	2.5770
One-sided Pr > Z	0.0050
Two-sided Pr > |Z|	0.0100

Output 4.17 Measures of Association

Statistic	Value	ASE
Gamma	-0.5309	0.1626
Kendall's Tau-b	-0.2622	0.0905
Stuart's Tau-c	-0.2500	0.0917
Somers' D C|R	-0.2014	0.0726
Somers' D R|C	-0.3413	0.1171
Pearson Correlation	-0.2763	0.0966
Spearman Correlation	-0.2761	0.0955
Lambda Asymmetric C|R	0.0000	0.0000
Lambda Asymmetric R|C	0.0000	0.0000
Lambda Symmetric	0.0000	0.0000
Uncertainty Coefficient C|R	0.0733	0.0510
Uncertainty Coefficient R|C	0.0392	0.0276
Uncertainty Coefficient Symmetric	0.0511	0.0357

There is a significant correlation between risk perception and adolescent usage for the table with father's usage, and there is similarity of the Pearson correlations according to father's usage, -0.267 for Yes and -0.226 for No. The Cochran-Armitage Z statistic has the value 2.5770 and a two-sided p-value of 0.0100. Note that exact p-values are available for the trend test for sparse data.

Section 4.5 discusses the analysis of the combined tables.

4.5 Sets of $s \times 2$ Tables

4.5.1 Correlation Statistic

Mantel (1963) also proposed a statistic for the association of two variables that were ordinal for a combined set of strata, based on assigning scores $\{a\}$ and $\{c\}$ to the columns and rows of the tables: extended Mantel-Haenszel correlation statistic

$$Q_{\text{CSMH}} = \frac{\left\{\sum_{h=1}^{q} n_h(\bar{f}_h - E\{\bar{f}_h|H_0\})\right\}^2}{\sum_{h=1}^{q} n_h^2 \text{var}\{\bar{f}_h|H_0\}}$$

$$= \frac{\left\{\sum_{h=1}^{q} n_h(v_{hc}v_{ha})^{1/2} r_{ca,h}\right\}^2}{\sum_{h=1}^{q} [n_h^2 v_{hc}v_{ha}/(n_h - 1)]}$$

Q_{CSMH} is called the extended Mantel-Haenszel correlation statistic. It approximately follows the chi-square distribution with one degree of freedom when the combined strata sample sizes are sufficiently large:

$$\sum_{h=1}^{q} n_h \geq 40$$

4.5.2 Analysis of Smokeless Tobacco Data

The following SAS statements request that Mantel-Haenszel correlation statistics be computed for the smokeless tobacco data. Two TABLES statements are included to specify analyses that use both integer scores and modified ridit scores.

```
proc freq;
   weight count;
   tables f_usage*risk*usage /cmh;
   tables f_usage*risk*usage /cmh scores=modridit;
run;
```

Output 4.18 contains the results for the integer scores, and Output 4.19 contains the results for the modified ridit scores.

Output 4.18 Results for Integer Scores

Summary Statistics for risk by usage
Controlling for f_usage

Cochran-Mantel-Haenszel Statistics (Based on Table Scores)				
Statistic	Alternative Hypothesis	DF	Value	Prob
1	Nonzero Correlation	1	40.6639	<.0001
2	Row Mean Scores Differ	2	41.0577	<.0001
3	General Association	2	41.0577	<.0001

Total Sample Size = 574

Output 4.19 Results for Modified Ridit Scores

Summary Statistics for risk by usage
Controlling for f_usage

Cochran-Mantel-Haenszel Statistics (Modified Ridit Scores)				
Statistic	Alternative Hypothesis	DF	Value	Prob
1	Nonzero Correlation	1	39.3048	<.0001
2	Row Mean Scores Differ	2	41.0826	<.0001
3	General Association	2	41.0577	<.0001

Total Sample Size = 574

Q_{CSMH} is Statistic 1 in the table, labeled the "Nonzero Correlation" statistic. It takes the value 40.6639 for integer scores, and it takes the value 39.3048 for modified ridit scores. Both results are similar, with strongly significant statistics; often, different sets of scores produce essentially the same results.

4.5.3 Pain Data Analysis

Clinical trials not only investigate measures of efficacy, or how well a drug works for its designed purpose, but they also address the matter of adverse effects, or whether the drug has harmful side effects. Table 4.6 contains data from a study concerned with measuring the adverse effects of a pain relief treatment that was given at five different dosages, including placebo, to patients with one of two diagnoses. Investigators were interested in whether there was a trend in the proportions with adverse effects.

Table 4.6 Adverse Effects for Pain Treatment

	Diagnosis			
	I		II	
	Adverse Effects		Adverse Effects	
Treatment	No	Yes	No	Yes
Placebo	26	6	26	6
Dosage1	26	7	12	20
Dosage2	23	9	13	20
Dosage3	18	14	1	31
Dosage4	9	23	1	31

The following SAS statements request a Q_{CSMH} statistic from PROC FREQ, using both integer scores and modified ridit scores. First, a TABLES statement requests the table of treatment by response pooled over the two diagnoses. Note the use of the ORDER=DATA option in the PROC statement. If this option was omitted, the levels of TREATMENT would be ordered incorrectly, with placebo being placed last instead of first.

```
data pain;
   input diagnosis $ treatment $ response $ count @@;
   datalines;
I  placebo  no 26 I  placebo  yes  6
I  dosage1  no 26 I  dosage1  yes  7
I  dosage2  no 23 I  dosage2  yes  9
I  dosage3  no 18 I  dosage3  yes 14
I  dosage4  no  9 I  dosage4  yes 23
II placebo  no 26 II placebo  yes  6
II dosage1  no 12 II dosage1  yes 20
II dosage2  no 13 II dosage2  yes 20
II dosage3  no  1 II dosage3  yes 31
II dosage4  no  1 II dosage4  yes 31
;

proc freq order=data;
   weight count;
   tables treatment*response / chisq;
   tables diagnosis*treatment*response / chisq cmh;
   tables diagnosis*treatment*response / scores=modridit cmh;
run;
```

As shown in Output 4.21, Q_{CS} for the combined table is strongly significant, with a value of 65.4730 and 1 df.

Output 4.20 Results for Combined Diagnoses

Frequency Percent Row Pct Col Pct	Table of treatment by response		
		response	
treatment	no	yes	Total
placebo	52 16.15 81.25 33.55	12 3.73 18.75 7.19	64 19.88
dosage1	38 11.80 58.46 24.52	27 8.39 41.54 16.17	65 20.19
dosage2	36 11.18 55.38 23.23	29 9.01 44.62 17.37	65 20.19
dosage3	19 5.90 29.69 12.26	45 13.98 70.31 26.95	64 19.88
dosage4	10 3.11 15.63 6.45	54 16.77 84.38 32.34	64 19.88
Total	155 48.14	167 51.86	322 100.00

Statistic	DF	Value	Prob
Chi-Square	4	68.0752	<.0001
Likelihood Ratio Chi-Square	4	73.2533	<.0001
Mantel-Haenszel Chi-Square	1	65.4730	<.0001
Phi Coefficient		0.4598	
Contingency Coefficient		0.4178	
Cramer's V		0.4598	

Output 4.21 contains the statistics for the individual tables. Q_{CS} takes the value 22.8188 for Diagnosis I and the value 52.3306 for Diagnosis II.

Output 4.21 Results for Table 1, Table 2

Statistic	DF	Value	Prob
Chi-Square	4	26.6025	<.0001
Likelihood Ratio Chi-Square	4	26.6689	<.0001
Mantel-Haenszel Chi-Square	1	22.8188	<.0001
Phi Coefficient		0.4065	
Contingency Coefficient		0.3766	
Cramer's V		0.4065	

Statistic	DF	Value	Prob
Chi-Square	4	60.5073	<.0001
Likelihood Ratio Chi-Square	4	68.7446	<.0001
Mantel-Haenszel Chi-Square	1	52.3306	<.0001
Phi Coefficient		0.6130	
Contingency Coefficient		0.5226	
Cramer's V		0.6130	

Output 4.22 contains the stratified analysis results. Integer scores produce a Q_{CSMH} of 71.7263, and modified ridit scores produce a Q_{CSMH} of 71.6471. These statistics are clearly significant. The proportion of patients with adverse effects is correlated with level of dosage; higher dosages produce more reports of adverse effects.

Output 4.22 Combined Results

Summary Statistics for treatment by response
Controlling for diagnosis

Cochran-Mantel-Haenszel Statistics (Based on Table Scores)				
Statistic	Alternative Hypothesis	DF	Value	Prob
1	Nonzero Correlation	1	71.7263	<.0001
2	Row Mean Scores Differ	4	74.5307	<.0001
3	General Association	4	74.5307	<.0001

Cochran-Mantel-Haenszel Statistics (Modified Ridit Scores)				
Statistic	Alternative Hypothesis	DF	Value	Prob
1	Nonzero Correlation	1	71.6471	<.0001
2	Row Mean Scores Differ	4	74.5307	<.0001
3	General Association	4	74.5307	<.0001

Output 4.23 and Output 4.24 displays measures of association for the individual tables.

Output 4.23 Table 1 Measures

Statistic	Value	ASE
Gamma	0.5313	0.0935
Kendall's Tau-b	0.3373	0.0642
Stuart's Tau-c	0.4111	0.0798
Somers' D C|R	0.2569	0.0499
Somers' D R|C	0.4427	0.0837
Pearson Correlation	0.3776	0.0714
Spearman Correlation	0.3771	0.0718
Lambda Asymmetric C|R	0.2373	0.0837
Lambda Asymmetric R|C	0.1250	0.0662
Lambda Symmetric	0.1604	0.0621
Uncertainty Coefficient C|R	0.1261	0.0467
Uncertainty Coefficient R|C	0.0515	0.0191
Uncertainty Coefficient Symmetric	0.0731	0.0271

Output 4.24 Table 2 Measures

Statistic	Value	ASE
Gamma	0.7871	0.0571
Kendall's Tau-b	0.5114	0.0473
Stuart's Tau-c	0.6080	0.0638
Somers' D C|R	0.3800	0.0399
Somers' D R|C	0.6883	0.0606
Pearson Correlation	0.5719	0.0527
Spearman Correlation	0.5718	0.0529
Lambda Asymmetric C|R	0.3774	0.0842
Lambda Asymmetric R|C	0.1875	0.0668
Lambda Symmetric	0.2431	0.0637
Uncertainty Coefficient C|R	0.3369	0.0625
Uncertainty Coefficient R|C	0.1327	0.0256
Uncertainty Coefficient Symmetric	0.1904	0.0362

Somers' D for Diagnosis I has the value 0.2569, which takes the value 0.3800 for Diagnosis II, indicating somewhat stronger evidence for association in the latter case.

4.6 Relationships between Sets of Tables

Suppose you transposed the rows and columns of Table 4.6. You would obtain the results displayed in Table 4.7:

Table 4.7 Adverse Effects for Pain Treatment

Diagnosis	Adverse Effects	Placebo	Dosage1	Dosage2	Dosage3	Dosage4
I	No	26	26	23	18	9
I	Yes	6	7	9	14	23
II	No	26	12	13	1	1
II	Yes	6	20	20	31	31

Furthermore, suppose you analyzed these tables as two $2 \times r$ tables, making the response variable the row variable and the grouping variable the column variable. The first TABLES statement requests an analysis for this table, and the second TABLES statement requests the earlier analysis again.

```
proc freq order=data;
   weight count;
   tables diagnosis*response*treatment / cmh;
   tables diagnosis*treatment*response / cmh;
run;
```

Output 4.25 contains the results.

Output 4.25 Combined Results

Summary Statistics for response by treatment
Controlling for diagnosis

Cochran-Mantel-Haenszel Statistics (Based on Table Scores)				
Statistic	Alternative Hypothesis	DF	Value	Prob
1	Nonzero Correlation	1	71.7263	<.0001
2	Row Mean Scores Differ	1	71.7263	<.0001
3	General Association	4	74.5307	<.0001

Output 4.25 *continued*

Cochran-Mantel-Haenszel Statistics (Based on Table Scores)				
Statistic	Alternative Hypothesis	DF	Value	Prob
1	Nonzero Correlation	1	71.7263	<.0001
2	Row Mean Scores Differ	4	74.5307	<.0001
3	General Association	4	74.5307	<.0001

Q_{SMH} and Q_{CSMH} are identical in the first table, which contains the results for the $2 \times 2 \times 5$ analysis. One degree of freedom is needed to compare the mean differences across two groups, in the case of Q_{SMH}, and one degree of freedom is needed to assess correlation, in the case of Q_{CSMH}. The second table in Output 4.25 contains the results for the $2 \times 5 \times 2$ analysis. Only Q_{SMH} has the value 71.7263.

In Chapter 6, the Mantel-Haenszel statistic is extended to sets of $s \times r$ tables. The mean score statistic for the case of more than two groups has $(s - 1)$ degrees of freedom, since you are comparing mean differences across s groups. Thus, Q_S for the $2 \times r$ table is a special case of the more general mean score statistic and has $(s - 1) = (2 - 1) = 1$ degree of freedom. When $s = 2$, Q_{SMH} and Q_{CSMH} take the same value with table scores and can be used interchangeably. Thus, transposing the Table 4.6 data and computing these statistics produced identical mean score and correlation statistics, since the transposed data produced a mean score statistic with one degree of freedom.

Similarly, when $s = 2$, Q_S and Q_{CS} take the same value. This is why, in Section 4.2, you are able to use the Mantel-Haenszel statistic produced by the CHISQ option of PROC FREQ. That statistic is actually Q_{CS}, but for $2 \times r$ tables it is also the mean score statistic.

Table 4.8 summarizes the Mantel-Haenszel statistics for the tables discussed in this chapter; it also lists the labels associated with these statistics in PROC FREQ output.

Table 4.8 Summary of Extended Mantel-Haenszel Statistics

Table Dimensions	Statistic	DF	Corresponding PROC FREQ MH Label
2×2	Q_{MH}	1	Nonzero Correlation Row Mean Scores Differ General Association
$2 \times r$	Q_{SMH}	1	Nonzero Correlation Row Mean Scores Differ
$s \times 2$	Q_{CSMH}	1	Nonzero Correlation

4.7 Exact Analysis of Association for the $s \times 2$ Table

The data in Table 4.9 come from a study of mice who were exposed to a bacterial challenge and treated with the drugs carbenicillin or cefotaxmine (Bowdre, et al. 1983). Surviving mice were assessed at intervals of 6 to 24 hours.

Table 4.9 Mice Surviving Exposure to Vibrio Vulnificus

Hours	Carbenicillin	Cefotaxime	Total	Ranks	Logranks
0–6	1	1	2	1.5	0.909
6–12	3	1	4	4.5	0.709
12–18	5	1	6	9.5	0.334
18–24	1	0	1	13	0.234
24–30	1	2	3	15	−0.099
30–48	0	2	2	17.5	−0.433
48–72	1	1	2	19.5	−0.933
72–96	0	1	1	21	−1.433
> 96	0	1	1	22	−2.433
Total	12	10	22		

Researchers were interested in whether one of the drugs had a better outcome over time. Note that this table has response outcomes down the rows and treatments across the columns. Analysis is concerned with whether the mean number of survival hours is different for the two drugs, and thus Q_S is the appropriate statistic. However, as was discussed in the previous section, since 1 df of freedom is required to compare the mean survival hours for the two drug groups, the 1 df statistic Q_{CS} will be identical to Q_S. Logrank scores are indicated since many of the mice (in both treatment groups) only survived for the low end of the time scale.

In addition, the counts are very small and do not satisfy the asymptotic requirements for Q_{CS}. However, there is an exact counterpart to Q_{CS} which is appropriate. Logranks have to be input directly in order for PROC FREQ to use them to compute Mantel-Haenszel statistics.

Table 4.9 displays the rank and logrank scores computed for each row. In addition, the scores used in the algorithm for the exact computations in PROC FREQ need to be presented in ascending order. The order does not make a difference in these test statistics. The following DATA step inputs the data into SAS data set MICE and sorts the resulting data set.

```
data mice;
input LogRank Treatment $ count @@;
datalines;
0.909  Ca 1 0.909  Ce 1
0.709  Ca 3 0.709  Ce 1
0.334  Ca 5 0.334  Ce 1
0.234  Ca 1 0.234  Ce 0
-0.099 Ca 1 -0.099 Ce 2
-0.433 Ca 0 -0.433 Ce 2
-0.933 Ca 1 -0.933 Ce 1
-1.433 Ca 0 -1.433 Ce 1
```

```
    -2.433 Ca 0 -2.433 Ce 1
    ;

proc sort data=mice;
   by LogRank;
run;
```

The following PROC FREQ statements request both the asymptotic Q_{CS} and the exact analysis for both Wilcoxon rank scores and logrank scores. The SCOROUT option requests that the scores used in the analysis be printed.

```
proc freq;
   weight count;
   tables LogRank*Treatment / norow nocol nopct scorout chisq;
   tables LogRank*Treatment / noprint scores=rank scorout chisq;
   exact mhchi;
run;
```

Output 4.26 displays the crosstabulations for these data. Since the data were sorted according to logrank score, they appear in reverse order from Table 4.9.

Output 4.26 Frequencies for Mice Challenge

Frequency	Table of LogRank by Treatment			
		Treatment		
	LogRank	Ca	Ce	Total
	-2.433	0	1	1
	-1.433	0	1	1
	-0.933	1	1	2
	-0.433	0	2	2
	-0.099	1	2	3
	0.234	1	0	1
	0.334	5	1	6
	0.709	3	1	4
	0.909	1	1	2
	Total	12	10	22

Output 4.27 displays the row scores used in the analysis. (The SCOROUT option produces the column scores, too, but they are not displayed here.) The scores are identical to the logrank scores presented in Table 4.9.

Output 4.27 Logrank Scores

Row Scores	
LogRank	Score
-2.433	-2.433
-1.433	-1.433
-0.933	-0.933
-0.433	-0.433
-0.099	-0.099
0.234	0.234
0.334	0.334
0.709	0.709
0.909	0.909

The correlation Mantel-Haenszel statistic based on the logrank scores is displayed in Output 4.28. Based on the logrank scores, it has the value 4.0569. The asymptotic $p = 0.0440$, and the exact $p = 0.0367$.

Output 4.28 MH Chi-Square Test for Logrank Scores

Mantel-Haenszel Chi-Square Test	
Chi-Square	4.0569
DF	1
Asymptotic Pr > ChiSq	0.0440
Exact Pr >= ChiSq	0.0367

Output 4.29 displays the results for the Wilcoxon rank scores.

Output 4.29 MH Chi-Square Test for Rank Scores

Mantel-Haenszel Chi-Square Test (Rank Scores)	
Chi-Square	3.5118
DF	1
Asymptotic Pr > ChiSq	0.0609
Exact Pr >= ChiSq	0.0625

Here, $Q_{CS} = 3.5118$, and the asymptotic $p = 0.0609$ and the exact $p = 0.0625$. Thus, the choice of scores can make a difference in your conclusion. Logrank scores generally work better than Wilcoxon rank scores in these low count situations with relatively greater differences between treatments for the longer survival times. In addition, note that the exact analysis has the larger p-value for the rank scores, but it has the smaller p-value for the logrank scores. However, in both cases, the exact result is more accurate for the small sample size.

Chapter 5
The $s \times r$ Table

Contents

5.1	Introduction		**107**
5.2	Association		**108**
	5.2.1	Tests for General Association	108
	5.2.2	Mean Score Test	112
	5.2.3	Correlation Test	115
5.3	Exact Tests for Association		**119**
	5.3.1	General Association	119
	5.3.2	Test of Correlation	122
5.4	Measures of Association		**124**
	5.4.1	Ordinal Measures of Association	124
	5.4.2	Exact Tests for Ordinal Measures of Association	127
	5.4.3	Nominal Measures of Association	129
5.5	Observer Agreement		**131**
	5.5.1	Computing the Kappa Statistic	131
	5.5.2	Exact p-values for the Kappa Statistic	134
5.6	Test for Ordered Differences		**136**

5.1 Introduction

Previous chapters address the concepts of association and measures of association in 2×2 tables, $2 \times r$ tables, and $s \times 2$ tables. This chapter extends these concepts to the general $s \times r$ table. The main difference from these earlier chapters is that scale of measurement is always a consideration: the statistics you choose depend on whether the rows and columns of the table are nominally or ordinally scaled. This is true for investigating whether association exists and for summarizing the degree of association. Section 5.2 addresses tests for association, and Section 5.4 addresses measures of association.

Often, subjects or experimental units are observed by two or more researchers, and the question of interest is how closely their evaluations agree. Such studies are called *observer agreement* studies. The columns of the resulting table are the classifications of one observer, and the rows are the classifications of the other observer. Subjects are cross-classified into table cells according to their observed profiles. Observer agreement is discussed in Section 5.5. Sometimes you are interested in ordered alternatives to the hypothesis of no association. Section 5.6 discusses the Jonckheere-Terpstra test for ordered differences.

Exact *p*-values are available for many tests of association and measures of association. The Fisher exact test for the $s \times r$ table and exact *p*-values for several chi-square statistics are discussed in Section 5.3. Exact *p*-values are also discussed for measures of association, observer agreement, and the Jonckheere-Terpstra test in those respective sections. The exact *p*-value computations for the actual test statistics such as chi-square statistics, which take only nonnegative values, are based on the sum of the exact probabilities for those tables where the test statistic is greater than or equal to the one you observed. The tables you consider are those with the same margins as the table you observe.

For tests where you might consider one-sided or two-sided alternative hypotheses, such as for the kappa coefficient, the computation is a bit more involved. For one-sided tests, the FREQ procedure computes the right-sided *p*-value when the observed value of the test statistic is greater than its expected value, and it computes the left-sided *p*-value when the test statistic is less than or equal to its expected value. In each case, the *p*-value is the sum of the probabilities for tables that have a more extreme test statistic than the observed one. The two-sided *p*-value is computed as the sum of the one-sided *p*-value and the area in the other tail of the distribution for the statistic that is at least as far from the expected value. See Agresti (1992) for a review of the strategies for exact *p*-value computations for table statistics.

5.2 Association

5.2.1 Tests for General Association

Table 5.1 contains data from a study concerning the distribution of party affiliation in a city suburb. The interest was whether there was an association between registered political party and neighborhood.

Table 5.1 Distribution of Parties in Neighborhoods

Party	Neighborhood			
	Bayside	Highland	Longview	Sheffeld
Democrat	221	160	360	140
Independent	200	291	160	311
Republican	208	106	316	97

For these data, both row and column variables are nominally scaled; there is no inherent ordering of the response values for either neighborhood or political party. Thus, the alternative to the null hypothesis of no association is general association, defined as heterogeneous patterns of distribution of the response (column) levels across the row levels. Table 5.2 represents the general $s \times r$ table.

Table 5.2 s × r Contingency Table

Group	Response Variable Categories				Total
	1	2	...	r	
1	n_{11}	n_{12}	...	n_{1r}	n_{1+}
2	n_{21}	n_{22}	...	n_{2r}	n_{2+}
⋮	⋮	⋮		⋮	⋮
s	n_{s1}	n_{s2}	...	n_{sr}	n_{s+}
Total	n_{+1}	n_{+2}	...	n_{+r}	n

One test statistic for the hypothesis of no general association is the Pearson chi-square. This statistic is defined the same as for the 2 × 2 table, except that the summation for i is from 1 to s and the summation for j is from 1 to r.

$$Q_P = \sum_{i=1}^{s} \sum_{j=1}^{r} \frac{(n_{ij} - m_{ij})^2}{m_{ij}}$$

where

$$m_{ij} = E\{n_{ij}|H_0\} = \frac{n_{i+}n_{+j}}{n}$$

is the expected value of the frequencies in the ith row and jth column.

If the sample size is sufficiently large, that is, all expected cell counts $m_{ij} \geq 5$, then Q_P approximately has the chi-square distribution with $(s-1)(r-1)$ degrees of freedom. In the case of the 2 × 2 table, $r = 2$ and $s = 2$ so that Q_P has 1 df.

Just as for 2 × 2 tables, the randomization statistic Q can be written

$$Q = \left(\frac{n-1}{n}\right) Q_P$$

and it also has an approximate chi-square distribution with $(s-1)(r-1)$ degrees of freedom under the null hypothesis.

For more detail, recall from Chapter 2, "The 2 × 2 Table," that the derivation of Q depends on the assumption of fixed marginal totals such that the table frequencies have a hypergeometric distribution. For the $s \times r$ table, the distribution is multivariate hypergeometric under the null hypothesis of no association.

You can write the probability distribution as

$$Pr\{n_{ij}\} = \frac{\prod_{i=1}^{s} n_{i+}! \prod_{j=1}^{r} n_{+j}!}{n! \prod_{i=1}^{s} \prod_{j=1}^{r} n_{ij}!}$$

The covariance structure under H_0 is

$$\text{Cov}\{n_{ij}, n_{i'j'}|H_0\} = \frac{m_{ij}(n\delta_{ii'} - n_{i'+})(n\delta_{jj'} - n_{+j'})}{n(n-1)}$$

where $\delta_{kk'} = 1$ if $k = k'$ and $\delta_{kk'} = 0$ if $k \neq k'$.

Q is computed from the quadratic form

$$Q = (\mathbf{n} - \mathbf{m})'\mathbf{A}'(\mathbf{A}\mathbf{V}\mathbf{A}')^{-1}\mathbf{A}(\mathbf{n} - \mathbf{m})$$

where $\mathbf{n} = (n_{11}, n_{12}, \ldots, n_{1r}, \ldots, n_{s1}, \ldots, n_{sr})'$ is the compound vector of observed frequencies, $\mathbf{m}$ is the corresponding vector of expected frequencies, $\mathbf{V}$ is the covariance matrix, and $\mathbf{A}$ is a matrix of coefficients defined such that $\mathbf{A}\mathbf{V}\mathbf{A}'$ is nonsingular. The symbol $\otimes$ denotes the left-hand Kronecker product (the matrix on the left of the $\otimes$ multiplies each element in the matrix on the right).

The usual choice for $\mathbf{A}$ for testing general association is

$$\mathbf{A} = \begin{bmatrix} \mathbf{I}_{(r-1)}, \mathbf{0}_{(r-1)} \end{bmatrix} \otimes \begin{bmatrix} \mathbf{I}_{(s-1)}, \mathbf{0}_{(s-1)} \end{bmatrix}$$

where $\mathbf{I}_{(u-1)}$ is the $(u-1) \times (u-1)$ identity matrix and $\mathbf{0}_{(u-1)}$ is a $(u-1)$ vector of 0s.

For example, for a 2×3 table,

$$\mathbf{A} = \begin{bmatrix} 1 & 0 & 0 & 0 & 0 & 0 \\ 0 & 1 & 0 & 0 & 0 & 0 \end{bmatrix}$$

Generating Q_P and Q requires no new PROC FREQ features. The CHISQ option in the TABLES statement produces Q_P, and the CMH option produces Q. The following statements produce these statistics for the neighborhood data. A mosaic plot is also requested with the PLOTS=MOSAICPLOT option.

```
data neighbor;
   length party $ 11 neighborhood $ 10;
   input party $ neighborhood $ count @@;
   datalines;
democrat      longview    360  democrat     bayside  221
democrat      sheffeld    140  democrat     highland 160
republican    longview    316  republican   bayside  208
republican    sheffeld     97  republican   highland 106
independent   longview    160  independent  bayside  200
independent   sheffeld    311  independent  highland 291
;

ods graphics on;
proc freq ;
   weight count;
   tables party*neighborhood /
      plots=mosaicplot chisq cmh nocol nopct;
run;
ods graphics off;
```

Output 5.1 contains the frequency table.

Output 5.1 Frequency Table

Frequency Row Pct	Table of party by neighborhood				
		neighborhood			
party	bayside	highland	longview	sheffeld	Total
democrat	221 25.09	160 18.16	360 40.86	140 15.89	881
independent	200 20.79	291 30.25	160 16.63	311 32.33	962
republican	208 28.61	106 14.58	316 43.47	97 13.34	727
Total	629	557	836	548	2570

Output 5.2 displays the frequencies graphically with a mosaic plot.

Output 5.2 Frequency Plot

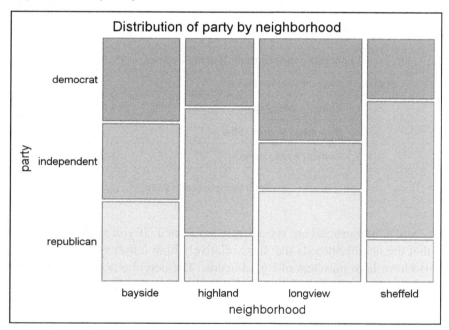

The mosaic plot uses tiles proportional to cell frequencies to provide a visual representation of a contingency table. First, the horizontal space is divided by sizes corresponding to the relative count size of the levels of the variable on the x-axis, and then the vertical spaces are divided up according to the relative sizes of the counts of the levels of the variable on the y-axis. Different colors or shading is used to differentiate the different categories.

The mosaic plot for the neighborhood data seems to indicate that political party is not distributed the same across the various neighborhoods. Highland and Sheffeld have large numbers of independents.

Output 5.3 displays the table statistics. $Q_P = 273.9188$ with 6 df, $p < 0.0001$.

Output 5.3 Pearson Chi-Square

Statistic	DF	Value	Prob
Chi-Square	6	273.9188	<.0001
Likelihood Ratio Chi-Square	6	282.3266	<.0001
Mantel-Haenszel Chi-Square	1	0.8124	0.3674
Phi Coefficient		0.3265	
Contingency Coefficient		0.3104	
Cramer's V		0.2308	

Output 5.4 contains the MH statistics. PROC FREQ computes Q as the extended Mantel-Haenszel statistic for one stratum. Q is the "General Association" statistic, with a value of 273.8122 and 6 df. Notice how close the values of Q and Q_P are for these data; this is expected since the sample size is large (2,570).

Output 5.4 Randomization Q

Summary Statistics for party by neighborhood

Cochran-Mantel-Haenszel Statistics (Based on Table Scores)				
Statistic	Alternative Hypothesis	DF	Value	Prob
1	Nonzero Correlation	1	0.8124	0.3674
2	Row Mean Scores Differ	2	13.8938	0.0010
3	General Association	6	273.8122	<.0001

Total Sample Size = 2570

Political party and neighborhood are statistically associated. If you study the column percentages, you can see that the neighborhoods that have relatively high numbers of Democrats (Bayside and Longview) also have high numbers of Republicans. The neighborhoods that have relatively high numbers of Independents (Highland and Sheffeld) also have low numbers of both Democrats and Republicans.

5.2.2 Mean Score Test

The data in Table 5.3 come from a study on headache pain relief. A new treatment was compared with the standard treatment and a placebo. Researchers measured the number of hours of substantial relief from headache pain.

Table 5.3 Pain Study Data

Treatment	Hours of Relief				
	0	1	2	3	4
Placebo	6	9	6	3	1
Standard	1	4	6	6	8
Test	2	5	6	8	6

Clearly, number of hours of relief is an ordinally scaled response measure. While Q and Q_P are good strategies for detecting general association, they aren't as good as other strategies when the response variable is ordinally scaled and the alternative to no association is location shifts. Section 4.2 discusses the mean score test for a $2 \times r$ table. Scores are assigned to the levels of the response variable, and row mean scores are computed. The statistic Q_S is then derived. Q_S also applies to $s \times r$ tables, in which case it has $(s-1)$ degrees of freedom since you are comparing mean scores across s groups.

For more detail, the statistic Q_S is derived from the same general quadratic form as Q discussed in Section 5.2. You choose **A** so that it assigns scores to the response levels and then compares the resulting linear functions of scores for $(s-1)$ groups to their expected values. **A** is the $(s-1) \times sr$ matrix:

$$\mathbf{A} = \begin{bmatrix} \mathbf{a}' & \underline{0}' & \cdots & \underline{0}' & \underline{0}' \\ \underline{0}' & \mathbf{a}' & \cdots & \underline{0}' & \underline{0}' \\ \vdots & \vdots & & & \vdots \\ \underline{0}' & \underline{0}' & \cdots & \mathbf{a}' & \underline{0}' \end{bmatrix}$$

For example, if the actual values were used as scores for the columns in Table 5.3, then $\mathbf{a}' = (0\ 1\ 2\ 3\ 4)$.

It is interesting to note that Q_S can be written in a one-way analysis-of-variance form

$$Q_S = \frac{(n-1) \sum_{i=1}^{s} n_{i+} (\bar{f}_i - \mu_\mathbf{a})^2}{n v_\mathbf{a}}$$

where, as discussed in Section 4.2,

$$\bar{f}_i = \sum_{j=1}^{r} \frac{a_j n_{ij}}{n_{i+}}$$

and $\mu_\mathbf{a}$ is its expected value:

$$\mu_\mathbf{a} = E\{\bar{f}_i | H_0\} = \sum_{j=1}^{r} \frac{a_j n_{+j}}{n}$$

$$v_\mathbf{a} = \sum_{j=1}^{r} (a_j - \mu_\mathbf{a})^2 \left(\frac{n_{+j}}{n}\right)$$

See Section 4.2.3 for choices of scoring systems. For the pain data, integer scores make sense. The following PROC FREQ statements request crosstabulations and the mean score test Q_S for the pain data.

```
data pain;
   input treatment $ hours count @@;
   datalines;
placebo  0 6 placebo  1 9 placebo  2 6 placebo  3 3 placebo  4 1
standard 0 1 standard 1 4 standard 2 6 standard 3 6 standard 4 8
test     0 2 test     1 5 test     2 6 test     3 8 test     4 6
;

proc freq;
   weight count;
   tables treatment*hours/ cmh nocol nopct;
run;
```

The following PROC MEANS statements request mean hours by treatment.

```
proc means;
   freq count;
   class treatment;
   var hours;
run;
```

Output 5.5 contains the frequency table produced by PROC FREQ.

Output 5.5 Frequency Table

Frequency Row Pct	Table of treatment by hours					
	hours					
treatment	0	1	2	3	4	Total
placebo	6 24.00	9 36.00	6 24.00	3 12.00	1 4.00	25
standard	1 4.00	4 16.00	6 24.00	6 24.00	8 32.00	25
test	2 7.41	5 18.52	6 22.22	8 29.63	6 22.22	27
Total	9	18	18	17	15	77

Output 5.6 displays the mean hours of relief by type of treatment as well as the standard deviations.

Output 5.6 Means and Standard Errors

Analysis Variable : hours						
treatment	N Obs	N	Mean	Std Dev	Minimum	Maximum
placebo	5	5	1.3600000	2.7276363	0	4.0000000
standard	5	5	2.6400000	2.9899833	0	4.0000000
test	5	5	2.4074074	3.1827079	0	4.0000000

Patients on standard and test treatments reported 2.640 and 2.407 average hours of relief, respectively, while patients on placebo reported 1.36 hours of relief on average.

Output 5.7 displays the summary statistics.

Output 5.7 Mean Score Statistic

Summary Statistics for treatment by hours

Cochran-Mantel-Haenszel Statistics (Based on Table Scores)				
Statistic	Alternative Hypothesis	DF	Value	Prob
1	Nonzero Correlation	1	8.0668	0.0045
2	Row Mean Scores Differ	2	13.7346	0.0010
3	General Association	8	14.4030	0.0718

Total Sample Size = 77

Q_S is the "Row Mean Scores Differ" statistic. $Q_S = 13.7346$, with 2 df, and is clearly significant. Note that Q for these data takes the value 14.403, which has a p-value of 0.0718 with 8 df. In fact, there are a number of cells whose expected values are ≤ 5, so the chi-square approximation for the test for general association may not even be valid. However, since the row totals of the table are all greater than 20 and each row has counts ≥ 5 for both outcomes ≤ 1 and ≥ 2, there is sufficient sample size for Q_S. This is an example of how taking advantage of the ordinality of the data is not only the more appropriate approach, but it might be the only possible Mantel-Haenszel strategy due to sample size constraints.

5.2.3 Correlation Test

Sometimes, both the row variable and the column variable are ordinally scaled. This is common when you are studying responses that are evaluated on an ordinal scale and what is being compared are different dosage levels, which are also ordinally scaled. Consider the data in Table 5.4. A water treatment company is studying water additives and investigating how they affect clothes washing. The treatments studied were no treatment (plain water), the standard treatment, and a double dose of the standard treatment, called super. Washability was measured as low, medium, and high.

Table 5.4 Washability Data

Treatment	Washability			Total
	Low	Medium	High	
Water	27	14	5	46
Standard	10	17	26	53
Super	5	12	50	67

As discussed in Section 4.2, the appropriate statistic to investigate association for this situation is one that takes advantage of the ordinality of both the row variable and the column variable and tests the null hypothesis of no association against the alternative of linear association. In Chapter

4, "Sets of $2 \times r$ and $s \times 2$ Tables," the test statistic Q_{CS} was developed for the $s \times 2$ table and was shown to have one degree of freedom. A similar strategy applies to the $s \times r$ table. You assign scores both to the levels of the response variable and to the levels of the grouping variable to obtain Q_{CS}, which is approximately chi-square with one degree of freedom. Thus, whether the table is 2×2, $s \times 2$, or $s \times r$, Q_{CS} always has one degree of freedom. (See Section 4.6 for a related discussion.)

For more detail, this statistic is also derived from the general quadratic form

$$Q = (\mathbf{n} - \mathbf{m})' \mathbf{A}' (\mathbf{A} \mathbf{V} \mathbf{A}')^{-1} \mathbf{A} (\mathbf{n} - \mathbf{m}) = (n-1) r_{ac}^2$$

where r_{ac} is the Pearson correlation. You obtain Q_{CS} by choosing $\mathbf{A}$ to be

$$\mathbf{A} = [\mathbf{a}' \otimes \mathbf{c}'] = [a_1 c_1, \ldots, a_r c_1, \ldots, a_r c_s]$$

where $\mathbf{a}' = (a_1, a_2, \ldots, a_r)$ are scores for the response levels and $\mathbf{c}' = (c_1, c_2, \ldots, c_s)$ are scores for the levels of the grouping variable. $\mathbf{A}$ has dimension $1 \times sr$.

The following PROC FREQ statements produce the correlation statistic for the washability data. It is of interest to use both integer scores and modified ridit scores and compare the results. The following statements request both integer scores (the default) and modified ridit scores. The ORDER= option maintains the desired order of the levels of the rows and columns; it is the same as the order in which the variable values are encountered in the DATA step. The NOPRINT option suppresses the printing of the individual tables.

```
data wash;
   input treatment $ washability $ count @@;
   datalines;
water low 27 water medium 14 water high 5
standard low 10 standard medium 17 standard high 26
super low 5 super medium 12 super high 50
;

proc freq order=data;
   weight count;
   tables treatment*washability / chisq cmh nocol nopct;
   tables treatment*washability / scores=modridit cmh
                                  noprint nocol nopct;
run;
```

Output 5.8 displays the frequency table.

Output 5.8 Frequency Table

Frequency Row Pct	Table of treatment by washability			
		washability		
treatment	low	medium	high	Total
water	27 58.70	14 30.43	5 10.87	46
standard	10 18.87	17 32.08	26 49.06	53
super	5 7.46	12 17.91	50 74.63	67
Total	42	43	81	166

Output 5.9 displays the measures of association. The Pearson correlation has the value 0.5538 for the table scores.

Output 5.9 Frequency Table

Statistics for Table of treatment by washability

Statistic	Value	ASE
Gamma	0.6974	0.0636
Kendall's Tau-b	0.4969	0.0553
Stuart's Tau-c	0.4803	0.0545
Somers' D C|R	0.4864	0.0542
Somers' D R|C	0.5077	0.0572
Pearson Correlation	0.5538	0.0590
Spearman Correlation	0.5479	0.0596
Lambda Asymmetric C|R	0.2588	0.0573
Lambda Asymmetric R|C	0.2727	0.0673
Lambda Symmetric	0.2663	0.0559
Uncertainty Coefficient C|R	0.1668	0.0389
Uncertainty Coefficient R|C	0.1609	0.0372
Uncertainty Coefficient Symmetric	0.1638	0.0380

The CHISQ option always produces the correlation statistic Q_{CS}. Compare its value, $Q_{CS} = 50.6016$ (displayed in Output 5.10), with the statistic displayed under "Nonzero Correlation" in Output 5.11. These statistics are the same. Thus, you don't need to specify CMH to obtain Q_{CS} for a single table. For a 2 × 2 table, Q_{CS} is equivalent to Q and Q_S; for a 2 × r table, Q_{CS} is equivalent to Q_S.

Note that $Q_{CS} = 50.6016$ is indeed equal to $(n-1)r_{ac}^2 = 165 \times (0.5538)^2$.

Output 5.10 Chi-Square Statistics

Statistics for Table of treatment by washability

Statistic	DF	Value	Prob
Chi-Square	4	55.0879	<.0001
Likelihood Ratio Chi-Square	4	58.0366	<.0001
Mantel-Haenszel Chi-Square	1	50.6016	<.0001
Phi Coefficient		0.5761	
Contingency Coefficient		0.4992	
Cramer's V		0.4073	

Output 5.11 Q_{CS} for Integer Scores

Summary Statistics for treatment by washability

Cochran-Mantel-Haenszel Statistics (Based on Table Scores)				
Statistic	Alternative Hypothesis	DF	Value	Prob
1	Nonzero Correlation	1	50.6016	<.0001
2	Row Mean Scores Differ	2	52.7786	<.0001
3	General Association	4	54.7560	<.0001

Total Sample Size = 166

Q_{CS} is clearly significant. Washability increases with the degree of additive to the water. Output 5.12 displays Q_{CS} for the modified ridit scores. It has the value 49.541, which is clearly significant.

Output 5.12 Q_{CS} for Modified Ridit Scores

Summary Statistics for treatment by washability

Cochran-Mantel-Haenszel Statistics (Modified Ridit Scores)				
Statistic	Alternative Hypothesis	DF	Value	Prob
1	Nonzero Correlation	1	49.5410	<.0001
2	Row Mean Scores Differ	2	52.5148	<.0001
3	General Association	4	54.7560	<.0001

Total Sample Size = 166

5.3 Exact Tests for Association

5.3.1 General Association

In some cases, the samples sizes are not sufficient for the chi-square statistics discussed earlier in this chapter to be valid (several $m_{ij} \leq 5$). An alternative strategy for these situations is the Fisher exact test for $s \times r$ tables. This method follows the same principles as Fisher's exact test for the 2×2 table, except that the probabilities that are summed are taken from the multivariate hypergeometric distribution. Mehta and Patel (1983) describe a network algorithm for obtaining exact p-values that works much faster and more efficiently than direct enumeration; Baglivo, Olivier, and Pagano (1988), Cox and Plackett (1980), and Pagano and Halvorsen (1981) have also done work in this area. Besides Fisher's exact test, which produces the exact p-value for the table, exact p-values are available for general association tests such as Q, Q_P, and Q_L.

Consider Table 5.5. A marketing research firm organized a focus group to consider issues of new car marketing. Members of the group included people who had purchased a car from a local dealer in the last month. Researchers were interested in whether there was an association between the type of car bought and the manner in which group members found out about the car in the media. Cars were classified as sedans, sporty, and utility. The types of media included television, magazines, newspapers, and radio.

Table 5.5 Car Marketing Data

Type of Car	Advertising Source				Total
	TV	Magazine	Newspaper	Radio	
Sedan	4	0	0	2	6
Sporty	0	3	3	4	10
Utility	5	5	2	2	14

It is clear that the data do not meet the requirements for the usual tests of association via the Pearson chi-square or the randomization chi-square. There are a number of zero cells and a number of other cells whose expected values are less than 5. Under these circumstances, the exact test for no association is an appropriate strategy.

The following SAS statements produce the exact test for the car marketing data. Recall that Fisher's exact test is produced automatically for 2×2 tables with the CHISQ option; to generate the exact test for $s \times r$ tables, you need to specify the EXACT option in the TABLES statement. This generates the usual statistics produced with the CHISQ option and the exact test. Since the ORDER= option isn't specified, the columns of the resulting table will be ordered alphabetically. No ordering is assumed for this test, so this does not matter.

```
data market;
   length AdSource $ 9. ;
   input car $ AdSource $ count @@;
   datalines;
sporty    paper 3 sporty    radio 4 sporty    tv 0 sporty    magazine 3
sedan     paper 0 sedan     radio 2 sedan     tv 4 sedan     magazine 0
```

```
utility paper 2 utility radio 2 utility tv 5 utility magazine 5
;

proc freq;
   weight count;
   table car*AdSource / norow nocol nopct;
   exact fisher pchi lrchi;
run;
```

Output 5.13 contains the frequency table.

Output 5.13 Car Marketing Frequency Table

Frequency	Table of car by AdSource				
	AdSource				
car	magazine	paper	radio	tv	Total
sedan	0	0	2	4	6
sporty	3	3	4	0	10
utility	5	2	2	5	14
Total	8	5	8	9	30

Output 5.14 displays the Fisher exact p-value for the table, in addition to the asymptotic and exact results for Q_P and Q_L. For these data, the exact p-value for the table is $p = 0.0473$. Note that Q_P is 11.5984 with 6 df, $p = 0.0716$, and Q_L has the value 16.3095 with $p = 0.0122$. Since the alternative hypothesis is general association, there are no left-tail or right-tail analogies to what Fisher's exact test for 2×2 tables provides since the alternative hypothesis can be directional association.

The exact p-value for Q_P is 0.0664, and the exact p-value for Q_L is 0.0272. Thus, with the exact computations, Q_P became somewhat stronger and Q_L became somewhat weaker but still significant at the $\alpha = 0.05$ level. For this table, you would typically use the Fisher exact p-value as your indication of the strength of the association and consider the association to be significant at the 0.05 level of significance.

While you can't directly produce an exact p-value for the general association Q (that is, the test produced by PROC FREQ with the CMH option for the one-stratum case), the exact distribution for Q is identical to the exact distribution for Q_P. Thus, the exact p-value for Q_P is the same as the exact p-value for Q. This is because

$$Q = \left(\frac{n-1}{n}\right) Q_P$$

for general association.

Output 5.14 Exact Test Results

Pearson Chi-Square Test	
Chi-Square	11.5984
DF	6
Asymptotic Pr > ChiSq	0.0716
Exact Pr >= ChiSq	0.0664

Likelihood Ratio Chi-Square Test	
Chi-Square	16.3095
DF	6
Asymptotic Pr > ChiSq	0.0122
Exact Pr >= ChiSq	0.0272

Fisher's Exact Test	
Table Probability (P)	2.545E-05
Pr <= P	0.0473

5.3.1.1 Notes on Exact Computations

Even though the network algorithms used to produce these exact tests are very fast compared to direct enumeration, exact methods are computationally intensive. The memory requirements and CPU time requirements can be quite high. As the sample size becomes larger, the test is likely to become computationally infeasible. For most situations when the sample size is moderately large, asymptotic methods are valid. An exception would be data that have marked sparseness in the row and column marginal totals. The exact test is mainly useful when significance is suggested by the approximate results of Q_P and Q_L. Also, in these situations, the computations are not overly lengthy. Computations are lengthy when the p-value is somewhere around 0.5, and in this situation, the exact p-value is usually not needed.

When SAS is performing exact computations, it prints a message to the log stating that you can press the system interrupt key if you want to terminate the computations. In addition, you can specify the MAXTIME= option in the EXACT statement to request, in seconds, a length of time after which the procedure is to stop exact computations.

There are some data for which computing the exact p-values is going to be very memory- and time- intensive and yet the asymptotic tests are not quite justifiable. You can request Monte Carlo estimation for these situations by specifying MC as an EXACT statement option. PROC FREQ uses Monte Carlo methods to estimate the exact p-value and give a confidence interval for the estimate. See Agresti, Wackerly, and Boyett (1979) for more detail. With PROC FREQ, you can specify the number of samples (the default is 10,000) and the random number seed. See the *SAS/STAT User's Guide* for more information about the exact computational algorithms used in the FREQ procedure.

5.3.2 Test of Correlation

Section 5.2 discusses Mantel-Haenszel test statistics for the evaluation of general association, location shifts, and correlation. You can obtain exact *p*-values for the correlation test with SAS for the case when both the rows and columns of your table are ordinally scaled.

Consider the data in Table 5.6 from a study on a new drug for a skin disorder. Subjects were randomly assigned to one of four dosage levels and, after a suitable period of time, the affected skin area was examined and classified on a four-point scale ranging from 0 for poor to 3 for excellent.

Table 5.6 Skin Disorder Data

Dose in Mg	Response			
	Poor	Fair	Good	Excellent
25	1	1	1	0
50	1	2	1	1
75	0	0	2	2
100	0	0	7	0

Since both the rows and columns can be considered to be on an ordinal scale, the type of association involved is linear and the correlation Mantel-Haenszel statistic is suitable. However, note that there are several zero cells, many other cells with counts of 1 or 2, and a total sample size of 19. This is on the border of too small for the asymptotic MH test, which requires an overall sample size of at least 20. In addition, if you collapse this table into various 2×2 tables, many of the resulting cell counts are less than 5; for the asymptotic MH correlation test you generally want any cell count of a collapsed 2×2 table to be 5 or larger.

However, you can compute an exact *p*-value, which is the sum of the exact *p*-values associated with the tables where the test statistic is larger than the statistic for the table you observe.

The following DATA step creates SAS data set DISORDER.

```
data disorder;
   input dose outcome count @@;
   datalines;
 25 0 1    25 1 1    25 2 1    25 3 0
 50 0 1    50 1 2    50 2 1    50 3 1
 75 0 0    75 1 0    75 2 2    75 3 2
100 0 0   100 1 0   100 2 7   100 3 0
;
```

Specifying the EXACT statement with the MHCHI keyword produces both the asymptotic and the exact MH test.

```
proc freq;
   weight count;
   tables dose*outcome / nocol norow nopct measures;
   exact mhchi;
run;
```

Output 5.15 displays the frequency table for the skin disorder data.

Output 5.15 Skin Disorder Data

Frequency	Table of dose by outcome				
		outcome			
dose	0	1	2	3	Total
25	1	1	1	0	3
50	1	2	1	1	5
75	0	0	2	2	4
100	0	0	7	0	7
Total	2	3	11	3	19

Output 5.16 displays the measures of association, including the Pearson correlation which has a value of 0.4673.

Output 5.16 Skin Disorder Data

Statistics for Table of dose by outcome

Statistic	Value	ASE
Gamma	0.4194	0.2170
Kendall's Tau-b	0.3264	0.1721
Stuart's Tau-c	0.2881	0.1625
Somers' D C|R	0.2977	0.1633
Somers' D R|C	0.3578	0.1843
Pearson Correlation	0.4673	0.1497
Spearman Correlation	0.4192	0.1962
Lambda Asymmetric C|R	0.1250	0.3508
Lambda Asymmetric R|C	0.4167	0.1423
Lambda Symmetric	0.3000	0.2100
Uncertainty Coefficient C|R	0.4104	0.0931
Uncertainty Coefficient R|C	0.3484	0.1004
Uncertainty Coefficient Symmetric	0.3768	0.0959

Output 5.17 displays Q_{CS} and both the asymptotic and exact p-values.

Output 5.17 Exact Results for Correlation MH

Mantel-Haenszel Chi-Square Test	
Chi-Square	3.9314
DF	1
Asymptotic Pr > ChiSq	0.0474
Exact Pr >= ChiSq	0.0488

With a Q_{CS} value of 3.9314 and 1 df, the chi-square approximation provides a significant *p*-value of 0.0474. The exact *p*-value is a little higher (0.0488), but is still significant at $\alpha = 0.05$. These data clearly have an association that is detected with a linear correlation statistic.

5.4 Measures of Association

Analysts are sometimes interested in assessing the strength of association in the $s \times r$ table. Although there is no counterpart to the odds ratios in 2×2 tables, there are several measures of association available, and, as you might expect, their choice depends on the scale of measurement.

5.4.1 Ordinal Measures of Association

If the data in the table have an interval scale or have scores that are equally spaced, then the Pearson correlation coefficient is an appropriate measure of association, and one that is familiar to most readers.

If the data do not lie on an obvious scale but are ordinal in nature, then other measures apply. The Spearman rank correlation coefficient is produced by substituting ranks as variable values for the Pearson correlation coefficient. Other measures are based on the classification of all possible pairs of subjects in the table as concordant or discordant pairs. If a pair is *concordant*, then the subject that ranks higher on the row variable also ranks higher on the column variable. If a pair is *discordant*, then the subject that ranks higher on the row variable also ranks lower on the column variable. The pair can also be tied on the row and column variables.

The gamma, Kendall's tau-*b*, Stuart's tau-*c*, and Somer's *D* statistics are all based on concordant and discordant pairs; that is, they use the relative ordering on the levels of the variables to determine whether association is negative, positive, or present at all. For example, gamma is estimated by

$$\hat{\gamma} = \frac{(C - D)}{(C + D)}$$

where C is the total number of concordant pairs and D is the total number of discordant pairs.

These measures, like the Pearson correlation coefficient, take values between -1 and 1. They differ mainly in their strategies for adjusting for ties and sample size. Somer's *D* depends on which variable is considered to be explanatory (the grouping variable—adjustments for ties are made only on it). Somer's *D*, Stuart's tau-*c*, and Kendall's tau-*b* generally express less strength of association than gamma.

Asymptotic standard errors are available for these measures. Although the measure of association is always valid, these standard errors are valid only if the sample size is large. Very conservative guidelines are the usual requirements for the Pearson chi-square that the expected cell counts are 5 or greater. A more realistic guideline is to collapse the $s \times r$ table to a 2×2 table by choosing cutpoints and then adding the appropriate rows and columns. Think of this as a line under one row and beside one column; the 2×2 table is the result of summing the cells in the resulting quadrants. The sample size is adequate if each of the cells of this 2×2 table is 5 or greater.

If the sample size is adequate, then the measure of association is approximately normally distributed and you can form the confidence intervals of interest. For example,

$$\text{measure} \pm 1.96 \times \text{ASE}$$

forms the bounds of a 95% confidence interval. See the *SAS/STAT User's Guide* for more information on these ordinal measures of association.

Measures of association are produced in the PROC FREQ output by the MEASURES option in the TABLES statement. In addition, you can request confidence limits by specifying the CL option. The following statements produce measures of association for the washability data listed in Table 5.4. The SCORES=RANK option in the second TABLES statement requests that rank scores be used in calculating Pearson's correlation coefficient.

```
data wash;
   input treatment $ washability $ count @@;
   datalines;
water    low 27 water    medium 14 water    high  5
standard low 10 standard medium 17 standard high 26
super    low  5 super    medium 12 super    high 50
;

proc freq order=data;
   weight count;
   tables treatment*washability / measures noprint nocol nopct cl;
   tables treatment*washability / measures scores=rank noprint cl;
run;
```

Output 5.18 contains the table produced by the first PROC FREQ invocation. All of the measures of ordinal association indicate a positive association. Note also that the Somer's *D* statistics, Kendall's tau-*b*, and Stuart's tau-*c* all have smaller values than gamma. Somer's D statistic has two forms: Somer's $DC|R$ means that the column variable is considered the dependent (response) variable, and Somer's $DR|C$ means that the row variable is considered the response variable.

Output 5.18 Measures of Association

Statistics for Table of treatment by washability

Statistic	Value	ASE	95% Confidence Limits	
Gamma	0.6974	0.0636	0.5728	0.8221
Kendall's Tau-b	0.4969	0.0553	0.3885	0.6053
Stuart's Tau-c	0.4803	0.0545	0.3734	0.5872
Somers' D C|R	0.4864	0.0542	0.3802	0.5926
Somers' D R|C	0.5077	0.0572	0.3956	0.6197
Pearson Correlation	0.5538	0.0590	0.4382	0.6693
Spearman Correlation	0.5479	0.0596	0.4311	0.6648
Lambda Asymmetric C|R	0.2588	0.0573	0.1465	0.3711
Lambda Asymmetric R|C	0.2727	0.0673	0.1409	0.4046
Lambda Symmetric	0.2663	0.0559	0.1567	0.3759
Uncertainty Coefficient C|R	0.1668	0.0389	0.0906	0.2431
Uncertainty Coefficient R|C	0.1609	0.0372	0.0880	0.2339
Uncertainty Coefficient Symmetric	0.1638	0.0380	0.0893	0.2383

Output 5.19 contains the output produced by the second PROC FREQ invocation. The only difference is that rank scores were used in the calculation of Pearson's correlation coefficient. When rank scores are used, Pearson's correlation coefficient is equivalent to Spearman's correlation, as illustrated in the output. (However, the asymptotic standard errors are not equivalent.)

Output 5.19 Rank Scores for Pearson's Correlation

Statistics for Table of treatment by washability

Statistic	Value	ASE	95% Confidence Limits	
Gamma	0.6974	0.0636	0.5728	0.8221
Kendall's Tau-b	0.4969	0.0553	0.3885	0.6053
Stuart's Tau-c	0.4803	0.0545	0.3734	0.5872
Somers' D C|R	0.4864	0.0542	0.3802	0.5926
Somers' D R|C	0.5077	0.0572	0.3956	0.6197
Pearson Correlation (Rank Scores)	0.5479	0.0591	0.4322	0.6637
Spearman Correlation	0.5479	0.0596	0.4311	0.6648
Lambda Asymmetric C|R	0.2588	0.0573	0.1465	0.3711
Lambda Asymmetric R|C	0.2727	0.0673	0.1409	0.4046
Lambda Symmetric	0.2663	0.0559	0.1567	0.3759
Uncertainty Coefficient C|R	0.1668	0.0389	0.0906	0.2431
Uncertainty Coefficient R|C	0.1609	0.0372	0.0880	0.2339
Uncertainty Coefficient Symmetric	0.1638	0.0380	0.0893	0.2383

5.4.2 Exact Tests for Ordinal Measures of Association

The only difference is that rank scores were used in the calculation of Pearson's correlation coefficient. When rank scores are used, Pearson's correlation coefficient is equivalent to Spearman's correlation, as illustrated in the output. (However, the asymptotic standard errors are not equivalent.)

In addition to estimating measures of association, you can also test whether a particular measure is equal to zero. In the case of the correlation coefficients, you can produce exact p-values for this test. Thus, you have access to exact methods in the evaluation of the correlation coefficients.

Table 5.7 displays data that a recreation supervisor collected from her girls' soccer league coaches. Hearing complaints about too-intense parental involvement, she surveyed each coach to see whether they considered the parental interference to be of low, medium, or high intensity for the three different grade level leagues. Interference was considered to be parents questioning their child's playing time or position, questioning referee calls during the games, or yelling very specific instructions to the children on the team. She was interested in whether interference was associated with league grade. Since both grade level and interference level lie on an ordinal scale, the Spearman rank correlation coefficient is an appropriate statistic to consider.

Table 5.7 Soccer Coach Interviews

Grades	Parental Interference		
	Low	Medium	High
1–2	3	1	0
3–4	3	2	1
5–6	1	3	2

Since there were four teams in the league for Grades 1 and 2 and six teams each in the leagues for Grades 2 and 4 and for Grades 5 and 6, the counts are necessarily small. If you apply various cutpoints to produce collapsed 2 ×2 tables as suggested previously to determine whether the asymptotic confidence intervals for the measures of association would be valid, you determine that no cutpoints exist so that each component cell is ≥ 5. However, you can apply exact methods to get an exact p-value for the hypothesis that the Spearman rank correlation coefficient is equal to zero.

The following DATA step inputs the soccer data into SAS data set SOCCER.

```
data soccer;
   input grades $ degree $ count @@;
   datalines;
1-2 low 3 1-2 medium 1 1-2 high 0
3-4 low 3 3-4 medium 2 3-4 high 1
5-6 low 1 5-6 medium 3 5-6 high 2
;
```

In order to produce the exact p-value for the Spearman's rank test, you specify an EXACT statement that includes the keyword SCORR. The ORDER=DATA option is specified in the PROC statement to ensure that the columns and rows maintain the correct ordering.

```
proc freq order=data;
   weight count;
   tables grades*degree / nocol nopct norow;
   exact scorr;
run;
```

Output 5.20 displays the frequency table for the soccer data.

Output 5.20 Soccer Frequency Table

Frequency	Table of grades by degree			
		degree		
grades	low	medium	high	Total
1-2	3	1	0	4
3-4	3	2	1	6
5-6	1	3	2	6
Total	7	6	3	16

Output 5.21 contains the Spearman correlation coefficient, which has the value 0.4878, indicating the possibility of modest correlation.

Output 5.21 Spearman Correlation Coefficient

Spearman Correlation Coefficient	
Correlation (r)	0.4878
ASE	0.1843
95% Lower Conf Limit	0.1265
95% Upper Conf Limit	0.8491

The results for the hypothesis test that the correlation is equal to zero are listed in Output 5.22. The exact two-sided p-value is 0.0637, a borderline result, although you would not reject the hypothesis at a strict $\alpha = 0.05$ level of confidence. Note that the asymptotic test results in a two-sided p-value of 0.0092. However, with these counts, you could not justify the use of the asymptotic test.

Output 5.22 Hypothesis Test for Spearman's Rank Test

Test of H0: Correlation = 0	
ASE under H0	0.1872
Z	2.6055
One-sided Pr > Z	0.0046
Two-sided Pr > |Z|	0.0092
Exact Test	
One-sided Pr >= r	0.0354
Two-sided Pr >= |r|	0.0637

Note that you can also test whether the asymptotic statistics produced by the MEASURES option are equal to zero. You request such tests with the TEST statement in the FREQ procedure; you can also test hypotheses concerning the kappa statistics discussed in Section 5.5.

See the *SAS/STAT User's Guide* for more information.

5.4.3 Nominal Measures of Association

Measures of association when one or both variables are nominally scaled are more difficult to define, since you can't think of association in these circumstances as negative or positive in any sense. However, indices of association in the nominal case have been constructed, and most are based on mimicking R-squared in some fashion. One such measure is the uncertainty coefficient, and another is the lambda coefficient. More information about these statistics can be obtained in the *SAS/STAT User's Guide*, including the appropriate references. Agresti (2002) also discusses some of these measures.

The following PROC FREQ invocation produces nominal measures of association for the neighborhood data.

```
data neighbor;
   length party $ 11 neighborhood $ 10;
   input party $ neighborhood $ count @@;
   datalines;
democrat      longview    360  democrat     bayside  221
democrat      sheffeld    140  democrat     highland 160
republican    longview    316  republican   bayside  208
republican    sheffeld     97  republican   highland 106
independent   longview    160  independent  bayside  200
independent   sheffeld    311  independent  highland 291
;

proc freq ;
   weight count;
   tables party*neighborhood / chisq measures nocol nopct;
run;
```

Output 5.23 displays the resulting table.

Output 5.23 Nominal Measures of Association

Statistics for Table of party by neighborhood

Statistic	Value	ASE
Gamma	-0.0183	0.0226
Kendall's Tau-b	-0.0130	0.0161
Stuart's Tau-c	-0.0137	0.0169
Somers' D C|R	-0.0138	0.0170
Somers' D R|C	-0.0123	0.0152
Pearson Correlation	-0.0178	0.0190
Spearman Correlation	-0.0150	0.0189
Lambda Asymmetric C|R	0.0871	0.0120
Lambda Asymmetric R|C	0.1374	0.0177
Lambda Symmetric	0.1113	0.0119
Uncertainty Coefficient C|R	0.0401	0.0046
Uncertainty Coefficient R|C	0.0503	0.0058
Uncertainty Coefficient Symmetric	0.0446	0.0051

You should ignore the ordinal measures of association here since the data are not ordinally scaled.

There are three versions of both the lambda coefficient and the uncertainty coefficient: column variable as the response variable, row variable as the response variable, and a symmetric version. Obviously, this makes a difference in the resulting statistic.

5.5 Observer Agreement

5.5.1 Computing the Kappa Statistic

For many years, researchers in medicine, epidemiology, psychiatry, and psychological measurement and testing have been aware of the importance of observer error as a major source of measurement error. In many cases, different observers, or even the same observer at a different time, may examine an X-ray or perform a physical examination and reach different conclusions. It is important to evaluate observer agreement, both to understand the possible contributions to measurement error and as part of the evaluation of testing new instruments and procedures. See Landis et al. (2011) for a review of methods for assessing observer agreement.

Often, the data collected as part of an observer agreement study form a contingency table, where the column levels represent the ratings of one observer and the row levels represent the ratings of another observer. Each cell represents one possible profile of the observers' ratings. The cells on the diagonal represent the cases where the observers agree.

Consider Table 5.8. These data come from a study concerning the diagnostic classification of multiple sclerosis patients. Patients from Winnipeg and New Orleans were classified into one of four diagnostic classes by both a Winnipeg neurologist and a New Orleans neurologist. Table 5.8 contains the data for the Winnipeg patients (Landis and Koch 1977).

Table 5.8 Ratings of Neurologists

New Orleans Neurologist	Winnipeg Neurologist			
	1	2	3	4
1	38	5	0	1
2	33	11	3	0
3	10	14	5	6
4	3	7	3	10

Certainly one way to assess the association between these two raters is to compute the usual measures of association. However, while measures of association can reflect the strength of the predictable relationship between two raters or observers, they don't target how well they agree. Agreement can be considered a special case of association—to what degree do different observers classify a particular subject into the identical category? All measures of agreement target the diagonal cells of a contingency table in their computations, and some measures take into consideration how far away from the diagonal elements other cells fall.

Suppose π_{ij} is the probability of a subject being classified in the ith category by the first observer and the jth category by the second observer. Then

$$\Pi_o = \sum \pi_{ii}$$

is the probability that the observers agree. If the ratings are independent, then the probability of agreement is

$$\Pi_e = \sum \pi_{i+} \pi_{+i}$$

So, $\Pi_o - \Pi_e$ is the amount of agreement beyond that expected by chance. The *kappa coefficient* (Cohen 1960) is defined as

$$\kappa = \frac{\Pi_o - \Pi_e}{1 - \Pi_e}$$

Since $\Pi_o = 1$ when agreement is perfect (all non-diagonal elements are zero), κ equals 1 when agreement is perfect, and κ equals 0 when the agreement equals that expected by chance. The closer the value is to 1, the more agreement there is in the table. It is possible to obtain negative values, but that rarely occurs. Note that κ is analogous to the intraclass correlation coefficient obtained from ANOVA models for quantitative measurements; it can be used as a measure of reliability of multiple determinations on the same subject (Fleiss and Cohen 1973, Fleiss 1975).

You might be interested in distinguishing degrees of agreement in a table, particularly if the categories are ordered in some way. For example, you might want to take into account those disagreements that are just one category away. A weighted form of the kappa statistic allows you to assign weights, or scores, to the various categories so that you can incorporate such considerations into the construction of the test statistic.

Weighted κ is written

$$\kappa_w = \frac{\sum \sum w_{ij} \pi_{ij} - \sum \sum w_{ij} \pi_{i+} \pi_{+j}}{1 - \sum \sum_{ij} w_{ij} \pi_{i+} \pi_{+j}}$$

where w_{ij} represents weights with values between 0 and 1. One possible set of weights is

$$w_{ij} = 1 - \frac{|\text{score}(i) - \text{score}(j)|}{\text{score}(\text{dim}) - \text{score}(1)}$$

where score(i) is the score for the ith row, score(j) is the score for the jth column, and dim is the dimension of an $s \times s$ table. This scoring system puts more weight on those cells closest to the diagonal. These weights are known as Cicchetti-Allison weights (Cicchetti and Allision 1969) and are the default weights for the weighted kappa statistic in PROC FREQ. Fleiss-Cohen weights are also available (Fleiss and Cohen 1973).

The following SAS statements generate kappa statistics for the Winnipeg data. To produce measures of agreement, you specify AGREE in the TABLES statement. When ODS Graphics is enabled and kappa statistics are requested, PROC FREQ also produces an agreement plot.

```
data classify;
   input no_rater w_rater count @@;
   datalines;
1 1 38 1 2   5 1 3 0 1 4   1
2 1 33 2 2  11 2 3 3 2 4   0
3 1 10 3 2  14 3 3 5 3 4   6
4 1  3 4 2   7 4 3 3 4 4  10
;
```

```
ods graphics on;
proc freq;
   weight count;
   tables no_rater*w_rater / agree norow nocol nopct;
run;
ods graphics off;
```

Output 5.24 contains the table.

Output 5.24 Winnipeg Data

Frequency	Table of no_rater by w_rater					
		w_rater				
	no_rater	1	2	3	4	Total
	1	38	5	0	1	44
	2	33	11	3	0	47
	3	10	14	5	6	35
	4	3	7	3	10	23
	Total	84	37	11	17	149

Output 5.25 displays the kappa statistics.

Output 5.25 Kappa Statistics

Kappa Statistics				
Statistic	Value	ASE	95% Confidence Limits	
Simple Kappa	0.2079	0.0505	0.1091	0.3068
Weighted Kappa	0.3797	0.0517	0.2785	0.4810

$\hat{\kappa}$ has the value 0.2079. This is indicative of slight agreement. Values of 0.4 or higher are considered to indicate moderate agreement, and values of 0.8 or higher indicate excellent agreement. The asymptotic standard error is also printed, as are confidence bounds. Since the confidence bounds do not contain the value 0, you can reject the hypothesis that κ is 0 for these data (no agreement) at the $\alpha = 0.05$ level of significance.

Using the default scores, $\hat{\kappa}_w$ takes the value 0.3797. This means that if you consider disagreement close to the diagonals less heavily than disagreement further away from the diagonals, you get higher agreement. $\hat{\kappa}$ treats all off-diagonal cells the same. When $\hat{\kappa}_w$ is high (say $\hat{\kappa}_w \geq 0.6$ for moderate sample size), it might be preferable to produce confidence bounds on a transformed scale (such as logarithms or the Fisher z transformation) and then exponentiate to compute the limits.

Output 5.26 contains Bowker's test of symmetry (Bowker 1948).

Output 5.26 Symmetry Test

Test of Symmetry	
Statistic (S)	46.7492
DF	6
Pr > S	<.0001

The null hypothesis of this test is that the square table is symmetric (the cell probabilities p_{ij} and p_{ji} are equal). When you have a 2 × 2 table, the test is the same as McNemar's test.

Output 5.27 displays the agreement plot, which provides a visual impression of the strength of the agreement (Bangdiwala and Bryan 1987).

Output 5.27 Agreement Plot

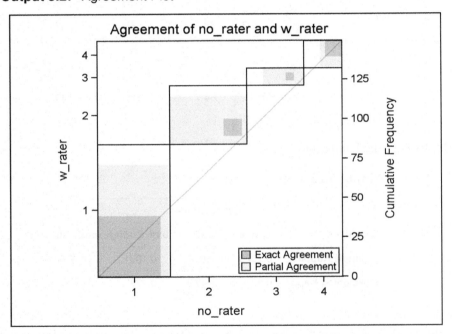

The large rectangles represent the marginal totals, that is, n_{i+} by n_{+i}, or the maximum possible agreement. The dark squares represent agreement, and their size is n_{ii} by n_{ii}. For example, the biggest square represents the 38 subjects rated 1 by both raters. The lighter shade represents partial agreement. The comparison of shaded portions of the rectangles to the entire rectangles provides an impression of the strength of the agreement.

5.5.2 Exact *p*-values for the Kappa Statistic

Exact *p*-values are also available for the kappa statistic. Consider the data in Table 5.9. Elderly residents of a midwestern community enrolled in a pilot program that provided resources to seniors and also sought to identify seniors who required additional living assistance. Researchers tested a tool for in-home evaluation of a resident's agility. The test rated the ease with which basic tasks could be performed and provided an overall rating of ability on a four-point scale ranging from 1 for poor to 4 for excellent. Two raters evaluated the same 24 people.

Table 5.9 Ratings of Social Workers

Rater One	Rater Two			
	1	2	3	4
1	4	0	1	0
2	0	2	6	1
3	1	0	2	1
4	0	2	1	3

The table includes numerous 0 and 1 counts, too many for the asymptotic requirements to be fulfilled. However, rater agreement can still be assessed with the use of exact *p*-values for the test of null agreement with kappa statistics.

The following PROC FREQ statements produce the desired exact results. You specify the keyword KAPPA in the EXACT statement to generate a table with both the asymptotic and exact results.

```
data pilot;
   input rater1 rater2 count @@;
   datalines;
1 1 4 1 2 0 1 3 1 1 4 0
2 1 0 2 2 2 2 3 6 2 4 1
3 1 1 3 2 0 3 3 2 3 4 1
4 1 0 4 2 2 4 3 1 4 4 3
;

proc freq;
   weight count;
   tables rater1*rater2 /norow nocol nopct;
   exact kappa;
run;
```

Output 5.28 contains the table of ratings.

Output 5.28 Pilot Data

Frequency	Table of rater1 by rater2				
		rater2			
rater1	1	2	3	4	Total
1	4	0	1	0	5
2	0	2	6	1	9
3	1	0	2	1	4
4	0	2	1	3	6
Total	5	4	10	5	24

Output 5.29 contains the estimate of the kappa coefficient, which is 0.2989 with asymptotic 95% confidence limits of (0.0469, 0.5509). The table "Test of H0: Kappa=0" presents both the asymptotic test for the hypothesis that a constructed Z statistic is equal to zero as well as the exact p-value for that test. The exact p-value is 0.0088, both one-sided and two-sided. There is some agreement between raters.

Output 5.29 Exact Results for Kappa Test

Simple Kappa Coefficient	
Kappa (K)	0.2989
ASE	0.1286
95% Lower Conf Limit	0.0469
95% Upper Conf Limit	0.5509

Test of H0: Kappa = 0	
ASE under H0	0.1066
Z	2.8032
One-sided Pr > Z	0.0025
Two-sided Pr > |Z|	0.0051
Exact Test	
One-sided Pr >= K	0.0088
Two-sided Pr >= |K|	0.0088

5.6 Test for Ordered Differences

Sometimes you have a contingency table in which the columns represent an ordinal outcome and the rows are either nominal or ordinal. One test of interest is whether there are location shifts in the mean response; this is evaluated with the mean score test as discussed in Section 5.2.2. However, you might also be interested in testing against an ordered alternative; that is, are the mean scores strictly increasing (or decreasing) across the levels of the row variable? The Jonckheere-Terpstra test is designed to test the null hypothesis that the distribution of the ordered responses is the same across the various rows of the table. This test detects whether there are differences in

$$d_1 \le d_2 \le \cdots \le d_s \text{ or } d_s \ge d_{s-1} \ge \cdots \ge d_1$$

where d_i represents the ith group effect.

The Jonckheere-Terpstra test is a nonparametric test that is based on sums of Mann-Whitney test statistics; the asymptotic p-values are produced by using the normal approximation for the distribution of the standardized test statistic. See the *SAS/STAT User's Guide* for more computational detail, and see Pirie (1983) and Hollander and Wolfe (1973) for more information on the Jonckheere-Terpstra test.

Table 5.10 displays the dumping syndrome data, which have appeared frequently in the categorical data analysis literature, beginning with Grizzle, Starmer, and Koch (1969). Investigators conducted a randomized clinical trial in four hospitals, where patients were assigned to one of four surgical procedures for the treatment of severe duodenal ulcers. The treatments include:

v + d: vagotomy and drainage

v + a: vagotomy and antrectomy (removal of 25% of gastric tissue)

v + h: vagatomy and hemigastrectomy (removal of 50% of gastric tissue)

gre: gastric resection (removal of 75% of gastric tissue)

The response measured was the severity (none, slight, moderate) of the dumping syndrome, which is expected to increase directly with the proportion of gastric tissue removed. This response, an adverse effect of surgery, can be considered ordinally scaled, as can operation. Investigators wanted to determine whether type of operation was associated with severity of dumping syndrome, after adjusting for hospital. This analysis is performed in the next chapter.

Table 5.10 Dumping Syndrome Data

Hospital	Operation	Severity of Symptoms			Total
		None	Slight	Moderate	
1	v + d	23	7	2	32
1	v + a	23	10	5	38
1	v + h	20	13	5	38
1	gre	24	10	6	40
2	v + d	18	6	1	25
2	v + a	18	6	2	26
2	v + h	13	13	2	28
2	gre	9	15	2	26
3	v + d	8	6	3	17
3	v + a	12	4	4	20
3	v + h	11	6	2	19
3	gre	7	7	4	18
4	v + d	12	9	1	22
4	v + a	15	3	2	20
4	v + h	14	8	3	25
4	gre	13	6	4	23

Ignoring hospital, there is interest in determining whether the responses are ordered the same across the operations. The Jonckheere-Terpstra test is appropriate here. The following SAS statements input these data.

```
data operate;
   input hospital trt $ severity $ wt @@;
   datalines;
1 v+d none 23    1 v+d slight  7    1 v+d moderate 2
1 v+a none 23    1 v+a slight 10    1 v+a moderate 5
1 v+h none 20    1 v+h slight 13    1 v+h moderate 5
1 gre none 24    1 gre slight 10    1 gre moderate 6
2 v+d none 18    2 v+d slight  6    2 v+d moderate 1
2 v+a none 18    2 v+a slight  6    2 v+a moderate 2
2 v+h none 13    2 v+h slight 13    2 v+h moderate 2
2 gre none  9    2 gre slight 15    2 gre moderate 2
3 v+d none  8    3 v+d slight  6    3 v+d moderate 3
3 v+a none 12    3 v+a slight  4    3 v+a moderate 4
3 v+h none 11    3 v+h slight  6    3 v+h moderate 2
3 gre none  7    3 gre slight  7    3 gre moderate 4
4 v+d none 12    4 v+d slight  9    4 v+d moderate 1
4 v+a none 15    4 v+a slight  3    4 v+a moderate 2
4 v+h none 14    4 v+h slight  8    4 v+h moderate 3
4 gre none 13    4 gre slight  6    4 gre moderate 4
;
```

The following PROC FREQ statements request the Jonckheere-Terpstra test by specifying the JT option in the TABLES statement. The order of the table columns is very important for such a test; in this PROC FREQ invocation, the ORDER=DATA option in the PROC statement produces the desired order. The CHISQ option with SCORES=RANKS is specified for comparison.

```
proc freq order=data;
   weight wt;
   tables trt*severity / chisq scores=rank norow nocol nopct jt;
run;
```

Output 5.30 contains the contingency table of treatment by severity.

Output 5.30 Dumping Syndrome Data

Frequency	Table of trt by severity			
		severity		
trt	none	slight	moderate	Total
v+d	61	28	7	96
v+a	68	23	13	104
v+h	58	40	12	110
gre	53	38	16	107
Total	240	129	48	417

The correlation Mantel-Haenszel statistic for rank scores has the value 6.6405 ($p = 0.0100$), as shown in Output 5.31. It addresses association in a similar manner to the Jonckheere-Terpstra test.

5.6. Test for Ordered Differences

Output 5.31 Chi-Square Statistics

Statistics for Table of trt by severity

Statistic	DF	Value	Prob
Chi-Square	6	10.5419	0.1036
Likelihood Ratio Chi-Square	6	10.8782	0.0922
MH Chi-Square (Rank Scores)	1	6.6405	0.0100
Phi Coefficient		0.1590	
Contingency Coefficient		0.1570	
Cramer's V		0.1124	

Output 5.32 displays the results of the Jonckheere-Terpstra test.

Output 5.32 Jonckheere-Terpstra

Jonckheere-Terpstra Test	
Statistic	35697.0000
Z	2.5712
One-sided Pr > Z	0.0051
Two-sided Pr > |Z|	0.0101

The value of the actual Jonckheere-Terpstra statistic is 35,697. The corresponding Z-statistic has the value 2.5712 with a two-sided p-value of 0.0101 (very close to the results of the MH test). At a 0.05 α level, you would conclude that there are significant differences among groups in their respective ordering for the ordered response variable that is represented by the columns of the table.

An exact version of the Jonckheere-Terpstra test is available. You simply specify the option JT in an EXACT statement.

Chapter 6

Sets of $s \times r$ Tables

Contents

6.1	Introduction		**141**
6.2	General Mantel-Haenszel Methodology		**142**
	6.2.1	General Association Statistic	144
	6.2.2	Mean Score Statistic	144
	6.2.3	Correlation Statistic	145
	6.2.4	Summary	145
6.3	Mantel-Haenszel Applications		**146**
	6.3.1	Dumping Syndrome Data	146
	6.3.2	Shoulder Harness Data	148
	6.3.3	Learning Preference Data	151
6.4	Advanced Topic: Application to Repeated Measures		**155**
	6.4.1	Introduction	155
	6.4.2	Dichotomous Response: Two Time Points (McNemar's Test)	157
	6.4.3	Dichotomous Response: Three Repeated Measurements	161
	6.4.4	Ordinal Response	164
	6.4.5	Ordinal Response with Missing Data	168

6.1 Introduction

Previous chapters address stratified analysis as the assessment of association in sets of 2×2 tables, $2 \times r$ tables where the response variable, represented in the table columns, is ordinally scaled, and $s \times 2$ tables where the groups for the row variable are ordinally scaled. Such analyses are special cases of the analysis of sets of $s \times r$ tables, which includes the cases where the row and column variables are both nominally scaled, the row variable is nominally scaled and the column variable is ordinally scaled, and the row variable and the column variable are both ordinally scaled. The Mantel-Haenszel strategy can be extended to handle these situations. It provides statistics that detect general association, mean score differences, and linear correlation as alternatives to the null hypothesis of no association; the choice of statistic depends on the scale of the row and column variables.

The general idea of stratified analyses is that you control for the effects of factors that are part of the research design, such as medical centers or hospitals in a randomized clinical trial, or factors that represent a prespecified poststudy stratification to adjust for explanatory variables that are thought to be related to the response variable. This is a common strategy for retrospective and observational studies. As mentioned in previous chapters, the Mantel-Haenszel strategy potentially removes the confounding influence of the explanatory variables that comprise the stratification and provides a gain of power for detecting association by comparing like subjects. In some sense, the strategy is similar to adjustment for blocks in a two-way analysis of variance for randomized blocks; it is also similar to covariance adjustment for a categorical explanatory variable.

Historically, the principle of combining information across strata was identified by Cochran (1954): this was in the context of combining differences of proportions from binomial distributions. Mantel and Haenszel (1959) refined the procedure to apply to hypergeometric distributions and produced a statistic to which central limit theory was more applicable for the combined strata. Thus, only the overall sample size needed to be reasonably large. The Mantel-Haenszel statistic proved more useful than Cochran's method. (Cochran's influence is the reason why the FREQ procedure output is labeled "Cochran-Mantel-Haenszel Statistics"; current literature tends to use the terms "extended Mantel-Haenszel statistics" and "Mantel-Haenszel statistics.")

Mantel (1963) discussed extensions to the MH strategy, including strategies for sets of $2 \times r$ tables, sets of $s \times 2$ tables, and the correlation statistic for $s \times r$ tables. The method was further elaborated by Landis, Heyman, and Koch (1978) to encompass the family of Mantel-Haenzsel statistics, which included the statistics for general association, nonparametric ANOVA (mean score), the correlation statistic, and other special cases. Kuritz, Landis, and Koch (1988) present a useful overview of the Mantel-Haenszel strategy, and so do Landis et al. (1998).

The Mantel-Haenszel strategy requires minimal assumptions. The methods it encompasses are based on randomization considerations; the only assumptions required are the randomization of the subjects to levels of the row variable. This can be done explicitly, such as for randomized clinical trials; implicitly, via hypothesis; or conditionally, such as for retrospective studies or observational data. The minimal assumptions often allow you to perform hypothesis tests on data that do not meet the more rigorous assumptions concerning random sampling or underlying distributions that are required for statistical modeling. However, the conclusions of the analysis may be restricted to the study population at hand, versus inference to a larger population. Most often, a complete analysis includes the applications of these minimal assumption methods to perform hypothesis tests and then statistical modeling to describe more completely the variation in the data.

Another advantage of the Mantel-Haenszel strategy is the fact that sample size requirements are based on total frequencies, or quantities summed across tables, rather than on individual cell sizes. This is partly because the Mantel-Haenszel methods are targeted at detecting average effects across strata; they are often called methods of assessing average partial association.

Section 6.2 discusses the formulation of the Mantel-Haenszel statistics in matrix terminology. Section 6.3 illustrates the use of the Mantel-Haenszel strategy for several applications. Finally, Section 6.4 addresses the advanced topic of the use of the Mantel-Haenszel strategy for repeated measurements analysis.

6.2 General Mantel-Haenszel Methodology

Table 6.1 represents the generic $s \times r$ table in a set of q $s \times r$ tables.

6.2. General Mantel-Haenszel Methodology

Table 6.1 hth $s \times r$ Contingency Table

Group	Response Variable Categories				Total
	1	2	...	r	
1	n_{h11}	n_{h12}	...	n_{h1r}	n_{h1+}
2	n_{h21}	n_{h22}	...	n_{h2r}	n_{h2+}
$\vdots$	$\vdots$	$\vdots$		$\vdots$	$\vdots$
s	n_{hs1}	n_{hs2}	...	n_{hsr}	n_{hs+}
Total	n_{h+1}	n_{h+2}	...	n_{h+r}	n_h

Under the assumption that the marginal totals n_{hi+} and n_{h+j} are fixed, the overall null hypothesis of no partial association can be stated as follows:

> For each of the levels of the stratification variable $h = 1, 2, \ldots, q$, the response variable is distributed at random with respect to the groups (row variable levels).

Suppose $\mathbf{n}'_h = (n_{h11}, n_{h12}, \ldots, n_{h1r}, \ldots, n_{hs1}, \ldots, n_{hsr})$, where n_{hij} is the number of subjects in the hth stratum in the ith group in the jth response category. The probability distribution for the vector $\mathbf{n}_h$ under H_0 can be written

$$Pr\{\mathbf{n}_h | H_0\} = \frac{\prod_{i=1}^{s} n_{hi+}! \prod_{j=1}^{r} n_{h+j}!}{n_h! \prod_{i=1}^{s} \prod_{j=1}^{r} n_{hij}!}$$

For the hth stratum, suppose that $p_{hi+} = n_{hi+}/n_h$ denotes the marginal proportion of subjects belonging to the ith group, and suppose that $p_{h+j} = n_{h+j}/n_h$ denotes the marginal proportion of subjects classified as belonging to the jth response category. These proportions can be denoted in vector notation as

$$\mathbf{p}'_{h*+} = (p_{h1+}, \ldots, p_{hs+})$$

$$\mathbf{p}'_{h+*} = (p_{h+1}, \ldots, p_{h+r})$$

Then,

$$E\{n_{hij} | H_0\} = m_{hij} = n_h p_{hi+} p_{h+j}$$

and the expected value of $\mathbf{n}_h$ can be written

$$E\{\mathbf{n}_h | H_0\} = \mathbf{m}_h = n_h [\mathbf{p}_{h+*} \otimes \mathbf{p}_{h*+}]$$

where $\otimes$ denotes the left-hand Kronecker product (the matrix on the left of the $\otimes$ multiplies each element of the matrix on the right).

The variance of $\mathbf{n}_h$ under H_0 is

$$V_h = \text{Var}\{\mathbf{n}_h | H_0\} = \frac{n_h^2}{(n_h - 1)} \{[\mathbf{D}_{\mathbf{p}_{h+*}} - \mathbf{p}_{h+*} \mathbf{p}'_{h+*}] \otimes [\mathbf{D}_{\mathbf{p}_{h*+}} - \mathbf{p}_{h*+} \mathbf{p}'_{h*+}]\}$$

where $\mathbf{D}_{\mathbf{p}_{h+*}}$ and $\mathbf{D}_{\mathbf{p}_{h*+}}$ are diagonal matrices with elements of the vectors $\mathbf{p}_{h+*}$ and $\mathbf{p}_{h*+}$ as the main diagonals.

The general form of the extended Mantel-Haenszel statistic for $s \times r$ tables is

$$Q_{\text{EMH}} = \left\{ \sum_{h=1}^{q} (\mathbf{n}_h - \mathbf{m}_h)' \mathbf{A}_h' \right\} \left\{ \sum_{h=1}^{q} \mathbf{A}_h \mathbf{V}_h \mathbf{A}_h' \right\}^{-1} \left\{ \sum_{h=1}^{q} \mathbf{A}_h (\mathbf{n}_h - \mathbf{m}_h) \right\}$$

where $\mathbf{A}_h$ is a matrix that specifies the linear functions of the $\{\mathbf{n}_h - \mathbf{m}_h\}$ at which the test statistic is directed. Choices of the $\{\mathbf{A}_h\}$ provide stratification-adjusted counterparts to the randomization chi-square statistic Q, the mean score statistic Q_S, and the correlation statistic Q_{CS} that are discussed in Chapter 5, "The $s \times r$ Table."

6.2.1 General Association Statistic

When both the row and column variables are nominally scaled, the alternative hypothesis of interest is that of general association, where the pattern of distribution of the response levels across the row levels is heterogeneous. This is the most general alternative hypothesis and is always valid, no matter how the row and column variables are scaled.

In this case,

$$\mathbf{A}_h = \{[\mathbf{I}_{(r-1)}, \mathbf{0}_{(r-1)}] \otimes [\mathbf{I}_{(s-1)}, \mathbf{0}_{(s-1)}]\}$$

which, applied to $(\mathbf{n}_h - \mathbf{m}_h)$, produces the differences between the observed and expected frequencies under H_0 for the $(s-1)(r-1)$ cells of the table after eliminating the last row and column. This results in Q_{GMH}, which is approximately chi-square with $(s-1)(r-1)$ degrees of freedom. Q_{GMH} is often called the test of general association.

6.2.2 Mean Score Statistic

When the response levels are ordinally scaled, you can assign scores to them to compute row mean scores. In this case, the alternative hypothesis to the null hypothesis of no association is that there are location shifts for these mean scores across the levels of the row variables.

Here,

$$\mathbf{A}_h = \mathbf{a}_h' \otimes [\mathbf{I}_{(s-1)}, \mathbf{0}_{(s-1)}]$$

where $\{\mathbf{a}_h\} = (a_{h1}, a_{h2}, \ldots, a_{hr})$ specifies scores for the jth response level in the hth stratum, from which the means

$$\bar{y}_{hi} = \sum_{j=1}^{r} (a_{hj} n_{hij} / n_{hi+})$$

are created for comparisons of the s populations across the strata.

This produces the extended Mantel-Haenszel Q_{SMH}, which is approximately chi-square with $(s-1)$ degrees of freedom under H_0. Q_{SMH} is called the mean score statistic and is the general form of the Q_{SMH} statistic for $2 \times r$ tables discussed in Chapter 4, "Sets of $2 \times r$ and $s \times 2$ Tables," where $(s - 1) = 1$. If marginal rank or ridit scores are used, with midranks assigned for ties, Q_{SMH} is equivalent to an extension of the Kruskal-Wallis ANOVA test on ranks to account for strata and the Friedman ANOVA test on ranks to account for more than one subject per group within strata. See Chapter 7, "Nonparametric Methods," for further discussion about nonparametric tests that are special cases of Mantel-Haenszel strategies.

6.2.3 Correlation Statistic

When both the response variable (columns) and the row variable (groups) are ordinally scaled, you can assign scores to both the response levels and the row levels in the hth stratum. The alternative hypothesis to no association in this situation is a linear trend on the mean scores across the levels of the row variable. In this case,

$$\mathbf{A}_h = [\mathbf{a}'_h \otimes \mathbf{c}'_h]$$

where the $\{\mathbf{a}_h\}$ are defined as before and the $\{\mathbf{c}_h\} = (c_{h1}, c_{h2}, \ldots, c_{hs})$ specify a set of scores for the ith level of the row variable in the hth stratum. This produces the differences between the observed and expected sum of products of the row and column scores with the frequencies n_{hij}, so that the resulting test statistic is directed at detecting correlation.

This test statistic is Q_{CSMH}, which is approximately chi-square with one degree of freedom under H_0. It is the general form of Q_{CSMH} discussed in Chapter 4 for stratified $s \times 2$ tables where the row variable is ordinally scaled. It has increased power relative to Q_{GMH} or Q_{SMH} for linear association alternatives to the null hypothesis of no association.

6.2.4 Summary

Table 6.2 summarizes the various types of extended Mantel-Haenszel statistics.

Table 6.2 Extended Mantel-Haenszel Statistics

MH Statistic	Alternative Hypothesis	SAS Output Label	Degrees of Freedom	Scale Requirements	Nonparametric Equivalents
Q_{GMH}	general association	General Association	$(s-1) \times (r-1)$	none	
Q_{SMH}	mean score location shifts	Row Means Scores Differ	$(s-1)$	column variable ordinal	Kruskal-Wallis
Q_{CSMH}	linear association	Nonzero Correlation	1	row and column variable ordinal	Spearman correlation

6.3 Mantel-Haenszel Applications

The Mantel-Haenszel strategy has applications in many different settings, including a number of different sampling frameworks. Chapter 3 demonstrates the use of this strategy for analyzing sets of 2×2 tables, and Chapter 4 demonstrates the use of the strategy for sets of $2 \times r$ and $s \times 2$ tables. If you haven't read these chapters, you should review them since they contain many general remarks about the application of Mantel-Haenszel methods. Section 6.3 illustrates the use of these methods for sets of $s \times r$ tables, including examples from clinical trials, observational studies, and prospective studies.

6.3.1 Dumping Syndrome Data

The previous chapter described the dumping syndrome data, which are displayed in Table 5.10. The response measured was the severity (none, slight, moderate) of the dumping syndrome, which is expected to increase directly with the proportion of gastric tissue removed. This response, an adverse effect of surgery, can be considered ordinally scaled, as can operation. Investigators wanted to determine if whether type of operation was associated with severity of dumping syndrome, after adjusting for hospital.

Since both the row and column variables are ordinally scaled, you can use the correlation statistic Q_{CSMH} to assess the null hypothesis of no association against the alternative that type of operation and severity of response are linearly associated.

The following SAS statements input the data into the SAS data set OPERATE and request the MH analysis. Note the use of the option ORDER=DATA, as well as the request for both integer scores (the default table scores) and standardized midrank scores (SCORES=MODRIDIT).

```
data operate;
   input hospital trt $ severity $ wt @@;
   datalines;
1 v+d none 23    1 v+d slight  7   1 v+d moderate 2
1 v+a none 23    1 v+a slight 10   1 v+a moderate 5
1 v+h none 20    1 v+h slight 13   1 v+h moderate 5
1 gre none 24    1 gre slight 10   1 gre moderate 6
2 v+d none 18    2 v+d slight  6   2 v+d moderate 1
2 v+a none 18    2 v+a slight  6   2 v+a moderate 2
2 v+h none 13    2 v+h slight 13   2 v+h moderate 2
2 gre none  9    2 gre slight 15   2 gre moderate 2
3 v+d none  8    3 v+d slight  6   3 v+d moderate 3
3 v+a none 12    3 v+a slight  4   3 v+a moderate 4
3 v+h none 11    3 v+h slight  6   3 v+h moderate 2
3 gre none  7    3 gre slight  7   3 gre moderate 4
4 v+d none 12    4 v+d slight  9   4 v+d moderate 1
4 v+a none 15    4 v+a slight  3   4 v+a moderate 2
4 v+h none 14    4 v+h slight  8   4 v+h moderate 3
4 gre none 13    4 gre slight  6   4 gre moderate 4
;
```

```
proc freq order=data;
   weight wt;
   tables hospital*trt*severity / cmh;
   tables hospital*trt*severity / cmh scores=modridit;
run;
```

Output 6.1 contains the results for the extended Mantel-Haenszel analysis using integer scores. Q_{CSMH} takes the value 6.3404, which is significant at the $\alpha = 0.05$ level; note that the statistics for general association, Q_{GMH}, and mean score differences, Q_{SMH}, are not significant at the $\alpha = 0.05$ level of significance. This is an example of the utility of taking advantage of the correlation statistic when it is appropriate; its greater power against the alternative hypothesis of linear association has detected significant evidence against the null hypothesis.

Output 6.1 Table Scores

Summary Statistics for trt by severity
Controlling for hospital

Cochran-Mantel-Haenszel Statistics (Based on Table Scores)				
Statistic	Alternative Hypothesis	DF	Value	Prob
1	Nonzero Correlation	1	6.3404	0.0118
2	Row Mean Scores Differ	3	6.5901	0.0862
3	General Association	6	10.5983	0.1016

Total Sample Size = 417

Output 6.2 contains the results for the standardized midrank scores. $Q_{CSMH} = 6.9266$, with $p = 0.0085$. As with the integer scores, the other statistics do not detect as much evidence against the null hypothesis of no association. Since the response variable levels are subjective and undoubtedly not equally spaced, the analysis of standardized midrank scores may provide the most appropriate test.

Output 6.2 Standardized Midrank Scores

Summary Statistics for trt by severity
Controlling for hospital

Cochran-Mantel-Haenszel Statistics (Modified Ridit Scores)				
Statistic	Alternative Hypothesis	DF	Value	Prob
1	Nonzero Correlation	1	6.9266	0.0085
2	Row Mean Scores Differ	3	7.6370	0.0541
3	General Association	6	10.5983	0.1016

Total Sample Size = 417

This analysis shows that, adjusting for hospital, there is a clear monotonic association between degree of gastric tissue removal and severity of dumping syndrome. The greater the degree of gastric tissue removal, the worse the dumping syndrome.

6.3.2 Shoulder Harness Data

The following data were collected in a study of shoulder harness usage in observations for a sample of North Carolina cars (Hochberg, Stutts, and Reinfurt 1977).

Table 6.3 Shoulder Harness Data

Area	Location	Larger Cars		Medium		Smaller Cars		Total
		No	Yes	No	Yes	No	Yes	
Coast	Urban	174	69	134	56	150	54	637
Coast	Rural	52	14	31	14	25	17	153
Piedmont	Urban	127	62	94	63	112	93	551
Piedmont	Rural	35	29	32	30	46	34	206
Mountains	Urban	111	26	120	47	145	68	517
Mountains	Rural	62	31	44	32	85	43	297

For these data, researchers were interested in whether there was an association between the size of car and shoulder harness usage, after controlling for geographic area and location. First, there is interest in looking at the pooled table of car size × usage. Then, a Mantel-Haenszel analysis is requested for a stratification consisting of the combinations of levels of area and location, resulting in six strata. Finally, Mantel-Haenszel analyses are requested for the association of size with usage stratified on area and location, singly. Standardized midrank scores are specified.

The following SAS statements request these analyses. Note that the NOPRINT option is specified in the last two TABLES statements to suppress the printing of tables.

```
data shoulder;
    input area $ location $ size $ usage $ count @@;
    datalines;
coast     urban large  no 174 coast     urban large  yes 69
coast     urban medium no 134 coast     urban medium yes 56
coast     urban small  no 150 coast     urban small  yes 54
coast     rural large  no  52 coast     rural large  yes 14
coast     rural medium no  31 coast     rural medium yes 14
coast     rural small  no  25 coast     rural small  yes 17
piedmont  urban large  no 127 piedmont  urban large  yes 62
piedmont  urban medium no  94 piedmont  urban medium yes 63
piedmont  urban small  no 112 piedmont  urban small  yes 93
piedmont  rural large  no  35 piedmont  rural large  yes 29
piedmont  rural medium no  32 piedmont  rural medium yes 30
piedmont  rural small  no  46 piedmont  rural small  yes 34
mountain  urban large  no 111 mountain  urban large  yes 26
mountain  urban medium no 120 mountain  urban medium yes 47
mountain  urban small  no 145 mountain  urban small  yes 68
mountain  rural large  no  62 mountain  rural large  yes 31
mountain  rural medium no  44 mountain  rural medium yes 32
mountain  rural small  no  85 mountain  rural small  yes 43
;
```

```
proc freq;
   weight count;
   tables size*usage / cmh scores=modridit;
   tables area*location*size*usage / cmh scores=modridit;
   tables area*size*usage / noprint cmh scores=modridit;
   tables location*size*usage / noprint cmh scores=modridit;
run;
```

Output 6.3 contains the frequency table for size cross-classified by shoulder harness usage.

Output 6.3 Pooled Table

Frequency Percent Row Pct Col Pct	Table of size by usage		
		usage	
size	no	yes	Total
large	561 23.76 70.83 35.53	231 9.78 29.17 29.54	792 33.55
medium	455 19.27 65.28 28.82	242 10.25 34.72 30.95	697 29.52
small	563 23.85 64.56 35.66	309 13.09 35.44 39.51	872 36.93
Total	1579 66.88	782 33.12	2361 100.00

Output 6.4 contains the Mantel-Haenszel results for this table. Q_{CSMH} is valid for these data since SIZE is ordinally scaled, and the response is dichotomous; it indicates that there is a strong association between size of car and shoulder harness usage ($Q_{CS} = 7.1169$). By looking at the row percentages in the table cells, you can see that drivers of small and medium sized cars exhibit a greater tendency to use shoulder harnesses than do the drivers of large cars.

Output 6.4 Pooled Table

Summary Statistics for size by usage

Cochran-Mantel-Haenszel Statistics (Modified Ridit Scores)

Statistic	Alternative Hypothesis	DF	Value	Prob
1	Nonzero Correlation	1	7.1169	0.0076
2	Row Mean Scores Differ	2	8.5013	0.0143
3	General Association	2	8.5013	0.0143

Total Sample Size = 2361

This association holds when you control for area and location. Output 6.5 contains the frequency table for rural locations in the coast region (the other tables are not reproduced here).

Output 6.5 Table for AREA=coast and LOCATION=rural

Frequency Percent Row Pct Col Pct	Table 1 of size by usage		
	Controlling for area=coast location=rural		
	usage		
size	no	yes	Total
large	52 33.99 78.79 48.15	14 9.15 21.21 31.11	66 43.14
medium	31 20.26 68.89 28.70	14 9.15 31.11 31.11	45 29.41
small	25 16.34 59.52 23.15	17 11.11 40.48 37.78	42 27.45
Total	108 70.59	45 29.41	153 100.00

Output 6.6 displays the Mantel-Haenszel results for the stratified analysis where the strata are all combinations of area and location. $Q_{CSMH} = 6.6398$, which is strongly significant. Controlling for area and location, shoulder harness usage is clearly associated with size of car.

Output 6.6 Stratified by Area and Location

Summary Statistics for size by usage
Controlling for area and location

Cochran-Mantel-Haenszel Statistics (Modified Ridit Scores)				
Statistic	Alternative Hypothesis	DF	Value	Prob
1	Nonzero Correlation	1	6.6398	0.0100
2	Row Mean Scores Differ	2	8.4226	0.0148
3	General Association	2	8.4258	0.0148

Total Sample Size = 2361

Output 6.7 and Output 6.8 contain the Mantel-Haenszel results for the association of size and shoulder harness usage, controlling for area and location singly. $Q_{CSMH} = 6.5097$ and 7.0702, respectively. Controlling only for area or location, the significant association between shoulder harness and size of car remains evident. Q_{GMH} and Q_{SMH} are also significant for the preceding analyses, but most of the information is contained in the correlation statistic Q_{CSMH}.

However, you should use caution in interpreting the mean score statistic for modified ridit scores when the outcome is a dichotomous response. Ordinarily, you would want the values 0 and 1 to be maintained in such an analysis; by using modified ridit scores you are effectively assigning different values from 0 and 1 to the columns, and these scores will be different in the different strata.

Also, the fact that Q_{GMH} and Q_{SMH} have very close or the same values in Output 6.6, Output 6.7, and Output 6.8 is an artifact. However, these statistics are identical for sets of $s \times 2$ tables when integer scores are used.

Output 6.7 Stratified by Area

Summary Statistics for size by usage
Controlling for area

Cochran-Mantel-Haenszel Statistics (Modified Ridit Scores)				
Statistic	Alternative Hypothesis	DF	Value	Prob
1	Nonzero Correlation	1	6.5097	0.0107
2	Row Mean Scores Differ	2	8.1203	0.0172
3	General Association	2	8.1203	0.0172

Total Sample Size = 2361

Output 6.8 Stratified by Location

Summary Statistics for size by usage
Controlling for location

Cochran-Mantel-Haenszel Statistics (Modified Ridit Scores)				
Statistic	Alternative Hypothesis	DF	Value	Prob
1	Nonzero Correlation	1	7.0702	0.0078
2	Row Mean Scores Differ	2	8.5794	0.0137
3	General Association	2	8.5789	0.0137

Total Sample Size = 2361

6.3.3 Learning Preference Data

In this study, educational researchers compared three different approaches to mathematics instruction for third graders. During the year, students were rotated through three different styles: a self-instructional mode that was largely based on computer use, a team approach in which students solved problems in groups of four students, and a traditional class approach. Researchers were interested in both how other school programs influenced the effectiveness of the styles and how they influenced the students' perceptions of the different styles. Table 6.4 displays data that reflect the students' preferences of styles, cross-classified by the school program they are in: Regular, which is a regular school schedule, and After, which supplements the regular school day with an afternoon school program that involves the same classmates. The study included three different schools.

Table 6.4 School Program Data

School	Program	Learning Style Preference		
		Self	Team	Class
1	Regular	10	17	26
1	After	5	12	50
2	Regular	21	17	26
2	After	16	12	36
3	Regular	15	15	16
3	After	12	12	20

The question of interest is whether students' learning style preference is associated with their school program, after adjusting for any effects of individual school. There may be some ordinality to the response measure, in the sense of increasing group participation, but that isn't robust when you try to distinguish the team approach from the classroom approach. Thus, the appropriate extended Mantel-Haenszel statistic for the stratified analysis of these data is the test for general association. Since $(s-1)(r-1)$ for these data is equal to 2, Q_{GMH} has two degrees of freedom.

The following SAS statements request the appropriate analysis.

```
data school;
   input school program $ style $ count @@;
   datalines;
1 regular  self 10  1 regular  team 17  1 regular  class  26
1 after    self  5  1 after    team 12  1 after    class  50
2 regular  self 21  2 regular  team 17  2 regular  class  26
2 after    self 16  2 after    team 12  2 after    class  36
3 regular  self 15  3 regular  team 15  3 regular  class  16
3 after    self 12  3 after    team 12  3 after    class  20
;

proc freq;
   weight count;
   tables school*program*style / cmh chisq;
run;
```

Output 6.9 contains the results for the stratified analysis. Q_{GMH} has a value of 10.9577, with 2 df, with a p-value of 0.0042. School program and learning style preference are strongly associated. For these data, the general association statistic is most appropriate. The other statistics printed in this table are not applicable since the scale of the row and column variables of these tables do not justify their use. Note that since the ORDER=DATA option is not specified, the columns and rows of the tables are arranged alphabetically. This has no bearing on the general association statistic. However, if you wanted to order the rows and columns of the table as displayed in Table 6.4, then you would use ORDER=DATA.

Output 6.9 Stratified Analysis

Summary Statistics for program by style
Controlling for school

Cochran-Mantel-Haenszel Statistics (Based on Table Scores)				
Statistic	Alternative Hypothesis	DF	Value	Prob
1	Nonzero Correlation	1	9.0072	0.0027
2	Row Mean Scores Differ	1	9.0072	0.0027
3	General Association	2	10.9577	0.0042

Total Sample Size = 338

Output 6.10, Output 6.11, and Output 6.12 contain the results for the individual tables. Note that most of the association seems to be occurring in School 1, judging by Q_P. Nevertheless, methods for logistic regression for nominal response variables support absence of school*program interaction and thereby homogeneity of the association between learning program and learning style across schools (see Section 9.3.2 in Chapter 9).

Output 6.10 Results for School 1

Frequency Percent Row Pct Col Pct	Table 1 of program by style			
	Controlling for school=1			
		style		
program	class	self	team	Total
after	50 41.67 74.63 65.79	5 4.17 7.46 33.33	12 10.00 17.91 41.38	67 55.83
regular	26 21.67 49.06 34.21	10 8.33 18.87 66.67	17 14.17 32.08 58.62	53 44.17
Total	76 63.33	15 12.50	29 24.17	120 100.00

Statistics for Table 1 of program by style
Controlling for school=1

Output 6.10 *continued*

Statistic	DF	Value	Prob
Chi-Square	2	8.5913	0.0136
Likelihood Ratio Chi-Square	2	8.6385	0.0133
Mantel-Haenszel Chi-Square	1	6.4209	0.0113
Phi Coefficient		0.2676	
Contingency Coefficient		0.2585	
Cramer's V		0.2676	

Output 6.11 Results for School 2

Frequency Percent Row Pct Col Pct	Table 2 of program by style			
	Controlling for school=2			
	style			
program	class	self	team	Total
after	36 28.13 56.25 58.06	16 12.50 25.00 43.24	12 9.38 18.75 41.38	64 50.00
regular	26 20.31 40.63 41.94	21 16.41 32.81 56.76	17 13.28 26.56 58.62	64 50.00
Total	62 48.44	37 28.91	29 22.66	128 100.00

Statistics for Table 2 of program by style
Controlling for school=2

Statistic	DF	Value	Prob
Chi-Square	2	3.1506	0.2069
Likelihood Ratio Chi-Square	2	3.1641	0.2056
Mantel-Haenszel Chi-Square	1	2.7062	0.1000
Phi Coefficient		0.1569	
Contingency Coefficient		0.1550	
Cramer's V		0.1569	

Output 6.12 Results for School 3

Frequency Percent Row Pct Col Pct	Table 3 of program by style			
	Controlling for school=3			
		style		
program	class	self	team	Total
after	20 22.22 45.45 55.56	12 13.33 27.27 44.44	12 13.33 27.27 44.44	44 48.89
regular	16 17.78 34.78 44.44	15 16.67 32.61 55.56	15 16.67 32.61 55.56	46 51.11
Total	36 40.00	27 30.00	27 30.00	90 100.00

Statistics for Table 3 of program by style
Controlling for school=3

Statistic	DF	Value	Prob
Chi-Square	2	1.0672	0.5865
Likelihood Ratio Chi-Square	2	1.0690	0.5860
Mantel-Haenszel Chi-Square	1	0.8259	0.3635
Phi Coefficient		0.1089	
Contingency Coefficient		0.1083	
Cramer's V		0.1089	

6.4 Advanced Topic: Application to Repeated Measures

6.4.1 Introduction

The Mantel-Haenszel strategy has a useful application to the analysis of repeated measurements data. Such data occur when measurements are obtained over time, when responses from experimental units are measured under multiple conditions (such as multiple teeth in the same subject), and when multiple measurements are obtained from the same experimental unit (such as from two or more observers.) Using repeated measurements enables comparisons among different times or conditions to avoid being obscured by subject-to-subject variability.

By specifying the appropriate tables for the data, you construct a setting in which Mantel-Haenszel methods can address the hypothesis of no association between a repeated measurement factor (such as time or condition) and a response variable, adjusting for the effect of subject. This type of analysis might be sufficient, or there might also be interest in statistical modeling of the repeated measurements data, which is discussed in Chapter 13, "Weighted Least Squares," and Chapter 14, "Generalized Estimating Equations."

Consider the general situation in which t measurements of a univariate response variable Y are obtained from each of n experimental units. One common application is to longitudinal studies, in which repeated measurements are obtained at t time points for each subject. In other applications, the responses from each experimental unit are measured under multiple conditions rather than at multiple time points. In some settings in which repeated measures data are obtained, the independent experimental units are not individual subjects. For example, in a matched case-control study, the experimental units are matched sets, and responses are obtained from the individual members of each set. In a toxicological study, the experimental units might be litters; responses are then obtained from the multiple newborns in each litter. In a genetic study, experimental units might be defined by families; responses are then obtained from the members of each family.

Although interest focuses primarily on the situation in which Y is categorical, the response might be either continuous or categorical. Let y_{ij} denote the response for subject i at time (or condition) j. The resulting data are commonly displayed in an $n \times t$ data matrix, as shown in Table 6.5.

Table 6.5 One-Sample Repeated Measures Data

Subject	Time Point				
	1	...	j	...	t
1	y_{11}	...	y_{1j}	...	y_{1t}
$\vdots$	$\vdots$	$\ddots$	$\vdots$	$\ddots$	$\vdots$
i	y_{i1}	...	y_{ij}	...	y_{it}
$\vdots$	$\vdots$	$\ddots$	$\vdots$	$\ddots$	$\vdots$
n	y_{n1}	...	y_{nj}	...	y_{nt}

Alternatively, suppose c denotes the number of distinct values of Y and suppose indicator variables

$$n_{ijk} = \begin{cases} 1 & \text{if subject } i \text{ is classified in response category } k \text{ at time } j \\ 0 & \text{otherwise} \end{cases}$$

for $i = 1, \ldots, n$; $j = 1, \ldots, t$; and $k = 1, \ldots, c$. In this case, the data from subject i can be displayed in a $t \times c$ contingency table, as shown in Table 6.6. Thus, the data from a one-sample repeated measures study can be viewed as a set of n independent two-way contingency tables, where each table has t rows and c columns.

Table 6.6 Contingency Table Layout for Subject i

Time Point	Response Category			Total
	1	...	c	
1	n_{i11}	...	n_{i1c}	n_{i1+}
$\vdots$	$\vdots$	$\ddots$	$\vdots$	$\vdots$
t	n_{it1}	...	n_{itc}	n_{it+}
Total	n_{i+1}	...	n_{i+c}	n_i

If the response variable Y is categorical with a limited number of possible values, the number of columns in each table, c, will be relatively small. On the other hand, if Y is a continuous variable, the number of distinct values of Y may be very large. The most extreme case results when each of the n subjects has a unique response at each time. In this situation, c is equal to nt and every column marginal total n_{i+k} is equal to zero or one.

When the data are complete, the total sample size for each of the n tables is $n_i = t$ and every row marginal total n_{ij+} is equal to 1. In this case, each row of Table 6.6 has exactly one n_{ijk} value equal to 1 and the remaining values are equal to 0. This situation occurs when the outcome variable is measured once at every time point for each subject.

However, if a particular subject has a missing response at one or more time points, the corresponding row of the subject's table will have each n_{ijk} value equal to 0 and the marginal total n_{ij+} will consequently equal 0. In this case, the total sample size n_i equals t minus the number of missing observations.

Based on the framework displayed in Table 6.6, Mantel-Haenszel statistics can be used to test the null hypothesis of no association between the row dimension (time) and the column dimension (response), adjusted for subject. Under the assumption that the marginal totals $\{n_{ij+}\}$ and $\{n_{i+k}\}$ of each table are fixed, the null hypothesis is that, for each subject, the response variable Y is distributed at random with respect to the t time points. As discussed in Landis et al. (1988), this null hypothesis is precisely the interchangeability hypothesis of Madansky (1963). Interchangeability states that all permutations of responses across conditions within a subject are equally likely. In turn, the hypothesis of interchangeability implies *marginal homogeneity* in the distribution of Y across the t time points; that is, the marginal distribution of Y is the same at each of the t time points.

Although the interchangeability hypothesis is a somewhat stronger condition than marginal homogeneity, the general association statistic Q_{GMH}, mean score statistic Q_{SMH}, and correlation statistic Q_{CSMH} are directed at alternatives that correspond to various types of departures from marginal homogeneity. The following examples demonstrate the use of MH statistics in testing marginal homogeneity for repeated measures.

6.4.2 Dichotomous Response: Two Time Points (McNemar's Test)

A running shoe company produces a new model of running shoe that includes a harder material for the insert that corrects for overpronation. However, the company is concerned that the material will induce heel tenderness as a result of some loss of cushioning on the strike of each step. It conducted a study on 87 runners who used the new shoe for a month. Researchers asked the participants whether they experienced occasional heel tenderness before and after they used the new shoe.

The data was collected as one observation per time period, that is, two measurements were collected for each subject, and they included the time period (before or after) and whether heel tenderness was experienced (yes or no). Table 6.7 contains the contingency table that summarizes these data.

Table 6.7 Heel Tenderness for Runners

Before	After No	Yes	Total
No	48	15	63
Yes	5	19	24

However, you can think of the measurements for each subject as being one of four 2×2 tables, corresponding to the four cells of Table 6.7. These tables are displayed in Table 6.8 through Table 6.11. Each subject's set of responses can be represented by one of these tables.

Table 6.8 (No, No) Configuration Table (48)

Time	Heel Tenderness No	Yes	Total
Before	1	0	1
After	1	0	1

Table 6.9 (No, Yes) Configuration Table (15)

Time	Heel Tenderness No	Yes	Total
Before	1	0	1
After	0	1	1

Table 6.10 (Yes, No) Configuration Table (5)

Time	Heel Tenderness No	Yes	Total
Before	0	1	1
After	1	0	1

Table 6.11 (Yes, Yes) Configuration Table (19)

Time	Heel Tenderness No	Yes	Total
Before	0	1	1
After	0	1	1

You can determine whether there is an association between the response and time for before and after responses by performing a stratified analysis where each subject constitutes a stratum. There are 87 tables altogether: 48 with the (no, no) configuration, 15 with the (no, yes) configuration, 5 with the (yes, no) configuration, and 19 with the (yes, yes) configuration.

If you study Table 6.7, you can see that these data effectively have the matched pairs framework that was discussed in Section 2.7. In fact, the Mantel-Haenszel statistic for the described analysis is equivalent to McNemar's test. The following analysis demonstrates the Mantel-Haenszel approach to analyzing these repeated measurements data. The same strategy is followed when the tables involved have dimensions greater than 2 × 2 and there is no straightforward alternative strategy such as McNemar's test.

The following SAS statements input the running shoes data. The data are in case record form: one observation per time point per subject. Thus, there are 174 observations altogether.

```
data pump;
   input subject time $ response $ @@;
   datalines;
1 before no    1 after no    2 before no    2 after no
3 before no    3 after no    4 before no    4 after no
5 before no    5 after no    6 before no    6 after no
7 before no    7 after no    8 before no    8 after no
9 before no    9 after no   10 before no   10 after no
11 before no  11 after no   12 before no   12 after no
13 before no  13 after no   14 before no   14 after no
15 before no  15 after no   16 before no   16 after no
17 before no  17 after no   18 before no   18 after no
19 before no  19 after no   20 before no   20 after no
21 before no  21 after no   22 before no   22 after no
23 before no  23 after no   24 before no   24 after no
25 before no  25 after no   26 before no   26 after no
27 before no  27 after no   28 before no   28 after no
29 before no  29 after no   30 before no   30 after no
31 before no  31 after no   32 before no   32 after no
33 before no  33 after no   34 before no   34 after no
35 before no  35 after no   36 before no   36 after no
37 before no  37 after no   38 before no   38 after no
39 before no  39 after no   40 before no   40 after no
41 before no  41 after no   42 before no   42 after no
43 before no  43 after no   44 before no   44 after no
45 before no  45 after no   46 before no   46 after no
47 before no  47 after no   48 before no   48 after no
49 before no  49 after yes  50 before no   50 after yes
51 before no  51 after yes  52 before no   52 after yes
53 before no  53 after yes  54 before no   54 after yes
55 before no  55 after yes  56 before no   56 after yes
57 before no  57 after yes  58 before no   58 after yes
59 before no  59 after yes  60 before no   60 after yes
61 before no  61 after yes  62 before no   62 after yes
63 before no  63 after yes  64 before yes  64 after no
65 before yes 65 after no   66 before yes  66 after no
67 before yes 67 after no   68 before yes  68 after no
69 before yes 69 after yes  70 before yes  70 after yes
71 before yes 71 after yes  72 before yes  72 after yes
73 before yes 73 after yes  74 before yes  74 after yes
75 before yes 75 after yes  76 before yes  76 after yes
77 before yes 77 after yes  78 before yes  78 after yes
79 before yes 79 after yes  80 before yes  80 after yes
81 before yes 81 after yes  82 before yes  82 after yes
83 before yes 83 after yes  84 before yes  84 after yes
```

```
85 before yes 85 after yes 86 before yes 86 after yes
87 before yes 87 after yes
;
```

The next statements request the Mantel-Haenszel analysis. Since the data are in case record form, no WEIGHT statement is required. Since 87 tables are to be computed, the NOPRINT option is specified so that the tables are not printed.

```
proc freq;
    tables subject*time*response/ noprint cmh;
run;
```

Output 6.13 contains the Mantel-Haenszel results. Q_{MH} has the value 5.0000 with $p = 0.0253$. This is clearly significant. Runners reported more heel tenderness with the new running shoes than with their old running shoes.

Output 6.13 Mantel-Haenszel Results

Summary Statistics for time by response
Controlling for subject

	Cochran-Mantel-Haenszel Statistics (Based on Table Scores)			
Statistic	Alternative Hypothesis	DF	Value	Prob
1	Nonzero Correlation	1	5.0000	0.0253
2	Row Mean Scores Differ	1	5.0000	0.0253
3	General Association	1	5.0000	0.0253

Total Sample Size = 174

The CMH option always produces the "Estimates of Relative Risk" table and the "Breslow-Day Test for Homogeneity" for sets of 2×2 tables. However, Q_{BD} is not valid here because the data are paired, and neither is the logit estimator of the confidence interval for the odds ratio (the MH estimator of the confidence interval can be problematic as well.) To obtain the odds ratio and its appropriate confidence interval, you can fit a conditional logistic regression model with the subject as the strata variable and time as the main effect, a technique demonstrated in Section 10.3 of Chapter 10.

Another way of obtaining these results for sets of 2×2 tables is to input the original 2×2 table and specify the AGREE option to obtain McNemar's test.

```
data shoes;
    input before $ after $ count;
    datalines;
yes yes 19
yes no   5
no  yes 15
no  no  48
;

proc freq;
    weight count;
```

```
        tables before*after / agree;
    run;
```

Output 6.14 contains the resulting frequency table and McNemar's test. $Q_M = 5.0000$, the same value as was obtained for Q_{MH}.

Output 6.14 Frequency Table and McNemar's Test

Frequency Percent Row Pct Col Pct	Table of before by after			
		after		
before		no	yes	Total
no		48 55.17 76.19 90.57	15 17.24 23.81 44.12	63 72.41
yes		5 5.75 20.83 9.43	19 21.84 79.17 55.88	24 27.59
Total		53 60.92	34 39.08	87 100.00

McNemar's Test	
Statistic (S)	5.0000
DF	1
Pr > S	0.0253

Sample Size = 87

Recall that McNemar's test did not make use of the diagonal cells—the (no, no) and (yes, yes) cells. Thus, if you repeated the Mantel-Haenszel analysis and eliminated the tables corresponding to the (no, no) and (yes, yes) configurations, you would obtain identical results.

6.4.3 Dichotomous Response: Three Repeated Measurements

Grizzle, Starmer, and Koch (1969) analyze data in which 46 patients were each treated with three drugs (A, B, and C). The response to each drug was recorded as favorable (F) or unfavorable (U). Table 6.12 summarizes the eight possible combinations of favorable or unfavorable response for the three drugs and the number of patients with each response pattern.

Table 6.12 Drug Response Data

Drug			Frequency
A	B	C	
F	F	F	6
F	F	U	16
F	U	F	2
F	U	U	4
U	F	F	2
U	F	U	4
U	U	F	6
U	U	U	6

The objective of the analysis is to determine whether the three drugs have similar probabilities for favorable response. Thus, the null hypothesis is interchangeability (that is, no association between drug and response for each patient), which implies equality of the marginal probabilities of a favorable response for the three drugs across patients. This hypothesis can be tested using the general association statistic Q_{GMH}. The data in Table 6.12 must first be restructured so that there are forty-six 3×2 contingency tables, one for each of the 46 patients. For example, Table 6.13 shows the underlying table for a patient who responded favorably to drugs A and C and unfavorably to drug B.

Table 6.13 Sample Contingency Table for a Single Patient

Drug	Response		Total
	F	U	
A	1	0	1
B	0	1	1
C	1	0	1
Total	2	1	3

To apply the Mantel-Haenszel strategy to this data, you have to create a SAS data set that contains $46 \times 3 = 138$ observations (one observation per measurement) and three variables that represent patient, drug, and measurement, respectively. If the data are supplied in frequency count form, they must be rearranged. The following SAS statements read the data in frequency form, as displayed in Table 6.12, and rearrange them into the form displayed in Table 6.13. Thus, three observations are created for each patient, one for each drug. Each of the observations in data set DRUG2 contains an arbitrary patient identifier (numbered from 1 to 46), the drug code (A, B, or C), and the response (F or U).

Finally, the FREQ procedure computes the MH statistics that assess the association of drug and response, adjusting for patient. The NOPRINT option of the TABLES statement suppresses the printing of the 46 individual contingency tables. You almost always use this option when analyzing repeated measures data using MH methods.

```
data drug;
   input druga $ drugb $ drugc $ count;
   datalines;
```

```
   F F F  6
   F F U 16
   F U F  2
   F U U  4
   U F F  2
   U F U  4
   U U F  6
   U U U  6
   ;
   run;
   data drug2; set drug;
      keep patient drug response;
      retain patient 0;
      do i=1 to count;
      patient=patient+1;
      drug='A';   response=druga;   output;
      drug='B';   response=drugb;   output;
      drug='C';   response=drugc;   output;
      end;

   proc freq;
      tables patient*drug*response / noprint cmh;
   run;
```

Output 6.15 displays the results from PROC FREQ. Since the response is dichotomous, the general association and mean score statistics both have 2 df. With table scores, their values are identical. Since the repeated measures factor (drug) is not ordered, the correlation statistic does not apply.

Output 6.15 Test of Marginal Homogeneity

Summary Statistics for drug by response
Controlling for patient

Cochran-Mantel-Haenszel Statistics (Based on Table Scores)				
Statistic	Alternative Hypothesis	DF	Value	Prob
1	Nonzero Correlation	1	6.3529	0.0117
2	Row Mean Scores Differ	2	8.4706	0.0145
3	General Association	2	8.4706	0.0145

Total Sample Size = 138

The value of Q_{GMH} is 8.4706. With reference to the approximate chi-square distribution with 2 df, there is a clearly significant association between drug and response. This test is the same as Cochran's Q statistic (Cochran 1950). In order to summarize the nature of the association, it is helpful to report the estimated marginal probabilities of a favorable response for Drugs A, B, and C. These can be computed from Table 6.12 and are equal to 28/46 = 0.61, 28/46 = 0.61, and 16/46 = 0.35, respectively. It is evident that the marginal proportion for Drug C differs considerably from that of Drugs A and B. Drugs A and B have a much greater probability of favorable response than Drug C.

6.4.4 Ordinal Response

The same Mantel-Haenszel strategy is appropriate when the repeated measurements response variable is ordinally scaled. In this case, the statistic of interest is Q_{SMH}, the mean score statistic.

Macknin, Mathew, and Medendorp (1990) studied the efficacy of steam inhalation in the treatment of common cold symptoms. Thirty patients with colds of recent onset (symptoms of nasal drainage, nasal congestion, and sneezing for three days or less) received two 20-minute steam inhalation treatments. On four successive days, these patients self-assessed the severity of nasal drainage on a four-point ordinal scale (0 = no symptoms, 1 = mild symptoms, 2 = moderate symptoms, 3 = severe symptoms). Table 6.14 displays the resulting data.

Table 6.14 Nasal Drainage Data

Patient ID	Study Day				Patient ID	Study Day			
	1	2	3	4		1	2	3	4
1	1	1	2	2	16	2	1	1	1
2	0	0	0	0	17	1	1	1	1
3	1	1	1	1	18	2	2	2	2
4	1	1	1	1	19	3	1	1	1
5	0	2	2	0	20	1	1	2	1
6	2	0	0	0	21	2	1	1	2
7	2	2	1	2	22	2	2	2	2
8	1	1	1	0	23	1	1	1	1
9	3	2	1	1	24	2	2	3	1
10	2	2	2	3	25	2	0	0	0
11	1	0	1	1	26	1	1	1	1
12	2	3	2	2	27	0	1	1	0
13	1	3	2	1	28	1	1	1	1
14	2	1	1	1	29	1	1	1	0
15	2	3	3	3	30	3	3	3	3

The objective of the study was to determine whether nasal drainage becomes less severe following steam inhalation treatment. Thus, the relevant null hypothesis is that the distribution of the symptom severity scores is the same on each of the four study days for each patient. Since there are only four possible values of the response variable, the assumptions for the usual parametric methods are not directly applicable. In addition, the sample size is too small to justify analysis of the full 4^4 contingency table obtained by the joint cross-classification of the four-level response variable on four days. Thus, randomization model MH methods seem appropriate.

Although the general association statistic Q_{GMH} can be considered for this example, its use of 9 df would have low power to detect departures from marginal homogeneity in a sample of only 30 patients. Since the response is ordinal, the mean score statistic Q_{SMH}, with 3 df, can be used to compare the average symptom scores across the four days. The adequacy of the sample size to support the use of this statistic may also be questionable. Alternatively, since the repeated measures factor (study day) is also ordinal, you could test for a linear trend over study day for symptom severity using the correlation statistic Q_{CSMH}.

Both Q_{SMH} and Q_{CSMH} require that scores be assigned to the values of the repeated measures and response variables. Since study day is quantitative, it is natural to use the scores 1–4 for this variable. If it is reasonable to assume that the symptom severity ratings are equally spaced, the actual scores 0–3 can be used. You could also assign scores that incorporate unequal spacing between the four levels of symptom severity.

Another possibility is to use rank scores for the symptom severity ratings. In PROC FREQ, the SCORES=RANK option of the TABLES statement uses rank scores for both the row and column variables. However, since each patient contributes exactly one observation on each of the four days, the rank scores for study day are also 1, 2, 3, and 4. Thus, this option only affects the scoring of the symptom severity levels. The SCORES=RIDIT and SCORES=MODRIDIT options compute rank scores that are standardized by a function of the stratum-specific sample size. Since the sample sizes in the 30 underlying 4×4 contingency tables are all equal to 4, the results from the SCORES=RANK, SCORES=RIDIT, and SCORES=MODRIDIT options would be identical.

The following SAS statements read in the data in case record form with responses for all days on the same record and rearrange it so that there are four observations per patient.

```
data cold;
   keep id day drainage;
   input id day1-day4;
   day=1; drainage=day1; output;
   day=2; drainage=day2; output;
   day=3; drainage=day3; output;
   day=4; drainage=day4; output;
   datalines;
 1 1 1 2 2
 2 0 0 0 0
 3 1 1 1 1
 4 1 1 1 1
 5 0 2 2 0
 6 2 0 0 0
 7 2 2 1 2
 8 1 1 1 0
 9 3 2 1 1
10 2 2 2 3
11 1 0 1 1
12 2 3 2 2
13 1 3 2 1
14 2 1 1 1
15 2 3 3 3
16 2 1 1 1
17 1 1 1 1
18 2 2 2 2
19 3 1 1 1
20 1 1 2 1
21 2 1 1 2
22 2 2 2 2
23 1 1 1 1
24 2 2 3 1
25 2 0 0 0
26 1 1 1 1
27 0 1 1 0
```

```
28 1 1 1 1
29 1 1 1 0
30 3 3 3 3
;
```

You can generate the mean scores at days 1–4 by using PROC GLM and the LSMEANS statement. In addition, you can use the ESTIMATE statement to determine the direction of the mean difference between days from the first day to the fourth day.

```
proc glm;
    class day;
    model drainage=id day;
    lsmeans day;
    estimate 'direction' day -3 -1 1 3 / divisor=6;
run;
```

Output 6.16 displays the mean symptom severity scores. The observed mean scores at days 1–4 are 1.50, 1.37, 1.37, and 1.17; thus, symptom severity is decreasing over time.

Output 6.16 Mean Severity Scores

Least Squares Means

day	drainage LSMEAN
1	1.50000000
2	1.36666667
3	1.36666667
4	1.16666667

Output 6.17 confirms this with an weighted average pairwise difference of −0.1666 from day 1 to day 4.

Output 6.17 Distance Measure Estimate

Dependent Variable: drainage

Parameter	Estimate	Standard Error	t Value	Pr > |t|
direction	-0.16666667	0.11652958	-1.43	0.1554

PROC FREQ then computes the MH statistics, first using equally spaced table scores and then using rank scores.

```
proc freq;
    tables id*day*drainage / cmh2 noprint;
    tables id*day*drainage / cmh2 noprint scores=rank;
run;
```

Output 6.18 displays the MH statistics based on table scores, and Output 6.19 displays the corresponding results using rank scores. Using the default table scores, the test statistic that the mean symptom severity scores are the same at all four days is not statistically significant ($Q_{SMH} = 4.9355$, $p = 0.1766$). However, there is a statistically significant trend between study day and nasal drainage severity ($Q_{CSMH} = 4.3548$, $p = 0.0369$). The observed mean scores at days 1–4 are 1.50, 1.37, 1.37, and 1.17; thus, symptom severity is decreasing over time.

Output 6.18 MH Tests Using Table Scores

Summary Statistics for day by drainage
Controlling for id

Cochran-Mantel-Haenszel Statistics (Based on Table Scores)				
Statistic	Alternative Hypothesis	DF	Value	Prob
1	Nonzero Correlation	1	4.3548	0.0369
2	Row Mean Scores Differ	3	4.9355	0.1766

Total Sample Size = 120

Output 6.19 MH Tests Using Rank Scores

Summary Statistics for day by drainage
Controlling for id

Cochran-Mantel-Haenszel Statistics (Based on Rank Scores)				
Statistic	Alternative Hypothesis	DF	Value	Prob
1	Nonzero Correlation	1	2.6825	0.1015
2	Row Mean Scores Differ	3	3.3504	0.3407

Total Sample Size = 120

In this example, the use of rank scores leads to a less clear conclusion regarding the statistical significance of the correlation statistic ($Q_{CSMH} = 2.6825$, $p = 0.1015$). Some authors recommend the routine use of rank scores in preference to the arbitrary assignment of scores (for example, Fleiss 1986, pp. 83–84). However, as demonstrated by Graubard and Korn (1987), rank scores can be a poor choice when the column margin is far from uniformly distributed. This occurs because rank scores also assign a spacing between the levels of the categories. This spacing is generally not known by the analyst and may not be as powerful as other spacings for certain patterns of differences among distributions. Graubard and Korn (1987) recommend that you specify the scores whenever possible. If the choice of scores is not apparent, they recommend integer (or equally spaced) scores.

When there is no natural set of scores, Agresti (2002, p. 88) recommends that the data be analyzed using several reasonably assigned sets of scores to determine whether substantive conclusions depend on the choice of scores. This type of sensitivity analysis seems especially appropriate in this example, since the results that assume equally spaced scores differ from those obtained using rank scores. For example, the scores 0, 1, 3, 5 assume that the moderate category is equally spaced between the mild and severe categories, while none and mild are closer together. Another possibility would be 0, 1, 2, 4; this choice places severe symptoms further from the other three categories. These alternative scoring specifications are easily implemented by redefining the values of the DRAINAGE variable in the DATA step and then using the default table scores, which are just the input numeric values for drainage.

Note that since the general association statistic does not use scores, the value of Q_{GMH} is the same in both analyses.

6.4.5 Ordinal Response with Missing Data

Researchers at the C. S. Mott Children's Hospital in Ann Arbor, Michigan, investigated the effect of pulse duration on the development of acute electrical injury during transesophageal atrial pacing in animals. In brief, this procedure involves placing a pacemaker in the esophagus. Each of the 14 animals available for experimentation then received atrial pacing at pulse durations of 2, 4, 6, 8, and 10 milliseconds (ms), with each pulse delivered at a separate site in the esophagus for 30 minutes. The response variable, lesion severity, was classified according to depth of injury by histologic examination using an ordinal staging scale from 0 to 5 (0 = no lesion, 5 = acute inflammation of extraesophageal fascia). Table 6.15 displays the resulting data (missing observations are denoted by –). Landis et al. (1988) previously analyzed the data from the first 11 animals.

Table 6.15 Lesion Severity Data

	Pulse Duration (ms)				
ID	2	4	6	8	10
6	0	0	5	0	3
7	0	3	3	4	5
8	0	3	4	3	2
9	2	2	3	0	4
10	0	0	4	4	3
12	0	0	0	4	4
13	0	4	4	4	0
15	0	4	0	0	0
16	0	3	0	1	1
17	–	–	0	1	0
19	0	0	1	1	0
20	–	0	0	2	2
21	0	0	2	3	3
22	–	0	0	3	0

The investigators were primarily interested in determining the extent to which increasing the pulse duration from 2 to 10 ms tends to increase the severity of the lesion. In an experiment in which five repeated measurements of a six-category ordinal response are obtained from only 14 experimental units, the choice of statistical methodology is limited. The study is further complicated by the fact that 3 of the 14 animals have incomplete data.

The general association statistic Q_{GMH} has 20 df in this case ($s = 5$, $r = 6$). In addition to the fact that Q_{GMH} will not have a chi-square distribution when the sample size is so small relative to the degrees of freedom, there will be very low power to detect general departures from the null hypothesis of interchangeability. Although the alternative of location shift for mean responses across the five pulse durations can be addressed using the mean score statistic Q_{SMH}, this statistic does not take into account the ordering of the pulse durations. The 1 df correlation statistic Q_{CSMH} specifically focuses on the narrow alternative of a monotone relationship between lesion severity and pulse duration. This test addresses the objective of the investigators and is also best justified given the small sample size.

The following SAS statements read in the data in the format shown in Table 6.15 and rearrange them so that each subject has five observations, one for each pulse duration.

```
data animals;
   keep id pulse severity;
   input id sev2 sev4 sev6 sev8 sev10;
   pulse=2;   severity=sev2;   output;
   pulse=4;   severity=sev4;   output;
   pulse=6;   severity=sev6;   output;
   pulse=8;   severity=sev8;   output;
   pulse=10;  severity=sev10;  output;
   datalines;
 6 0 0 5 0 3
 7 0 3 3 4 5
 8 0 3 4 3 2
 9 2 2 3 0 4
10 0 0 4 4 3
12 0 0 0 4 4
13 0 4 4 4 0
15 0 4 0 0 0
16 0 3 0 1 1
17 . . 0 1 0
19 0 0 1 1 0
20 . 0 0 2 2
21 0 0 2 3 3
22 . 0 0 3 0
;
```

The following PROC GLM statements produce the lesion severity means for each pulse duration level in addition to the average pairwise difference as the pulse duration level increases.

```
proc glm;
   class pulse;
   model severity = id pulse;
   lsmeans pulse;
   estimate 'direction' pulse -4 -2 0 2 4 / divisor=10;
run;
```

Output 6.20 reports the mean lesion severity scores, which range from 0.0237 for a pulse duration of 2 to 1.967 for a pulse duration of 10, with the highest severity mean of 2.1814 for a pulse duration of 8. Note that these means are computed for the 11 animals with complete data.

Output 6.20 Mean Severity Scores

Least Squares Means

pulse	severity LSMEAN
2	0.02377386
4	1.47047331
6	1.89576987
8	2.18148415
10	1.96719844

Output 6.21 reports the average pairwise weighted distance as pulse duration increases, which is 0.9196.

Output 6.21 Distance Measure Estimate

Dependent Variable: severity

Parameter	Estimate	Standard Error	t Value	Pr > |t|
direction	0.91957200	0.26734944	3.44	0.0011

The next statements request the MH statistics using table scores and all three types of rank scores. The CMH2 option specifies that only the Mantel-Haenszel statistics Q_{SMH} and Q_{CSMH} be computed. (The CMH1 option specifies that only the correlation statistic Q_{CSMH} be computed.)

```
proc freq;
    tables id*pulse*severity / noprint cmh2;
    tables id*pulse*severity / noprint cmh2 scores=rank;
    tables id*pulse*severity / noprint cmh2 scores=ridit;
    tables id*pulse*severity / noprint cmh2 scores=modridit;
run;
```

Output 6.22 displays the results using the default table scores. In this case, lesion severity is scored using the integers $0, \ldots, 5$ in computing both the mean score statistic Q_{SMH} and the correlation statistic Q_{CSMH}. In addition, pulse duration is scored as 2, 4, 6, 8, or 10 in computing Q_{CSMH}. The correlation statistic Q_{CSMH} shows a highly significant monotone association (trend) between pulse duration and lesion severity; the results from the mean score statistic are also statistically significant.

Output 6.22 MH Tests Using Table Scores

Summary Statistics for pulse by severity
Controlling for id

Cochran-Mantel-Haenszel Statistics (Based on Table Scores)				
Statistic	Alternative Hypothesis	DF	Value	Prob
1	Nonzero Correlation	1	8.8042	0.0030
2	Row Mean Scores Differ	4	12.3474	0.0149

Effective Sample Size = 66
Frequency Missing = 4

Output 6.23, Output 6.24, and Output 6.25 display the corresponding results using rank, ridit, and modified ridit scores, respectively. In this example, the values of the mean score and correlation statistics differ slightly among the three types of rank statistics. This is due to the fact that the sample sizes are no longer the same across the 14 tables (due to the occurrence of missing data).

Output 6.23 MH Tests Using Rank Scores

Summary Statistics for pulse by severity
Controlling for id

Cochran-Mantel-Haenszel Statistics (Based on Rank Scores)				
Statistic	Alternative Hypothesis	DF	Value	Prob
1	Nonzero Correlation	1	9.9765	0.0016
2	Row Mean Scores Differ	4	13.6796	0.0084

Effective Sample Size = 66
Frequency Missing = 4

Output 6.24 MH Tests Using Ridit Scores

Summary Statistics for pulse by severity
Controlling for id

Cochran-Mantel-Haenszel Statistics (Based on Ridit Scores)				
Statistic	Alternative Hypothesis	DF	Value	Prob
1	Nonzero Correlation	1	10.0335	0.0015
2	Row Mean Scores Differ	4	14.2628	0.0065

Effective Sample Size = 66
Frequency Missing = 4

Output 6.25 MH Tests Using Modified Ridit Scores

Summary Statistics for pulse by severity
Controlling for id

Cochran-Mantel-Haenszel Statistics (Modified Ridit Scores)				
Statistic	Alternative Hypothesis	DF	Value	Prob
1	Nonzero Correlation	1	10.1102	0.0015
2	Row Mean Scores Differ	4	14.1328	0.0069

Effective Sample Size = 66
Frequency Missing = 4

As shown in Table 6.15, 3 of the 14 animals had incomplete data. Table 6.16 through Table 6.18 display the underlying contingency tables for these strata (ID numbers 17, 20, and 22). Although each of these three tables has one or more rows with a marginal total of zero, the remaining rows provide useful information about the association between pulse duration and lesion severity.

Table 6.16 Contingency Table for ID 17

Pulse Duration	Lesion Severity						Total
	0	1	2	3	4	5	
2	0	0	0	0	0	0	0
4	0	0	0	0	0	0	0
6	1	0	0	0	0	0	1
8	0	1	0	0	0	0	1
10	1	0	0	0	0	0	1
Total	2	1	0	0	0	0	3

Table 6.17 Contingency Table for ID 20

Pulse Duration	Lesion Severity						Total
	0	1	2	3	4	5	
2	0	0	0	0	0	0	0
4	1	0	0	0	0	0	1
6	1	0	0	0	0	0	1
8	0	0	1	0	0	0	1
10	0	0	1	0	0	0	1
Total	2	0	2	0	0	0	4

Table 6.18 Contingency Table for ID 22

Pulse Duration	Lesion Severity						Total
	0	1	2	3	4	5	
2	0	0	0	0	0	0	0
4	1	0	0	0	0	0	1
6	1	0	0	0	0	0	1
8	0	0	0	1	0	0	1
10	1	0	0	0	0	0	1
Total	3	0	0	1	0	0	4

The following statements exclude these three animals from the analysis and compute the test statistics for the subset of complete cases. In this case, all three types of rank scores produce the same results; thus, only the SCORES=RANK option is used. The WHERE clause is used to delete observations whose ID is equal to 17, 20, or 22.

```
proc freq data=animals;
   where id notin(17,20,22);
   tables id*pulse*severity / noprint cmh2;
   tables id*pulse*severity / noprint cmh2 scores=rank;
run;
```

Output 6.26 and Output 6.27 display the results from the analysis of complete cases.

Output 6.26 MH Tests Using Table Scores: Complete Cases Only

Summary Statistics for pulse by severity
Controlling for id

Cochran-Mantel-Haenszel Statistics (Based on Table Scores)				
Statistic	Alternative Hypothesis	DF	Value	Prob
1	Nonzero Correlation	1	7.5610	0.0060
2	Row Mean Scores Differ	4	11.5930	0.0206

Total Sample Size = 55

Output 6.27 MH Tests Using Rank Scores: Complete Cases Only

Summary Statistics for pulse by severity
Controlling for id

Cochran-Mantel-Haenszel Statistics (Based on Rank Scores)				
Statistic	Alternative Hypothesis	DF	Value	Prob
1	Nonzero Correlation	1	8.5099	0.0035
2	Row Mean Scores Differ	4	12.2637	0.0155

Total Sample Size = 55

The value of each of the test statistics is somewhat smaller than the corresponding value that is computed using all available data. Thus, the partial data from the incomplete cases strengthen the evidence in favor of the existence of a significant trend between pulse duration and lesion severity.

Chapter 7

Nonparametric Methods

Contents

7.1	Introduction	175
7.2	Kruskal-Wallis Test	176
7.3	Friedman's Chi-Square Test	178
7.4	Aligned Ranks Test for Randomized Complete Blocks	181
7.5	Analyzing Incomplete Data	182
7.6	Rank Analysis of Covariance	185

7.1 Introduction

Parametric methods of statistical inference require you to assume that your data come from some underlying distribution whose general form is known, such as the normal, binomial, Poisson, or Weibull distribution. Statistical methods for estimation and hypothesis testing are then based on these assumptions. The focus is on estimating parameters and testing hypotheses about them.

In contrast, nonparametric statistical methods make few assumptions about the underlying distribution from which the data are sampled. One of their main advantages is that inference is not focused on specific population parameters, and it is thus possible to test hypotheses that are more general than statements about parameters. For example, nonparametric methods allow you to test whether two distributions are the same without having to test hypotheses concerning population parameters. Nonparametric procedures can also be used when the underlying distribution is unknown or when parametric assumptions are not valid.

The main disadvantage is that a nonparametric test is generally less powerful than the corresponding parametric test when the assumptions are satisfied. However, for many of the commonly used nonparametric methods, the decrease in power is not large.

Most of this book concentrates on the analysis of categorical response variables that are measured on nominal or ordinal scales. This chapter focuses on the analysis of continuous response variables with the use of nonparametric statistical methods. The reason for considering these methods is that many of the commonly used nonparametric tests, such as the Kruskal-Wallis, Spearman correlation, and Friedman tests, can be computed using Mantel-Haenszel procedures. While previous chapters have shown how to use Mantel-Haenszel procedures to analyze two-way tables and sets of two-way tables, this chapter shows how to use the same procedures to perform nonparametric analyses of continuous response variables.

7.2 Kruskal-Wallis Test

The Kruskal-Wallis (1952) test is a nonparametric test of the null hypothesis that the distribution of a response variable is the same in multiple independently sampled populations. The test requires an ordinally scaled response variable and is sensitive to the alternative hypothesis that there is a location difference among the populations. The Kruskal-Wallis test can be used whenever a one-way analysis of variance (ANOVA) model is appropriate. The Kruskal-Wallis test is a generalization of the Wilcoxon-Mann-Whitney test to three or more groups (Wilcoxon 1945, Mann and Whitney 1947).

When the sample sizes in the groups are small, you should use tables of the exact distribution of the test statistic. Alternatively, you can carry out exact tests of significance for small sample sizes. If there are at least five observations per group, the p-value can be approximated using the asymptotic chi-square distribution with $s - 1$ degrees of freedom, where s is the number of groups. The approximate test is simply the Mantel-Haenszel mean score statistic for the special case of one stratum when rank scores are used.

Table 7.1 displays data from a study of antecubital vein cortisol levels at time of delivery in pregnant women (Cawson et al. 1974). The investigators wanted to determine if median cortisol levels differed among three groups of women, all of whom had delivery between 38 and 42 weeks gestation. The data were obtained before the onset of labor at elective Caesarean section (Group I), at emergency Caesarean section during induced labor (Group II), or at the time of vaginal or Caesarean delivery in women in whom spontaneous labor occurred (Group III).

Table 7.1   Antecubital Vein Cortisol Levels at Time of Delivery

Group I		Group II		Group III	
Patient	Level	Patient	Level	Patient	Level
1	262	1	465	1	343
2	307	2	501	2	772
3	211	3	455	3	207
4	323	4	355	4	1048
5	454	5	468	5	838
6	339	6	362	6	687
7	304				
8	154				
9	287				
10	356				

The following statements create a SAS data set containing the data of Table 7.1 and request the Mantel-Haenszel mean score statistic comparing the mean rank scores in the three groups of subjects.

```
data cortisol;
    input group $ subject cortisol;
    datalines;
I    1   262
I    2   307
I    3   211
I    4   323
I    5   454
I    6   339
I    7   304
I    8   154
I    9   287
I   10   356
II   1   465
II   2   501
II   3   455
II   4   355
II   5   468
II   6   362
III  1   343
III  2   772
III  3   207
III  4  1048
III  5   838
III  6   687
;

proc freq;
    tables group*cortisol / noprint cmh2 scores=rank;
run;
```

The Kruskal-Wallis statistic, labeled "Row Mean Scores Differ" in Output 7.1, is equal to 9.2316 with 2 df, which corresponds to a p-value of 0.0099. Thus, the cortisol level distributions differ among the three groups of patients. Since there are more than two groups, the Mantel-Haenszel correlation statistic, labeled "Nonzero Correlation," does not produce the same results as the Kruskal-Wallis test. The correlation statistic uses rank scores to test the null hypothesis that there is no association between group and cortisol level, versus the alternative hypothesis of a monotone association between the two variables. Thus, this statistic is only valid if the three groups are ordered (which might be realistic for this example in terms of the timing for cortisol level determination).

Output 7.1 Kruskal-Wallis Test Using PROC FREQ

Summary Statistics for group by cortisol

Cochran-Mantel-Haenszel Statistics (Based on Rank Scores)				
Statistic	Alternative Hypothesis	DF	Value	Prob
1	Nonzero Correlation	1	8.2857	0.0040
2	Row Mean Scores Differ	2	9.2316	0.0099

Total Sample Size = 22

The Kruskal-Wallis test can also be computed by using the WILCOXON option in the NPAR1WAY procedure.

```
proc nparlway wilcoxon;
   class group;
   var cortisol;
run;
```

The Kruskal-Wallis test displayed in Output 7.2 is identical to the value shown in Output 7.1. The NPAR1WAY procedure gives additional results showing that the mean rank scores in Groups II and III are nearly equivalent and are substantially greater than the mean rank score in Group I.

Output 7.2 Kruskal-Wallis Test Using PROC NPAR1WAY

Wilcoxon Scores (Rank Sums) for Variable cortisol Classified by Variable group					
group	N	Sum of Scores	Expected Under H0	Std Dev Under H0	Mean Score
I	10	69.0	115.0	15.165751	6.900000
II	6	90.0	69.0	13.564660	15.000000
III	6	94.0	69.0	13.564660	15.666667

Kruskal-Wallis Test	
Chi-Square	9.2316
DF	2
Pr > Chi-Square	0.0099

7.3 Friedman's Chi-Square Test

Friedman's test (1937) is a nonparametric method for analyzing a randomized complete block design. This type of study design is applicable when interest is focused on one particular factor, but there are other factors whose effects you want to control. The experimental units are first divided into blocks (groups) in such a way that units within a block are relatively homogeneous. The size of each block is equal to the number of treatments or conditions under study. The treatments are then assigned at random to the experimental units within each block so that each treatment is given once and only once per block. The basic design principle is to partition the experimental units in such a way that background variability between blocks is maximized so that the variability within blocks is minimized.

The standard parametric ANOVA methods for analyzing randomized complete block designs require the assumption that the experimental errors are normally distributed. The Friedman test, which does not require this assumption, depends only on the ranks of the observations within each block and is sometimes called the two-way analysis of variance by ranks.

For small randomized complete block designs, the exact distribution of the Friedman test statistic should be used; for example, Odeh et al. (1977) tabulate the critical values of the Friedman test for up to six blocks and up to six treatments. Alternatively, you can carry out exact tests of significance for small sample sizes. As the number of blocks increases, the distribution of the Friedman statistic approaches that of a chi-square random variable with $s - 1$ degrees of freedom, where s is the number of treatments. The approximate test is simply the Mantel-Haenszel mean score statistic for the special case of rank scores and one subject per treatment group in each block.

Table 7.2 displays data from an experiment designed to determine if five electrode types performed similarly (Berry 1987). In this study, all five types were applied to the arms of 16 subjects and the resistance was measured. Each subject is a block in this example, and all five treatments are applied once and only once per block.

Table 7.2 Electrical Resistance Data

Subject	Electrode Type				
	1	2	3	4	5
1	500	400	98	200	250
2	660	600	600	75	310
3	250	370	220	250	220
4	72	140	240	33	54
5	135	300	450	430	70
6	27	84	135	190	180
7	100	50	82	73	78
8	105	180	32	58	32
9	90	180	220	34	64
10	200	290	320	280	135
11	15	45	75	88	80
12	160	200	300	300	220
13	250	400	50	50	92
14	170	310	230	20	150
15	66	1000	1050	280	220
16	107	48	26	45	51

The following statements read in one record per subject and create a SAS data set containing one observation per electrode per subject. The Mantel-Haenszel mean score statistic is then computed using rank scores, where the 16 subjects define 16 strata.

```
data electrod;
   input subject resist1-resist5;
   type=1;  resist=resist1;  output;
   type=2;  resist=resist2;  output;
   type=3;  resist=resist3;  output;
   type=4;  resist=resist4;  output;
   type=5;  resist=resist5;  output;
   datalines;
1   500   400    98   200   250
2   660   600   600    75   310
3   250   370   220   250   220
4    72   140   240    33    54
5   135   300   450   430    70
6    27    84   135   190   180
7   100    50    82    73    78
8   105   180    32    58    32
9    90   180   220    34    64
10  200   290   320   280   135
11   15    45    75    88    80
12  160   200   300   300   220
13  250   400    50    50    92
14  170   310   230    20   150
15   66  1000  1050   280   220
16  107    48    26    45    51
;

proc freq;
   tables subject*type*resist / noprint cmh2 scores=rank;
run;
```

Output 7.3 displays the results. The value of the test statistic is 5.4522 with 4 df. The p-value of 0.2440 indicates that there is little evidence of a statistically significant difference among the five types of electrodes.

Output 7.3 Friedman Test

Summary Statistics for type by resist
Controlling for subject

Cochran-Mantel-Haenszel Statistics (Based on Rank Scores)				
Statistic	Alternative Hypothesis	DF	Value	Prob
1	Nonzero Correlation	1	2.7745	0.0958
2	Row Mean Scores Differ	4	5.4522	0.2440

Total Sample Size = 80

In experimental situations with more than one subject per group in each block, PROC FREQ can be used to compute generalizations of the Friedman test. The general principle is that the strata are defined by the blocks, and the treatments or groups define the rows of each table.

7.4 Aligned Ranks Test for Randomized Complete Blocks

When the number of blocks or treatments is small, the Friedman test has relatively low power. This results from the fact that the test statistic is based on ranking the observations within each block, which provides comparisons only of the within-block responses. Thus, direct comparison of responses in different blocks is not meaningful, due to variation between blocks. If the blocks are small, there are too few comparisons to permit an effective overall comparison of the treatments. As an example, the Friedman test reduces to the sign test if there are only two treatments. This disadvantage becomes less serious as the number of treatments increases or as the number of subjects per block increases for a fixed number s of treatments.

An alternative to the Friedman test is to use *aligned ranks*. The basic idea is to make the blocks more comparable by subtracting from each observation within a block some estimate of the location of the block, such as the average or median of the observations. The resulting differences are called *aligned observations*. Instead of separately ranking the observations within each block, you rank the complete set of aligned observations relative to each other. Thus, the ranking scheme is the same as that used in computing the Kruskal-Wallis statistic. The resulting ranks are called aligned ranks.

The aligned rank test was introduced by Hodges and Lehmann (1962). Koch and Sen (1968) considered four cases of interest in the analysis of randomized complete block experiments and independently proposed the aligned rank procedure for their Case IV. Apart from the fact that one set of aligned ranks is used instead of separate within-block ranks, the computation of the aligned rank statistic is the same as for the Friedman test.

The exact distribution of the test statistic is cumbersome to compute. In addition, tables are not feasible since the distribution depends on the way the aligned ranks are distributed over the blocks. However, the null distribution of the test statistic is approximately chi-square with $s - 1$ degrees of freedom, where s is the number of treatments (or block size when there is one observation per treatment in each block). Tardif (1980, 1981, 1985) studied the asymptotic efficiency and other aspects of aligned rank tests in randomized block designs.

In section 7.3, Friedman's test was used to analyze data from an experiment designed to determine if five electrode types performed similarly (Table 7.2). Using the SAS data set created in section 7.3, the following statements compute the aligned rank statistic.

```
proc standard mean=0;
   by subject;
   var resist;
proc rank;
   var resist;
proc freq;
   tables subject*type*resist / noprint cmh2;
run;
```

The STANDARD procedure standardizes the observations within each block (subject) to have mean zero. Thus, the subject-specific sample mean is subtracted from each response. The RANK procedure computes a single set of rankings for the combined aligned observations. Using the resulting aligned ranks as scores, the FREQ procedure computes the aligned rank statistic.

Output 7.4 displays the results. The test statistic is equal to 13.6003 with 4 df. With reference to the chi-square distribution with four degrees of freedom, there is a clearly significant difference among the five electrode types. Recall that the Friedman test (Output 7.3) was not statistically significant. Thus, this example illustrates the potentially greater power of the aligned ranks test.

Output 7.4 Aligned Ranks Test

Summary Statistics for type by resist
Controlling for subject

Cochran-Mantel-Haenszel Statistics (Based on Table Scores)				
Statistic	Alternative Hypothesis	DF	Value	Prob
1	Nonzero Correlation	1	4.9775	0.0257
2	Row Mean Scores Differ	4	13.6003	0.0087

Total Sample Size = 80

7.5 Analyzing Incomplete Data

Table 7.3 displays artificial data collected for the purpose of determining if pH level alters action potential characteristics following administration of a drug (Harrell 1989). The response variable of interest (Vmax) was measured at up to four pH levels for each of 25 patients. While at least two measurements were obtained from each patient, only three patients provided data at all four pH levels.

Table 7.3 Action Potential Data

Patient	pH Level				Patient	pH Level			
	6.5	6.9	7.4	7.9		6.5	6.9	7.4	7.9
1		284	310	326	14	204	234	268	
2			261	292	15			258	267
3		213	224	240	16		193	224	235
4		222	235	247	17	185	222	252	263
5			270	286	18		238	301	300
6			210	218	19		198	240	
7		216	234	237	20		235	255	
8		236	273	283	21		216	238	
9	220	249	270	281	22		197	212	219
10	166	218	244		23		234	238	
11	227	258	282	286	24			295	281
12	216		284		25			261	272
13			257	284					

Even though numerous responses are missing, Mantel-Haenszel statistics can still be used to determine if the average Vmax differs among the four pH values (Q_{SMH}) and if there is a trend between Vmax and pH (Q_{CSMH}). This approach offers the advantage of not requiring any assumptions concerning the distribution of Vmax. In addition, the MH methodology accommodates the varying numbers of observations per patient (under the assumption that missing values are missing completely at random and the test statistic is specified with either table scores or ranks).

The following SAS statements read in the data in the format shown in Table 7.3 and restructure the data set for use by the FREQ procedure.

```
data ph_vmax;
   keep subject ph vmax;
   input subject vmax1-vmax4;
   ph=6.5;   vmax=vmax1;   output;
   ph=6.9;   vmax=vmax2;   output;
   ph=7.4;   vmax=vmax3;   output;
   ph=7.9;   vmax=vmax4;   output;
   datalines;
1   .   284 310 326
2   .    .  261 292
3   .   213 224 240
4   .   222 235 247
5   .    .  270 286
6   .    .  210 218
7   .   216 234 237
8   .   236 273 283
9  220  249 270 281
10 166  218 244  .
11 227  258 282 286
12 216   .  284  .
13  .    .  257 284
14 204  234 268  .
15  .    .  258 267
16  .   193 224 235
17 185  222 252 263
18  .   238 301 300
19  .   198 240  .
20  .   235 255  .
21  .   216 238  .
22  .   197 212 219
23  .   234 238  .
24  .    .  295 281
25  .    .  261 272
;
```

The following PROC GLM statements produce the means for each pH level as well as the average pairwise difference as the pH level increases.

```
proc glm;
   class ph;
   model vmax = subject ph;
   lsmeans ph;
   estimate 'direction' ph -1 0 0 1 / divisor=3;
run;
```

Output 7.5 and Output 7.6 contain these results. The mean Vmax ranges from 202.78 at the 6.5 pH level to 266.96 at the 7.9 pH level. The average weighted difference of means from lower to higher pH level is 21.39.

Output 7.5 Mean Scores

Least Squares Means

ph	vmax LSMEAN
6.5	202.781663
6.9	227.587950
7.4	256.122018
7.9	266.959134

Output 7.6 Difference Measure

Dependent Variable: vmax

Parameter	Estimate	Standard Error	t Value	Pr > |t|
direction	21.3924902	4.10791831	5.21	<.0001

The following statements compute the MH mean score and correlation statistics. The CMH2 option is used since it is not possible (or sensible) to compute the general association statistic Q_{GMH}. Since both pH and Vmax are quantitative variables, the default table scores are used. In addition, the trend is also assessed by using modified ridit scores.

```
proc freq;
   tables subject*ph*vmax / noprint cmh2;
   tables subject*ph*vmax / noprint cmh2 scores=modridit;
run;
```

Output 7.7 shows that the mean Vmax differs significantly among the four pH levels ($Q_{SMH} = 27.7431$, 3 df, $p < 0.0001$). In addition, there is a highly significant linear trend between pH and Vmax ($Q_{CSMH} = 27.3891$, 1 df, $p < 0.0001$).

Output 7.7 MH Mean Score and Correlation Tests: Table Scores

Summary Statistics for ph by vmax
Controlling for subject

Cochran-Mantel-Haenszel Statistics (Based on Table Scores)				
Statistic	Alternative Hypothesis	DF	Value	Prob
1	Nonzero Correlation	1	27.3891	<.0001
2	Row Mean Scores Differ	3	27.7431	<.0001

Effective Sample Size = 66
Frequency Missing = 34

WARNING: 34% of the data are missing.

The mean score and correlation statistics are even more significant when modified ridit scores are used (Output 7.8). Note that Vmax tends to progressively increase with pH for almost all patients (patients 18 and 24 are the exception).

Output 7.8 MH Mean Score and Correlation Tests: Modified Ridit Scores

Summary Statistics for ph by vmax
Controlling for subject

Cochran-Mantel-Haenszel Statistics (Modified Ridit Scores)				
Statistic	Alternative Hypothesis	DF	Value	Prob
1	Nonzero Correlation	1	35.3818	<.0001
2	Row Mean Scores Differ	3	34.7945	<.0001

Effective Sample Size = 66
Frequency Missing = 34

WARNING: 34% of the data are missing.

In this example, the column variable of each table is continuous and the row variable, although quantitative, has only four possible values. Thus, both Q_{SMH} and Q_{CSMH} can be used. The MH approach to the analysis of one-sample repeated measures can also be very useful when the row and column variables are both continuous. In this case, only Q_{CSMH} can be used. This can be specified by using the CMH1 option in the TABLES statement.

The methodology is also applicable when there are multiple groups (samples). However, the observations are viewed as a single group when comparing conditions within subjects. Note that if the data come from a balanced, incomplete block design, this analysis is equivalent to the Durbin test (Durbin 1951), which is a rank test used to test the null hypothesis of no differences among treatments in a balanced incomplete block design. The Durbin test reduces to the Friedman test if the number of treatments equals the number of experimental units per block.

7.6 Rank Analysis of Covariance

The analysis of covariance (ANCOVA) is a standard statistical methodology that combines the features of analysis of variance (ANOVA) and linear regression to determine if there is a difference in some response variable between two or more groups. The basic idea is to augment the ANOVA model containing the group effects with one or more additional categorical or quantitative variables that are related to the response variable. These additional variables *covary* with the response and so are called covariables or covariates.

One of the main uses of ANCOVA is to increase precision in randomized experiments by using the relationship between the response variable and the covariates to reduce the error variability in comparing treatment groups. In this setting, ANCOVA often results in more powerful tests, shorter confidence intervals, and a reduction in the sample size required to establish differences among treatment groups.

The validity of classical parametric ANCOVA depends on several assumptions, including normality of error terms, equality of error variances for different treatments, equality of slopes for the different treatment regression lines, and linearity of regression. For situations in which these assumptions may not be satisfied, Quade (1967) proposed the use of rank analysis of covariance. This technique can be combined with the randomization model framework of extended Mantel-Haenszel statistics to carry out nonparametric comparisons between treatment groups, after adjusting for the effects of one or more covariates. The methodology, which has been described by Koch et al. (1982, 1990), can easily be implemented using SAS.

The methodology can also be modified for the situation in which there are multiple strata. Table 7.4 displays data from an experiment to evaluate the effectiveness of topically applied stannous fluoride and acid phosphate fluoride in reducing the incidence of dental caries, as compared with a placebo treatment of distilled water (Cartwright, Lindahl, and Bawden 1968; Quade 1982). These data are from 69 female children from three centers who completed the two-year study. The stannous fluoride, acid phosphate fluoride, and distilled water treatment groups are denoted by SF, APF, and W. The columns labeled B and A represent the number of decayed, missing, or filled teeth (DMFT) before and after the study, respectively. In this example, the response to be compared among the three groups is the number of DMFT after treatment; the number of DMFT before treatment is used as a covariate. In addition, the analysis is stratified by center.

Table 7.4 Dental Caries Data

Center 1				Center 2				Center 3							
ID	Grp	B	A	ID	Grp	B	A	ID	Grp	B	A	ID	Grp	B	A
1	W	7	11	1	W	10	14	1	W	2	4	18	APF	10	12
2	W	20	24	2	W	13	17	2	W	13	18	19	APF	7	11
3	W	21	25	3	W	3	4	3	W	9	12	20	APF	13	12
4	W	1	2	4	W	4	7	4	W	15	18	21	APF	5	8
5	W	3	7	5	W	4	9	5	W	13	17	22	APF	1	3
6	W	20	23	6	SF	15	18	6	W	2	5	23	APF	8	9
7	W	9	13	7	SF	6	8	7	W	9	12	24	APF	4	5
8	W	2	4	8	SF	4	6	8	SF	4	6	25	APF	4	7
9	SF	11	13	9	SF	18	19	9	SF	10	14	26	APF	14	14
10	SF	15	18	10	SF	11	12	10	SF	7	11	27	APF	8	10
11	APF	7	10	11	SF	9	9	11	SF	14	15	28	APF	3	5
12	APF	17	17	12	SF	4	7	12	SF	7	10	29	APF	11	12
13	APF	9	11	13	SF	5	7	13	SF	3	6	30	APF	16	18
14	APF	1	5	14	SF	11	14	14	SF	9	12	31	APF	8	8
15	APF	3	7	15	SF	4	6	15	SF	8	10	32	APF	0	1
				16	APF	4	4	16	SF	19	19	33	APF	3	4
				17	APF	7	7	17	SF	10	13				
				18	APF	0	4								
				19	APF	3	3								
				20	APF	0	1								
				21	APF	8	8								

The following SAS statements read in the variables CENTER, ID, GROUP, BEFORE, and AFTER, whose values are displayed in Table 7.4.

7.6. Rank Analysis of Covariance

```
data caries;
   input center id group $ before after @@;
   datalines;
1  1 W      7 11    1  2 W    20 24    1  3 W    21 25    1  4 W     1  2
1  5 W      3  7    1  6 W    20 23    1  7 W     9 13    1  8 W     2  4
1  9 SF    11 13    1 10 SF   15 18    1 11 APF   7 10    1 12 APF  17 17
1 13 APF    9 11    1 14 APF   1  5    1 15 APF   3  7    2  1 W    10 14
2  2 W     13 17    2  3 W     3  4    2  4 W     4  7    2  5 W     4  9
2  6 SF    15 18    2  7 SF    6  8    2  8 SF    4  6    2  9 SF   18 19
2 10 SF    11 12    2 11 SF    9  9    2 12 SF    4  7    2 13 SF    5  7
2 14 SF    11 14    2 15 SF    4  6    2 16 APF   4  4    2 17 APF   7  7
2 18 APF    0  4    2 19 APF   3  3    2 20 APF   0  1    2 21 APF   8  8
3  1 W      2  4    3  2 W    13 18    3  3 W     9 12    3  4 W    15 18
3  5 W     13 17    3  6 W     2  5    3  7 W     9 12    3  8 SF    4  6
3  9 SF    10 14    3 10 SF    7 11    3 11 SF   14 15    3 12 SF    7 10
3 13 SF     3  6    3 14 SF    9 12    3 15 SF    8 10    3 16 SF   19 19
3 17 SF    10 13    3 18 APF  10 12    3 19 APF   7 11    3 20 APF  13 12
3 21 APF    5  8    3 22 APF   1  3    3 23 APF   8  9    3 24 APF   4  5
3 25 APF    4  7    3 26 APF  14 14    3 27 APF   8 10    3 28 APF   3  5
3 29 APF   11 12    3 30 APF  16 18    3 31 APF   8  8    3 32 APF   0  1
3 33 APF    3  4
;
```

The next statements produce standardized ranks for the covariate BEFORE and the response variable AFTER in each of the three centers. Standardized ranks are used to adjust for the fact that the number of patients differs among centers.

```
proc rank nplus1 ties=mean out=ranks;
   by center;
   var before after;
run;
```

The NPLUS1 option in the RANK procedure requests fractional ranks by using the denominator $n + 1$, where n is the center-specific sample size. The TIES=MEAN option requests that tied values receive the mean of the corresponding ranks (midranks). Since TIES=MEAN is the default for PROC RANK, this option was not specified in the previous example. However, when fractional ranks are requested using either the FRACTION (denominator is n) or NPLUS1 (denominator is $n + 1$) options, the TIES=HIGH option is the default. Thus, you must specify both the NPLUS1 and TIES=MEAN options.

PROC REG is then used to fit separate linear regression models for the three centers. In each model, the standardized ranks of the AFTER and BEFORE variables are used as the dependent and independent variables, respectively. The following statements request these models and output the corresponding residuals into an output data set named RESIDUAL.

```
proc reg noprint;
   by center;
   model after=before;
   output out=residual r=resid;
run;
```

Finally, the stratified mean score test, using the values of the residuals as scores, compares the three groups.

```
proc freq;
   tables center*group*resid / noprint cmh2;
run;
```

Output 7.9 displays the results. The difference among the three treatment groups, after adjusting for the baseline number of DMFT and center, is clearly significant (row mean score chi-square = 17.5929, 2 df, $p = 0.0002$).

Output 7.9 Results of Stratified Rank Analysis of Covariance

Summary Statistics for group by resid
Controlling for center

Cochran-Mantel-Haenszel Statistics (Based on Table Scores)				
Statistic	Alternative Hypothesis	DF	Value	Prob
1	Nonzero Correlation	1	17.1716	<.0001
2	Row Mean Scores Differ	2	17.5929	0.0002

Total Sample Size = 69

The rank analysis of covariance strategy described in this section is generally limited to randomized clinical trials, since the covariables should have similar distributions in the groups being compared. In this example, patients were randomly assigned to one of the three treatment groups, and Cartwright, Lindahl, and Bawden (1968) reported that the groups were comparable with respect to both the number of DMFT at baseline and other baseline variables. Although the patients in the APF group appear to have fewer DMFT at baseline than the patients in the SF and W groups (the corresponding medians are 7, 9, and 9, respectively), there is insufficient evidence to conclude that the distributions are significantly different (Kruskal-Wallis chi-square = 4.4 with 2 df, $p = 0.1100$); thus, rank analysis of covariance methods are appropriate.

You can apply these strategies to data from an observational (nonrandomized) study to adjust for sources of bias. However, such results should be interpreted cautiously unless you can make a strong case that the data collection was representative of a randomized scheme.

Chapter 8
Logistic Regression I: Dichotomous Response

Contents

8.1	Introduction		**190**
8.2	Dichotomous Explanatory Variables		**191**
	8.2.1	Logistic Model	191
	8.2.2	Model Fitting	192
	8.2.3	Goodness of Fit	194
	8.2.4	Using PROC LOGISTIC	194
	8.2.5	Interpretation of Main Effects Model	197
	8.2.6	Alternative Methods of Assessing Goodness of Fit	200
	8.2.7	Overdispersion	202
8.3	Using the CLASS Statement		**202**
	8.3.1	Analysis of Sentencing Data	202
	8.3.2	Goodness-of-Fit Statistics for Single Main Effect Model	206
	8.3.3	Deviation from the Mean Parameterization	207
8.4	Qualitative Explanatory Variables		**210**
	8.4.1	Model Fitting	211
	8.4.2	PROC LOGISTIC for Nominal Effects	212
	8.4.3	Testing Hypotheses about the Parameters	218
	8.4.4	Additional Graphics	220
	8.4.5	Fitting Models with Interactions	221
8.5	Continuous and Ordinal Explanatory Variables		**229**
	8.5.1	Goodness of Fit	229
	8.5.2	Fitting a Main Effects Model	230
8.6	A Note on Diagnostics		**235**
8.7	Alternatives to Maximum Likelihood Estimation		**238**
	8.7.1	Analyzing the Pre-Clinical Study Data	241
	8.7.2	Analysis of Completely Separated Data	242
	8.7.3	Analysis of Liver Function Data	244
	8.7.4	Exact Confidence Limits for Common Odds Ratios for 2×2 Tables	249
8.8	Using the GENMOD Procedure for Logistic Regression		**252**
	8.8.1	Performing Logistic Regression with the GENMOD Procedure	252
	8.8.2	Fitting Logistic Regression Models with PROC GENMOD	253
	Appendix A: Statistical Methodology for Dichotomous Logistic Regression		**257**

8.1 Introduction

The previous chapters discussed the investigation of statistical association, primarily by testing the hypothesis of no association between a set of groups and outcomes for a response with adjustment for a set of strata. Recall that Mantel-Haenszel strategies produced tests for specific alternatives to no association: general association, location shifts for means, and linear trends. This chapter shifts the focus to statistical models, methods aimed at describing the nature of the association in terms of a parsimonious number of parameters. Besides describing the variation in the data, statistical modeling allows you to address questions about association in terms of hypotheses concerning model parameters.

If certain realistic sampling assumptions are plausible, a statistical model can be used to make inferences from a study population to a larger target population. If you are analyzing a clinical trial that assigned its subjects to a randomized protocol, then you can generalize your results to the population from which the subjects were selected and possibly to a more general target population. If you are analyzing observational data, and you can argue that your study subjects are conceptually representative of some larger target population, then you may make inferences to that target population.

Logistic regression is a form of statistical modeling that is often appropriate for categorical outcome variables. It describes the relationship between a categorical response variable and a set of explanatory variables. The response variable is usually dichotomous, but it may be polytomous, that is, have more than two response levels. These multiple-level response variables can be nominally or ordinally scaled. This chapter addresses logistic regression when the response is dichotomous; typically the two outcomes are yes and no. Logistic regression with more than two response variable levels is covered in Chapter 9, "Logistic Regression II: Polytomous Response." Another kind of logistic regression is called conditional logistic regression and is often used for highly stratified data. Chapter 10, "Conditional Logistic Regression," describes this methodology.

Chapter 8 and Chapter 9 mainly focus on asymptotic methods that require an adequate sample size in order for model fit and effect assessment tests to be valid. However, sometimes your data are so sparse or have such small cell counts that these methods are not valid. This chapter also discusses exact logistic regression, which is an alternative strategy for these situations.

The explanatory variables in logistic regression can be categorical or continuous. Sometimes the term "logistic regression" is restricted to analyses that include continuous explanatory variables, and the term "logistic analysis" is used for those situations where all the explanatory variables are categorical. In this book, logistic regression refers to both cases. Logistic regression has applications in fields such as epidemiology, medical research, banking, market research, and social research. As you will see, one of its advantages is that model interpretation is possible through odds ratios, which are functions of model parameters.

A number of procedures in SAS/STAT software can be used to perform logistic regression, but this chapter focuses on the LOGISTIC procedure. It is designed primarily for logistic regression analysis, and it provides useful information such as odds ratio estimates and model diagnostics. The GENMOD procedure analyzes generalized linear models, of which logistic regression is a simple case. This chapter also provides a basic example of the use of the GENMOD procedure. Logistic regression is also available through the CATMOD, GLIMMIX, and PROBIT procedures.

8.2 Dichotomous Explanatory Variables

8.2.1 Logistic Model

Table 8.1 displays the coronary artery disease data that were analyzed in Chapter 3, "Sets of 2 × 2 Tables." Recall that the study population consists of people who visited a clinic on a walk-in basis and required a catheterization. The response, presence of coronary artery disease (CA), is dichotomous, as are the explanatory variables, sex and ECG. These data were analyzed in Section 3.3.2 with Mantel-Haenszel methods; also, odds ratios and the common odds ratio were computed. Recall that ECG was clearly associated with disease status, adjusted for gender.

Table 8.1 Coronary Artery Disease Data

Sex	ECG	Disease	No Disease	Total
Female	< 0.1 ST segment depression	4	11	15
Female	≥ 0.1 ST segment depression	8	10	18
Male	< 0.1 ST segment depression	9	9	18
Male	≥ 0.1 ST segment depression	21	6	27

Assume that these data arise from a stratified simple random sample so that presence of coronary artery disease is distributed binomially for each sex × ECG combination, that is, for each row of Table 8.1. These rows are called groups or subpopulations. You can then write a model for the probability, or the likelihood, of these data. The sex by ECG by disease status classification has the product binomial distribution

$$\Pr\{n_{hij}\} = \prod_{h=1}^{2} \prod_{i=1}^{2} \frac{n_{hi+}!}{n_{hi1}! n_{hi2}!} \theta_{hi}^{n_{hi1}} (1 - \theta_{hi})^{n_{hi2}}$$

The quantity θ_{hi} is the probability that a person of the hth sex with an ith ECG status has coronary artery disease, and n_{hi1} and n_{hi2} are the numbers of persons of the hth sex and ith ECG with and without coronary artery disease, respectively ($h = 1$ for females, $h = 2$ for males; $i = 1$ for ECG < 0.1, $i = 2$ for ECG ≥ 0.1; $j = 1$ for disease, $j = 2$ for no disease, and $n_{hi+} = (n_{hi1} + n_{hi2})$). You can apply the logistic model to describe the variation among the $\{\theta_{hi}\}$:

$$\theta_{hi} = \frac{1}{1 + \exp\{-(\alpha + \sum_{k=1}^{t} \beta_k x_{hik})\}}$$

Another form of this equation that is often used is

$$\theta_{hi} = \frac{\exp\{\alpha + \sum_{k=1}^{t} \beta_k x_{hik}\}}{1 + \exp\{\alpha + \sum_{k=1}^{t} \beta_k x_{hik}\}}$$

The quantity α is the intercept parameter; the $\{x_{hik}\}$ are the t explanatory variables for the hth sex and ith ECG; $k = 1, \ldots, t$; and the $\{\beta_k\}$ are the t regression parameters.

The matrix form of this equation is

$$\theta_{hi} = \frac{\exp(\alpha + \mathbf{x}'_{hi}\boldsymbol{\beta})}{1 + \exp(\alpha + \mathbf{x}'_{hi}\boldsymbol{\beta})}$$

where the quantity $\boldsymbol{\beta}$ is a vector of t regression parameters, and $\mathbf{x}_{hi}$ is a vector of explanatory variables corresponding to the hith group.

You can show that the odds of CA disease for the hith group is

$$\frac{\theta_{hi}}{1 - \theta_{hi}} = \exp\{\alpha + \sum_{k=1}^{t} \beta_k x_{hik}\}$$

By taking natural logarithms on both sides, you obtain a linear model for the logit:

$$\log\left\{\frac{\theta_{hi}}{1 - \theta_{hi}}\right\} = \alpha + \sum_{k=1}^{t} \beta_k x_{hik}$$

The logit is the log of an odds, so this model is for the log odds of coronary artery disease versus no coronary artery disease for the hith group. The log odds for the hith group can be written as the sum of an intercept and a linear combination of explanatory variable values multiplied by the appropriate parameter values. This result allows you to obtain the model-predicted odds ratios for variation in the x_{hik} by exponentiating model parameter estimates for the β_k, as explained below.

Besides taking the familiar linear form, the logistic model has the useful property that all possible values of $(\alpha + \mathbf{x}'_{hi}\boldsymbol{\beta})$ in $(-\infty, \infty)$ map into $(0, 1)$ for θ_{hi}. Thus, predicted probabilities produced by this model are constrained to lie between 0 and 1. This model produces no negative predicted probabilities and no predicted probabilities greater than 1. Maximum likelihood methods are generally used to estimate α and $\boldsymbol{\beta}$. PROC LOGISTIC uses the Fisher scoring method, which is equivalent to model fitting with iteratively weighted least squares. When the overall sample size $n = \sum_h \sum_i n_{hi}$ is sufficiently large, the resulting estimates for α and $\boldsymbol{\beta}$ have a multivariate normal distribution for which a consistent estimate of the corresponding covariance matrix is conveniently available. On this basis, confidence intervals and test statistics are straightforward to construct for inferences concerning α and $\boldsymbol{\beta}$. See Appendix A in this chapter for more methodological detail.

8.2.2 Model Fitting

A useful first model for the coronary disease data is one that includes main effects for sex and ECG. Since these effects are dichotomous, there are three parameters in this model, including the intercept.

You can write this main effects model as

$$\begin{bmatrix} \text{logit}(\theta_{11}) \\ \text{logit}(\theta_{12}) \\ \text{logit}(\theta_{21}) \\ \text{logit}(\theta_{22}) \end{bmatrix} = \begin{bmatrix} \alpha \\ \alpha \quad\quad + \beta_2 \\ \alpha + \beta_1 \\ \alpha + \beta_1 + \beta_2 \end{bmatrix} = \begin{bmatrix} 1 & 0 & 0 \\ 1 & 0 & 1 \\ 1 & 1 & 0 \\ 1 & 1 & 1 \end{bmatrix} \begin{bmatrix} \alpha \\ \beta_1 \\ \beta_2 \end{bmatrix}$$

This type of parameterization is often called *incremental effects* parameterization. It has a model matrix (also called a design matrix) composed of 0s and 1s. The quantity α is the log odds of coronary artery disease for females with an ECG of less than 0.1. Since females with ST segment depression less than 0.1 are described by the intercept, this group is known as the *reference cell* in this parameterization. The parameter β_1 is the increment in log odds for males, and β_2 is the increment in log odds for having an ECG of at least 0.1. Table 8.2 displays the probabilities and odds predicted by this model.

Table 8.2 Model-Predicted Probabilities and Odds

Sex	ECG	Pr{CA Disease}=θ_{hi}	Odds of CA Disease
Females	< 0.1	$e^{\alpha}/(1+e^{\alpha})$	e^{α}
Females	≥ 0.1	$e^{\alpha+\beta_2}/(1+e^{\alpha+\beta_2})$	$e^{\alpha+\beta_2}$
Males	< 0.1	$e^{\alpha+\beta_1}/(1+e^{\alpha+\beta_1})$	$e^{\alpha+\beta_1}$
Males	≥ 0.1	$e^{\alpha+\beta_1+\beta_2}/(1+e^{\alpha+\beta_1+\beta_2})$	$e^{\alpha+\beta_1+\beta_2}$

You can calculate the odds ratio for males versus females by forming the ratio of male odds of CA disease to female odds of CA disease for either low or high ECG (see Chapter 2, "The 2 × 2 Table," for a discussion of odds ratios):

$$\frac{e^{\alpha+\beta_1}}{e^{\alpha}} = e^{\beta_1} \quad \text{or} \quad \frac{e^{\alpha+\beta_1+\beta_2}}{e^{\alpha+\beta_2}} = e^{\beta_1}$$

Similarly, the odds ratio for high ECG versus low ECG is determined by forming the corresponding ratio of the odds of CA disease for either sex:

$$\frac{e^{\alpha+\beta_1+\beta_2}}{e^{\alpha+\beta_1}} = e^{\beta_2} \quad \text{or} \quad \frac{e^{\alpha+\beta_2}}{e^{\alpha}} = e^{\beta_2}$$

Thus, you can obtain odds ratios as functions of the model parameters in logistic regression. With incremental effects parameterization for a main effects model, you simply exponentiate the parameter estimates. However, unlike the odds ratios you calculate from individual 2 × 2 tables, these odds ratios have been adjusted for all other explanatory variables in the model.

8.2.3 Goodness of Fit

Once you have applied the model, you need to assess how well it fits the data, or how close the model-predicted values are to the corresponding observed values. Test statistics that assess fit in this manner are known as *goodness-of-fit statistics*. They address the differences between observed and predicted values, or their ratio, in some appropriate manner. Departures of the predicted proportions from the observed proportions should be essentially random. The test statistics have approximate chi-square distributions when the $\{n_{hij}\}$ are sufficiently large. If they are larger than a tolerable value, then you have an oversimplified model and you need to identify some other factors to better explain the variation in the data.

Two traditional goodness-of-fit tests are the Pearson chi-square, Q_P, and the likelihood ratio chi-square, Q_L, also known as the *deviance*.

$$Q_P = \sum_{h=1}^{2} \sum_{i=1}^{2} \sum_{j=1}^{2} (n_{hij} - m_{hij})^2 / m_{hij}$$

$$Q_L = \sum_{h=1}^{2} \sum_{i=1}^{2} \sum_{j=1}^{2} 2 n_{hij} \log\left(\frac{n_{hij}}{m_{hij}}\right)$$

where the m_{hij} are the model-predicted counts defined as

$$m_{hij} = \begin{cases} n_{hi+}\hat{\theta}_{hi} & \text{for } j=1 \\ n_{hi+}(1 - \hat{\theta}_{hi}) & \text{for } j=2 \end{cases}$$

The quantity $\hat{\theta}_{hi}$ is the estimate of θ_{hi} using the estimates of α and the β_k. If the model fits, both Q_P and Q_L are approximately distributed as chi-square with degrees of freedom equal to the number of rows in the table minus the number of parameters. For the main effects model being discussed, there are four rows in the table (four groups) and three parameters, including the intercept, and so Q_P and Q_L have $4 - 3 = 1$ degree of freedom. Sample size guidelines for these statistics to be approximately chi-square include

- each of the groups has at least 10 subjects ($n_{hi+} \geq 10$)
- 80% of the predicted counts (m_{hij}) are at least 5
- all other expected counts are greater than 2, with essentially no 0 counts

When the above guidelines do not apply, there is usually a tendency for the chi-square approximation to Q_P and Q_L to overstate lack of fit, and so tolerably small values for them are robustly interpretable as supporting goodness of fit. For a more rigorous evaluation of goodness of fit when the $\{n_{hij}\}$ are not large enough to justify chi-square approximations for Q_L and Q_P, exact methods for logistic regression are available (see Section 8.8).

8.2.4 Using PROC LOGISTIC

The LOGISTIC procedure was designed specifically to fit logistic regression models. You specify the response variable and the explanatory variables in a MODEL statement, and it fits the model via maximum likelihood estimation. PROC LOGISTIC produces the parameter estimates, their standard errors, and statistics to assess model fit. In addition, it also provides several model selection methods, puts predicted values and other statistics into output data sets, and includes a number of options for controlling the model-fitting process.

The following SAS code creates the data set CORONARY.

```
data coronary;
   input sex ecg ca count @@;
   datalines;
0 0 0   11  0 0 1    4
0 1 0   10  0 1 1    8
1 0 0    9  1 0 1    9
1 1 0    6  1 1 1   21
;
```

The variable CA is the response variable, and SEX and ECG are the explanatory variables. The variable SEX takes the value 0 for females and 1 for males, and ECG takes the value 0 for lower ST segment depression and 1 for higher ST segment depression. Thus, these variables provide the values for the model matrix. Such coding is known as *indicator-coding* or *dummy-coding*.

The variable CA takes the value 1 if CA disease is present and is 0 otherwise. By default, PROC LOGISTIC orders the response variable values alphanumerically so that, for these data, it bases its model on the probability of the smallest value, Pr{CA=0}, which is Pr{no coronary artery disease}. This means that it models the log odds of {no coronary artery disease}. If you want to change the basis of the model to be Pr{CA=1}, which is Pr{coronary artery disease}, you have to alter this default behavior. Data analysts usually want their models to be based on the probability of the event (disease, success), which is often coded as 1.

For a dichotomous response variable, the effect of reversing the order of the response values is to change the sign of the parameter estimates. Another effect will be that the odds ratio is the reciprocal of the desired one. Thus, if your estimates for the parameters have opposite signs from another logistic regression run, you have modeled opposite levels for the dichotomous response variable.

The next group of SAS statements invokes PROC LOGISTIC. Since the data are in frequency (count) form, you need to indicate that to the procedure. This is done with the FREQ statement, which is similar in use to the WEIGHT statement in PROC FREQ. (Note that a WEIGHT statement is available with the LOGISTIC procedure; however, it is used somewhat differently.) The main effects model is specified in the MODEL statement, which also includes the options SCALE=NONE and AGGREGATE. The EVENT='1' option for the response variable in the MODEL statement requests that the basis of the model be Pr{CA=1} or Pr{coronary artery disease}. (Another way to do this is to use the DESCENDING option in the PROC LOGISTIC statement. This reverses the default alphanumeric ordering of the response variable outcomes.)

The SCALE option produces goodness-of-fit statistics; the AGGREGATE option requests that PROC LOGISTIC treat each unique combination of the explanatory variable values as a distinct group in computing the goodness-of-fit statistics.

```
proc logistic;
   freq count;
   model ca(event='1')=sex ecg / scale=none aggregate;
run;
```

Output 8.1 displays the resulting "Response Profile" table. The response variable values are listed according to their PROC LOGISTIC *ordered values*, but the EVENT='1' option has made Pr{coronary artery disease} the basis of the model. It is always important to check the "Response Profile" table and note to ensure that PROC LOGISTIC is forming its model the way you desired.

Output 8.1 Response Profile

Response Profile		
Ordered Value	ca	Total Frequency
1	0	36
2	1	42

Probability modeled is ca=1.

Output 8.2 contains the goodness-of-fit statistics. Q_P has the value 0.2155, and Q_L has the value 0.2141. Compared to a chi-square distribution with 1 df, these relatively small values suggest that the model fits the data adequately. The note that the number of unique profiles is 4 means that these statistics are computed based on the 4 groups that are the rows of Table 8.1, the result of the AGGREGATE option.

Output 8.2 Goodness-of-Fit Statistics

Deviance and Pearson Goodness-of-Fit Statistics				
Criterion	Value	DF	Value/DF	Pr > ChiSq
Deviance	0.2141	1	0.2141	0.6436
Pearson	0.2155	1	0.2155	0.6425

Number of unique profiles: 4

Output 8.3 lists various criteria for assessing model fit through the quality of the explanatory capacity of the model; for $-2 \log L$ this is done by testing whether the explanatory variables are jointly significant relative to the chi-square distribution. AIC and SC serve a similar purpose while adjusting for the number of explanatory variables in the model. All of these statistics are analogous to the overall F test for the model parameters in a linear regression setting. Generally, you are interested in the second column, which pertains to the model with the intercept and covariates. Refer to the *SAS/STAT User's Guide*, for more information on these statistics.

Output 8.3 Testing Joint Significance of the Explanatory Variables

Model Fit Statistics		
Criterion	Intercept Only	Intercept and Covariates
AIC	109.669	101.900
SC	112.026	108.970
-2 Log L	107.669	95.900

8.2.5 Interpretation of Main Effects Model

With the satisfactory goodness of fit, it is appropriate to examine the parameter estimates from the model. Note that these results apply only to the population consisting of those persons who visited this medical clinic and required catheterization. The "Analysis of Maximum Likelihood Estimates" table in Output 8.4 lists the estimated model parameters, their standard errors, Wald chi-square tests, and p-values. A Wald test is a statistic that takes the form of the squared value ratio for the estimate to its standard error; it follows an approximate chi-square distribution when the sample size is sufficiently large. Wald statistics are easy to compute and are based on normal theory; however, their statistical properties are somewhat less optimal than those of the likelihood ratio statistics for small samples. Moreover, when there is concern for the statistical properties of results from small samples, exact methods can be helpful; see Section 8.8.

Output 8.4 Main Effects Model: ANOVA Table

Analysis of Maximum Likelihood Estimates					
Parameter	DF	Estimate	Standard Error	Wald Chi-Square	Pr > ChiSq
Intercept	1	-1.1747	0.4854	5.8571	0.0155
sex	1	1.2770	0.4980	6.5750	0.0103
ecg	1	1.0545	0.4980	4.4844	0.0342

The variable SEX is significant compared to a significance level of 0.05, with a Wald statistic (usually denoted Q_W) of 6.5750. The variable ECG is also significant, with $Q_W = 4.4844$.

The model equation can be written as follows:

$$\text{logit}(\theta_{hi}) = -1.1747 + 1.2770 \, \text{SEX} + 1.0545 \, \text{ECG}$$

Table 8.3 lists the parameter interpretations, and Table 8.4 displays the predicted logits and odds of coronary disease.

Table 8.3 Interpretation of Parameters

Parameter	Estimate	Standard Error	Interpretation
α	−1.1747	0.485	log odds of coronary disease for females with ECG < 0.1
β_1	1.2770	0.498	increment to log odds for males
β_2	1.0545	0.498	increment to log odds for ECG $\geq$ 0.1

Table 8.4 Model-Predicted Logits and Odds of CA Disease

Sex	ECG	Logit	Odds of Coronary Artery Disease
Female	< 0.1	$\hat{\alpha} = -1.1747$	$e^{\hat{\alpha}} = e^{-1.1747} = 0.3089$
Female	≥ 0.1	$\hat{\alpha} + \hat{\beta}_2 = -0.1202$	$e^{\hat{\alpha}+\hat{\beta}_2} = e^{-0.1202} = 0.8867$
Male	< 0.1	$\hat{\alpha} + \hat{\beta}_1 = 0.1023$	$e^{\hat{\alpha}+\hat{\beta}_1} = e^{0.1023} = 1.1077$
Male	≥ 0.1	$\hat{\alpha} + \hat{\beta}_1 + \hat{\beta}_2 = 1.1568$	$e^{\hat{\alpha}+\hat{\beta}_1+\hat{\beta}_2} = e^{1.1568} = 3.1797$

The odds ratio for males compared to females is the ratio of the predicted odds of CA disease for males versus females, which, on page 193, was shown to be

$$e^{\hat{\beta}_1} = e^{1.2770} = 3.586$$

Men in the study have 3.6 times higher odds for coronary artery disease than women in the study, controlling for ECG status. The odds ratio for ECG $\geq$ 0.1 versus ECG < 0.1 is the ratio of the predicted odds of CA disease for high ECG versus low ECG, which was shown to be

$$e^{\hat{\beta}_2} = e^{1.0545} = 2.871$$

Those persons with ECG $\geq$ 0.1 have 2.9 times the odds of coronary artery disease as those with ECG < 0.1, controlling for gender. This quantity is very similar to the common odds ratio estimates computed by PROC FREQ and displayed in Section 3.3.2 ($\hat{\psi}_{MH} = 2.847$ and $\hat{\psi}_L = 2.859$).

Output 8.5 contains the adjusted odds ratios and their 95% Wald confidence limits. The point estimates have the values calculated above. Neither of the confidence limits includes the value 1 in agreement with the statistical significance of each factor relative to the hypothesis of no association.

Output 8.5 Confidence Limits for Odds Ratios

Odds Ratio Estimates			
Effect	Point Estimate	95% Wald Confidence Limits	
sex	3.586	1.351	9.516
ecg	2.871	1.082	7.618

Predicted values are easily produced. The OUTPUT statement specifies that predicted values for the first ordered value (CA=1) be put into the variable PROB and output into the SAS data set PREDICT along with the variables from the input data set. You can print these values with the PRINT procedure. Note that PROC LOGISTIC also provides a SCORE statement for generating predicted values for other data sets.

This time, the DESCENDING option in the PROC LOGISTIC statement determines the internal order of the response variable values. It specifies that the model be based on the highest ordered response variable value, which is CA=1.

```
proc logistic descending;
   freq count;
   model ca=sex ecg;
   output out=predict pred=prob;
run;
proc print data=predict;
run;
```

The data set PREDICT contains model-predicted values for each observation in the input data set. The created variable named PROB contains these predicted values; the created variable _LEVEL_ tells you that they are the predicted values for the first ordered value, or Pr{coronary artery disease}. Observations 7 and 8 display the predicted value 0.76075 for males with high ECG.

Output 8.6 Predicted Values Output Data Set

Obs	sex	ecg	ca	count	_LEVEL_	prob
1	0	0	0	11	1	0.23601
2	0	0	1	4	1	0.23601
3	0	1	0	10	1	0.46999
4	0	1	1	8	1	0.46999
5	1	0	0	9	1	0.52555
6	1	0	1	9	1	0.52555
7	1	1	0	6	1	0.76075
8	1	1	1	21	1	0.76075

In conclusion, the main effects model is satisfactory. Being male and having ECG ≥ 0.1 are risk indicators for the presence of coronary artery disease for these data. If you can make the argument that this convenience sample is representative of a target group of coronary artery disease patients, possibly those persons who visit clinics on a walk-in basis, then these results may also apply to that population.

8.2.6 Alternative Methods of Assessing Goodness of Fit

There are other strategies available for assessing goodness of fit; these are based on fitting an appropriate expanded model and then evaluating whether the contribution of the additional terms is nonsignificant. If so, you then conclude that the original model has an adequate fit. You can compute likelihood ratio tests for the significance of the additional terms by taking the difference in the log likelihood for both models ($-2 \log L$ in the "Model Fit Statistics" table); this difference has an approximate chi-square distribution with degrees of freedom equal to the difference in the number of parameters in the models. You can also examine the Wald statistic for the additional parameters in order to assess goodness of fit.

For these data, the expanded model would be the one that contains the main effects for sex and ECG plus their interaction. The desired likelihood ratio statistic tests the significance of the interaction term and thus serves as a goodness-of-fit test for the main effects model.

You can write this model as

$$\begin{bmatrix} \text{logit}(\theta_{11}) \\ \text{logit}(\theta_{12}) \\ \text{logit}(\theta_{21}) \\ \text{logit}(\theta_{22}) \end{bmatrix} = \begin{bmatrix} \alpha \\ \alpha + \beta_2 \\ \alpha + \beta_1 \\ \alpha + \beta_1 + \beta_2 + \beta_3 \end{bmatrix} = \begin{bmatrix} 1 & 0 & 0 & 0 \\ 1 & 0 & 1 & 0 \\ 1 & 1 & 0 & 0 \\ 1 & 1 & 1 & 1 \end{bmatrix} \begin{bmatrix} \alpha \\ \beta_1 \\ \beta_2 \\ \beta_3 \end{bmatrix}$$

The model matrix column corresponding to β_3, the interaction term, is constructed by multiplying the columns for β_1 and β_2 together. Note that this model is a *saturated* model, since there are as many parameters as there are logit functions being modeled.

The following SAS code fits this model. Since PROC LOGISTIC includes a complete model-building facility, you simply cross SEX and ECG in the MODEL statement to specify their interaction. The interaction term in the resulting model matrix has the value 1 if both SEX and ECG are 1; otherwise, it is 0. The ODS SELECT statement is used to restrict the output to the fit statistics and the parameter estimates.

```
ods select FitStatistics ParameterEstimates;
proc logistic descending;
   freq count;
   model ca=sex ecg sex*ecg;
run;
```

The resulting tables titled "Model Fit Statistics" and "Analysis of Maximum Likelihood Estimates" follow.

Output 8.7 Results for Saturated Model

Model Fit Statistics		
Criterion	Intercept Only	Intercept and Covariates
AIC	109.669	103.686
SC	112.026	113.112
-2 Log L	107.669	95.686

Output 8.8 Results for Saturated Model

Analysis of Maximum Likelihood Estimates					
Parameter	DF	Estimate	Standard Error	Wald Chi-Square	Pr > ChiSq
Intercept	1	-1.0116	0.5839	3.0018	0.0832
sex	1	1.0116	0.7504	1.8172	0.1776
ecg	1	0.7885	0.7523	1.0985	0.2946
sex*ecg	1	0.4643	1.0012	0.2151	0.6428

The value for −2(log likelihood) is 95.686 for the saturated model; this is the value for −2 Log L listed under "Intercept and Covariates." The value for the main effects model is 95.900 (see Output 8.3), yielding a difference of 0.214. This difference is the likelihood ratio test value, with 1 df (4 parameters for the expanded model − 3 parameters for the main effects model). Compared with a chi-square distribution with 1 df, the nonsignificance of this statistic supports the adequacy of the main effects model. Note that you can always compute a likelihood ratio test in this manner for the contribution of a particular model term or a set of model terms when you have a nested model, that is, the reduced model is a subset of the model effects of the full model.

This likelihood ratio test value is the same as the deviance reported for the main effects model in Output 8.2. This is because the deviance statistic is effectively comparing the log likelihood for the main effects model with that for the saturated model.

Note that the value of the Wald statistic is 0.2151 for the interaction listed in the "Analysis of Maximum Likelihood Estimates" table. Both the likelihood ratio statistic and the Wald statistic are evaluating the same hypothesis: whether or not the interaction explains any of the variation among the different log odds beyond that explained by the main effects. They support goodness of fit of the main effects model by indicating nonsignificance of the interaction between sex and ECG. The Wald statistic and the likelihood ratio statistic are essentially equivalent for large samples.

8.2.7 Overdispersion

Sometimes a logistic model is considered reasonable, but the goodness-of-fit statistics indicate that too much variation remains (usually the deviance or deviance/df is examined). This condition is known as *overdispersion*, and it occurs when the data do not follow a binomial distribution well; the condition is also known as heterogeneity.

You can adjust for the overdispersion by scaling the covariance matrix to account for it. This involves the additional estimation of a dispersion parameter, often called a scaling parameter. PROC LOGISTIC allows you to specify a scaling parameter through the use of the SCALE= option; this explains why the SCALE=NONE option is used to generate the goodness-of-fit statistics, including the deviance, when no scale adjustment is desired. McCullagh and Nelder (1989) and Collett (2003) discuss overdispersion comprehensively. The *SAS/STAT User's Guide* describes these options in detail. Another method for addressing overdispersion is discussed in Chapter 15 in the context of methods involving generalized estimating equations.

8.3 Using the CLASS Statement

In the previous example, PROC LOGISTIC used the values of the explanatory variables to construct the model matrix. These values were already coded as 0s and 1s. However, that was a relatively simplistic case and more often the data set includes a variety of values for the explanatory and response variables, including variables with character values. The LOGISTIC procedure handles character-valued response variables by creating ordered values based on the alphanumeric order of the response variable values by default and offers numerous other options to control how it manages them. Its CLASS statement provides a choice of parameterization schemes. The next example illustrates how PROC LOGISTIC handles character-valued response variables and how the CLASS statement simplifies the use of classification variables in your model.

8.3.1 Analysis of Sentencing Data

Table 8.5 displays data based on a study on prison sentencing for persons convicted of a burglary or larceny. Investigators collected information on whether there was a prior arrest record and whether the crime was a nonresidential burglary, residential burglary, or something else—usually some sort of larceny. Here, type of crime is divided into nonresidential burglary versus all others. Sentence was recorded as to whether the offender was sent to prison.

Table 8.5 Sentencing Data

Type	Prior Arrest	Prison	No Prison	Total
Nonresidential	Some	42	109	151
Nonresidential	None	17	75	92
Other	Some	33	175	208
Other	None	53	359	412

Assume that these data arise from a stratified simple random sample so that sentencing is distributed binomially for each offense type × prior arrest record combination, that is, for each row of Table 8.5. The type of offense by prior arrest status by sentence classification has the product binomial distribution.

$$\Pr\{n_{hij}\} = \prod_{h=1}^{2} \prod_{i=1}^{2} \frac{n_{hi+}!}{n_{hi1}! n_{hi2}!} \theta_{hi}^{n_{hi1}} (1 - \theta_{hi})^{n_{hi2}}$$

The quantity θ_{hi} is the probability that a person arrested for a crime of type h with an ith prior arrest record receives a prison sentence, and n_{hi1} and n_{hi2} are the number of persons of the hth type and ith prior record who did and did not receive prison sentences, respectively (h=1 for nonresidential, h=2 for other; i=1 for prior arrest, i=2 for no arrest).

Similar to the previous example, a useful preliminary model for the sentencing data is one that includes main effects for type of offense and prior arrest record. There are three parameters in this model. The parameter α is the intercept, β_1 is the increment in log odds for committing a nonresidential burglary, and β_2 is the increment in log odds for having a prior arrest record. The probabilities and odds predicted by this model have identical structure to those presented in Table 8.2, replacing the first column with the values Nonresidential and Other and replacing the second column with the values Some and None. The model matrix is identical to the one displayed on page 193.

The following DATA step creates the SAS data set SENTENCE.

```
data sentence;
   input type $ prior $ sentence $ count @@;
   datalines;
nrb     some   y   42  nrb     some   n  109
nrb     none   y   17  nrb     none   n   75
other   some   y   33  other   some   n  175
other   none   y   53  other   none   n  359
;
```

The variable SENTENCE is the response variable, and TYPE and PRIOR are the explanatory variables. Note that SENTENCE is character valued, with values 'y' for prison sentence and 'n' for no prison sentence. PROC LOGISTIC orders these values alphabetically by default so that it bases its model on the probability of the value 'n', or Pr{no prison sentence}. If you want to change the basis of the model to be Pr{prison sentence}, you have to alter this default behavior.

The following group of SAS statements invoke PROC LOGISTIC. Note that since the desired model is based on Pr{prison sentence}, the EVENT= 'y' option for the response variable is specified.

```
proc logistic;
   class type prior(ref=first) / param=ref;
   freq count;
   model sentence(event='y') = type prior / scale=none aggregate;
run;
```

You list your classification variables in the CLASS statement. If you desire the incremental effects parameterization, you specify the option PARAM=REF after a '/'. The procedure provides a number of other parameterizations as well, including the default effect (deviation from the mean) parameterization and the less than full rank parameterization used in PROC GLM and PROC GENMOD. The incremental effects parameterization is a full rank parameterization.

By default, PROC LOGISTIC uses the last ordered value of the explanatory variable as the reference level and assigns it the value 0. If you want another value to be the reference level, you specify it with the REF= option after a slash (/) or after each individual variable, enclosed in parentheses. Here, REF=FIRST indicates that the 'none' level of PRIOR is the reference. You could also specify this value explicitly with the REF='none' option. Since 'other' is the last alphanumeric value in TYPE, it becomes the reference value for that effect, which is desired.

The "Response Profile" table indicates that the model is based on Pr{prison sentence}.

Output 8.9 Response Profiles

Response Profile		
Ordered Value	sentence	Total Frequency
1	n	718
2	y	145

Probability modeled is sentence='y'.

The "Class Level Information" table informs you how the model matrix is constructed. The design variables are the values associated with the explanatory variable levels. Since you want PRIOR='some' and TYPE='nrb' to be the incremental effects, the design variables take the value 1 for those levels.

Output 8.10 Class Level Information

Class Level Information		
Class	Value	Design Variables
type	nrb	1
	other	0
prior	none	0
	some	1

The goodness-of-fit statistics $Q_L = 0.5076$ and $Q_P = 0.5025$ indicate an adequate model fit. Note that if these statistics have values that are dissimilar, it is an indication that sample sizes in the groups are not large enough to support their use as goodness-of-fit statistics.

Output 8.11 Goodness of Fit

Deviance and Pearson Goodness-of-Fit Statistics				
Criterion	Value	DF	Value/DF	Pr > ChiSq
Deviance	0.5076	1	0.5076	0.4762
Pearson	0.5025	1	0.5025	0.4784

Number of unique profiles: 4

Since there are CLASS variables in the model, PROC LOGISTIC prints out the "Type 3 Analysis of Effects" table. These are Wald tests for the effects. Since both TYPE and PRIOR have 1 df, these tests are the same as for the parameter estimates in Output 8.13.

Output 8.12 Type III Analysis of Effects

Type 3 Analysis of Effects			
Effect	DF	Wald Chi-Square	Pr > ChiSq
type	1	9.0509	0.0026
prior	1	3.3127	0.0687

The variable TYPE is clearly significant, with $Q_W = 9.0509$. The variable PRIOR nearly approaches significance, with $Q_W = 3.3127$ and $p = 0.0687$. While some analysts might delete any effects that do not meet their designated 0.05 significance level, it is sometimes reasonable to keep modestly suggestive effects in the model to avoid potential bias for estimates of the other effects. In fact, for main effects models where presumably each explanatory variable chosen has some potential basis for its inclusion, many analysts keep all effects in the model, regardless of their significance. The model still appropriately describes the data, and it is easier to compare with other researchers' models where those nonsignificant effects may prove to be more important. And finally, sometimes the variables reflect the study design, in which case you would always include them.

Output 8.13 Main Effects Model

Analysis of Maximum Likelihood Estimates						
Parameter		DF	Estimate	Standard Error	Wald Chi-Square	Pr > ChiSq
Intercept		1	-1.9523	0.1384	199.0994	<.0001
type	nrb	1	0.5920	0.1968	9.0509	0.0026
prior	some	1	0.3469	0.1906	3.3127	0.0687

However, you may want to consider removing modest or clearly nonsignificant effects if some of them are redundant; that is, they are reflecting essentially the same factor. This can induce collinearity, and sometimes the association of explanatory variables with each other may mask the true effect. The additional model terms lead to poorer quality of the individual parameter estimates since they will be less precise (higher standard errors). In this case, PRIOR is kept in the model.

The model equation can be written as follows:

$$\text{logit}(\theta_{hi}) = -1.9523 + 0.5920 \text{ TYPE} + 0.3469 \text{ PRIOR}$$

The "Analysis of Maximum Likelihood Estimates" table for this model is displayed in Output 8.13. The estimates of the βs are printed as well as standard errors and significance tests. Output 8.14 displays the odds ratio estimates and confidence limits. The odds ratios are 1.808 ($e^{0.5920}$) for type of offense and 1.415 ($e^{0.3469}$) for prior arrest record. Thus, those persons committing a nonresidential burglary have nearly twice the odds of receiving prison sentences as those committing another offense. Those with a prior arrest record are about 40% more likely to receive a prison sentence than those with no prior record.

Output 8.14 Odds Ratio Estimates and Confidence Limits

Odds Ratio Estimates			
Effect	Point Estimate	95% Wald Confidence Limits	
type nrb vs other	1.808	1.229	2.658
prior some vs none	1.415	0.974	2.056

8.3.2 Goodness-of-Fit Statistics for Single Main Effect Model

Suppose that you did decide to fit the model with a single main effect, TYPE, and you wanted to generate the appropriate goodness-of-fit statistics for that model. Using the SCALE=NONE and AGGREGATE options would not work for this model, since the AGGREGATE option creates groups on which to base the goodness-of-fit statistic according to the values of the explanatory variables. Since there is just one dichotomous explanatory variable remaining in the model, only two groups would be created. To produce the groups consistent with the sampling framework, you need to specify AGGREGATE=(TYPE PRIOR), where the list of variables inside the parentheses are those whose unique values determine the rows of Table 8.5.

The following statements request the main effects model. The ODS SELECT statement restricts the output to the goodness-of-fit information.

```
ods select GoodnessOfFit;
proc logistic;
   class type prior (ref=first) / param=ref;
   freq count;
   model sentence(event='y') = type / scale=none aggregate=(type prior);
run;
```

Output 8.15 includes the goodness-of-fit statistics. Note the SAS message that there are 4 unique covariate profiles; this tells you that the correct groups were formed and that the statistics are based on the intended subpopulations.

Output 8.15 Single Effect Model

Deviance and Pearson Goodness-of-Fit Statistics				
Criterion	Value	DF	Value/DF	Pr > ChiSq
Deviance	3.8086	2	1.9043	0.1489
Pearson	3.7527	2	1.8763	0.1532

Number of unique profiles: 4

Since $Q_L = 3.8086$ and $Q_P = 3.7527$, both with 2 df and p-values of about 0.15, this single main effect model has a satisfactory fit.

8.3.3 Deviation from the Mean Parameterization

The preceding example used incremental effects parameterization, also called reference cell parameterization. However, that is not the default parameterization for the LOGISTIC procedure. If you do not specify the PARAM= option, you will obtain *deviation from the mean* parameterization, also known as *effect* parameterization. You can specify this explicitly in PROC LOGISTIC with the option PARAM=EFFECT in the CLASS statement.

In this parameterization, also a full rank parameterization like the incremental effects parameterization, the effects are differential rather than incremental. This model is written as follows:

$$\begin{bmatrix} \text{logit}(\theta_{11}) \\ \text{logit}(\theta_{12}) \\ \text{logit}(\theta_{21}) \\ \text{logit}(\theta_{22}) \end{bmatrix} = \begin{bmatrix} \alpha + \beta_1 + \beta_2 \\ \alpha + \beta_1 - \beta_2 \\ \alpha - \beta_1 + \beta_2 \\ \alpha - \beta_1 - \beta_2 \end{bmatrix} = \begin{bmatrix} 1 & 1 & 1 \\ 1 & 1 & -1 \\ 1 & -1 & 1 \\ 1 & -1 & -1 \end{bmatrix} \begin{bmatrix} \alpha \\ \beta_1 \\ \beta_2 \end{bmatrix}$$

Here, α is the average log odds (across the four populations) of a prison sentence, β_1 is the average differential change in log odds for whether a nonresidential burglary was committed, and β_2 is the differential change in log odds for having a prior arrest record. β_1 is an added amount for a nonresidential burglary and a subtracted amount for other burglary. β_2 is an added amount for a prior arrest record and a subtracted amount for no previous arrest record. The formulas for the model-predicted probabilities and odds for this parameterization are listed in Table 8.6.

Table 8.6 Model-Predicted Probabilities and Odds

Type	Prior Arrest	Pr{Prison}	Odds of Prison
Nonresidential	Some	$e^{\alpha+\beta_1+\beta_2}/(1+e^{\alpha+\beta_1+\beta_2})$	$e^{\alpha+\beta_1+\beta_2}$
Nonresidential	None	$e^{\alpha+\beta_1-\beta_2}/(1+e^{\alpha+\beta_1-\beta_2})$	$e^{\alpha+\beta_1-\beta_2}$
Other	Some	$e^{\alpha-\beta_1+\beta_2}/(1+e^{\alpha-\beta_1+\beta_2})$	$e^{\alpha-\beta_1+\beta_2}$
Other	None	$e^{\alpha-\beta_1-\beta_2}/(1+e^{\alpha-\beta_1-\beta_2})$	$e^{\alpha-\beta_1-\beta_2}$

The odds of a prison sentence for nonresidential burglary (nrb) versus other is obtained by forming the ratio of the odds for nrb versus other for either prior arrest level. Using some prior arrest, this is computed as

$$\frac{e^{\alpha+\beta_1+\beta_2}}{e^{\alpha-\beta_1+\beta_2}} = e^{2\beta_1}$$

The odds of a prison sentence for some arrest record versus none is obtained by forming the ratio of the odds for some prior arrest versus no prior arrest for either level of burglary type. Using nrb, this is computed as

$$\frac{e^{\alpha+\beta_1+\beta_2}}{e^{\alpha+\beta_1-\beta_2}} = e^{2\beta_2}$$

Thus, with this parameterization for a two-level explanatory variable, you need to exponentiate twice the parameter estimates to calculate the odds ratios, instead of simply exponentiating them, as was true for the reference cell model. However, this is taken care of by the LOGISTIC procedure.

The following SAS statements request an analysis of the sentencing data with the differential effects parameterization.

```
ods select ClassLevelInfo GoodnessOfFit
         ParameterEstimates OddsRatios;
proc logistic data=sentence;
   class type prior(ref='none');
   freq count;
   model sentence(event='y')= type prior / scale=none aggregate;
run;
```

Since not all of the output from the LOGISTIC procedure is desired, the ODS SELECT statement is used to request that only specific tables be generated. Since no PARAM= option is specified, the differential effects parameterization is used.

The "Class Level Information" table details the way in which the parameterization is constructed. The values of the CLASS variables are ordered alphanumerically, and the first ordered value gets the value 1 and the second gets the value -1, as illustrated in the case of variable TYPE. Since REF='none' was specified for variable PRIOR, the -1 is assigned to 'none' as the reference level, and the 1 is assigned to the value 'some'.

Output 8.16 Class Level Information

Class	Value	Design Variables
type	nrb	1
	other	-1
prior	none	-1
	some	1

Next, the goodness-of-fit statistics Q_P and Q_L have the values 0.5025 and 0.5076 respectively, the same as in the analysis with the incremental effects parameterization. In both cases, the test is assessing the same effect. A geometric way of looking at this is to say that the sets of explanatory variables for the two parameterizations span the same space, and so their estimated parameters produce the same predicted values.

Output 8.17 Goodness of Fit

Deviance and Pearson Goodness-of-Fit Statistics				
Criterion	Value	DF	Value/DF	Pr > ChiSq
Deviance	0.5076	1	0.5076	0.4762
Pearson	0.5025	1	0.5025	0.4784

Number of unique profiles: 4

Output 8.18 displays the "Analysis of Maximum Likelihood Estimates" table.

Output 8.18 Analysis of Maximum Likelihood Estimates

Analysis of Maximum Likelihood Estimates						
Parameter		DF	Estimate	Standard Error	Wald Chi-Square	Pr > ChiSq
Intercept		1	-1.4828	0.0951	243.2458	<.0001
type	nrb	1	0.2960	0.0984	9.0509	0.0026
prior	some	1	0.1735	0.0953	3.3127	0.0687

However, the parameter estimates are very different. This is because they represent very different quantities. The intercept is now the average log odds (across the four populations) of a prison sentence and the other parameters are the differential changes in the log odds for prior arrest and type of offense.

Output 8.19 displays the "Odds Ratio Estimates" table.

Output 8.19 Odds Ratio Estimates

Odds Ratio Estimates		
Effect	Point Estimate	95% Wald Confidence Limits
type nrb vs other	1.808	1.229 2.658
prior some vs none	1.415	0.974 2.056

The estimate for the odds ratio for a prison sentence comparing nonresidential burglary to other is $e^{2\beta_1} = 1.808$, which is the exponentiation of 2×0.2960; thus, PROC LOGISTIC has computed the odds ratio correctly. Similarly, the odds ratio for a prison sentence comparing prior arrest record to no arrest record is 1.415. The confidence limits for these point estimates are (1.229, 2.658) and (0.974, 2.056), respectively.

8.4 Qualitative Explanatory Variables

The previous examples have been concerned with analyses of dichotomous outcomes when the explanatory variables were also dichotomous. However, explanatory variables can be nominal (qualitative) with three or more levels, ordinal, or continuous. Logistic regression allows for any combination of these types of explanatory variables. This section is concerned with handling explanatory variables that are qualitative and contain three or more levels.

The following data come from a study on urinary tract infections (Koch, Imrey, et al. 1985). Investigators applied three treatments to patients who had either a complicated or uncomplicated diagnosis of urinary tract infection. Since complicated cases of urinary tract infections are difficult to cure, investigators were interested in whether the pattern of treatment differences is the same across diagnoses: did the diagnosis status of the patients affect the relative effectiveness of the three treatments? This is the same as determining whether there is a treatment × diagnosis interaction. Diagnosis is a dichotomous explanatory variable and treatment is a nominal explanatory variable consisting of levels for treatments A, B, and C. Table 8.7 displays the data.

Table 8.7 Urinary Tract Infection Data

Diagnosis	Treatment	Cured	Not Cured	Proportion Cured
Complicated	A	78	28	0.736
Complicated	B	101	11	0.902
Complicated	C	68	46	0.596
Uncomplicated	A	40	5	0.889
Uncomplicated	B	54	5	0.915
Uncomplicated	C	34	6	0.850

These data can be assumed to arise from a stratified simple random sample so that the response (cured or not cured) is distributed binomially for each diagnosis × treatment combination, that is, for each row of Table 8.7. The diagnosis by treatment classification has the product binomial distribution.

$$\Pr\{n_{hij}\} = \prod_{h=1}^{2} \prod_{i=1}^{3} \frac{n_{hi+}!}{n_{hi1}! n_{hi2}!} \theta_{hi}^{n_{hi1}} (1-\theta_{hi})^{n_{hi2}}$$

The quantity θ_{hi} is the probability that a person with the hth diagnosis receiving the ith treatment is cured, and n_{hi1} and n_{hi2} are the numbers of patients of the hth diagnosis and ith treatment who were and were not cured, respectively ($h = 1$ for complicated, $h = 2$ for uncomplicated; $i = 1$ for treatment A, $i = 2$ for treatment B, $i = 3$ for treatment C). You can then apply the logistic model to describe the variation among the $\{\theta_{hi}\}$. This is the same likelihood function as in the previous example except that i takes on the values 1, 2, and 3 instead of 1, 2.

8.4.1 Model Fitting

Since there is interest in the interaction term, the preliminary model includes main effects and their interaction (saturated model). There is one parameter for the intercept (α), which is the reference parameter corresponding to the log odds of being cured if you have an uncomplicated diagnosis and are getting treatment C. The parameter β_1 is the increment for complicated diagnosis. The effect for treatment consists of two parameters: β_2 is the incremental effect for treatment A, and β_3 is the incremental effect for treatment B.

There is no particular reason to choose a parameterization that includes incremental effects for treatments A and B; you could choose to parameterize the model by including incremental effects for treatments A and C. Often, data analysts choose the reference parameter to be the control group, with incremental effects representing various exposure effects. However, it's important to note that an effect with L levels must be represented by $(L - 1)$ parameters.

The interaction effect is made up of two additional parameters, β_4 and β_5, which represent the interaction terms for complicated diagnosis and treatment A, and complicated diagnosis and treatment B, respectively. When you are creating interaction terms from two effects, you create a number of terms equal to the product of the number of terms for both effects.

You can write this saturated model in matrix formulation as

$$\begin{bmatrix} \text{logit}(\theta_{11}) \\ \text{logit}(\theta_{12}) \\ \text{logit}(\theta_{13}) \\ \text{logit}(\theta_{21}) \\ \text{logit}(\theta_{22}) \\ \text{logit}(\theta_{23}) \end{bmatrix} = \begin{bmatrix} \alpha + \beta_1 + \beta_2 & + \beta_4 & \\ \alpha + \beta_1 & + \beta_3 & + \beta_5 \\ \alpha + \beta_1 & & \\ \alpha & + \beta_2 & \\ \alpha & + \beta_3 & \\ \alpha & & \end{bmatrix} = \begin{bmatrix} 1 & 1 & 1 & 0 & 1 & 0 \\ 1 & 1 & 0 & 1 & 0 & 1 \\ 1 & 1 & 0 & 0 & 0 & 0 \\ 1 & 0 & 1 & 0 & 0 & 0 \\ 1 & 0 & 0 & 1 & 0 & 0 \\ 1 & 0 & 0 & 0 & 0 & 0 \end{bmatrix} \begin{bmatrix} \alpha \\ \beta_1 \\ \beta_2 \\ \beta_3 \\ \beta_4 \\ \beta_5 \end{bmatrix}$$

Note that if you had parameterized the model so that there were three columns for treatment effects, each consisting of 1s corresponding to those logits representing the respective treatments, the columns would add up to a column of 1s. This would be redundant with the column of 1s for the intercept, and so PROC LOGISTIC would set the parameter corresponding to the third column of the effect equal to zero, since it is a linear combination of other columns. You could of course fit this model by creating indicator variables both for the incremental effects and for their interactions. You would need two indicator variables for the incremental effects for treatment A and treatment B, one indicator variable for complicated diagnosis, and two indicator variables for the interaction of diagnosis and treatment. However, you can perform this analysis much more easily by using a CLASS statement.

8.4.2 PROC LOGISTIC for Nominal Effects

The following DATA step creates SAS data set UTI.

```
data uti;
   input diagnosis : $13. treatment $ response $ count @@;
   datalines;
complicated    A  cured  78  complicated    A not 28
complicated    B  cured 101  complicated    B not 11
complicated    C  cured  68  complicated    C not 46
uncomplicated  A  cured  40  uncomplicated  A not  5
uncomplicated  B  cured  54  uncomplicated  B not  5
uncomplicated  C  cured  34  uncomplicated  C not  6
;
```

Since this model is saturated, the goodness-of-fit statistics don't apply; there are no available degrees of freedom because the number of groups and the number of parameters are the same (6). PROC LOGISTIC prints out near-zero values and zero df for saturated models. However, fitting this model does allow you to determine whether there is an interaction effect. Fitting the reduced model without the interaction terms and taking the difference in the likelihood allows you to determine whether the interaction is meaningful with a likelihood ratio test. The following PROC LOGISTIC statements fit the full model.

```
proc logistic;
   freq count;
   class diagnosis treatment /param=ref;
   model response = diagnosis|treatment;
run;
```

Output 8.20 contains -2 Log L for the full model, which is 447.556.

Output 8.20 Log Likelihood for the Full Model

Model Fit Statistics		
Criterion	Intercept Only	Intercept and Covariates
AIC	494.029	459.556
SC	498.194	484.549
-2 Log L	492.029	447.556

Output 8.21 contains the "Type 3 Analysis of Effects" table.

Output 8.21 Analysis of Effects

Type 3 Analysis of Effects			
Effect	DF	Wald Chi-Square	Pr > ChiSq
diagnosis	1	7.7653	0.0053
treatment	2	1.0069	0.6045
diagnosis*treatment	2	2.6384	0.2674

The Wald statistic for the interaction effect has the value 2.6384 with $p = 0.2674$ for 2 df, which is clearly not significant. This test also serves as a goodness-of-fit test for the main effects model, which is fit next.

The SCALE=NONE and AGGREGATE options are added to the MODEL statement to produce goodness-of-fit statistics.

```
proc logistic;
   freq count;
   class diagnosis treatment;
   model response = diagnosis treatment /
      scale=none aggregate;
run;
```

Output 8.22 contains the -2 Log L for the reduced model, which is 450.071.

Output 8.22 Log Likelihood for the Reduced Model

Model Fit Statistics		
Criterion	Intercept Only	Intercept and Covariates
AIC	494.029	458.071
SC	498.194	474.733
-2 Log L	492.029	450.071

The difference between 447.556 (full) and 450.071 (reduced) is 2.515; since the difference in the number of parameters in these models is 2, the value 2.515 should be compared to a chi-square distribution with 2 df. This computation is easily performed with the DATA step.

```
data;
   p=1-probchi(2.515,2);
   put p= ;
run;
```

The result is $p = 0.2844$. Thus, the likelihood ratio test for the hypothesis that the additional terms in the expanded model are zero cannot be rejected. The interaction between treatment and diagnosis is not significant, as was seen in the Wald test for the interaction. The likelihood ratio test also serves as a test for the adequacy of the main effects model.

Output 8.23 contains the goodness-of-fit statistics Q_P and Q_L for the reduced model. Note that Q_L has the same value as the likelihood ratio statistic; thus, you could have simply fit the main effects model and used Q_L as the test for interaction, knowing that the two omitted terms were the two interaction terms.

Output 8.23 Goodness-of-Fit Statistics

Deviance and Pearson Goodness-of-Fit Statistics				
Criterion	Value	DF	Value/DF	Pr > ChiSq
Deviance	2.5147	2	1.2573	0.2844
Pearson	2.7574	2	1.3787	0.2519

Number of unique profiles: 6

Output 8.24 displays the parameter estimates from the main effects model.

Output 8.24 Parameter Estimates

Analysis of Maximum Likelihood Estimates						
Parameter		DF	Estimate	Standard Error	Wald Chi-Square	Pr > ChiSq
Intercept		1	1.6528	0.1557	112.7189	<.0001
diagnosis	complicated	1	-0.4808	0.1499	10.2885	0.0013
treatment	A	1	-0.1304	0.1696	0.5914	0.4419
treatment	B	1	0.8456	0.1970	18.4336	<.0001

Output 8.25 contains the odds ratio estimates and their confidence limits, which are the 95% Wald confidence limits. None of these limits contain the value 1, indicating that there are significant treatment and diagnosis effects.

Output 8.25 Odds Ratio Estimates

Odds Ratio Estimates			
Effect	Point Estimate	95% Wald Confidence Limits	
diagnosis complicated vs uncomplicated	0.382	0.212	0.688
treatment A vs C	1.795	1.069	3.011
treatment B vs C	4.762	2.564	8.847

You have 4.8 times higher odds of being cured if you get treatment B compared with treatment C, and 1.8 times higher odds of being cured if you get treatment A compared to treatment C. You have 0.38 times lower odds of being cured if you have a complicated diagnosis as compared to an uncomplicated diagnosis (or a 62% reduction in odds); you have $(1/0.382) = 2.6$ times as high odds of being cured if you have uncomplicated diagnosis compared with complicated diagnosis. Note that all of these odds ratios have been adjusted for the other explanatory variable.

To confirm what these odds ratios represent, consider the model-predicted probabilities and odds listed in Table 8.8. Taking the ratio of odds for complicated diagnosis and treatment A versus complicated diagnosis and treatment C yields e^{β_2}. A similar exercise for treatment B yields e^{β_3}. To determine the odds ratio for complicated diagnosis to uncomplicated diagnosis, take the ratio of the odds for complicated to uncomplicated diagnosis at any level of treatment. You should get e^{β_1}.

Table 8.8 Model-Predicted Probabilities and Odds

Diagnosis	Treatment	Pr{Cured}	Odds of Cured
Complicated	A	$e^{\alpha+\beta_1+\beta_2}/(1+e^{\alpha+\beta_1+\beta_2})$	$e^{\alpha+\beta_1+\beta_2}$
Complicated	B	$e^{\alpha+\beta_1+\beta_3}/(1+e^{\alpha+\beta_1+\beta_3})$	$e^{\alpha+\beta_1+\beta_3}$
Complicated	C	$e^{\alpha+\beta_1}/(1+e^{\alpha+\beta_1})$	$e^{\alpha+\beta_1}$
Uncomplicated	A	$e^{\alpha+\beta_2}/(1+e^{\alpha+\beta_2})$	$e^{\alpha+\beta_2}$
Uncomplicated	B	$e^{\alpha+\beta_3}/(1+e^{\alpha+\beta_3})$	$e^{\alpha+\beta_3}$
Uncomplicated	C	$e^{\alpha}/(1+e^{\alpha})$	e^{α}

You may also want the odds ratio for the comparison of treatment A and treatment B, because, as pointed out above, the odds ratios produced by default are for treatments A and B relative to treatment C. The ODDSRATIO statement produces odds ratios for every pairwise difference of the levels of a classification variable. In addition, if you have enabled ODS Graphics, then you will also produce a graph that compares the odds ratios and their confidence intervals.

PROC LOGISTIC can also produce confidence limits for the odds ratios that are likelihood-ratio based. These are also known as profile likelihood confidence intervals. They are particularly desirable when the sample sizes are only moderately large rather than very large. (You can also request profile likelihood confidence intervals for the regression parameters with the CLPARM=PL option in the MODEL statement.)

The following SAS statements request the odds ratios. A separate ODDSRATIO statement is required for each variable; the results will be combined. The CL=BOTH option specifies both the default Wald and profile likelihood confidence intervals for the odds ratios. With ODS Graphics enabled, PROC LOGISTIC produces a graph of the odds ratios and confidence intervals when the ODDSRATIO statement is used. Specifying it with the PLOTS= option in the PROC LOGISTIC statement allows you to request that Log 2 be the basis of the x-axis. The PLOTS=EFFECT option specifies an effect plot.

```
ods graphics on;
proc logistic data=uti
        plots(only)=(effect(clband yrange=(.5,1) x=treatment*diagnosis)
                            oddsratio(logbase=2));
    freq count;
    class diagnosis treatment;
    model response = diagnosis treatment /
    scale=none aggregate;
run;
ods graphics off;
```

Output 8.26 displays the output produced by the ODDSRATIO statements.

Output 8.26 Confidence Limits for Odds Ratios

Odds Ratio Estimates and Wald Confidence Intervals			
Label	Estimate	95% Confidence Limits	
treatment A vs B	0.377	0.197	0.721
treatment A vs C	1.795	1.069	3.011
treatment B vs C	4.762	2.564	8.847
diagnosis complicated vs uncomplicated	0.382	0.212	0.688

Odds Ratio Estimates and Profile-Likelihood Confidence Intervals			
Label	Estimate	95% Confidence Limits	
treatment A vs B	0.377	0.193	0.711
treatment A vs C	1.795	1.074	3.031
treatment B vs C	4.762	2.615	9.085
diagnosis complicated vs uncomplicated	0.382	0.206	0.672

The odds ratios for comparing diagnostic levels and comparing treatments A and B with treatment C are the same as reported in Output 8.25. In addition, the odds ratio comparing treatment A to treatment B is 0.377, or treatment B has $(1/0.377) = 2.65$ times as high odds of resulting in a cure as treatment A. If you compare the Wald and profile likelihood confidence intervals in Output 8.26, you will find that they are similar for these data.

Output 8.27 displays the odds ratios comparison plot.

Output 8.27 Odds Ratios for UTI Analysis

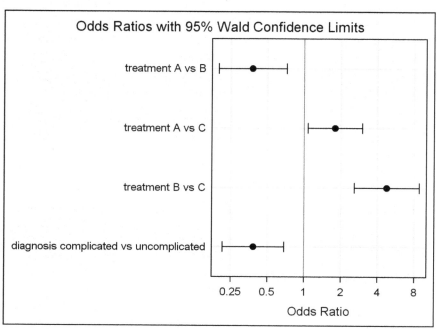

Output 8.28 displays the predicted probabilities by Diagnosis and Treatment.

Output 8.28 Predicted Probabilities for Cured by Diagnosis and Treatment

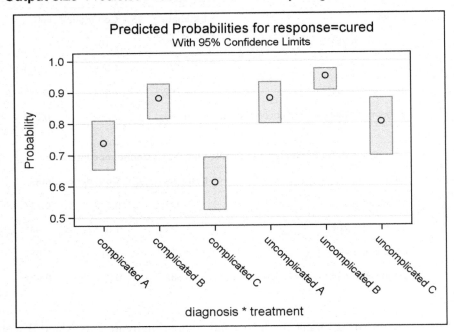

This plot clearly shows that patients getting treatment B did best for both complicated and uncomplicated diagnosis.

8.4.3 Testing Hypotheses about the Parameters

In the previous analysis, the overall effect for treatment was significant and so were the individual incremental effects parameters for treatment A and treatment B. Since the confidence interval for the odds ratio comparing treatment A and treatment B did not contain the value 1, you may also conclude that the effect of treatment A is different from the effect for treatment B. But you may also be interested in performing a formal test for the difference of these effects. You can request such a comparison test with the CONTRAST statement in PROC LOGISTIC.

In order to assess whether any of the treatments are similar, linear combinations of the parameters are tested to see if they are significantly different from zero.

$$H_0: \mathbf{L}\boldsymbol{\beta} = \mathbf{0}$$

By choosing the appropriate elements of **L**, you can construct linear combinations of the parameters that will produce the test of interest. The Wald statistic for a given linear combination **L** is computed as

$$Q_W = (\mathbf{L}\hat{\boldsymbol{\beta}})'(\mathbf{L}\mathbf{V}(\hat{\boldsymbol{\beta}})\mathbf{L}')^{-1}(\mathbf{L}\hat{\boldsymbol{\beta}})$$

where $\hat{\boldsymbol{\beta}}$ is the vector of parameter estimates. Q_W follows the chi-square distribution with degrees of freedom equal to the number of linearly independent rows of **L**.

The test for whether treatment A is equivalent to treatment B is expressed as

$$H_0: \beta_2 - \beta_3 = 0$$

which corresponds to $\mathbf{L} = \{1 - 1\}$ for $\boldsymbol{\beta} = \{\beta_2, \beta_3\}$. The test for whether treatment A is equivalent to treatment C is expressed as

$$H_0: \beta_2 = 0$$

which corresponds to $\mathbf{L}= \{1\ 0\ \}$ for $\boldsymbol{\beta} = \beta_2$, since β_2 is an incremental effect for treatment A in reference to treatment C. If β_2 equals zero, then treatment A is the same as treatment C, and the intercept represents the logit for uncomplicated diagnosis for either treatment A or treatment C.

You follow the same logic to see whether treatment B is equivalent to treatment C, and this corresponds to $\mathbf{L}=\{0\ 1\}$ for $\boldsymbol{\beta} = \beta_3$.

To compute the Wald test for the joint effect of treatment A and treatment B relative to treatment C (or the equality of treatments A, B, and C to one another), you test the hypothesis

$$H_0: \beta_2 = \beta_3 = 0$$

This is the hypothesis tested in the "Type III Analysis of Effects" table. The $\mathbf{L}$ for this third contrast is

$$\begin{bmatrix} 1 & 0 \\ 0 & 1 \end{bmatrix}$$

You specify these hypotheses in the CONTRAST statement. You list each hypothesis in a different statement, providing a name for the test within quotes. This can be up to 256 characters long. You then list the effect variable name and provide the coefficients for the $\mathbf{L}$ matrix. The following CONTRAST statements request the test comparing A and B, the individual test for A, and the joint test for A, B, and C.

```
ods select ContrastTest ContrastEstimate;
proc logistic data=uti;
   freq count;
   class diagnosis treatment /param=ref;
   model response = diagnosis treatment;
   contrast 'A versus B' treatment 1 -1
        / estimate=exp;
   contrast 'A' treatment 1 0;
   contrast 'joint test' treatment 1 0,
                          treatment 0 1;
run;
```

The ESTIMATE= option in the CONTRAST statement requests the estimate of the linear combination $\mathbf{L}\boldsymbol{\beta}$. The ESTIMATE=EXP option requests that the estimate be produced and exponentiated. Recall that the odds ratio for being cured for treatment A compared to treatment B is $e^{\beta_2-\beta_3}$. Thus, the ESTIMATE=EXP option produces the estimate of the odds ratio comparing treatment A and treatment B.

Output 8.29 contains the results. With a Wald chi-square of 8.6919 and a *p*-value of 0.0032, clearly treatments A and B are significantly different. The joint test statistic has the value 24.6219, which is the same as displayed in the "Type III Analysis of Effects" for the treatment effect.

Output 8.29 Contrast Test Results

Contrast	DF	Wald Chi-Square	Pr > ChiSq
A versus B	1	8.6919	0.0032
A	1	4.9020	0.0268
joint test	2	24.6219	<.0001

Output 8.30 contains the results of the contrast estimation.

Output 8.30 Contrast Estimation Results

Contrast Estimation and Testing Results by Row							
Contrast	Type	Row	Estimate	Standard Error	Alpha	Confidence Limits	
A versus B	EXP	1	0.3768	0.1247	0.05	0.1969	0.7210

Contrast Estimation and Testing Results by Row					
Contrast	Type	Row	Wald Chi-Square	Pr > ChiSq
A versus B	EXP	1	8.6919	0.0032

The point estimate for the odds ratio is 0.3768, as previously reported in Output 8.26.

8.4.4 Additional Graphics

Other useful graphics are available as well. The PLOTS= option in the PROC LOGISTIC statement provides a number of graphs plus the means to fine-tune them. The following PLOTS option requests an EFFECT plot where treatment is on the x-axis, predicted probabilities are displayed for each level of DIAGNOSIS, error bars are included, and the predicted values are connected with a line.

```
ods graphics on;
proc logistic data=uti plots(only)=effect(x=treatment
             sliceby=diagnosis clbar
             connect yrange=(0.5));
   freq count;
```

```
    class diagnosis treatment /param=ref;
    model response = diagnosis treatment;
    oddsratio treatment / cl=pl;
    oddsratio treatment / cl=pl;
  run;
  ods graphics off;
```

Output 8.31 displays the predicted probabilities for treatment and diagnosis; this is a useful plot for summarizing the results of this analysis.

Output 8.31 Predicted Probabilities of Cured for Treatment and Diagnosis

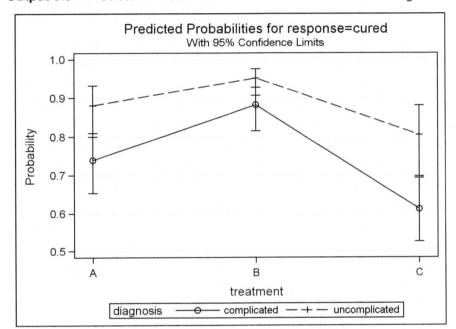

Those persons with uncomplicated diagnosis do better than those with complicated diagnosis for all treatments. Persons who received treatment B did the best, and persons receiving treatment A did better than those persons receiving treatment C.

8.4.5 Fitting Models with Interactions

The data displayed in Table 8.9 are from a cross-sectional prevalence study conducted in 1973 to investigate textile worker complaints about respiratory symptoms experienced while working in the mills (Higgins and Koch 1977). Investigators were interested in whether occupational environment was related to the prevalence of respiratory ailments associated with the disease byssinosis.

The study provides estimates of prevalence for a multi-way classification; these estimates are useful to identify subgroups with higher prevalence and thereby with a potentially higher need for health services that might be helpful for managing the byssinosis symptoms. In this case, the cross-sectional data only need to be representative of some corresponding population where comparable prevalences may apply. In other words, some of the more restrictive assumptions necessary to make inferences to a target population aren't required. More simply, note that the explanatory variables simply help to identify the types of patients with higher prevalences and thereby in potentially greater need for health services to address their byssinosis symptoms.

Table 8.9 Byssinosis Complaints

Workplace Condition	Years of Employment	Smoking	Complaints Yes	Complaints No
Dusty	< 10	Yes	30	203
Dusty	< 10	No	7	119
Dusty	≥ 10	Yes	57	161
Dusty	≥ 10	No	11	81
Not Dusty	< 10	Yes	14	1340
Not Dusty	< 10	No	12	1004
Not Dusty	≥ 10	Yes	24	1360
Not Dusty	≥ 10	No	10	986

If you can assume that there is a justifiable target population, then it becomes reasonable to think of these frequencies as coming from some stratified simple random sampling scheme, so that the table is distributed as product multinomial. With the dichotomous outcome, logistic regression is an obvious approach.

The following SAS statements create data set BYSS.

```
data byss;
    input work $ years $ smoke $ status $ count @@;
datalines ;
dusty  <10  yes  yes  30  dusty  <10   yes  no   203
dusty  <10  no   yes   7  dusty  <10   no   no   119
dusty  >=10 yes  yes  57  dusty  >=10  yes  no   161
dusty  >=10 no   yes  11  dusty  >=10  no   no    81
not    <10  yes  yes  14  not    <10   yes  no  1340
not    <10  no   yes  12  not    <10   no   no  1004
not    >=10 yes  yes  24  not    >=10  yes  no  1360
not    >=10 no   yes  10  not    >=10  no   no   986
;
```

The following PROC LOGISTIC statements fit the model that contains all pairwise interactions. The EVENT=LAST option in the MODEL statement means that the model is based on the probability of byssinosis symptoms. The REF=FIRST option for variables YEARS and SMOKE in the CLASS statement means that less than ten years of employment and not smoking will be the reference levels for those variables, respectively; the default REF=LAST option specifies the reference level for WORK (non-dusty workplace). The '@2' symbol with the single bar notation specifies that main effects and pairwise interactions are included in the model.

```
proc logistic data=byss;
    freq count;
    class work years(ref=first) smoke(ref=first) /param=ref;
    model status(event=last) = work|years|smoke@2 /
        scale=none aggregate;
run;
```

The response profiles in Output 8.32 show that 165 subjects had byssinosis symptoms and 5254 subjects did not have symptoms.

Output 8.32 Response Profiles

Response Profile		
Ordered Value	status	Total Frequency
1	no	5254
2	yes	165

Probability modeled is status='yes'.

The Pearson and deviance goodness-of-fit statistics displayed in Output 8.33 have p-values greater than 0.4, indicating adequate fit.

Output 8.33 Goodness-of-Fit Statistics

Deviance and Pearson Goodness-of-Fit Statistics				
Criterion	Value	DF	Value/DF	Pr > ChiSq
Deviance	0.6943	1	0.6943	0.4047
Pearson	0.6905	1	0.6905	0.4060

Number of unique profiles: 8

Output 8.34 displays the "Type 3 Analysis of Effects" table, which PROC LOGISTIC prints by default. For explanatory variables that have two levels, the results are the same as for the "Parameter Estimates" table since each effect is represented by a single parameter.

Output 8.34 Type 3 Analysis of Effects

Type 3 Analysis of Effects			
Effect	DF	Wald Chi-Square	Pr > ChiSq
work	1	23.9781	<.0001
years	1	0.0085	0.9267
work*years	1	2.3264	0.1272
smoke	1	0.0100	0.9202
work*smoke	1	3.2242	0.0726
years*smoke	1	0.9101	0.3401

Since the YEARS*SMOKE interaction appears to be unimportant with $p = 0.3401$, the model excluding that term is fit. The EFFECTPLOT statement requests a plot of the predicted log odds at all levels of SMOKE and YEARS for dusty and non-dusty workplace. The ODDSRATIO statement produces odds ratios for workplace, and an odds ratio plot is also requested with the PLOTS= option.

```
ods graphics on;
proc logistic plots(only)=(oddsratio(logbase=2));
   freq count;
   class work years(ref=first) smoke(ref=first) /param=ref;
   model status(event=last) = work years smoke
                     work*years work*smoke
                     /scale=none aggregate;
   effectplot  interaction (x=work) / at(smoke=all years=all) link noobs;
   oddsratios work;
run;
ods graphics off;
```

Output 8.35 displays the goodness-of-fit statistics. Reducing the model results in a Q_P with a p-value of 0.4487 and a Q_L with a p-value of 0.4490.

Output 8.35 Goodness-of-Fit Statistics

Deviance and Pearson Goodness-of-Fit Statistics				
Criterion	Value	DF	Value/DF	Pr > ChiSq
Deviance	1.6016	2	0.8008	0.4490
Pearson	1.6027	2	0.8013	0.4487

Number of unique profiles: 8

The global fit statistics are displayed in Output 8.36.

Output 8.36 Global Tests

Testing Global Null Hypothesis: BETA=0			
Test	Chi-Square	DF	Pr > ChiSq
Likelihood Ratio	284.3172	5	<.0001
Score	543.9026	5	<.0001
Wald	299.1096	5	<.0001

Output 8.37 displays the model parameter estimates.

Output 8.37 Parameter Estimates

Analysis of Maximum Likelihood Estimates							
Parameter			DF	Estimate	Standard Error	Wald Chi-Square	Pr > ChiSq
Intercept			1	-4.6446	0.2598	319.6394	<.0001
work	dusty		1	1.7936	0.3833	21.9032	<.0001
years	>=10		1	0.2651	0.2622	1.0221	0.3120
smoke	yes		1	0.2387	0.2696	0.7843	0.3758
work*years	dusty	>=10	1	0.6014	0.3444	3.0491	0.0808
work*smoke	dusty	yes	1	0.7047	0.3857	3.3387	0.0677

In the reference cell parameterization, the intercept represents the log odds of symptoms for persons in the non-dusty workplace who didn't smoke and worked less than ten years. The coefficient for workplace addresses the role of workplace for those who didn't smoke and who had less than ten years of employment, and it is strongly significant. The coefficient for smoking pertains to smoking in the non-dusty workplace, and the coefficient for years of employment pertains to at least ten years of employment in the non-dusty workplace; both of these have *p*-values in the 0.3 to 0.4 range.

Both the WORK*YEARS and WORK*SMOKE interactions are modestly influential, with *p*-values of 0.0808 and 0.0677, respectively, for 1 df. These interactions are kept in the model.

A main effect (no interaction) means that the variable's effect has roughly the same influence at all levels of the second variable. Pairwise interactions occur when one variable's effect depends on the level of a second variable. Sometimes, one variable has a measurable effect at one level of a second variable but virtually no effect at a different level of that variable. That appears to be the case here, as both the WORK*YEARS and the WORK*SMOKING interactions can be explained by the interplay of the dusty workplace with smoking and years of employment. You do need to keep the smoking and years of employment terms in the model when they are also contained within interactions.

Output 8.38 provides a visual interpretation of this model.

Output 8.38 InteractionPanel

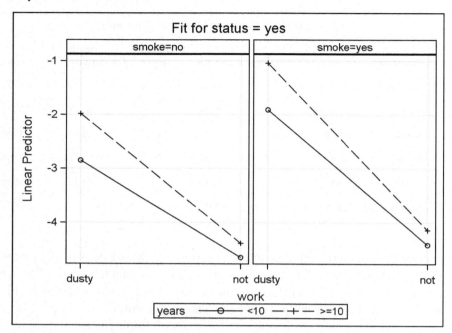

The predicted log odds (logit) is higher for dusty workplace. Its effect depends on the level of smoking and years of employment, as the predicted log odds is much higher for smokers and for subjects with ten or more years of employment. And the predicted log odds is higher again for that group of subjects who are smokers and have ten or more years of employment.

You can produce odds ratios when the model includes interactions, and they can provide a useful framework for interpretation. Consider the model at hand:

$$\text{logit}(\theta_h) = \alpha + \{\text{dusty workplace}\} * \beta_1 + \{>=10 \text{ years}\} * \beta_2 + \{\text{smoking}\} * \beta_3$$
$$+ \{\text{work*years: dusty and} >=10 \text{ years}\} * \beta_4 + \{\text{work*smoke: dusty and smoking}\} * \beta_5$$

where $h = 1, \ldots, 8$ represents the eight rows of Table 8.9, and { } indicates the presence of that level of the explanatory variables(s).

In order to produce the odds ratio comparing the odds of byssinosis symptoms for a dusty workplace to a non-dusty workplace in a main effects model, you would exponentiate the parameter estimate corresponding to workplace, which is β_1 in this model. But having interactions in the model means that the workplace effect depends on the levels of the other variables for smoking and years of employment, and so does the corresponding odds ratios. You need to compute a set of odds ratios for workplace—for each combination of the levels of the variables that interact with it.

To form the complete set of odds ratios estimates for workplace, you form the ratio of the odds for Dusty to Not Dusty workplace for each of the four combinations of the levels of Years of Employment (<10 and ≥ 10) and Smoking (Yes and No). Table 8.10 contains these odds based on the model.

Table 8.10 Byssinosis Complaints

Workplace Condition	Years of Employment	Smoking	Odds of Symptoms
Dusty	< 10	Yes	$e^{\alpha+\beta_1+\beta_3+\beta_5}$
Dusty	< 10	No	$e^{\alpha+\beta_1}$
Dusty	≥ 10	Yes	$e^{\alpha+\beta_1+\beta_2+\beta_3+\beta_4+\beta_5}$
Dusty	≥ 10	No	$e^{\alpha+\beta_1+\beta_2+\beta_4}$
Not Dusty	< 10	Yes	$e^{\alpha+\beta_3}$
Not Dusty	< 10	No	e^{α}
Not Dusty	≥ 10	Yes	$e^{\alpha+\beta_2+\beta_3}$
Not Dusty	≥ 10	No	$e^{\alpha+\beta_2}$

So e^{β_1} is the ratio of the odds in the second line of the table to the odds in the sixth line in the table, and thus it is the odds of symptoms for a dusty workplace compared to the odds of symptoms in the non-dusty workplace for subjects who didn't smoke and who worked for less than ten years. And $e^{1.7936} = 6.011$ when you supply $\hat{\beta}_1$.

To determine the odds ratio at <10 years of employment and smoking, you form the ratio for the odds listed in the first and fifth lines in Table 8.10:

$$e^{\alpha+\beta_1+\beta_3+\beta_5}/e^{\alpha+\beta_3} = e^{\beta_1+\beta_5}$$

Plugging in the parameter estimates, you obtain

$$e^{1.7936+0.7047} = 12.1618$$

Thus, subjects who worked in a dusty workplace had 12.1618 times the odds of symptoms in a non-dusty workplace if they smoked and had less than ten years on the job.

When you specify the ODDSRATIO in the LOGISTIC procedure for a variable that is included in interactions, PROC LOGISTIC computes the appropriate set of odds ratio estimates. Output 8.39 contains the estimates and the Wald-based confidence intervals for the odds ratio comparing the odds of symptom for dusty workplace versus non-dusty workplace for each combination of the levels of SMOKE and YEARS.

Output 8.39 Odds Ratios

Odds Ratio Estimates and Wald Confidence Intervals			
Label	Estimate	95% Confidence Limits	
work dusty vs not at years=<10 smoke=no	6.011	2.836	12.741
work dusty vs not at years=<10 smoke=yes	12.163	6.926	21.359
work dusty vs not at years=>=10 smoke=no	10.968	5.430	22.156
work dusty vs not at years=>=10 smoke=yes	22.192	13.652	36.073

For non-smokers with less than ten years of employment, those subjects in a dusty workplace had 6.011 times the odds of symptoms compared to those subjects in a non-dusty workplace. However, this effect increases for both smoking (12.163) or ten or more years of employment (10.968). If the subject both smoked and had ten or more years of employment, then the odds ratio is 22.192. These increasing odds ratio estimates clearly demonstrate the interaction effect of smoking and years of employment with the dusty workplace.

Output 8.40 compares these odds ratio estimates graphically.

Output 8.40 Odds Ratios

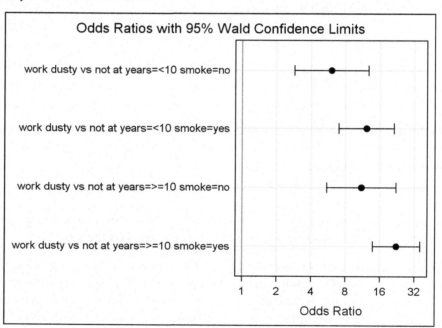

8.5 Continuous and Ordinal Explanatory Variables

8.5.1 Goodness of Fit

Frequently, some or all of the explanatory variables in a logistic regression analysis are continuous. Analysis strategies are the same as those described in previous sections, except in the evaluation of goodness of fit.

The following data are from the same study on coronary artery disease as previously analyzed; in addition, the continuous variable AGE is an explanatory variable. The variable ECG is now treated as an ordinal variable, with values 0, 1, and 2. ECG is coded 0 if the ST segment depression is less than 0.1, 1 if it equals 0.1 or higher but less than 0.2, and 2 if the ST segment depression is greater than or equal to 0.2. The variable AGE is age in years.

```
data coronary;
   input sex ecg age ca @@  ;
   datalines;
0 0 28 0    1 0 42 1    0 1 46 0    1 1 45 0
0 0 34 0    1 0 44 1    0 1 48 1    1 1 45 1
0 0 38 0    1 0 45 0    0 1 49 0    1 1 45 1
0 0 41 1    1 0 46 0    0 1 49 0    1 1 46 1
0 0 44 0    1 0 48 0    0 1 52 0    1 1 48 1
0 0 45 1    1 0 50 0    0 1 53 1    1 1 57 1
0 0 46 0    1 0 52 1    0 1 54 1    1 1 57 1
0 0 47 0    1 0 52 1    0 1 55 0    1 1 59 1
0 0 50 0    1 0 54 0    0 1 57 1    1 1 60 1
0 0 51 0    1 0 55 0    0 2 46 1    1 1 63 1
0 0 51 0    1 0 59 1    0 2 48 0    1 2 35 0
0 0 53 0    1 0 59 1    0 2 57 1    1 2 37 1
0 0 55 1    1 1 32 0    0 2 60 1    1 2 43 1
0 0 59 0    1 1 37 0    1 0 30 0    1 2 47 1
0 0 60 1    1 1 38 1    1 0 34 0    1 2 48 1
0 1 32 1    1 1 38 1    1 0 36 1    1 2 49 0
0 1 33 0    1 1 42 1    1 0 38 1    1 2 58 1
0 1 35 0    1 1 43 0    1 0 39 0    1 2 59 1
0 1 39 0    1 1 43 1    1 0 42 0    1 2 60 1
0 1 40 0    1 1 44 1
;
```

Look at the values listed for AGE. While some observations share the same AGE value, most of these values are unique. Thus, there will be only one observation in most of the cells created by the cross-classification of the explanatory variable values. In fact, the SEX by ECG by AGE cross-classification produces 68 groups from these 78 observations. This means that the sample size requirement for the use of the Pearson chi-square goodness-of-fit test and the likelihood ratio goodness-of-fit test—that each predicted cell count tends to be at least 5—is not met. This is almost always the case when you have continuous explanatory variables.

There are several alternative strategies. First, you can fit the desired model, fit an appropriate expanded model with additional explanatory variables, and look at the differences in the log-likelihood ratio statistics. This difference is distributed as chi-square with degrees of freedom equal to the difference in degrees in freedom of the two models (given sufficiently large samples to support approximate normal estimates from the expanded model).

The second strategy is to examine the residual score statistic, Q_{RS} (Breslow and Day 1980). This criterion is directed at the extent to which the residuals from the model are linearly associated with other potential explanatory variables. If there is an association, this is an indication that these variables should also be included in the model. Thus, to compute the residual score statistic, you need to have access to the variables that make up the potential expansion. Q_{RS} is distributed as chi-square, with degrees of freedom equal to the difference in the number of parameters for the two models.

However, unlike computing the log-likelihood ratio statistic where you have to execute PROC LOGISTIC twice and form the difference of the log-likelihood ratio statistics, you can generate this score goodness-of-fit statistic with one invocation of PROC LOGISTIC. You do this by taking advantage of the LOGISTIC procedure's model-building capabilities. The SELECTION=FORWARD method adds variables to your model in the manner in which you specify, computing model assessment statistics for each of the successive models it fits. In addition, it prints a score statistic that assesses the joint contribution of the remaining model effects that have not yet been incorporated into the model. With the right choice of model effects in the MODEL statement, this is the score goodness-of-fit statistic. You can also generate the constituent one degree of freedom score tests by including the DETAILS option in the MODEL statement.

A third strategy is to compute an alternative goodness-of-fit statistic proposed by Hosmer and Lemeshow (1989). This test places subjects into deciles based on the model-predicted probabilities, then computes a Pearson chi-square test based on the observed and expected number of subjects in the deciles. The statistic is compared to a chi-square distribution with t degrees of freedom, where t is the number of decile groups minus 2. Depending on the number of observations, there may be fewer than ten groups. PROC LOGISTIC prints this statistic when you specify the LACKFIT option in the MODEL statement. You should note that this method may have low power for detecting departures from goodness of fit, and so some caution may be needed in its interpretation.

These strategies are implemented in the next section.

8.5.2 Fitting a Main Effects Model

A model of interest for these data is a main effects model with terms for sex, ECG, and age. To generate a score statistic, you need to choose the effects that constitute the expanded model. Your choice depends partially on the sample size. There should be at least 5 observations for the rarer outcome per parameter being considered in the expanded model. Some analysts would prefer at least 10. In this data set, there are 37 observations with no coronary artery disease and 41 observations with coronary artery disease. Thus, no coronary artery disease is the rarer event, and the quotient 37/5 suggests that no more than 7–8 parameters can be supported.

For these data, an appropriate expanded model consists of all second-order terms, which are the squared terms for age and ECG plus all pairwise interactions. This creates eight parameters beyond the intercept. One might also include the third-order terms, but their inclusion would result in too few observations per parameter for the necessary sample size requirements for these statistics. If there did happen to be substantial third-order variation, this approach would not be appropriate.

The following PROC LOGISTIC statements fit the main effects model and compute the score test. The first- and second-order terms are listed on the right-hand side of the MODEL statement, with CA as the response variable. SELECTION=FORWARD is specified as a MODEL statement option after a slash (/). The option INCLUDE=3 specifies that the first three terms listed in the MODEL statement be included in each fitted model, regardless of significance level. PROC LOGISTIC first fits this model, which is the main effects model, and then produces the score goodness-of-fit statistic.

```
proc logistic descending;
model ca=sex ecg age ecg*ecg age*age
          sex*ecg sex*age ecg*age /
          selection=forward include=3 details lackfit;
run;
```

Note that 1 is the first ordered value, since the DESCENDING option was specified in the PROC statement, so the model is based on Pr{coronary artery disease}.

Output 8.41 Response Profile

Ordered Value	ca	Total Frequency
1	1	41
2	0	37

Probability modeled is ca=1.

Model Convergence Status
Convergence criterion (GCONV=1E-8) satisfied.

After the "Response Profile" table, PROC LOGISTIC prints a list of the variables included in each model. Note that the score statistic printed in the table "Testing Global Null Hypothesis: BETA=0" is not the score goodness-of-fit statistic. This score statistic is strictly testing the hypothesis that the specified model effects are jointly equal to zero.

Output 8.42 Assessing Fit

Testing Global Null Hypothesis: BETA=0			
Test	Chi-Square	DF	Pr > ChiSq
Likelihood Ratio	21.1145	3	<.0001
Score	18.5624	3	0.0003
Wald	14.4410	3	0.0024

The "Residual Chi-Square Test" is printed after the "Association of Predicted Probabilities and Observed Responses" table. This is the score goodness-of-fit statistic.

Output 8.43 Residual Chi-Square

Residual Chi-Square Test		
Chi-Square	DF	Pr > ChiSq
2.3277	5	0.8022

Since the difference between the number of parameters for the expanded model and the main effects model is $9 - 4 = 5$, it has 5 degrees of freedom. Since $Q_{RS} = 2.3277$ and $p = 0.8022$, the main effects model fits adequately. The DETAILS option causes the "Analysis of Effects Eligible for Entry" table to be printed. These tests are the score tests for the addition of the single effects to the model. Each of these tests has one degree of freedom. As one might expect, all of these tests indicate that the single effects add little to the main effects model. Since the sample size requirements for the global test are very roughly met, the confirmation of goodness of fit with the single tests is reasonable, since sample size requirements for these individual expanded models are easily met.

Output 8.44 Analysis of Effects Eligible for Entry

Analysis of Effects Eligible for Entry			
Effect	DF	Score Chi-Square	Pr > ChiSq
ecg*ecg	1	0.3766	0.5394
age*age	1	0.7712	0.3798
sex*ecg	1	0.0352	0.8513
sex*age	1	0.0290	0.8647
ecg*age	1	0.8825	0.3475

Note that this testing process is conservative with respect to confirming model fit. Inadequate sample size may produce spuriously large chi-squares and correspondingly small p-values. However, this would mean that you decide that the fit is not adequate, and you search for another model. Small sample sizes will not misleadingly cause these methods to suggest that poor fit is adequate, although they would have the limitation of low power to detect real departures from a model.

You may have a concern with the evaluation of multiple tests to assess model goodness of fit. However, by requiring the global test and most single tests to be nonsignificant, the assessment of goodness of fit is more stringent. Also, the multiplicity can be evaluated relative to what might be expected by chance in an assessment of goodness of fit.

Output 8.45 and Output 8.46 displays the results produced by the LACKFIT option.

Output 8.45 Results from the LACKFIT Option

		ca = 1		ca = 0	
Group	Total	Observed	Expected	Observed	Expected
1	8	2	1.02	6	6.98
2	8	1	1.80	7	6.20
3	8	3	2.59	5	5.41
4	8	3	3.42	5	4.58
5	8	4	4.07	4	3.93
6	9	6	5.38	3	3.62
7	9	4	5.97	5	3.03
8	8	7	5.99	1	2.01
9	8	7	6.98	1	1.02
10	4	4	3.77	0	0.23

Partition for the Hosmer and Lemeshow Test

Output 8.46 Goodness-of-Fit Test

Hosmer and Lemeshow Goodness-of-Fit Test		
Chi-Square	DF	Pr > ChiSq
4.7766	8	0.7812

The Hosmer and Lemeshow statistic has a value of 4.7766 with 8 df; $p = 0.7812$. Thus, this measure also supports the model's adequacy for these data. The output also includes the observed and expected counts for each predicted probability decile for each value of the response variable. This criterion can also be used as a measure of goodness of fit for the strictly qualitative explanatory variable situation.

The satisfactory goodness-of-fit statistics make it reasonable to examine the main effects parameter estimates. The following PROC LOGISTIC statements fit this model. The UNITS statement enables you to specify the units of change for which you want the odds ratios computed.

```
proc logistic descending;
   model ca=sex ecg age;
   units age=10;
run;
```

Output 8.47 Main Effects Parameter Estimates

Analysis of Maximum Likelihood Estimates					
Parameter	DF	Estimate	Standard Error	Wald Chi-Square	Pr > ChiSq
Intercept	1	-5.6418	1.8061	9.7572	0.0018
sex	1	1.3564	0.5464	6.1616	0.0131
ecg	1	0.8732	0.3843	5.1619	0.0231
age	1	0.0929	0.0351	7.0003	0.0081

Output 8.48 Odds Ratio Estimates

Odds Ratio Estimates		
Effect	Point Estimate	95% Wald Confidence Limits
sex	3.882	1.330 11.330
ecg	2.395	1.127 5.086
age	1.097	1.024 1.175

The parameter estimates are all significant at the 0.05 level, as judged by the accompanying Wald statistics. Thus, the estimated equation for the log odds is

$$\text{logit}(\theta_{hi}) = -5.6418 + 1.3564\,\text{SEX} + 0.8732\,\text{ECG} + 0.0929\,\text{AGE}$$

Presence of coronary artery disease is positively associated with age and ST segment depression, and it is more likely for males in this population. The odds ratio listed for SEX, 3.882, is the odds of coronary disease presence for males relative to females adjusted for age and ST segment depression. The value listed for ECG, 2.395, is the extent to which the odds of coronary artery disease presence is higher per level increase in ST segment depression. The value 1.097 for AGE is the extent to which the odds are higher each year.

A more desirable statistic may be the extent to which the odds of coronary artery disease increase per ten years of age; instead of exponentiating the parameter estimate 0.0929, you compute $e^{10 \times 0.0929}$ to obtain 2.53. Thus, the odds of coronary artery disease increase by a factor of 2.53 every ten years.

Output 8.49 displays this estimate as produced by the UNITS statement in PROC LOGISTIC.

Output 8.49 Odds Ratios for Units of 10

Odds Ratios		
Effect	Unit	Estimate
age	10.0000	2.531

However, note that this model is useful for prediction only for persons in the walk-in population who fall into the age range of those in this study—ages 28 to 60.

8.6 A Note on Diagnostics

While goodness-of-fit statistics can tell you how well a particular model fits the data, they tell you little about the lack of fit, or where a particular model fails to fit the data. Measures called regression diagnostics have long been useful tools to assess lack of fit for linear regression models, and in the 1980s researchers proposed similar measures for the analysis of binary data. In particular, work by Pregibon (1981) provided the theoretical basis of extending diagnostics used in linear regression to logistic regression. Both Hosmer and Lemeshow (2000) and Collett (2003) include lengthy discussions on model-checking for logistic regression models; Collett includes many references for work in this area. Standard texts on regression analysis like Draper and Smith (1998) discuss model-checking strategies for linear regression; Cook and Weisberg (1982) discuss residual analysis and diagnostics extensively.

This section presents a basic description of a few diagnostic tools and an example of their application with the urinary tract data set. The Pearson and deviance chi-square tests are two measures that assess overall model fit. It makes some sense that by looking at the individual components of these statistics, which are functions of the observed group counts and their model-predicted values, you will gain insight into a model's lack of fit.

Suppose that you have s groups, $i = 1, \ldots, s$, and n_i total subjects for the ith group. If y_i is the number of events (success, yes) for the ith group, and $\hat{\theta}_i$ denotes the predicted probability of success for the ith group, then define the ith residual as

$$r_i = \frac{y_i - n_i \hat{\theta}_i}{\sqrt{n_i \hat{\theta}_i (1 - \hat{\theta}_i)}}$$

These residuals are known as Pearson residuals, since the sum of their squares is Q_P. They compare the differences between observed counts and their predicted values, scaled by the observed count's standard deviation. By examining the r_i, you can determine how well the model fits the individual groups. Often, the residual values are considered to be indicative of lack of fit if they exceed 2 in size.

Similarly, the deviance residual is a component of the deviance statistic. The deviance residual is written

$$d_i = \text{sgn}(y_i - \hat{y}_i) \left[2y_i \log\left(\frac{y_i}{\hat{y}_i}\right) + 2(n_i - y_i) \log\left(\frac{n_i - y_i}{n_i - \hat{y}_i}\right) \right]^{\frac{1}{2}}$$

where $\hat{y}_i = n_i \hat{\theta}_i$. The sum of squares of the d_i values is the deviance statistic.

The Pearson and deviance residuals can be standardized to have approximately unit variances. Another type of residual is the likelihood residual, which is a weighted combination of the standardized Pearson and deviance residuals. Refer to Collett (2003) and the *SAS/STAT User's Guide* for details. The standardized deviance residuals and the likelihood residuals are recommended as they rank extreme observations well and are reasonably well approximated by a standard normal distribution when the numbers in each group are large enough (Collett 2003).

Residuals are often presented in tabular form; however, graphical display aids their inspection. One simple plot is called an *index plot*, in which the residuals are plotted against the corresponding observation number, the index. By examining these plots, you can determine if there are unusually large residuals, possibly indicative of outliers, or systematic patterns of variation, possibly indicative of a poor model choice.

Residuals are examined for the urinary tract data. As you will recall, the main effects model was considered to have an adequate fit. The INFLUENCE option requests that PROC LOGISTIC provide regression diagnostics.

The data are input differently than they were in Section 8.4. The variable RESPONSE is now the number of cures in a group, and the variable TRIALS is the total number of patients in that group, the sum of those who were cured and those who were not. The *events/trials* MODEL statement syntax allows you to specify the response as a ratio of two variables, the *events* variable and the *trials* variable. When the response is specified this way, developed to support the binomial trials framework, the residuals are calculated using an n_i that is based on the group size, which is desired. (If you specify a single response, called *actual model* syntax, when you compute residuals, the residuals are calculated using a group size of 1.)

```
data uti2;
    input diagnosis : $13. treatment $ response trials;
datalines;
complicated      A   78   106
complicated      B   101  112
complicated      C   68   114
uncomplicated    A   40   45
uncomplicated    B   54   59
uncomplicated    C   34   40
;
```

The following statements request the standardized residual plots with the STDRED suboption of the PLOTS=INFLUENCE option in the PROC LOGISTIC statement. The LABEL option displays the case number for diagnostic plots, and the UNPACK option suppresses the paneling of plots (the default is to combine multiple plots in a panel display).

```
ods graphics on;
proc logistic data=uti2 plots(label)=influence(unpack stdres);
    class diagnosis treatment / param=ref;
    model response/trials = diagnosis treatment;
run;
ods graphics off;
```

The graph in Output 8.50 displays the likelihood residuals; the residuals are in the acceptance region.

Output 8.50 Likelihood Residuals Plot

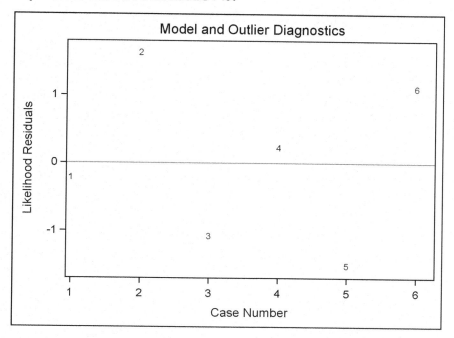

The standardized deviance residuals plot (not displayed here) is very similar.

Note that if the model didn't fit—the goodness-of-fit statistics were inadequate, possibly resulting from a missing effect—then you would likely see residual values higher than 2 for at least one group.

Residuals need to be used cautiously when the data contain continuous explanatory variables so that most of the group sizes are 1. This is for the same reason that Q_P and the deviance are inappropriate—the sample size requirements for approximate chi-square distributions are not met. However, these residuals are often considered useful as a rough indicator of model fit in this situation, and they are often examined.

Note that the residual score statistic can address sums of residuals in regions of interest via indicator variables with the value 1 for records in the region of interest and 0 elsewhere. Many such indicators can have assessment for diagnostic purposes. Also, the use of interactions with the score statistic partly performs such an assessment.

Other types of diagnostics include changes in the Q_P and deviance when the ith observation is excluded; the ith leverage; and distances between estimated parameters and the estimated parameters when the ith observation is excluded (or fit perfectly by addition of an indicator variable with the value 1 for the ith observation and the value 0 otherwise). In addition, there are a variety of graphs that have been devised to assist in evaluating model adequacy. Refer to the *SAS/STAT User's Guide* for information on what diagnostics are provided by the LOGISTIC procedure.

8.7 Alternatives to Maximum Likelihood Estimation

If you perform enough logistic regressions, you will encounter data for which maximum likelihood estimation does not produce a unique solution for the parameters; you do not obtain convergence. In addition, for data with small cell counts, large sample theory may not be applicable and thus tests based on the asymptotic normality of the maximum likelihood estimates may be unreliable. This section describes strategies for these situations, including exact logistic regression and the Firth bias reduction method.

To gain insight into the possible data configurations that result in non-convergence, consider Table 8.11:

Table 8.11 Infinite Odds Ratio Example

Factor	Response=Yes	Response=No
1	15	0
2	0	34

Computing the odds ratio for these data results in the quantity

$$\frac{a \times d}{b \times c} = \frac{15 \times 34}{0 \times 0}$$

which is infinite. Since the odds ratio is e^β, where β is the parameter for the factor, this means that β is infinite. You could request the exact odds ratio from the FREQ procedure, which would produce the 95% lower and upper confidence limits of 87.1520 and ∞, respectively.

The LOGISTIC procedure determines whether the input data have a configuration that leads to infinite parameter estimates. If convergence is not attained within eight iterations, PROC LOGISTIC computes the probability of allocating each observation to the correct response group. If this probability is equal to 1 for all observations, there is said to be *complete separation* of data points (this occurs if all the observations having unique covariate profiles have the same response outcome—for example, all the Factor=1 subjects responded yes, and all the Factor=2 subjects responded no). If complete separation is found, the iterative process is halted and a warning message is printed.

If nearly all the observations have a probability of 1 of being allocated to the correct response group, then the data configuration may be one of *quasi-complete separation*. (For quasi-complete separation to occur, the dispersion matrix also becomes unbounded.) Iteration also stops when this condition is detected, and a warning message is printed, since the parameter estimates are also infinite.

If neither of these conditions exists for the data, then they are said to be *overlapping*. The data points overlap, so observations with the same covariate profile have all possible responses. Maximum likelihood estimates exist and are unique for overlapping configurations. The problems of complete separation and quasi-complete separation generally occur for small data sets. Usually quasi-complete separation does not occur if you have a continuous explanatory variable; complete separation can always occur. Refer to Albert and Anderson (1984) for more information about infinite parameters and the data configurations that produce them; refer to Silvapulle (1981) for a discussion of the necessary and sufficient conditions for the existence of maximum likelihood estimators in binomial response models.

As an illustration of data that exhibit quasi-complete separation, consider Table 8.12.

Table 8.12 Pre-Clinical Study Data

Treatment A	Treatment B	No	Yes
0	0	0	0
0	1	2	0
1	0	0	8
1	1	6	21

The following statements creates SAS data set PRECLINICAL.

```
data preclinical;
    input treatA $ treatB $ response $ count @@;
datalines;
no   no   yes 0   no   no   no 0
yes  no   yes 8   yes  no   no 0
no   yes  yes 0   no   yes  no 2
yes  yes  yes 21  yes  yes  no 6
;
```

Logistic regression is requested with the LOGISTIC procedure:

```
proc logistic descending;
   freq count;
   class treatA treatB /param=ref reference=first;
   model response = treatA treatB;
run;
```

Output 8.51 contains the results from PROC LOGISTIC. Since there is quasi-complete separation, the maximum likelihood solution may not exist.

Output 8.51 Quasi-Complete Separation Note

Response Profile		
Ordered Value	response	Total Frequency
1	yes	29
2	no	8

Probability modeled is response='yes'.

Model Convergence Status
Quasi-complete separation of data points detected.

The LOGISTIC procedure proceeds, and it prints results based on the last maximum likelihood iteration (as well as a warning). If you examine the statistics listed in the "Global Tests" table in Output 8.52, you will see that they vary widely. The Wald chi-square has a value of 0.0089, and the Likelihood Ratio chi-square takes the value 10.0294. These statistics should be very similar when sample sizes are large enough for model fitting with maximum likelihood. When they are not similar, it is a red flag, regardless of whether the procedure also prints out a warning message.

Output 8.52 Global Tests

Testing Global Null Hypothesis: BETA=0			
Test	Chi-Square	DF	Pr > ChiSq
Likelihood Ratio	10.0294	2	0.0066
Score	9.4626	2	0.0088
Wald	0.0089	2	0.9956

Strategies do exist for situations where maximum likelihood estimation fails or small cell counts make the resulting maximum likelihood estimates fail to have their intended properties. Exact methods provide a way to compute parameter estimates, confidence intervals, and p-values for statistical tests using methodology based on exact distributions. The key is conditioning on the appropriate sufficient statistic.

The idea of computing parameter estimates, confidence intervals, and p-values for statistical tests using methodology based on exact distributions is not a new one. Cox (1970) suggested it decades ago, but it took some time for algorithmic advances in computing the exact distributions to make the strategy computationally practical. Refer to Tritchler (1984) and Hirji, Mehta, and Patel (1987) for more details regarding these algorithms. See Appendix C in Chapter 10 for a brief overview of the methodology involved in deriving exact conditional distributions and computing tests and point estimates.

SAS provides exact logistic regression with the LOGISTIC procedure (exact methods are also available in the GENMOD procedure). It provides an exact probability test and an exact score test for the hypotheses that parameters for the specified hypothesis are equal to zero; these tests produce an exact p-value, which is the probability of obtaining a more extreme statistic than the one observed and a mid p-value, which adjusts for the discreteness of the distribution. Simultaneous tests can be specified. You can also request the point estimates of the parameters and the exponentiated parameter estimates (usually representing an odds ratio); these come with one or two-sided confidence limits and one- or two-sided p-values for testing that the parameter estimate is zero.

Another alternative strategy is Firth's penalized likelihood method (Firth 1993, Heinze and Schemper 2002). This is a bias reduction method that adds a term to the usual log-likelihood function; when the resulting penalized likelihood is maximized, it shrinks the estimates towards zero. Firth's method is especially useful when you are dealing with continuous explanatory variables and exact methods may not be applicable. It always produces parameter estimates when the issue is complete or quasi-complete separation (Heinze and Schemper 2002).

8.7.1 Analyzing the Pre-Clinical Study Data

The data in Table 8.12 are analyzed with exact logistic regression.

The EXACT statement requests the exact analysis. Note that if you include the EXACTONLY option in the PROC statement, only the exact analysis is performed. You can include more than one EXACT statement, and exact analyses are performed for the variables listed in the statements. The tests are conditioned on any other variables included in the model. The JOINT option requests a joint test of the effect variables listed in the EXACT statement. The option ESTIMATE=BOTH specifies that point estimates for both the parameter and the exponentiated parameter be computed. Exponentiated parameters are computed for CLASS variables only if PARAM=REF is specified in the CLASS statement.

```
proc logistic descending exactonly;
   freq count;
   class treatA treatB /param=ref reference=first;
   model response = treatA treatB / alpha=.025;
   exact treatA treatB /joint estimate=both;
run;
```

Output 8.53 contains the exact test results. Both the exact conditional score test and the probability test are reported; in this instance, they have the same p-values. This will not always be the case. Since the exact probabilities are analogous to the exact p-values you obtain in a Fisher's exact test, where the p-value represents the sum of the more extreme table p-values, this may be preferred.

Output 8.53 Exact Tests

Exact Conditional Analysis

			\multicolumn{2}{c	}{p-Value}
Effect	Test	Statistic	Exact	Mid
Joint	Score	9.2069	0.0181	0.0143
	Probability	0.00767	0.0181	0.0143
treatA	Score	5.4444	0.0690	0.0345
	Probability	0.0690	0.0690	0.0345
treatB	Score	2.0843	0.2994	0.2082
	Probability	0.1824	0.2994	0.2082

For the treatment A effect, the exact p-value is 0.0690, and for the treatment B effect, the exact p-value is 0.2994. Again, these tests are conditioned on the other effects in the model. Note that, if an effect consists of two or more parameters, then this test is evaluating the hypothesis that all of the relevant parameters are equal to zero simultaneously. The joint effect of TreatA and TreatB has a p-value of 0.0181.

From these results, you could conclude that treatment A was nearly influential.

Since the ESTIMATE option was specified, PROC LOGISTIC also produces a table of parameter estimates. Output 8.54 displays the parameter estimates, and Output 8.55 displays the odds ratio estimates and their confidence limits. You would view any estimates with some caution.

Output 8.54 Exact Parameter Estimates

Exact Parameter Estimates							
Parameter		Estimate		Standard Error	97.5% Confidence Limits		Two-sided p-Value
treatA	yes	1.9547	*	.	-0.6498	Infinity	0.1379
treatB	yes	-1.0434	*	.	-Infinity	1.0473	0.3647

Note: * indicates a median unbiased estimate.

Output 8.55 Exact Odds Ratio Estimates

Exact Odds Ratios						
Parameter		Estimate		97.5% Confidence Limits		Two-sided p-Value
treatA	yes	7.062	*	0.522	Infinity	0.1379
treatB	yes	0.352	*	0	2.850	0.3647

Note: * indicates a median unbiased estimate.

The median unbiased estimates are produced instead of the maximum conditional likelihood statistic because the value of the observed sufficient statistic is at the extreme of the derived distribution. Note that the one-sided 97.5% confidence interval with a lower bound of 0 is also the two-sided 95% confidence interval.

8.7.2 Analysis of Completely Separated Data

Sometimes, exact analysis does not produce a solution when the data are completely separated. Table 8.13 displays data with the same number of zero counts as Table 8.12; however, the position of the zero counts results in complete separation of the data points with logistic regression.

Table 8.13 Completely Separated Data

Gender	Region	Yes	No
0	0	0	5
0	1	1	0
1	0	0	175
1	1	53	0

The following DATA step creates the SAS data set COMPLETE.

Analysis of Completely Separated Data

```
data complete;
   input gender region count response @@;
datalines;
0 0 0  1   0 0 5    0
0 1 1  1   0 1 0    0
1 0 0  1   1 0 175  0
1 1 53 1   1 1 0    0
;
```

If you perform the usual logistic regression analysis for these data, you will get a message that the data are completely separated and that the maximum likelihood solution does not exist.

The following statements request both an exact analysis for these data as well as one using the Firth penalized likelihood method. This is requested with the FIRTH option, and the CLPARM=PL option is also specified. This option should always be used with the FIRTH option since the profile likelihood-based confidence limits will be based on the penalized likelihood.

```
proc logistic data=complete descending;
   freq count;
   model response = gender region / firth clparm=pl;
   exact gender region;
run;
```

The exact results displayed in Output 8.56 are non-conclusive because the computations run into a degenerate distribution.

Output 8.56 Exact Tests

Exact Conditional Analysis

				p-Value	
Effect	Test	Statistic		Exact	Mid
gender	Score	.	#	.	.
	Probability	1.0000	#	1.0000	0.5000
region	Score	232.0		<.0001	<.0001
	Probability	5.44E-54		<.0001	<.0001

Note: # indicates that the conditional distribution is degenerate.

The Firth method does produce estimates, however, as displayed in Output 8.57.

Output 8.57 Penalized Parameter Estimates

Analysis of Penalized Maximum Likelihood Estimates					
Parameter	DF	Estimate	Standard Error	Wald Chi-Square	Pr > ChiSq
Intercept	1	-2.4001	1.6189	2.1978	0.1382
gender	1	-3.4599	2.1523	2.5843	0.1079
region	1	10.5320	2.0164	27.2817	<.0001

The penalized likelihood estimates and the profile likelihood confidence limits are displayed in Output 8.58.

Output 8.58 Penalized Parameter Estimates

Parameter Estimates and Profile-Likelihood Confidence Intervals		
Parameter	Estimate	95% Confidence Limits
Intercept	-2.4001	. -0.2218
gender	-3.4599	-8.7265 .
region	10.5320	7.5460 16.2653

These estimates should be used cautiously. However, the confidence interval for regions does convey the impression that region is an important effect. One way to evaluate the parameter estimates is to collapse the two tables in Table 8.13 into one 2×2 table and add 0.5 to each of the counts. The collapsing over gender is justified since gender appears to have no effect.

Table 8.14 Collapsed Over Gender Table

Region	Yes	No
0	0.5	180.5
1	54.5	0.5

If you compute the odds ratio for Table 8.14, you obtain $(0.5)(0.5)/(54.5)(180.5) = 0.00003$, which is about the same as the exponentiated parameter for region, or $e^{-10.5320}$. Thus, these estimates appear to be reasonable.

8.7.3 Analysis of Liver Function Data

Consider the data in Table 8.15 from a study on liver function outcomes for high-risk overdose patients in which antidote and historical control groups are compared. The data are stratified by time to hospital admission (Koch, Gillings, and Stokes 1980).

Table 8.15 Liver Function Outcomes

Time to Hospital	Antidote		Control	
	Severe	Not Severe	Severe	Not Severe
Early	6	12	6	2
Delayed	3	4	3	0
Late	5	1	6	0

The small counts in many cells—seven of the twelve cells have values less than 5—make the applicability of large sample theory somewhat questionable.

The following DATA step inputs the data.

```
data liver;
    input time $ group $ status $ count @@;
datalines;
early     antidote  severe  6 early    antidote not 12
early     control   severe  6 early    control  not  2
delayed   antidote  severe  3 delayed  antidote not  4
delayed   control   severe  3 delayed  control  not  0
late      antidote  severe  5 late     antidote not  1
late      control   severe  6 late     control  not  0
;
```

The following PROC LOGISTIC statements request a logistic regression analysis of the severity of the outcome with explanatory variables based on time to admission and treatment group. The early level for TIME is the reference level, and the control level is the reference level for GROUP. The PARAM=REF option requests incremental effects parameterization.

```
proc logistic descending;
    freq count;
    class time(ref='early') group(ref='control') /param=ref;
    model status = time group / clparm=wald;
run;
```

Output 8.59 contains the global fit statistics. Note the discrepancy between the Wald and the likelihood ratio test. The p-value for the former is more than ten times the p-value for the latter.

Output 8.59 Global Fit Statistics

Testing Global Null Hypothesis: BETA=0			
Test	Chi-Square	DF	Pr > ChiSq
Likelihood Ratio	16.3913	3	0.0009
Score	13.4256	3	0.0038
Wald	10.2488	3	0.0166

The parameter estimates are displayed in Output 8.60 for comparison's sake.

Output 8.60 MLE Estimates

Analysis of Maximum Likelihood Estimates						
Parameter		DF	Estimate	Standard Error	Wald Chi-Square	Pr > ChiSq
Intercept		1	1.4132	0.7970	3.1439	0.0762
time	delayed	1	0.7024	0.8344	0.7087	0.3999
time	late	1	2.5533	1.1667	4.7893	0.0286
group	antidote	1	-2.2170	0.8799	6.3480	0.0118

Odds ratios and their 95% confidence limits are displayed in Output 8.61.

Output 8.61 Odds Ratio Estimates

Odds Ratio Estimates		
Effect	Point Estimate	95% Wald Confidence Limits
time delayed vs early	2.019	0.393 10.359
time late vs early	12.849	1.305 126.471
group antidote vs control	0.109	0.019 0.611

You would not report the maximum likelihood results because of sample size concerns.

The following statements request an exact analysis. The option ESTIMATE=BOTH in the first EXACT statement specifies that point estimates for both the parameter and the exponentiated parameter be computed. The JOINTONLY option in the second EXACT statement requests a joint test for variables TIME and GROUP (and just the joint test).

```
proc logistic descending exactonly;
   freq count;
   class time(ref='early') group(ref='control') /param=ref;
   model status = time group / clparm=wald;
   exact 'Model 1' intercept time group /
       estimate=both;
   exact 'Joint Test' time group / jointonly;
run;
```

Output 8.62 contains the exact test results for the first EXACT statement.

Output 8.62 Exact Tests

Exact Conditional Analysis

Exact Conditional Tests for Model 1				
			p-Value	
Effect	Test	Statistic	Exact	Mid
Intercept	Score	3.4724	0.1150	0.0922
	Probability	0.0457	0.1150	0.0922
time	Score	6.0734	0.0442	0.0418
	Probability	0.00471	0.0442	0.0418
group	Score	7.1656	0.0085	0.0050
	Probability	0.00698	0.0085	0.0050

For the time effect, the exact p-value is 0.0442, and for the group effect, the exact p-value is 0.0085. Again, these tests are conditioned on the other effects in the model. If an effect consists of two or more parameters, then this test is evaluating the hypothesis that all of the relevant parameters are equal to zero simultaneously.

Output 8.63 contains the exact test results for the second EXACT statement.

Output 8.63 Exact Tests

Exact Conditional Tests for Joint Test				
			p-Value	
Effect	Test	Statistic	Exact	Mid
Joint	Score	13.1459	0.0027	0.0027
	Probability	0.000015	0.0015	0.0015

For the 'Joint' results, the score test produces an exact p-value of 0.0027, and the probability test produces an exact p-value of 0.0015.

Output 8.64 displays the parameter estimates and their 95% confidence limits. The parameter estimates are fairly similar to those based on the large sample approximate methods. The exact p-values reported for the effect parameters have different values from those reported for the exact conditional tests. This is because the exact p-values for the single parameters are twice the one-sided p-value, and they are constructed to be in harmony with the confidence intervals. For example, if the confidence interval contained the value zero, the p-value would be greater than 0.05 for a 95% confidence interval.

Output 8.64 Exact Parameter Estimates

Exact Parameter Estimates for Model 1						
Parameter		Estimate	Standard Error	95% Confidence Limits		Two-sided p-Value
Intercept		1.3695	0.7903	-0.2361	3.6386	0.1140
time	delayed	0.6675	0.8141	-1.2071	2.6444	0.6667
time	late	2.4388	1.1425	0.1364	6.4078	0.0331
group	antidote	-2.0992	0.8590	-4.5225	-0.3121	0.0154

Output 8.65 displays the odds ratio estimates and their 95% confidence limits.

Output 8.65 Exact Odds Ratio Estimates

Exact Odds Ratios for Model 1					
Parameter		Estimate	95% Confidence Limits		Two-sided p-Value
Intercept		3.934	0.790	38.037	0.1140
time	delayed	1.949	0.299	14.075	0.6667
time	late	11.460	1.146	606.546	0.0331
group	antidote	0.123	0.011	0.732	0.0154

Table 8.16 provides a comparison of the unconditional maximum likelihood estimates and the exact conditional estimates.

Table 8.16 Exact and Asymptotic Estimates

Variable	Inference Type	Estimate	Lower 95% CI Bound	Upper 95% CI Bound	p-value
Intercept	Asymptotic	1.4132	−0.1489	2.9754	0.0762
	Exact	1.3695	−0.2361	3.6386	0.1140
Delayed	Asymptotic	0.7024	−0.9330	2.3378	0.3999
	Exact	0.6675	−1.2071	2.6444	0.6667
Late	Asymptotic	2.5535	0.2666	4.8404	0.0286
	Exact	2.4387	0.1364	6.4078	0.0331
Antidote	Asymptotic	−2.2171	−3.9418	−0.4924	0.0118
	Exact	−2.0992	−4.5225	−0.3121	0.0154

For the exact computations performed with PROC LOGISTIC, the p-value listed is twice the one-sided p-value. For these data, you can see that exact logistic regression produces estimates that are different, although not substantially, from the maximum likelihood estimates. For each parameter, the p-values listed for the exact estimates are larger than those for the asymptotic estimates, and the confidence intervals are wider. Usually, the exact methods lead to more accurate results than the approximate methods. As a general rule, when the sample sizes are small and the approximate p-values are less than 0.10, it is a good idea to look at the exact results. If the approximate p-values are larger than 0.15, then the approximate methods are probably satisfactory in the sense that the exact results are likely to agree with them.

As an additional comparison, Firth's method is also applied to these data using the FIRTH option in PROC LOGISTIC (statements not shown here). The resulting parameter estimates and the 95% penalized likelihood confidence limits are displayed in Output 8.66.

Output 8.66 Penalized ML Estimates

Parameter Estimates and Profile-Likelihood Confidence Intervals				
Parameter		Estimate	95% Confidence Limits	
Intercept		1.2077	-0.0769	2.8718
time	delayed	0.6374	-0.9007	2.2523
time	late	2.1543	0.4031	4.5421
group	antidote	-1.9526	-3.7557	-0.5053

These parameter estimates are similar to those reported in Table 8.16; they are closer to the null than the exact parameter estimates, and the confidence intervals are narrower than those for the exact parameter estimates. In general, exact methods and, in particular, the exact conditional tests are recommended for small sample situations; however, the Firth penalized likelihood approach is a useful alternative, especially when the exact methods are computationally not feasible.

8.7.4 Exact Confidence Limits for Common Odds Ratios for 2 × 2 Tables

Section 3.4 in Chapter 3, "Sets of 2 × 2 Tables," describes how to use PROC FREQ to compute exact confidence limits for the average odds ratio in a set of 2 × 2 tables. You can also obtain them with exact logistic regression. You formulate the analysis as a regression where the column variable is the response variable and the row and stratification variables are the explanatory variables. Then, you condition on the stratification variable and estimate the odds ratio for the row variable. This odds ratio will be an average odds ratio.

Recall the exercise program data in Section 3.4, redisplayed in Table 8.17. The exercise counselor was interested in whether type of program was associated with good test results.

Table 8.17 Cardiovascular Test Outcomes

Location	Program	Good	Not Good	Total
Downtown	Office	12	5	17
Downtown	Home	3	5	8
	Total	15	11	26
Satellite	Office	6	1	7
Satellite	Home	1	3	4
	Total	7	4	11

Interest lies in computing an odds ratio comparing good results for the office program to good results for the home program. The sample sizes in these tables are too small to be able to justify asymptotic confidence limits for the odds ratio, so Section 3.4 demonstrated how to obtain exact confidence limits using the FREQ procedure. You can also obtain estimates of odds ratios and their exact confidence limits by performing an exact logistic regression.

The following DATA step inputs these data into SAS data set EXERCISE.

```
data exercise;
   input location $ program $ outcome $ count @@;
   datalines;
downtown office   good 12 downtown office not 5
downtown home     good 3 downtown  home    not 5
satellite office  good 6 satellite office  not 1
satellite home    good 1 satellite home    not 3
;
```

To perform the exact logistic regression, you put both LOCATION and PROGRAM in the CLASS statement and, in order to compare office to home, use the REF=FIRST option with the PROGRAM variable. The response variable is OUTCOME in the MODEL statement; since 'good' outcome is the first alphanumerically ordered value (the other is 'not'), the model is based on the probability of good outcome.

You then specify the EXACT statement, requesting exact tests for the variable PROGRAM and also specifying the ESTIMATE=BOTH option to obtain both the parameter estimate and the odds ratio estimate.

```
proc logistic;
   freq count;
   class location program(ref=first) /param=ref;
   model outcome = location program;
   exact program / estimate=both;
run;
```

Output 8.67 displays the results of the exact tests for exercise program. Both the score and probability tests have an exact p-value of 0.0307, indicating significance at the $\alpha = 0.05$ level.

Output 8.67 Exact Test Results

Exact Conditional Analysis

Exact Conditional Tests				
Effect	Test	Statistic	p-Value	
			Exact	Mid
program	Score	5.5739	0.0307	0.0215
	Probability	0.0183	0.0307	0.0215

Output 8.68 displays the exact parameter estimate and the exact odds ratio estimate comparing office program to home program; the odds ratio estimate takes the value 5.413 with 95% confidence limits of (1.049, 33.312). The confidence limits are identical to those displayed in Chapter 3. This estimate of the odds ratio is different from the estimate of 5.842 reported in the PROC FREQ output because that was the Mantel-Haenszel estimator.

Output 8.68 Exact Estimates

Exact Parameter Estimates						
Parameter		Estimate	Standard Error	95% Confidence Limits		Two-sided p-Value
program	office	1.6889	0.7435	0.0474	3.5059	0.0424

Exact Odds Ratios					
Parameter		Estimate	95% Confidence Limits		Two-sided p-Value
program	office	5.413	1.049	33.312	0.0424

Compare the exact results to those produced by the asymptotic analysis, which are displayed in Output 8.69.

Output 8.69 Odds Ratio for Asymptotic Analysis

Odds Ratio Estimates			
Effect	Point Estimate	95% Wald Confidence Limits	
location downtown vs satellit	0.758	0.151	3.803
program office vs home	6.111	1.331	28.062

The point estimate here is 6.111 and the 95% Wald confidence limits are (1.331, 28.062). Thus, using the exact method provides a more accurate picture than the inappropriate asymptotic method.

8.8 Using the GENMOD Procedure for Logistic Regression

The GENMOD procedure fits generalized linear models, and it performs logistic regression for correlated responses via the generalized estimating equations method (discussed in Chapter 15). Binary logistic regression is a simple case of the generalized linear model, and this section provides an introduction to using the GENMOD procedure for performing logistic regression.

8.8.1 Performing Logistic Regression with the GENMOD Procedure

Generalized linear models are a generalization of the general linear model that is fit by the GLM procedure. Generalized linear models include not only classical linear models but logistic and probit models for binary data, and Poisson regression and negative binomial regression models for count data. You can also fit loglinear models for multinomial data indirectly through computational equivalences with fitting Poisson regression models. You can generate many other statistical models by the appropriate selection of a *link function* and the probability distribution of the response.

A generalized linear model has three components:

- a random sample of independent response variable $\{y_i\}$ with some probability distribution, $i = 1, 2, \ldots, n$
- a set of explanatory variables $\mathbf{x}_i$ and parameter vector $\boldsymbol{\beta}$
- a monotonic link function g that describes how the expected value of y_i, μ_i, is related to $\mathbf{x}_i'\boldsymbol{\beta}$:

$$g(\mu_i) = \mathbf{x}_i'\boldsymbol{\beta}$$

You construct a generalized linear model by choosing the appropriate link function and response probability distribution. In the classical linear model, the probability distribution is the normal and the usual link function is the identity: $g(\mu) = \mu$. For logistic regression, the distribution is the binomial and the usual link function is the logit:

$$g(\mu) = \log\left(\frac{\mu}{1-\mu}\right)$$

For Poisson regression, the distribution is Poisson and the link function is $g(\mu) = \log(\mu)$.

The following section describes how to perform logistic regression using PROC GENMOD. See Chapter 12, "Poisson Regression and Related Loglinear Regression," for a discussion of Poisson regression and illustrations using the GENMOD procedure. For a comprehensive discussion of the generalized linear model, refer to McCullagh and Nelder (1989). For an introduction to the topic, refer to Dobson (1990) or Agresti (2007).

8.8.2 Fitting Logistic Regression Models with PROC GENMOD

Fitting logistic regression models with the GENMOD procedure is a relatively straightforward matter. PROC GENMOD includes a CLASS statement, so you simply list your classification variables in it, just as you now do in PROC LOGISTIC. The default parameterization is equivalent to the incremental effects parameterization implemented for most of the analyses performed in this chapter. In PROC GENMOD, the reference cell is the combination of the last sorted levels of the effects listed in the CLASS statement (and 0s for any continuous explanatory variables). Incremental effects parameters are estimated for the remaining levels. You can use the REF= option in the CLASS statement to change the reference levels. The default parameterization is illustrated with this example; note that you can also specify PARAM=REF in the CLASS statement in PROC GENMOD to produce the identical PARAM=REF parameterization utilized in the previous examples with PROC LOGISTIC.

Consider the urinary tract infection data analyzed in Section 8.4. If you sorted the values of TREATMENT and DIAGNOSIS, those observations that had an uncomplicated diagnosis and treatment C would become the reference cell.

The following statements produce an analysis using PROC GENMOD. You need to specify LINK=LOGIT and DIST=BINOMIAL to request logistic regression with PROC GENMOD. The TYPE3 option requests tests of effects for the model, and the AGGREGATE option generates the Pearson and deviance goodness-of-fit statistics.

```
proc genmod data=uti;
   freq count;
   class diagnosis treatment;
   model response = diagnosis treatment /
         link=logit dist=binomial type3 aggregate;
run;
```

Goodness-of-fit statistics are displayed in Output 8.70.

Output 8.70 Goodness of Fit

Criteria For Assessing Goodness Of Fit			
Criterion	DF	Value	Value/DF
Deviance	2	2.5147	1.2573
Scaled Deviance	2	2.5147	1.2573
Pearson Chi-Square	2	2.7574	1.3787
Scaled Pearson X2	2	2.7574	1.3787
Log Likelihood		-225.0355	
Full Log Likelihood		-13.4690	
AIC (smaller is better)		34.9379	
AICC (smaller is better)		35.0228	
BIC (smaller is better)		51.5996	

The table labeled "Criteria for Assessing Goodness of Fit" provides Q_P and Q_L, which take the value 2.5147 for 2 df. The Log Likelihood statistic has the value -225.0355. If you multiply this value by two and reverse the sign, you get the same value as -2LOG L displayed in the output for the same model in the PROC LOGISTIC output. Other criteria displayed are approximate chi-square statistics.

The estimates displayed in Output 8.71 are identical to those produced with PROC LOGISTIC for the same model. However, those levels that become the reference levels under incremental effects coding, uncomplicated diagnosis and treatment C, are assigned 0s for the parameter estimate and related statistics.

Output 8.71 Parameter Estimates

Analysis Of Maximum Likelihood Parameter Estimates							
Parameter		DF	Estimate	Standard Error	Wald 95% Confidence Limits		Wald Chi-Square
Intercept		1	1.4184	0.2987	0.8330	2.0038	22.55
diagnosis	complicated	1	-0.9616	0.2998	-1.5492	-0.3740	10.29
diagnosis	uncomplicated	0	0.0000	0.0000	0.0000	0.0000	.
treatment	A	1	0.5847	0.2641	0.0671	1.1024	4.90
treatment	B	1	1.5608	0.3160	0.9415	2.1800	24.40
treatment	C	0	0.0000	0.0000	0.0000	0.0000	.
Scale		0	1.0000	0.0000	1.0000	1.0000	

Analysis Of Maximum Likelihood Parameter Estimates		
Parameter		Pr > ChiSq
Intercept		<.0001
diagnosis	complicated	0.0013
diagnosis	uncomplicated	.
treatment	A	0.0268
treatment	B	<.0001
treatment	C	.
Scale		

Note: The scale parameter was held fixed.

The table labeled "LR Statistics For Type 3 Analysis" in Output 8.72 can be viewed as serving a similar role to that of an ANOVA table.

Output 8.72 Type 3 Analysis

LR Statistics For Type 3 Analysis			
Source	DF	Chi-Square	Pr > ChiSq
diagnosis	1	11.72	0.0006
treatment	2	28.11	<.0001

It includes likelihood ratio tests for each of the effects. The effect for treatment, which has three levels, has 2 df. The effect for diagnosis, with two levels, has 1 df. Both tests are clearly significant, with values of 28.11 and 11.72, respectively.

To assess whether any of the treatments are similar, linear combinations of the parameters are tested to see if they are significantly different from zero.

$$H_0: \mathbf{L}\boldsymbol{\beta} = \mathbf{0}$$

By choosing the right elements of **L**, you can construct linear combinations of the parameters that will produce the appropriate test. By default, PROC GENMOD computes a likelihood ratio test; on request, it can produce the corresponding Wald test. The likelihood ratio test for a contrast is twice the difference between the log likelihood of the current fitted model and the log likelihood of the model fitted under the constraint that the linear function of the parameters defined by the contrast is equal to zero.

The test for whether treatment A is equivalent to treatment B is expressed as

$$H_0: \beta_A = \beta_B$$

and the test for whether treatment A is equivalent to treatment C is expressed as

$$H_0: \beta_A = \beta_C$$

You request these tests with the CONTRAST statement in PROC GENMOD. The following CONTRAST statement is required to produce the first test. You place an identifying name for the test in quotes, name the effect variable, and then list the appropriate coefficients for **L**. These coefficients are listed according to the order in which the levels of the variable are known to PROC GENMOD. When you use a CONTRAST statement, or specify the ITPRINT, COVB, CORRB, WALDCI, or LRCI options in the MODEL statement, the GENMOD output includes information on what levels of effects the parameters represent.

The CONTRAST statement is very similar to the CONTRAST statement in PROC GLM, where you have to supply a coefficient for each level of the effect.

```
contrast 'A-B' treat 1 -1  0;
```

The following SAS statements produces the tests of interest.

```
proc genmod data=uti;
   freq count;
   class diagnosis treatment;
   model response = diagnosis treatment /
      link=logit dist=binomial;
   contrast 'treatment' treatment 1 0 -1 ,
                        treatment 0 1 -1;
   contrast 'A-B' treatment 1 -1  0;
   contrast 'A-C' treatment 1  0 -1;
run;
```

Output 8.73 contains the information about what the parameters represent.

Output 8.73 Parameter Information

Parameter Information			
Parameter	Effect	diagnosis	treatment
Prm1	Intercept		
Prm2	diagnosis	complicated	
Prm3	diagnosis	uncomplicated	
Prm4	treatment		A
Prm5	treatment		B
Prm6	treatment		C

Output 8.74 contains the results of the hypothesis tests.

Output 8.74 Contrasts

Contrast Results				
Contrast	DF	Chi-Square	Pr > ChiSq	Type
treatment	2	28.11	<.0001	LR
A-B	1	9.22	0.0024	LR
A-C	1	4.99	0.0255	LR

$Q_L = 9.22$ for the test of whether treatment A and treatment B are the same; $Q_L = 4.99$ for the test of whether treatment A and treatment C are the same; both of these are clearly significant at the $\alpha = 0.05$ level of significance. Note that these tests are similar to those displayed in the analysis performed in Section 8.4. If you execute these same statements using the WALD option, you will obtain identical results to the Wald tests obtained from PROC LOGISTIC in Section 8.4.

Appendix A: Statistical Methodology for Dichotomous Logistic Regression

Consider the relationship of a dichotomous outcome variable to a set of explanatory variables. Such situations can arise from clinical trials where the explanatory variables are treatment, stratification variables, and background covariables; another common source of such analyses are observational studies where the explanatory variables represent factors for evaluation and background variables.

The model for θ, the probability of the event, can be specified as follows:

$$\theta = \frac{\exp(\alpha + \sum_{k=1}^{t} \beta_k x_k)}{1 + \exp(\alpha + \sum_{k=1}^{t} \beta_k x_k)}$$

It follows that the odds are written

$$\frac{\theta}{1-\theta} = \exp(\alpha + \sum_{k=1}^{t} \beta_k x_k)$$

so the model for the logit is linear:

$$\log\left\{\frac{\theta}{1-\theta}\right\} = \alpha + \sum_{k=1}^{t} \beta_k x_k$$

The $\exp(\beta_k)$ are the odds ratios for unit changes in x_k, that is, the amount by which $\theta/(1-\theta)$ is multiplied per unit change in x_k.

You can apply the product binomial distribution when the data for the dichotomous outcome are from a sampling process equivalent to stratified simple random sampling from subpopulations according to the explanatory variables. Relative to this structure, the maximum likelihood estimates are obtained by iteratively solving the equations:

$$\sum_{i=1}^{s} n_{i+} \hat{\theta}_i (1, x_{i1}, \ldots, x_{it}) = \sum_{i=1}^{s} n_{i1} (1, x_{i1}, \ldots, x_{it})$$

where n_{i1} is the number of subjects who have the event corresponding to θ among n_i subjects with $(x_{i1}, \ldots, x_{it})$ status.

The quantity

$$\hat{\theta}_i = \frac{\exp\{\hat{\alpha} + \sum_{k=1}^{t} \hat{\beta}_k x_{ik}\}}{1 + \exp\{\hat{\alpha} + \sum_{k=1}^{t} \hat{\beta}_k x_{ik}\}}$$

is the model-predicted value for θ_i.

For sufficient sample size, the quantities $\hat{\alpha}$ and $\hat{\beta}_k$ have approximate multivariate normal distributions for which a consistent estimate of the covariance structure is available.

You can assess goodness of fit of the model with Pearson chi-square statistics when sample sizes are sufficiently large (80% of the $\{n_{i1}\}$ and the $\{n_i - n_{i1}\}$ are ≥ 5 and all others are ≥ 2).

$$Q_P = \sum_{i=1}^{s} \frac{(n_{i1} - n_{i+}\hat{\theta}_i)^2}{n_{i+}\hat{\theta}_i(1-\hat{\theta}_i)}$$

is approximately chi-square with $(s - 1 - t)$ degrees of freedom.

You can also use log-likelihood ratio statistics to evaluate goodness of fit by evaluating the need for a model to include additional explanatory variables.

In the setting where you have continuous explanatory variables, you cannot use Q_P to assess goodness of fit because you no longer have sufficient sample sizes n_{i+}. However, you can still apply the strategy of fitting an expanded model and then verifying that the effects not in the original model are nonsignificant. If the model matrix for the original model $\mathbf{X}$ has rank t, then the expanded model $[\mathbf{X}, \mathbf{W}]$ has rank $t + w$, where w is the rank of $\mathbf{W}$. You can evaluate the significance of $\mathbf{W}$ with the difference of the log-likelihood statistics for the models $\mathbf{X}$ and $[\mathbf{X}, \mathbf{W}]$.

$$Q_{LR} = \sum_{i=1}^{s}\sum_{j=1}^{2} 2n_{ij}\log\left(\frac{m_{ij,w}}{m_{ij}}\right)$$

where s is the total number of groups with at least one subject, m_{ij} is the predicted value of n_{ij} for model $\mathbf{X}$ ($m_{i1} = n_i\hat{\theta}_i$ and $m_{i2} = n_i(1-\hat{\theta}_i)$), and $m_{ij,w}$ is the predicted value of n_{ij} for model $[\mathbf{X}, \mathbf{W}]$. Q_{LR} has an approximate chi-square distribution with w degrees of freedom.

Another approach that doesn't involve fitting an expanded model is the score statistic for assessing the association of the residuals $(\mathbf{n}_{*1} - \mathbf{m}_{*1})$ with $\mathbf{W}$ via the linear functions $\mathbf{g} = \mathbf{W}'(\mathbf{n}_{*1} - \mathbf{m}_{*1})$. The score statistic is written

$$Q_S = \mathbf{g}'\{\mathbf{W}'[\mathbf{D}_\mathbf{v}^{-1} - \mathbf{D}_\mathbf{v}^{-1}\mathbf{X}_\mathbf{A}(\mathbf{X}_\mathbf{A}'\mathbf{D}_\mathbf{v}^{-1}\mathbf{X}_\mathbf{A})^{-1}\mathbf{X}_\mathbf{A}'\mathbf{D}_\mathbf{v}^{-1}]\mathbf{W}\}^{-1}\mathbf{g}$$

where $\mathbf{n}_{*1} = (n_{11}, n_{21}, \ldots, n_{s1})'$, $\mathbf{X}_\mathbf{A} = [\mathbf{1}, \mathbf{X}]$, $\mathbf{m}_{*1} = (m_{11}, m_{21}, \ldots, m_{s1})'$, and $\mathbf{D}_\mathbf{v}$ is a diagonal matrix with diagonal elements $v_i = [n_{i+}\hat{\theta}_i(1-\hat{\theta}_i)]^{-1}$. Q_S approximately has a chi-square distribution with w degrees of freedom when the total sample size is large enough to support an approximately multivariate normal distribution for the linear functions $[\mathbf{X}_\mathbf{A}', \mathbf{W}']\mathbf{n}_{*1}$.

Chapter 9
Logistic Regression II: Polytomous Response

Contents

9.1	Introduction	**259**
9.2	Ordinal Response: Proportional Odds Model	**260**
	9.2.1 Methodology	260
	9.2.2 Fitting the Proportional Odds Model with PROC LOGISTIC	263
	9.2.3 Multiple Qualitative Explanatory Variables	269
	9.2.4 Partial Proportional Odds Model	274
9.3	Nominal Response: Generalized Logits Model	**280**
	9.3.1 Methodology	280
	9.3.2 Fitting Models to Generalized Logits with PROC LOGISTIC	282
	9.3.3 Generalized Logit Model with Continuous Explanatory Variable	288
	9.3.4 Exact Methods for Generalized Logits Model	292

9.1 Introduction

Logistic regression most often involves modeling a dichotomous outcome, but it also applies to multilevel responses. The response might be ordinal (no pain, slight pain, substantial pain) or nominal (Democrats, Republicans, Independents). For ordinal response outcomes, you can model functions called *cumulative logits* by performing ordered logistic regression using the proportional odds model (McCullagh 1980). For nominal response outcomes, you form *generalized logits* and perform a logistic analysis similar to those described in the previous chapter, except that you model multiple logits per subpopulation. The analysis of generalized logits is a form of the loglinear model, discussed in Chapter 12, "Poisson Regression and Related Loglinear Models." The LOGISTIC procedure is used to model both cumulative logits and generalized logits.

9.2 Ordinal Response: Proportional Odds Model

9.2.1 Methodology

Consider the arthritis pain data in Table 9.1. Male and female subjects received an active or placebo treatment for their arthritis pain, and the subsequent extent of improvement was recorded as marked, some, or none (Koch and Edwards 1988).

Table 9.1 Arthritis Data

Sex	Treatment	Improvement			Total
		Marked	Some	None	
Female	Active	16	5	6	27
Female	Placebo	6	7	19	32
Male	Active	5	2	7	14
Male	Placebo	1	0	10	11

One possible analysis strategy is to create a dichotomous response variable by combining two of the response categories, basing a model on either Pr{marked improvement} versus Pr{some or no improvement} or Pr{marked or some improvement} versus Pr{no improvement}. However, since there is a natural ordering to these response levels, it makes sense to consider a strategy that takes advantage of this ordering.

Consider the quantities

$$\theta_{hi1} = \pi_{hi1}, \quad \theta_{hi2} = \pi_{hi1} + \pi_{hi2}$$

where π_{hi1} denotes the probability of marked improvement, π_{hi2} denotes the probability of some improvement, and π_{hi3} denotes the probability of no improvement. The $\{\theta_{hij}\}$ represent cumulative probabilities: θ_{hi1} is the probability of marked improvement, and θ_{hi2} is the probability of marked or some improvement ($h = 1$ for females, $h = 2$ for males; $i = 1$ for active treatment, $i = 2$ for placebo).

For a dichotomous response, you compute a logit function for each subpopulation. For a multilevel response, you create more than one logit function for each subpopulation. With ordinal data, you can compute *cumulative logits*, which are based on the cumulative probabilities. For three response levels, you compute two cumulative logits:

$$\text{logit}(\theta_{hi1}) = \log\left[\frac{\pi_{hi1}}{\pi_{hi2} + \pi_{hi3}}\right], \quad \text{logit}(\theta_{hi2}) = \log\left[\frac{\pi_{hi1} + \pi_{hi2}}{\pi_{hi3}}\right]$$

These cumulative logits are the log odds of marked improvement to none or some improvement and the log odds of marked or some improvement to no improvement, respectively. Both log odds focus on more favorable to less favorable response. The proportional odds model takes both of these odds into account.

Assuming that the data arise from a stratified simple random sample or are at least conceptually representative of a stratified population, they have the following likelihood:

$$\Pr\{n_{hij}\} = \prod_{h=1}^{2}\prod_{i=1}^{2} n_{hi+}! \prod_{j=1}^{3} \frac{\pi_{hij}^{n_{hij}}}{n_{hij}!}$$

where

$$\sum_{j=1}^{3} \pi_{hij} = 1$$

You could write a model that applies to both logits simultaneously for each combination of gender and treatment:

$$\text{logit}(\theta_{hik}) = \alpha_k + \mathbf{x}'_{hi}\boldsymbol{\beta}_k$$

where k indexes the two logits. This says that there are separate intercept parameters (α_k) and different sets of regression parameters ($\boldsymbol{\beta}_k$) for each logit.

If you take the difference in logits between two subpopulations for this model, you get

$$\text{logit}(\theta_{hik}) - \text{logit}(\theta_{hi'k}) = (\mathbf{x}_{hi} - \mathbf{x}_{hi'})'\boldsymbol{\beta}_k \text{ for } k = 1, 2$$

Thus, you would need to look at two differences in logits simultaneously to compare the response between two subpopulations. This is the same number of comparisons you would need to compare two subpopulations for a three-level nominal response, for example, in a test for association in a contingency table (that is, $r - 1$, where r is the number of response outcomes). Therefore, this general model doesn't take advantage of the ordinality of the data.

The proportional odds assumption is that $\boldsymbol{\beta}_k = \boldsymbol{\beta}$ for all k, simplifying the model to

$$\text{logit}(\theta_{hik}) = \alpha_k + \mathbf{x}'_{hi}\boldsymbol{\beta}$$

If you take the difference in logits for this model, you obtain the equations

$$\text{logit}(\theta_{hi1}) - \text{logit}(\theta_{hi'1}) = \log\left[\frac{\pi_{hi1}/(\pi_{hi2} + \pi_{hi3})}{\pi_{hi'1}/(\pi_{hi'2} + \pi_{hi'3})}\right] = (\mathbf{x}_{hi} - \mathbf{x}_{hi'})'\boldsymbol{\beta}$$

$$\text{logit}(\theta_{hi2}) - \text{logit}(\theta_{hi'2}) = \log\left[\frac{(\pi_{hi1} + \pi_{hi2})/\pi_{hi3}}{(\pi_{hi'1} + \pi_{hi'2})/\pi_{hi'3}}\right] = (\mathbf{x}_{hi} - \mathbf{x}_{hi'})'\boldsymbol{\beta}$$

This says that the log cumulative odds are proportional to the distance between the explanatory variable values and that the influence of the explanatory variables is independent of the cutpoint for the cumulative logit. In this case, there is a "cut" at marked improvement to form $\text{logit}(\theta_{hi1})$ and a cut at some improvement to form $\text{logit}(\theta_{hi2})$. This proportionality is what gives the proportional odds model its name. For a single continuous explanatory variable, the regression lines would be parallel to each other, their relative position determined by the values of the intercept parameter.

This model can also be stated as

$$\theta_{hik} = \frac{\exp(\alpha_k + \mathbf{x}'_{hi}\boldsymbol{\beta})}{1 + \exp(\alpha_k + \mathbf{x}'_{hi}\boldsymbol{\beta})}$$

and is written in summation notation as

$$\theta_{hik} = \frac{\exp\{\alpha_k + \sum_{g=1}^{t} \beta_g x_{hig}\}}{1 + \exp\{\alpha_k + \sum_{g=1}^{t} \beta_g x_{hig}\}}$$

where $g = (1, 2, \ldots, t)$ references the explanatory variables. This model is similar to the previous logistic regression models and is also fit with maximum likelihood methods. You can determine the values for π_{hij} from this model by performing the appropriate subtractions of the θ_{hik}.

$$\pi_{hi1} = \theta_{hi1}$$
$$\pi_{hi2} = \theta_{hi2} - \theta_{hi1}$$
$$\pi_{hi3} = 1 - \theta_{hi2}$$

The main effects model is an appropriate starting point for the analysis of the arthritis data. You can write this model in matrix notation as

$$\begin{bmatrix} \text{logit}(\theta_{111}) \\ \text{logit}(\theta_{112}) \\ \text{logit}(\theta_{121}) \\ \text{logit}(\theta_{122}) \\ \text{logit}(\theta_{211}) \\ \text{logit}(\theta_{212}) \\ \text{logit}(\theta_{221}) \\ \text{logit}(\theta_{222}) \end{bmatrix} = \begin{bmatrix} \alpha_1 & + \beta_1 + \beta_2 \\ & \alpha_2 + \beta_1 + \beta_2 \\ \alpha_1 & + \beta_1 \\ & \alpha_2 + \beta_1 \\ \alpha_1 & + \beta_2 \\ & \alpha_2 + \beta_2 \\ \alpha_1 & \\ & \alpha_2 \end{bmatrix} = \begin{bmatrix} 1 & 0 & 1 & 1 \\ 0 & 1 & 1 & 1 \\ 1 & 0 & 1 & 0 \\ 0 & 1 & 1 & 0 \\ 1 & 0 & 0 & 1 \\ 0 & 1 & 0 & 1 \\ 1 & 0 & 0 & 0 \\ 0 & 1 & 0 & 0 \end{bmatrix} \begin{bmatrix} \alpha_1 \\ \alpha_2 \\ \beta_1 \\ \beta_2 \end{bmatrix}$$

This is very similar to the models described in Chapter 8 except that there are two intercept parameters corresponding to the two cumulative logit functions being modeled for each group. The parameter α_1 is the intercept for the first cumulative logit, α_2 is the intercept for the second cumulative logit, β_1 is an incremental effect for females, and β_2 is an incremental effect for active treatment. Males on placebo constitute the reference cell.

Table 9.2 contains the cell probabilities for marked improvement and no improvement based on this model. Table 9.3 contains the odds. The cell probabilities for marked improvement are based on the model for the first logit function, and the probabilities for no improvement are based on the model for the second logit function (these probabilities are computed from $1 - \theta_{hi2}$). Since the probabilities for all three levels sum to 1, you can determine the cell probabilities for some improvement through subtraction.

The odds ratio for females versus males is e^{β_1}, and the odds ratio for active treatment versus placebo is e^{β_2}. The odds ratios are computed in the same manner as for the logistic regression analysis for a dichotomous response—you form the ratio of the appropriate odds.

Table 9.2 Formulas for Cell Probabilities

Sex	Treatment	Improvement	
		Marked	None
Female	Active	$e^{\alpha_1+\beta_1+\beta_2}/(1+e^{\alpha_1+\beta_1+\beta_2})$	$1/(1+e^{\alpha_2+\beta_1+\beta_2})$
Female	Placebo	$e^{\alpha_1+\beta_1}/(1+e^{\alpha_1+\beta_1})$	$1/(1+e^{\alpha_2+\beta_1})$
Male	Active	$e^{\alpha_1+\beta_2}/(1+e^{\alpha_1+\beta_2})$	$1/(1+e^{\alpha_2+\beta_2})$
Male	Placebo	$e^{\alpha_1}/(1+e^{\alpha_1})$	$1/(1+e^{\alpha_2})$

Table 9.3 Formulas for Model Odds

Sex	Treatment	Improvement	
		Marked Versus Some or None	Marked or Some Versus None
Female	Active	$e^{\alpha_1+\beta_1+\beta_2}$	$e^{\alpha_2+\beta_1+\beta_2}$
Female	Placebo	$e^{\alpha_1+\beta_1}$	$e^{\alpha_2+\beta_1}$
Male	Active	$e^{\alpha_1+\beta_2}$	$e^{\alpha_2+\beta_2}$
Male	Placebo	e^{α_1}	e^{α_2}

For example, when you compare the odds of marked improvement versus some or no improvement for active females versus active males, you obtain

$$\frac{e^{\alpha_1+\beta_1+\beta_2}}{e^{\alpha_1+\beta_2}} = e^{\beta_1}$$

As constrained by the proportional odds model, this is also the odds ratio for marked or some improvement versus no improvement.

9.2.2 Fitting the Proportional Odds Model with PROC LOGISTIC

PROC LOGISTIC fits the proportional odds model by default when the response variable has more than two levels. Thus, you need to ensure that you have an ordinal response variable because PROC LOGISTIC assumes that you do. The GENMOD, GLIMMIX, and PROBIT procedures also fit the proportional odds model with maximum likelihood estimation.

The following SAS statements create the data set ARTHRITIS. Note that these data are in the form of counts, so a variable named COUNT is created to contain the frequencies for each table cell. The variable IMPROVE is a character variable that takes the values marked, some, or none to indicate the subject's extent of improvement of arthritic pain. The variable SEX takes the values male and female, and the variable TREATMENT takes the values active and placebo.

```
data arthritis;
   input sex $ treatment $ improve $ count @@;
   datalines;
female active   marked 16 female active   some 5 female active   none  6
female placebo marked  6 female placebo some 7 female placebo none 19
male   active   marked  5 male   active   some 2 male   active   none  7
male   placebo marked  1 male   placebo some 0 male   placebo none 10
;
```

The use of PROC LOGISTIC is identical to previous invocations for dichotomous response logistic regression. The response variable is listed on the left-hand side of the equal sign and the explanatory variables are listed on the right-hand side. Since the ORDER=DATA option is specified in the PROC statement, the values for IMPROVE are ordered in the sequence in which PROC LOGISTIC encounters them in the data, which is marked, some, and none. (Another legitimate ordering would be none, some, and marked.) It is crucial to ensure that the ordering is correct when you are using ordinal data strategies. The procedure still performs an analysis if the response values are ordered incorrectly, but the results will be erroneous. The burden is on the user to specify the correct order and then to check the results.

The following statements requests that PROC LOGISTIC fit a proportional odds model.

```
proc logistic order=data;
   freq count;
   class treatment sex / param=reference;
   model improve = sex treatment / scale=none aggregate;
run;
```

The "Response Profile" table displayed in Output 9.1 shows that the response variable values are ordered correctly in terms of decreasing improvement. Thus, the cumulative logits modeled are based on more to less improvement. The procedure also prints out a note that a zero count observation has been encountered. For these data, this is not a problem since the total row counts are acceptably large. Computationally, zero counts are discarded. The model still produces predicted values for the cell that corresponds to the zero cell, males on placebo who showed some improvement.

Output 9.1 Response Profiles

Response Profile		
Ordered Value	improve	Total Frequency
1	marked	28
2	some	14
3	none	42

Probabilities modeled are cumulated over the lower Ordered Values.

The procedure next prints the "Class Level Information" table, which shows that the parameterization takes the form of incremental effects for active treatment and females.

Output 9.2 Class Levels

Class Level Information		
Class	Value	Design Variables
treatment	active	1
	placebo	0
sex	female	1
	male	0

Next, PROC LOGISTIC prints out a test for the appropriateness of the proportional odds assumption. The test performed is a score test that determines whether, if you fit a different set of explanatory variable parameters β_k for each logit function, those sets of parameters are equivalent. Thus, the model considered is

$$\text{logit}(\theta_{hik}) = \alpha_k + \mathbf{x}'_{hi}\boldsymbol{\beta}_k$$

The hypothesis tested is that there is a common parameter vector β instead of distinct β_k. The hypothesis can be stated as $\beta_k = \beta$ for all k. Thus, if you reject the null hypothesis, you reject the assumption of proportional odds and you need to consider a different approach. If the null hypothesis is not rejected, then the test supports the assumption of proportional odds. Since the test is comparing t parameters for the t explanatory variables across $(r-1)$ logits, where r is the number of response levels, it has $t * (r-2)$ degrees of freedom.

The sample size requirements for this test are moderately demanding; you need approximately five observations at each outcome at each level of each main effect, or roughly the same sample size as if you were fitting a generalized logit model. Small samples may artificially make the statistic large, meaning that any resulting significance needs to be interpreted cautiously. However, nonsignificant results are always informative.

The partial proportional odds model is an alternative model that can be fit when the proportionality assumption does not hold for all explanatory variables, but there is proportionality for some (Peterson and Harrell, 1990). Section 9.2.4 describes this approach See Koch, Amara, and Singer (1985) for another discussion of this model. When there appears to be no proportionality, you can fit the model with different parameters for each of the cumulative logits.

Output 9.3 displays the score test for the proportional odds assumption.

Output 9.3 Proportional Odds Test

Score Test for the Proportional Odds Assumption		
Chi-Square	DF	Pr > ChiSq
1.8833	2	0.3900

Q_{RS} takes the value 1.883 with 2 df. This is clearly nonsignificant, and so the assumption of proportional odds is a reasonable one for these data.

Output 9.4 contains the goodness-of-fit statistics. With values of 2.7121 and 1.9099, respectively, and 4 df, Q_L and Q_P support the adequacy of the model. The 4 df come from $(3-1)(4-1)-2 = 4$.

Output 9.4 Goodness-of-Fit Statistics

Deviance and Pearson Goodness-of-Fit Statistics				
Criterion	Value	DF	Value/DF	Pr > ChiSq
Deviance	2.7121	4	0.6780	0.6071
Pearson	1.9099	4	0.4775	0.7523

Number of unique profiles: 4

The tests for assessing model fit through explanatory capability are also supportive of the model; the likelihood ratio test has a value of 19.8865 with 2 df and the score test has a value of 17.8677 with 2 df, as displayed in Output 9.5.

Output 9.5 Global Tests

Testing Global Null Hypothesis: BETA=0			
Test	Chi-Square	DF	Pr > ChiSq
Likelihood Ratio	19.8865	2	<.0001
Score	17.8677	2	0.0001
Wald	16.7745	2	0.0002

You can also investigate goodness of fit by performing the score test for a set of additional terms not in the model. In this case, this effect would simply be the treatment × sex interaction. The following code requests that PROC LOGISTIC fit a main effects model and then perform a score test for the other effect listed in the MODEL statement, which is the interaction.

```
proc logistic order=data;
    freq count;
    class sex treatment / param=reference;
    model improve = sex treatment sex*treatment /
                    selection=forward start=2;
run;
```

The score test of interest is labeled "Residual Chi-Square" and is printed after the "Testing Global Null Hypothesis: BETA=0" table; it is displayed in Output 9.6. The value of the test statistic is 0.2801 (1 df since you are testing the addition of one term to the model) with $p = 0.5967$. This indicates that the main effects model is adequate.

Fitting the Proportional Odds Model with PROC LOGISTIC

Output 9.6 Score Statistic to Evaluate Goodness of Fit

Residual Chi-Square Test		
Chi-Square	DF	Pr > ChiSq
0.2801	1	0.5967

An alternative goodness-of-fit test is the difference in the likelihood ratios for the main effects model and the saturated model. Although the output is not displayed here, the difference in these statistics is $(150.029 - 149.721) = 0.308$. This is also clearly nonsignificant, compared to a chi-square distribution with 1 df. (Again, note that whenever you form a test statistic based on the difference in likelihoods, then the corresponding degrees of freedom are equal to the difference in the number of parameters for the two models.)

Output 9.7 contains the "Type 3 Analysis of Effects" table. Both sex and treatment are influential effects. Since these effects have 1 df each, the tests are the same as printed for the parameter estimates listed in Output 9.8.

Output 9.7 Type 3 Analysis of Effects

Type 3 Analysis of Effects			
Effect	DF	Wald Chi-Square	Pr > ChiSq
sex	1	6.2096	0.0127
treatment	1	14.4493	0.0001

Output 9.8 Parameter Estimates

Analysis of Maximum Likelihood Estimates						
Parameter		DF	Estimate	Standard Error	Wald Chi-Square	Pr > ChiSq
Intercept	marked	1	-2.6671	0.5997	19.7800	<.0001
Intercept	some	1	-1.8127	0.5566	10.6064	0.0011
sex	female	1	1.3187	0.5292	6.2096	0.0127
treatment	active	1	1.7973	0.4728	14.4493	0.0001

Table 9.4 displays the parameter interpretations.

Table 9.4 Parameter Estimates

Parameter	Estimate(SE)	Interpretation
α_1	−2.667(0.600)	log odds of marked improvement versus some or no improvement for males receiving placebo
α_2	−1.813(0.557)	log odds of marked or some improvement versus no improvement for males receiving placebo
β_1	1.319(0.529)	increment for both types of log odds due to female sex
β_2	1.797(0.473)	increment for both types of log odds due to active drug

Females have $e^{1.319} = 3.7$ times higher odds of showing improvement as males, both for marked improvement versus some or no improvement and for marked or some improvement versus no improvement. Those subjects receiving the active drug have $e^{1.8} = 6$ times higher odds of showing improvement as those on placebo, both for marked improvement versus some or no improvement and for some or marked improvement versus no improvement. These odds ratio estimates are displayed in Output 9.9.

Output 9.9 Odds Ratio Estimates

Odds Ratio Estimates		
Effect	Point Estimate	95% Wald Confidence Limits
sex female vs male	3.739	1.325 10.547
treatment active vs placebo	6.033	2.388 15.241

A graph of the predicted cumulative probabilities also provides a useful interpretation of the results of this analysis. The PLOTS=EFFECT option, with the POLYBAR and X=TREATMENT*SEX suboptions, specifies this plot from PROC LOGISTIC. The X= suboption specifies the cross-classication of the main effects for which you want to see predicted probabilities.

```
proc logistic order=data plots=effect(polybar x=treatment*sex);
   freq count;
   class sex treatment / param=reference;
   model improve = sex treatment sex*treatment /
                 selection=forward start=2;
run;
```

It's clear from the graph in Output 9.10 that the predicted probabilities of marked improvement and marked or some improvement, are highest for females and active treatment.

Output 9.10 Predicted Cumulative Probabilities

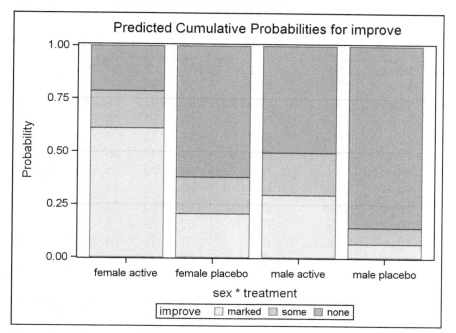

9.2.3 Multiple Qualitative Explanatory Variables

The inclusion of multiple explanatory variables in a proportional odds model produces no additional problems. The data in Table 9.5 are from an epidemiological study of chronic respiratory disease analyzed in Semenya and Koch (1980). Researchers collected information on subjects' exposure to general air pollution, exposure to pollution in their jobs, and whether they smoked. The response measured was chronic respiratory disease status. Subjects were assigned to one of four possible categories.

- Level I: no symptoms
- Level II: cough or phlegm less than three months a year
- Level III: cough or phlegm more than three months a year
- Level IV: cough and phlegm plus shortness of breath more than three months a year

Table 9.5 Chronic Respiratory Disease Data

Air Pollution	Job Exposure	Smoking Status	Response Level				Total
			I	II	III	IV	
Low	No	Non	158	9	5	0	172
Low	No	Ex	167	19	5	3	194
Low	No	Current	307	102	83	68	560
Low	Yes	Non	26	5	5	1	37
Low	Yes	Ex	38	12	4	4	58
Low	Yes	Current	94	48	46	60	248
High	No	Non	94	7	5	1	107
High	No	Ex	67	8	4	3	82
High	No	Current	184	65	33	36	318
High	Yes	Non	32	3	6	1	42
High	Yes	Ex	39	11	4	2	56
High	Yes	Current	77	48	39	51	215

The outcome is clearly ordinal, although there is no obvious distance between adjacent levels. You could combine response categories and fit the set of models that compared Level I versus Level II, III, and IV; Levels I and II versus Levels III and IV; and Levels I, II, and III versus Level IV. Note that if you did this, you would be computing models for the individual cumulative logits. The proportional odds model addresses these cumulative logits simultaneously by assuming that the slope parameters for the explanatory variables are the same regardless of the cumulative logit cutpoints.

From these data, you form three cumulative logits:

$$\text{logit}(\theta_{i1}) = \log\left[\frac{\pi_{i1}}{\pi_{i2} + \pi_{i3} + \pi_{i4}}\right]$$

$$\text{logit}(\theta_{i2}) = \log\left[\frac{\pi_{i1} + \pi_{i2}}{\pi_{i3} + \pi_{i4}}\right]$$

$$\text{logit}(\theta_{i3}) = \log\left[\frac{\pi_{i1} + \pi_{i2} + \pi_{i3}}{\pi_{i4}}\right]$$

where $i = 1, 2, \ldots, 12$ references the 12 populations determined by the levels of air pollution, job exposure, and smoking status, as ordered in Table 9.5. These cumulative logits are the log odds of a Level I response to a Level II, III, or IV response; the log odds of a Level I or II response to a Level III or IV response; and the log odds of a Level I, II, or III response to a Level IV response, respectively.

However, if you are more interested in the odds of more severe responses to less severe responses, you may want to order the cumulative logits in the opposite direction:

$$\text{logit}(\theta_{i1}) = \log\left[\frac{\pi_{i4}}{\pi_{i3} + \pi_{i2} + \pi_{i1}}\right]$$

$$\text{logit}(\theta_{i2}) = \log\left[\frac{\pi_{i4} + \pi_{i3}}{\pi_{i2} + \pi_{i1}}\right]$$

$$\text{logit}(\theta_{i3}) = \log\left[\frac{\pi_{i4} + \pi_{i3} + \pi_{i2}}{\pi_{i1}}\right]$$

You can generate this ordering in PROC LOGISTIC by using the DESCENDING option in the PROC statement, as shown in the following analysis.

The primary model of interest for these data is a main effects model. Besides three intercept terms α_1, α_2, and α_3 for the three cumulative logits, the main effects model includes the parameters β_1, β_2, β_3, and β_4 for incremental effects for air pollution exposure, job pollution exposure, ex-smoker status, and current smoking status, respectively.

The following SAS statements create the data set RESPIRE.

```
data respire;
    input air $ exposure $ smoking $ level count @@;
    datalines;
low  no  non  1 158  low  no  non  2   9
low  no  ex   1 167  low  no  ex   2  19
low  no  cur  1 307  low  no  cur  2 102
low  yes non  1  26  low  yes non  2   5
low  yes ex   1  38  low  yes ex   2  12
low  yes cur  1  94  low  yes cur  2  48
high no  non  1  94  high no  non  2   7
high no  ex   1  67  high no  ex   2   8
high no  cur  1 184  high no  cur  2  65
high yes non  1  32  high yes non  2   3
high yes ex   1  39  high yes ex   2  11
high yes cur  1  77  high yes cur  2  48
low  no  non  3   5  low  no  non  4   0
low  no  ex   3   5  low  no  ex   4   3
low  no  cur  3  83  low  no  cur  4  68
low  yes non  3   5  low  yes non  4   1
low  yes ex   3   4  low  yes ex   4   4
low  yes cur  3  46  low  yes cur  4  60
high no  non  3   5  high no  non  4   1
high no  ex   3   4  high no  ex   4   3
high no  cur  3  33  high no  cur  4  36
high yes non  3   6  high yes non  4   1
high yes ex   3   4  high yes ex   4   2
high yes cur  3  39  high yes cur  4  51
;
```

The following PROC LOGISTIC code requests a main effects proportional odds model. The MODEL statement generates a score statistic for the goodness of fit of the expanded model containing all pairwise interaction terms. The SCALE=NONE and AGGREGATE=(AIR EXPOSURE SMOKING) options request the goodness-of-fit tests based on the 12 subpopulations. The REF='no' option specified for the EXPOSURE variable in the CLASS statement causes no exposure to be the reference level.

```
proc logistic descending;
   freq count;
   class air exposure(ref='no') smoking / param=reference;
   model level = air exposure smoking
         air*exposure air*smoking exposure*smoking /
         selection=forward include=3 scale=none
         aggregate=(air exposure smoking);
run;
```

Output 9.11 shows the internal ordered values that PROC LOGISTIC uses. Since the response variable LEVEL has numeric values, the DESCENDING option causes PROC LOGISTIC to sort the values numerically, then reverses them to form the ordered values.

Output 9.11 Response Profile

Response Profile		
Ordered Value	level	Total Frequency
1	4	230
2	3	239
3	2	337
4	1	1283

Probabilities modeled are cumulated over the lower Ordered Values.

The score test for the proportional odds assumption takes the value $Q_{RS} = 12.0745$ ($p = 0.1479$) with 8 df ($4(4-2)$), as shown in Output 9.12. Thus, the proportional odds assumption is not contradicted.

Output 9.12 Test for Proportionality

Score Test for the Proportional Odds Assumption		
Chi-Square	DF	Pr > ChiSq
12.0745	8	0.1479

The three intercepts and four indicator variables representing the main effects are first entered into the model. The residual chi-square has a value of 2.7220 with 5 df and $p = 0.7428$, so this measure of goodness of fit suggests that the model-predicted cell proportions are acceptably close to the observed proportions.

Output 9.13 Assessment of Fit

Residual Chi-Square Test		
Chi-Square	DF	Pr > ChiSq
2.7220	5	0.7428

Output 9.14 displays the goodness-of-fit statistics. $Q_L = 29.9969$ and $Q_P = 28.0796$, both with $(r-1)(s-1) - t = 29$ df ($r = 4, s = 12, t = 4$). Model adequacy is again supported.

Output 9.14 Goodness-of-Fit Statistics

Deviance and Pearson Goodness-of-Fit Statistics				
Criterion	Value	DF	Value/DF	Pr > ChiSq
Deviance	29.9969	29	1.0344	0.4142
Pearson	28.0796	29	0.9683	0.5137

Number of unique profiles: 12

The "Type III Analysis of Effects" table displayed in Output 9.15 suggests a strong effect for job pollution exposure but no significant effect for outside air pollution ($p = 0.675$). The smoking effect is also highly significant.

Output 9.15 Type III Analysis of Effects

Type 3 Analysis of Effects			
Effect	DF	Wald Chi-Square	Pr > ChiSq
air	1	0.1758	0.6750
exposure	1	82.0603	<.0001
smoking	2	209.8507	<.0001

The parameter estimates are displayed in Output 9.16.

Output 9.16 Parameter Estimates

Analysis of Maximum Likelihood Estimates						
Parameter		DF	Estimate	Standard Error	Wald Chi-Square	Pr > ChiSq
Intercept	4	1	-3.8938	0.1779	479.2836	<.0001
Intercept	3	1	-2.9696	0.1693	307.7931	<.0001
Intercept	2	1	-2.0884	0.1633	163.5861	<.0001
air	high	1	-0.0393	0.0937	0.1758	0.6750
exposure	yes	1	0.8648	0.0955	82.0603	<.0001
smoking	cur	1	1.8527	0.1650	126.0383	<.0001
smoking	ex	1	0.4000	0.2019	3.9267	0.0475

The predicted odds ratios illustrate this model's conclusions. Persons with job exposure have $e^{0.8648} = 2.374$ times higher odds of having serious problems to less serious problems compared to persons not exposed on the job. Current smokers have $e^{1.8527} = 6.377$ times higher odds of having serious problems to less serious problems compared to nonsmokers. Both of these odds ratios have been adjusted for the other variables in the model.

Output 9.17 Odds Ratios

Odds Ratio Estimates		
Effect	Point Estimate	95% Wald Confidence Limits
air high vs low	0.961	0.800 1.155
exposure yes vs no	2.374	1.969 2.863
smoking cur vs non	6.377	4.615 8.812
smoking ex vs non	1.492	1.004 2.216

Note that if you fit the same model without reversing the order of the cumulative logits with the DESCENDING option, you fit an equivalent model. The intercepts will be in the opposite order and have opposite signs; that is, INTERCEP3 will have the value of this model's INTERCEP1 with the opposite sign. The parameters for the effects will have opposite signs, and the odds ratios will be inverted since they would represent the odds of less serious response to more serious response.

9.2.4 Partial Proportional Odds Model

Table 9.6 contains data from a study by a bicycling clothing manufacturer who wanted to assess their test glove. It was designed to combat the hand problems experienced by cyclists dealing with carpal tunnel syndrome. Cyclists wore either the company's standard gel glove or the new test glove for a week's worth of their standard rides. Then they reported whether they experienced major, moderate, or no relief from their usual symptoms (numbness and nerve pain) on the bike.

Table 9.6 Cycling Glove Data

Glove Type	Gender	Relief From Symptoms			Total
		Major	Moderate	None	
Test	Female	12	8	5	25
Test	Male	8	14	15	37
Gel	Female	5	5	9	19
Gel	Male	8	4	20	32

The data layout is similar to that for the arthritis data in Table 9.1: the response is ordinal in nature, and both explanatory variables have two levels. The proportional odds model is a reasonable one to consider.

The following DATA step inputs the cycling glove data.

```
data wrist;
   input glove $ gender $ relief $ count @@;
   datalines;
test  female  major     12  test female   moderate 8  test female   none  5
test  male major         8  test male     moderate 14 test male none 15
gel   female  major      5  gel female    moderate 5  gel female    none  9
gel   male major         8  gel male      moderate 4  gel male none 20
;
```

First, a saturated model is fit (not shown here). The interaction term was nonsignificant, so the following main effects model is fit.

```
proc logistic order=data;
   freq count;
   class glove gender / param=reference order=data;
   model relief= glove gender / scale=none aggregate;
run;
```

Output 9.18 displays the response profiles, listed from major to none, so the two cumulative logits modeled compare major symptom relief to moderate or no relief and major or moderate relief compared to no relief.

Output 9.18 Response Profile

Response Profile		
Ordered Value	relief	Total Frequency
1	major	33
2	moderate	31
3	none	49

Probabilities modeled are cumulated over the lower Ordered Values.

Output 9.19 shows the score test for the proportional odds assumption; it has the value 4.1734, 2 df, for $p = 0.1241$. Strictly speaking, that p-value would not lead you to reject the hypothesis of proportionality at the $\alpha = 0.05$ level, but it can be considered a marginal result. The partial proportional odds model may be appropriate in this situation if the parameters corresponding to one of the explanatory variables are consistent with proportional odds assumption but the parameters corresponding to the other explanatory variable are not consistent with the assumption.

Output 9.19 Proportional Odds Test

Score Test for the Proportional Odds Assumption		
Chi-Square	DF	Pr > ChiSq
4.1734	2	0.1241

If neither set of parameters is consistent with proportional odds, then you could fit a model for the cumulative logits with different parameters for each cumulative logit as previously discussed:

$$\text{logit}(\theta_{hik}) = \alpha_k + \mathbf{x}'_{hi}\boldsymbol{\beta}_k$$

where k indexes the two logits. This says that there are separate intercept parameters (α_k) and different sets of regression parameters ($\boldsymbol{\beta}_k$) for each cumulative logit.

But if proportionality holds for one set of coefficients, you could write the model as

$$\text{logit}(\theta_{hik}) = \alpha_k + \mathbf{x}'_{hi1}\boldsymbol{\beta}_k + \mathbf{x}'_{hi2}\boldsymbol{\gamma}$$

where $\mathbf{x}_{hi1}$ represents the explanatory variable with unequal slopes, $\mathbf{x}_{hi2}$ represents the explanatory variable with equal slopes, $\boldsymbol{\beta}_k$ represents the regression parameters for the $\mathbf{x}_{hi1}$, and $\boldsymbol{\gamma}$ represents the parameters for the $\mathbf{x}_{hi2}$.

The LOGISTIC procedure fits the general cumulative logits model as well as the partial proportional odds model. One way to determine whether you have partial proportional odds is to fit the general model and then perform contrast tests to see whether an effect's parameters are the same.

The following statements perform this analysis for the cycling glove data. The UNEQUALSLOPES option in the MODEL statement specifies the general model. Since two cumulative logits are being modeled, the model includes two intercept parameters, two parameters for the gender effect, and two parameters for the glove effect—each parameter corresponds to one of the two cumulative logits. The first TEST statement produces a contrast test to assess the equality of the two parameters for gender, and the second TEST statement produces a contrast test to assess the equality of the glove parameters. (You can determine the internal SAS names of the parameter effects by creating an OUTEST= SAS data set in the PROC LOGISTIC statement and printing it—not shown here).

```
proc logistic order=data;
   freq count;
```

```
    class glove gender / param=reference order=data;
    model relief = glove gender / link=clogit
                  scale=none aggregate unequalslopes;
    pogender: test genderfemale_major=genderfemale_moderate;
    poglove:  test glovetest_major=glovetest_moderate;
run;
```

Output 9.20 displays the goodness-of-fit statistics for this model, which are adequate.

Output 9.20 Goodness of Fit

Deviance and Pearson Goodness-of-Fit Statistics			
Criterion	Value	DF	Value/DF	Pr > ChiSq
Deviance	1.8487	2	0.9243	0.3968
Pearson	1.8365	2	0.9183	0.3992

Number of unique profiles: 4

Output 9.21 shows the results of the Type 3 tests. Both gender and glove type appear to be important factors.

Output 9.21 Type 3 Tests

Type 3 Analysis of Effects			
Effect	DF	Wald Chi-Square	Pr > ChiSq
glove	2	8.1812	0.0167
gender	2	5.1230	0.0772

The parameter estimates for the general model are displayed in Output 9.22. The parameter estimates for the gender effect appear to be similar for both the cumulative logit comparing major relief to moderate or no relief and the cumulative logit comparing any relief (major and moderate relief) compared to no relief.

Output 9.22 Parameter Estimates

Analysis of Maximum Likelihood Estimates						
Parameter		relief	DF	Estimate	Standard Error	Wald Chi-Square	Pr > ChiSq
Intercept		major	1	-1.3704	0.3799	13.0128	0.0003
Intercept		moderate	1	-0.6480	0.3362	3.7155	0.0539
glove	test	major	1	0.3209	0.4261	0.5672	0.4514
glove	test	moderate	1	1.0855	0.4022	7.2838	0.0070
gender	female	major	1	0.7189	0.4209	2.9177	0.0876
gender	female	moderate	1	0.9067	0.4143	4.7905	0.0286

Output 9.23 contains the results of formal tests of equal slopes for gender and glove type, considered separately. The *p*-value of 0.6307 for the gender effect indicates that the equal slopes assumption is viable, and the *p*-value of 0.0474 for glove type indicates that the equal slopes assumption is not viable.

Output 9.23 Proportional Odds Test

Linear Hypotheses Testing Results			
Label	Wald Chi-Square	DF	Pr > ChiSq
pogender	0.2311	1	0.6307
poglove	3.9326	1	0.0474

The following PROC LOGISTIC statements request a partial proportional odds model, where gender is handled with a single slope for both cumulative logits modeled and glove type is handled with a different parameter for each cumulative logit. This model is specified with the UNEQUALSLOPES=GLOVE option in the MODEL statement.

```
proc logistic order=data;
   freq count;
   class glove gender / param=reference order=data;
   model relief= glove gender / scale=none aggregate unequalslopes=glove;
run;
```

Output 9.24 provides the goodness of fit for the partial proportional odds model. The *p*-values of 0.3531 and 0.3500 for Q_L and Q_P, respectively, indicate a reasonable fit. The degrees of freedom are $(r-1)(s-1) - t$.

Output 9.24 Goodness of Fit

Deviance and Pearson Goodness-of-Fit Statistics				
Criterion	Value	DF	Value/DF	Pr > ChiSq
Deviance	2.0819	3	0.6940	0.5556
Pearson	2.0994	3	0.6998	0.5520

Number of unique profiles: 4

Output 9.25 displays the Type 3 Analysis of Effects results. Both gender and type of glove remain influential effects; the effect for gender has 1 df since that effect assumes the proportional odds assumption, and the effect for glove has 2 df since proportionality is not assumed (unequal slopes).

Output 9.25 Type 3 Tests

Type 3 Analysis of Effects			
Effect	DF	Wald Chi-Square	Pr > ChiSq
glove	2	7.9611	0.0187
gender	1	4.9395	0.0263

Output 9.26 shows the parameter estimates for the partial proportional odds model.

Output 9.26 Parameter Estimates

Analysis of Maximum Likelihood Estimates							
Parameter		relief	DF	Estimate	Standard Error	Wald Chi-Square	Pr > ChiSq
Intercept		major	1	-1.4269	0.3669	15.1215	0.0001
Intercept		moderate	1	-0.6123	0.3284	3.4773	0.0622
glove	test	major	1	0.3278	0.4269	0.5895	0.4426
glove	test	moderate	1	1.0672	0.3997	7.1292	0.0076
gender	female		1	0.8180	0.3680	4.9395	0.0263

Output 9.27 displays the odds ratios.

Output 9.27 Odds Ratios

Odds Ratio Estimates				
Effect	relief	Point Estimate	95% Wald Confidence Limits	
glove test vs gel	major	1.388	0.601	3.205
glove test vs gel	moderate	2.907	1.328	6.364
gender female vs male		2.266	1.101	4.661

There is only one odds ratio listed for gender since that effect was handled with just one parameter. Females reported 2.266 times higher odds of relief compared to males, for both major relief compared to moderate or no relief and for any relief compared to no relief. However, note that the Wald confidence interval barely excludes the value 1.

Two odds ratios are reported for glove type. When major relief is compared to moderate or no relief, people with the test glove had 1.33 times higher odds of reporting relief than people with the gel glove; however, this is not a significant result when you consider its 95% Wald confidence interval (0.601, 3.205). When any relief is compared to no relief, people with the test glove had 2.907 times higher odds of reporting relief than people with the gel glove.

To summarize this partial proportional odds model, gender was an influential effect—females reported relief more often than males—which needed to be accounted for in the model and this behavior held up for both cumulative logits. Test glove performed better than the gel glove, but that performance depended on the cumulative logit that was considered.

Peterson and Harrell (1990) described an unconstrained model and a constrained model, in which various constraints, such as linearity, are imposed on the parameters for an effect. The analysis described in this section is what is called an unconstrained model, where proportional odds can be imposed for some effects but not others, and that is the analysis provided by the UNEQUALSLOPES option. (The constrained model may be available in future releases of the LOGISTIC procedure.) See Koch, Amara, and Singer (1985) for other examples of data suitable for partial proportional odds analysis.

9.3 Nominal Response: Generalized Logits Model

When you have nominal response variables, you can also use logistic regression to model your data. Instead of fitting a model to cumulative logits, you fit a model to generalized logits. Table 9.7 redisplays the data analyzed in Section 6.3.3. Recall that schoolchildren in experimental learning settings were surveyed to determine which learning style they preferred. Investigators were interested in whether response was associated with school and their program, which was either a standard school day or also included after-hours care.

Table 9.7 School Program Data

School	Program	Learning Style Preference		
		Self	Team	Class
1	Regular	10	17	26
1	After	5	12	50
2	Regular	21	17	26
2	After	16	12	36
3	Regular	15	15	16
3	After	12	12	20

Since the levels of the response variable (self, team, and class) have no inherent ordering, the proportional odds model is not an appropriate mechanism for their analysis. You could form logits comparing self to team or class, or self or team to class, but that collapses the original structure of the response levels, which you might want to keep in your analysis. You can model a nominal response variable with more than two levels by performing a logistic analysis on the generalized logits.

9.3.1 Methodology

The generalized logit is defined as

$$\text{logit}_{hij} = \log\left[\frac{\pi_{hij}}{\pi_{hir}}\right]$$

for $j = 1, 2, \ldots, (r-1)$. A logit is formed for the probability of each succeeding category over the last response category.

Thus, the generalized logits for a three-level response like that displayed in Table 9.7 is

$$\text{logit}_{hi1} = \log\left[\frac{\pi_{hi1}}{\pi_{hi3}}\right], \quad \text{logit}_{hi2} = \log\left[\frac{\pi_{hi2}}{\pi_{hi3}}\right]$$

for $h = 1, 2, 3$ for the schools, $i = 1$ for regular program, and $i = 2$ for afterschool program.

The model you fit for generalized logits is the model discussed in Section 9.2.1.

$$\text{logit}_{hij} = \alpha_j + \mathbf{x}'_{hi}\boldsymbol{\beta}_j$$

where j indexes the two logits. This says that there are separate intercept parameters (α_j) and different sets of regression parameters ($\boldsymbol{\beta}_j$) for each logit. The matrix $\mathbf{x}_{hi}$ is the set of explanatory variable values for the hith group. Instead of estimating one set of parameters for one logit function, as in logistic regression for a dichotomous response variable, you are estimating sets of parameters for multiple logit functions. Whereas for the proportional odds model you estimated multiple intercept parameters for the cumulative logit functions but only one set of parameters corresponding to the explanatory variables, for the generalized logits model you are estimating multiple sets of parameters for both the intercept terms and the explanatory variables.

This model can also be stated as

$$\pi_{hij} = \frac{\exp(\alpha_j + \mathbf{x}'_{hi}\boldsymbol{\beta}_j)}{\sum_{j'=1}^{r} \exp(\alpha_{j'} + \mathbf{x}'_{hi}\boldsymbol{\beta}_{j'})}$$

where $\alpha_r = 0$ and $\boldsymbol{\beta}_r = 0$; and it is written in summation notation as

$$\pi_{hik} = \frac{\exp\{\alpha_j + \sum_{g=1}^{t} \beta_{jg} x_{hig}\}}{\sum_{j'=1}^{r} \exp\{\alpha_{j'} + \sum_{g=1}^{t} \beta_{j'g} x_{hig}\}}$$

where $g = (1, 2, \ldots, t)$ references the explanatory variables, $\alpha_r = 0$, and $\beta_{rg} = 0$ for all g. For the generalized logits model, however, there is a different $\boldsymbol{\beta}_j$ for each logit.

This poses no particular problems. Since there are multiple response functions being modeled per subpopulation, there are more degrees of freedom associated with each effect. Since the model matrix needs to account for multiple response functions, it takes a more complicated form. However, the modeling proceeds as usual; you fit your specified model, examine goodness-of-fit statistics, and possibly perform model reduction. Note that since you are predicting more than one response function per subpopulation, the sample size needs to be large enough to support the number of functions you are modeling. Sometimes, in those situations where there aren't enough data to justify the analysis of generalized logits, you will also encounter problems with parameter estimation and the software will print out notes about infinite parameters. In those situations, you can often simplify the response structure to a reasonable dichotomy and proceed with a binary logistic regression.

9.3.2 Fitting Models to Generalized Logits with PROC LOGISTIC

You fit the generalized logits model with the LOGISTIC procedure. The following SAS statements request the desired analysis. First, the data set SCHOOL is created, and then the LOGISTIC procedure is invoked. PROC LOGISTIC constructs two generalized logits per group from the levels of the variable STYLE; it creates six groups based on the unique values of the explanatory variables, SCHOOL and PROGRAM.

```
data school;
    input school program $ style $ count @@;
    datalines;
1 regular  self 10  1 regular  team 17  1 regular  class  26
1 after    self  5  1 after    team 12  1 after    class  50
2 regular  self 21  2 regular  team 17  2 regular  class  26
2 after    self 16  2 after    team 12  2 after    class  36
3 regular  self 15  3 regular  team 15  3 regular  class  16
3 after    self 12  3 after    team 12  3 after    class  20
;
```

The following PROC LOGISTIC statements perform this analysis. Incremental effects parameterization is requested. The LINK=GLOGIT option specifies that generalized logits are the response functions.

```
proc logistic order=data;
    freq count;
    class school (ref=first) program (ref=last) / param=ref;
    model style= program school school*program / link=glogit
        scale=none aggregate;
run;
```

Output 9.28 contains the response profiles. While the analysis does not require the response values to be ordered in any particular way, unlike the proportional odds model analyses, it is often useful to order the levels in a manner that facilitates interpretation. Since the ORDER=DATA option was specified in the PROC statement, the response variable levels are in the order self, team, and class. This means that generalized logits are formed for the probability of self with respect to class, and for the probability of team with respect to class.

Fitting Models to Generalized Logits with PROC LOGISTIC

Output 9.28 Response Profiles

Response Profile		
Ordered Value	style	Total Frequency
1	self	79
2	team	85
3	class	174

Logits modeled use style='class' as the reference category.

Output 9.29 displays the class level information. The reference logits are the ones corresponding to School 1 and the regular program.

Output 9.29 Class Level Information

Class Level Information			
Class	Value	Design Variables	
school	1	0	0
	2	1	0
	3	0	1
program	after	1	
	regular	0	

Since the model is saturated, with as many response functions being modeled as there are groups or subpopulations, the likelihood ratio test does not apply and PROC LOGISTIC prints missing values and 0 df, as seen in Output 9.30.

Output 9.30 Goodness Of Fit

Deviance and Pearson Goodness-of-Fit Statistics				
Criterion	Value	DF	Value/DF	Pr > ChiSq
Deviance	0.0000	0	.	.
Pearson	0.0000	0	.	.

Number of unique profiles: 6

The analysis of effects table is displayed in Output 9.31.

Output 9.31 Analysis of Effects Table

Type 3 Analysis of Effects			
Effect	DF	Wald Chi-Square	Pr > ChiSq
program	2	8.2669	0.0160
school	4	4.0142	0.4041
school*program	4	1.7439	0.7827

The school × program interaction is nonsignificant, with a Wald chi-square of 1.74 with 4 df. Note that the degrees of freedom for modeling two generalized logits are twice what you would expect for modeling one logit: instead of 1 df for the intercept you have 2 df; instead of 2 df for SCHOOL, which has three levels, you have 4 df. This is because you are simultaneously modeling two response functions instead of one; you are doubling the number of parameters being estimated since you have to estimate parameters for both logits. To determine the correct number of degrees of freedom for effects in models using generalized logits, multiply the number you would expect for modeling one logit (the usual logistic regression for a dichotomous outcome) by $r - 1$, where r is the number of response levels.

Since the interaction is nonsignificant, the main effects model is fit.

$$\begin{bmatrix} \text{logit}_{111} \\ \text{logit}_{112} \\ \text{logit}_{121} \\ \text{logit}_{122} \\ \text{logit}_{211} \\ \text{logit}_{212} \\ \text{logit}_{221} \\ \text{logit}_{222} \\ \text{logit}_{311} \\ \text{logit}_{312} \\ \text{logit}_{321} \\ \text{logit}_{322} \end{bmatrix} = \begin{bmatrix} 1 & 0 & 0 & 0 & 0 & 0 & 0 & 0 \\ 0 & 1 & 0 & 0 & 0 & 0 & 0 & 0 \\ 1 & 0 & 1 & 0 & 0 & 0 & 0 & 0 \\ 0 & 1 & 0 & 1 & 0 & 0 & 0 & 0 \\ 1 & 0 & 0 & 0 & 1 & 0 & 0 & 0 \\ 0 & 1 & 0 & 0 & 0 & 1 & 0 & 0 \\ 1 & 0 & 1 & 0 & 1 & 0 & 0 & 0 \\ 0 & 1 & 0 & 1 & 0 & 1 & 0 & 0 \\ 1 & 0 & 0 & 0 & 0 & 0 & 1 & 0 \\ 0 & 1 & 0 & 0 & 0 & 0 & 0 & 1 \\ 1 & 0 & 1 & 0 & 0 & 0 & 1 & 0 \\ 0 & 1 & 0 & 1 & 0 & 0 & 0 & 1 \end{bmatrix} \begin{bmatrix} \alpha_1 \\ \alpha_2 \\ \beta_1 \\ \beta_2 \\ \beta_3 \\ \beta_4 \\ \beta_5 \\ \beta_6 \end{bmatrix}$$

Essentially, this model matrix has the same structure as one for modeling a single response function, except that it models two response functions. Thus, the odd rows are for the first logit, and the even rows are for the second logit. Similarly, the odd columns correspond to parameters for the first logit, and the even columns correspond to parameters for the second logit. See Section 8.9 for further discussion.

Table 9.8 contains interpretations for these parameters.

Fitting Models to Generalized Logits with PROC LOGISTIC

Table 9.8 Parameter Interpretations

Model Parameter	Interpretation
α_1	logit_{hi1} for School 1 on Regular Program
α_2	logit_{hi2} for School 1 on Regular Program
β_1	Incremental Effect for After Hours for logit_{hi1}
β_2	Incremental Effect for After Hours for logit_{hi2}
β_3	Incremental Effect for School 2 for logit_{hi1}
β_4	Incremental Effect for School 2 for logit_{hi2}
β_5	Incremental Effect for School 3 for logit_{hi1}
β_6	Incremental Effect for School 3 for logit_{hi2}

The following statements produce the main effects model. The ODDSRATIO statements are added to compute odds ratios.

```
proc logistic order=data;
   freq count;
   class school (ref=first) program (ref=last) / param=ref;
   model style= program school  / link=glogit
      scale=none aggregate;
   oddsratio school;
   oddsratio program;
run;
```

Output 9.32 displays the goodness-of-fit statistics:

Output 9.32 Goodness of Fit

Deviance and Pearson Goodness-of-Fit Statistics				
Criterion	Value	DF	Value/DF	Pr > ChiSq
Deviance	1.7776	4	0.4444	0.7766
Pearson	1.7589	4	0.4397	0.7800

Number of unique profiles: 6

The deviance statistic has a value of 1.78 with 4 df, which is indicative of a good fit.

Output 9.33 contains the analysis of effects table for the main effects model.

Output 9.33 Analysis of Effects Table

Type 3 Analysis of Effects			
Effect	DF	Wald Chi-Square	Pr > ChiSq
program	2	10.9160	0.0043
school	4	14.8424	0.0050

The tests for the school and program effects are significant; SCHOOL has a Wald chi-square value of 14.84 with 4 df, and PROGRAM has a Wald chi-square value of 10.92 with 2 df.

The parameter estimates and tests for individual parameters are displayed in Output 9.34.

Output 9.34 Parameter Estimates

<td colspan="7" align="center">Analysis of Maximum Likelihood Estimates</td>
Parameter	style	DF	Estimate	Standard Error	Wald Chi-Square	Pr > ChiSq
Intercept		self	1	-1.2233	0.3154	15.0454	0.0001
Intercept		team	1	-0.5662	0.2586	4.7919	0.0286
program	after	self	1	-0.7474	0.2820	7.0272	0.0080
program	after	team	1	-0.7426	0.2706	7.5332	0.0061
school	2	self	1	1.0828	0.3539	9.3598	0.0022
school	2	team	1	0.1801	0.3172	0.3224	0.5702
school	3	self	1	1.3147	0.3839	11.7262	0.0006
school	3	team	1	0.6556	0.3395	3.7296	0.0535

Table 9.9 contains the parameter estimates arranged according to the logits they reference. This is often a useful way to display the results from an analysis of generalized logits.

Table 9.9 Coefficients from Final Model

Variable	logit(self/class) Coefficient	Standard Error	logit(team/class) Coefficient	Standard Error
Intercept	$-1.223\ (\hat{\alpha}_1)$	0.315	$-0.566\ (\hat{\alpha}_2)$	0.259
Program	$-0.747\ (\hat{\beta}_1)$	0.282	$-0.743\ (\hat{\beta}_2)$	0.271
School 2	$1.083\ (\hat{\beta}_3)$	0.354	$0.180\ (\hat{\beta}_4)$	0.317
School 3	$1.315\ (\hat{\beta}_5)$	0.384	$0.656\ (\hat{\beta}_6)$	0.340

School 3 has the largest incremental effect for school, particularly for the logit comparing self to class. Program has a nearly similar effect on both logits.

Odds ratios can also be used in models for generalized logits to facilitate model interpretation. Table 9.10 contains the odds corresponding to each logit function for each subpopulation in the data. However, unlike the proportional odds model where the form of the odds ratio was the same regardless of the cumulative logit being considered, the formulas for the odds ratio for the generalized logits model depend on which generalized logit is being considered. The third column, corresponding to the (self/team) logit, was generated by subtraction.

Table 9.10 Model-Predicted Odds

School	Program	Odds		
		Self/Class	Team/Class	Self/Team
1	Regular	e^{α_1}	e^{α_2}	$e^{\alpha_1-\alpha_2}$
1	After	$e^{\alpha_1+\beta_1}$	$e^{\alpha_2+\beta_2}$	$e^{\alpha_1-\alpha_2+\beta_1-\beta_2}$
2	Regular	$e^{\alpha_1+\beta_3}$	$e^{\alpha_2+\beta_4}$	$e^{\alpha_1-\alpha_2+\beta_3-\beta_4}$
2	After	$e^{\alpha_1+\beta_1+\beta_3}$	$e^{\alpha_2+\beta_2+\beta_4}$	$e^{\alpha_1-\alpha_2+\beta_1+\beta_3-\beta_2-\beta_4}$
3	Regular	$e^{\alpha_1+\beta_5}$	$e^{\alpha_2+\beta_6}$	$e^{\alpha_1-\alpha_2+\beta_5-\beta_6}$
3	After	$e^{\alpha_1+\beta_1+\beta_5}$	$e^{\alpha_2+\beta_2+\beta_6}$	$e^{\alpha_1-\alpha_2+\beta_1+\beta_5-\beta_2-\beta_6}$

To determine the odds ratio of self to class for school program, comparing after hours to regular program, you compute (for School 1)

$$\frac{e^{\alpha_1+\beta_1}}{e^{\alpha_1}} = e^{\beta_1}$$

Thus, the odds are $e^{-0.7474} = 0.4736$ times higher of choosing the self-learning style over the class learning style if students attended the after-hours program versus the regular program. Note that you obtain the same result if you do the comparison for School 2 or School 3. If you work through the exercise for the odds ratio of team to class, you find that the odds are $e^{(-.7426)} = 0.4759$ times higher of choosing the team learning style as the class learning style if students attended the after-hours program versus the regular program.

Comparing the odds ratio for School 1 compared to School 2 proceeds in the same manner. You form the ratio of the odds for School 1 regular program to School 2, regular program (after-hours program comparison would also work), to obtain (for self/class logit)

$$\frac{e^{\alpha_1}}{e^{\alpha_1+\beta_3}} = e^{-\beta_3}$$

Thus, the subjects from School 1 have $e^{-1.0828} = 0.339$ times the odds of choosing the self-learning style over the class learning style as those students from School 2.

You can also determine the odds ratios for the (self/team) logits. Form the ratio of the odds for the after-hours program to regular program for self-learning compared to team learning with

$$\frac{e^{\alpha_1-\alpha_2+\beta_1-\beta_2}}{e^{\alpha_1-\alpha_2}} = e^{\beta_1-\beta_2}$$

Thus, the students in the after-hours program have $e^{-0.0048} = 0.9942$ times the odds of choosing the self-learning style over the team learning style compared to those students in the regular program.

PROC LOGISTIC produces these odds ratios statistics for the generalized logits explicitly modeled when you use the ODDSRATIO statement; one statement is required for each variable. Comparisons are made to all levels of the CLASS variable.

Output 9.35 Odds Ratio Estimates

Odds Ratio Estimates and Wald Confidence Intervals			
Label	Estimate	95% Confidence Limits	
style self: school 2 vs 1	2.953	1.476	5.909
style team: school 2 vs 1	1.197	0.643	2.230
style self: school 3 vs 1	3.724	1.755	7.902
style team: school 3 vs 1	1.926	0.990	3.747
style self: school 2 vs 3	0.793	0.413	1.522
style team: school 2 vs 3	0.622	0.317	1.219
style self: program after vs regular	0.474	0.273	0.823
style team: program after vs regular	0.476	0.280	0.809

To summarize, School 1 made a difference for the (self/class) logit, but not for the (team/class) logit. Program was influential for both types of logits, as regular program participants had (1/0.474)=2.11 times the odds of choosing the self-learning styles as the class style, and the results for the (team/class) logit were almost identical.

To compute the odds ratios for (self/team) you would need to submit another PROC LOGISTIC run and explicitly model the (class/team) and (self/team) logits by not specifying the ORDER=DATA option in the PROC statement.

9.3.3 Generalized Logit Model with Continuous Explanatory Variable

The following data are from Table 6.1 in Agresti (2007). Researchers studied the food wild alligators chose to eat as their primary food in Lake George, Florida. The types of food were fish (F), invertebrate (I), and other (O). Invertebrates were generally apple snails, crayfish, and aquatic insects, while Other were mammal, plant material, stones and debris, and reptiles. Table 9.11 contains the choices as well as the length in meters of the alligators.

Generalized Logit Model with Continuous Explanatory Variable

Table 9.11   Alligator Size and Food Choice*

Length and Fish (F), Invertebrate (I), or Other (O) Food							
1.24 I	1.30 I	1.30 I	1.32 F	1.32 F	1.40 F	1.42 I	1.42 F
1.45 I	1.45 O	1.47 I	1.47 F	1.50 I	1.52 I	1.55 I	1.60 I
1.63 I	1.65 O	1.65 I	1.65 F	1.65 F	1.68 F	1.70 I	1.73 O
1.78 I	1.78 I	1.78 O	1.80 I	1.80 F	1.85 F	1.88 I	1.93 I
1.98 I	2.03 F	2.03 F	2.16 F	2.26 F	2.31 F	2.31 F	2.36 F
2.36 F	2.39 F	2.41 F	2.44 F	2.46 F	2.56 O	2.67 F	2.72 I
2.79 F	2.84 F	3.25 O	3.28 O	3.33 F	3.56 F	3.58 F	3.66 F
3.68 O	3.71 F	3.89 F					

*Reprinted by permission of John Wiley & Sons.

The choice of food is a nominal response since there is no logical order to the different choices. Thus, generalized logits regression is a reasonable strategy to determine whether size of the alligators impacted their choice of food. Unlike the previous example in which case the two explanatory variables were classification variables, size of alligator as measured in meters is a continuous explanatory variable.

The following SAS statements create the SAS data set ALLIGATOR.

```
data alligator;
    input length choice $ @@;
datalines;
1.24 I  1.30 I  1.30 I  1.32 F  1.32 F  1.40 F  1.42 I  1.42 F
1.45 I  1.45 O  1.47 I  1.47 F  1.50 I  1.52 I  1.55 I  1.60 I
1.63 I  1.65 O  1.65 I  1.65 F  1.65 F  1.68 F  1.70 I  1.73 O
1.78 I  1.78 I  1.78 O  1.80 I  1.80 F  1.85 F  1.88 I  1.93 I
1.98 I  2.03 F  2.03 F  2.16 F  2.26 F  2.31 F  2.31 F  2.36 F
2.36 F  2.39 F  2.41 F  2.44 F  2.46 F  2.56 O  2.67 F  2.72 I
2.79 F  2.84 F  3.25 O  3.28 O  3.33 F  3.56 F  3.58 F  3.66 F
3.68 O  3.71 F  3.89 F
;
```

The only difference in the approach when the explanatory variable is continuous is that the usual goodness-of-fit statistics Q_P and Q_L don't apply. If you created a cross-classification of choices and alligator size, only a few cells would have more than one animal, and the sample size requirement of these statistics that each cell contain a minimum of five subjects isn't met. However, you can assess goodness of fit by using the residual score statistic and creating an expanded model that includes the squared term for size.

The following PROC LOGISTIC statements produce a generalized logits analysis of the alligator data. The PLOTS=EFFECT option in the PROC LOGISTIC statement requests a prediction plot for each outcome by size. The ORDER=DATA option for the response variable CHOICE in the MODEL statement requests that the generalized logits be formed according to the order found in the data. Thus, the two generalized logits formed will compare fish to other and invertebrate to other. Both LENGTH and LENGTH*LENGTH are included as explanatory variables, and the SELECTION=FORWARD and INCLUDE=1 options are specified to produce the residual score statistic corresponding to the expanded model with the squared term in it.

Chapter 9: Logistic Regression II: Polytomous Response

The LINK=GLOGIT option in the MODEL statement is required to perform generalized logits regression.

```
ods graphics on;
proc logistic data=alligator plots=effect;
    model choice(order=data) = length length*length/
        selection=forward include=1 link=glogit;
run;
ods graphics off;
```

Output 9.36 displays the response profile. Most of the alligators chose fish for their primary food, while the next most popular choice was invertebrates. Eight of the alligators found other food source to be most appealing.

Output 9.36 Response Profile

Response Profile		
Ordered Value	choice	Total Frequency
1	I	20
2	F	31
3	O	8

Logits modeled use choice='O' as the reference category.

Output 9.37 contains the results for the residual score statistic. Q_{RS} takes the value 0.4260, with $p = 0.8081$ and 2 df, indicating that the main effect model with LENGTH is an adequate one.

Output 9.37 Residual Score Test

Residual Chi-Square Test		
Chi-Square	DF	Pr > ChiSq
0.4260	2	0.8081

The model fit statistics are displayed in Output 9.38.

Output 9.38 Fit Statistics

Model Fit Statistics		
Criterion	Intercept Only	Intercept and Covariates
AIC	119.142	106.341
SC	123.297	114.651
-2 Log L	115.142	98.341

Output 9.39 contains the parameter estimates. Length of alligator appears to be important for the (invertebrate/other) logit, but not for the (fish/other) logit.

Output 9.39 Parameter Estimates

\multicolumn{7}{c|}{Analysis of Maximum Likelihood Estimates}						
Parameter	choice	DF	Estimate	Standard Error	Wald Chi-Square	Pr > ChiSq
Intercept	I	1	5.6974	1.7938	10.0881	0.0015
Intercept	F	1	1.6177	1.3073	1.5314	0.2159
length	I	1	-2.4654	0.8997	7.5101	0.0061
length	F	1	-0.1101	0.5171	0.0453	0.8314

This is more evident when you examine the odds ratios displayed in Output 9.40.

Output 9.40 Odds Ratios

\multicolumn{3}{c|}{Odds Ratio Estimates and Wald Confidence Intervals}		
Label	Estimate	95% Confidence Limits
choice I: length	0.085	0.015 0.496
choice F: length	0.896	0.325 2.468

For invertebrate compared to other, other has (1/0.085)=11.76 greater odds of being chosen per meter increase in size of the alligator. For fish compared to other, other has (1/0.896) =1.12 times greater odds of being chosen per meter increase in size.

You can compute the odd ratio comparing fish to invertebrate by computing the implicit slope for the logit(fish/invertebrate) as

$$-(\beta_{\text{fish/other}} + \beta_{\text{invertebrate/other}}) = -(-2.4654 - 0.1101) = 2.3553$$

Thus, when fish are compared to invertebrate, fish have $e^{2.3553}$=10.54 times higher odds of being chose as the primary food source per meter increase in alligator length.

A graph of the model-predicted probabilities for each food choice by alligator length (Output 9.41) relays the results visually. Larger alligators clearly prefer to eat fish.

Output 9.41 Predicted Probabilities

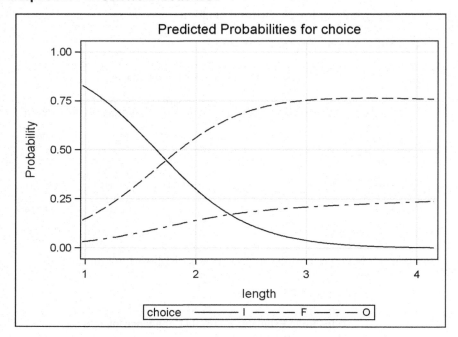

9.3.4 Exact Methods for Generalized Logits Model

Exact methods can also be applied to generalized logits regression. The following data come from a study in which a physical therapy practice experimented with different treatment modalities in a strengthening program. The therapists rotated patients through three programs for maintenance after basic strength had been attained. The strengthening programs included free weights, resistance bands, and weight machines. The data are displayed in Table 9.12.

Table 9.12 Physical Therapy Data

Gender	Status	Strengthening Choice		
		Machines	Free Weights	Bands
Males	Adult	2	13	3
Males	Senior	10	9	3
Females	Adult	3	9	1
Females	Senior	8	0	1

The possible outcomes are clearly nominal since there is no obvious ordinality across the choices of machine, free weights, or resistance bands. The generalized logit is a logical response function. The numerous table counts less than five, including a zero count, make it unlikely that an analysis based on asymptotics would be appropriate.

```
data pt;
   format therapy $11.;
   input gender $ age $ therapy $ count @@;
   datalines;
males   adult   machines      2   males   senior  machines       10
```

```
males    adult   freeweights 13    males   senior  freeweights   9
males    adult   bands        3    males   senior  bands         3
females  adult   machines     3    females senior  machines      8
females  adult   freeweights  9    females senior  freeweights   0
females  adult   bands        1    females senior  bands         1
;
```

The following PROC LOGISTIC statements request the generalized logits analysis.

```
proc logistic data=pt;
   freq count;
   class gender(ref='females') age(ref='senior') / param=ref;
   model therapy(ref='machines') = gender age / link=glogit scale=none aggregate;
run;
```

Output 9.42 contains the response profile. The logits formed compare free weights and bands to machines.

Output 9.42 Response Profile

Response Profile		
Ordered Value	therapy	Total Frequency
1	bands	8
2	freeweights	31
3	machines	23

Logits modeled use therapy='machines' as the reference category.

Output 9.43 contains the goodness-of-fit statistics.

Output 9.43 Goodness of Fit

Deviance and Pearson Goodness-of-Fit Statistics				
Criterion	Value	DF	Value/DF	Pr > ChiSq
Deviance	3.8087	2	1.9043	0.1489
Pearson	2.8044	2	1.4022	0.2461

Number of unique profiles: 4

Output 9.44 contains the global fit statistics for assessing goodness of fit through explanatory capacity. Their p-values are an indication that the sample sizes are too small for asymptotic methods; the p-value for the Wald statistic is more than ten times larger than the p-value for the likelihood ratio statistic (0.0122 versus 0.0008).

Output 9.44 Global Fit Statistics

Testing Global Null Hypothesis: BETA=0			
Test	Chi-Square	DF	Pr > ChiSq
Likelihood Ratio	18.9061	4	0.0008
Score	16.9631	4	0.0020
Wald	12.8115	4	0.0122

Output 9.45 and Output 9.46 contain the asymptotic results for comparison purposes.

Output 9.45 Parameter Estimates

Analysis of Maximum Likelihood Estimates							
Parameter		therapy	DF	Estimate	Standard Error	Wald Chi-Square	Pr > ChiSq
Intercept		bands	1	-2.6450	1.0166	6.7694	0.0093
Intercept		freeweights	1	-1.8774	0.7482	6.2969	0.0121
gender	males	bands	1	1.5261	1.0158	2.2570	0.1330
gender	males	freeweights	1	1.5758	0.7569	4.3347	0.0373
age	adult	bands	1	1.7416	0.9623	3.2754	0.0703
age	adult	freeweights	1	2.6472	0.7572	12.2212	0.0005

Output 9.46 Odds Ratios

Odds Ratio Estimates			
Effect	therapy	Point Estimate	95% Wald Confidence Limits
gender males vs females	bands	4.600	0.628 33.686
gender males vs females	freeweights	4.835	1.097 21.313
age adult vs senior	bands	5.707	0.865 37.629
age adult vs senior	freeweights	14.115	3.200 62.262

The following statements request the exact analysis. Variables GENDER and AGE are placed in the EXACT statement. The JOINT option requests a joint test for these effects, and the ESTIMATE=BOTH option requests both the parameter estimates and the exponentiated parameter estimates.

```
proc logistic data=pt;
   freq count;
   class gender(ref='females') age(ref='senior') / param=ref;
   model therapy(ref='machines') = gender age / link=glogit;
```

```
    exact gender age / joint estimate=both;
run;
```

Output 9.47 displays the exact test results. Age appears to be a significant effect, and gender appears to be a marginal effect.

Output 9.47 Exact Tests

Exact Conditional Analysis

			p-Value	
Effect	Test	Statistic	Exact	Mid
Joint	Score	16.6895	0.0013	0.0013
	Probability	7.115E-7	0.0010	0.0010
gender	Score	5.0830	0.0870	0.0835
	Probability	0.00697	0.0988	0.0953
age	Score	14.5093	0.0003	0.0003
	Probability	0.000032	0.0003	0.0003

Exact Conditional Tests

Output 9.48 contains the exact parameter estimates.

Output 9.48 Exact Parameter Estimates

Exact Parameter Estimates

Parameter		therapy	Estimate	Standard Error	95% Confidence Limits		Two-sided p-Value
gender	males	bands	1.4114	0.9710	-0.7079	4.0869	0.2715
gender	males	freeweights	1.5017	0.7376	-0.0692	3.4081	0.0641
age	adult	bands	1.6244	0.9252	-0.5146	3.9734	0.1673
age	adult	freeweights	2.5231	0.7317	0.9979	4.4303	0.0002

Output 9.49 displays the odds ratio estimates for the exact analysis. Age has a strong effect for the (freeweights/machines) logit; its effect for the (bands/machines) logit is marginal at best. You can also see that gender has a marginal influence when the (freeweight/bands) logit is considered, but no influence for the (bands/machines logit).

Output 9.49 Exact Odds Ratios

Exact Odds Ratios						
Parameter		therapy	Estimate	95% Confidence Limits		Two-sided p-Value
gender	males	bands	4.102	0.493	59.553	0.2715
gender	males	freeweights	4.489	0.933	30.209	0.0641
age	adult	bands	5.076	0.598	53.165	0.1673
age	adult	freeweights	12.467	2.713	83.960	0.0002

For these data, you should report the exact analysis results. In general, note that since the generalized logits analysis means many more parameters, exact methods may only handle relatively small models for relatively small data sets.

Chapter 10

Conditional Logistic Regression

Contents

10.1	Introduction	**297**
10.2	Paired Observations from a Highly Stratified Cohort Study	**298**
10.3	Clinical Trials Study Analysis	**300**
	10.3.1 Analysis Using the LOGISTIC Procedure	300
10.4	Crossover Design Studies	**308**
	10.4.1 Two-Period Crossover Design	309
	10.4.2 Three-Period Crossover Study	315
10.5	General Conditional Logistic Regression	**322**
	10.5.1 Analyzing Diagnostic Data	323
10.6	Paired Observations in a Retrospective Matched Study	**326**
	10.6.1 1:1 Conditional Logistic Regression	327
10.7	1:m Conditional Logistic Regression	**331**
10.8	Exact Conditional Logistic Regression in the Stratified Setting	**334**
	10.8.1 Printing More Digits	337
10.9	Appendix A: Theory for the Case-Control Retrospective Setting	**338**
10.10	Appendix B: Theory for General Conditional Logistic Regression	**340**
10.11	Appendix C: Theory for Exact Conditional Inference	**341**
10.12	Appendix D: ODS Macro	**343**

10.1 Introduction

The usual maximum likelihood approach to estimation in logistic regression is not always appropriate. As discussed in Chapter 8, "Logistic Regression I: Dichotomous Response," there may be insufficient sample size for logistic regression, particularly if the data are highly stratified and the strata contain small numbers of subjects. In these situations, you have a small sample size relative to the number of parameters being estimated since you will be estimating parameters for the stratification effects. For the maximum likelihood estimates to be valid, you need a large sample size relative to the number of parameters.

Often, highly stratified data come from a design with cluster sampling, that is, designs with two or more observations for each primary sampling unit or cluster. Common examples of such data are paired observations, such as fraternal twins (or litter mates), right and left sides of the body in a dermatology study, or two occasions for an expression of an opinion. Ordinary logistic regression

298 Chapter 10: Conditional Logistic Regression

may be inappropriate for such data, since you have insufficient sample size to estimate the stratum effect (family, litter, patient, respondent) without bias. However, by using conditioning arguments, you can eliminate the strata effects and estimate the other effects in which you are interested.

The appropriate form of logistic regression for these types of data is called *conditional logistic regression*. It takes the stratification into account by basing the maximum likelihood estimation of the model parameters on a conditional likelihood. You can fit these models in SAS with the LOGISTIC procedure. In the following sections, the conditional likelihood for paired observations from small clusters is derived, and the methodology is illustrated with data from a randomized clinical trial and a two-period crossover design study. Then, the more general stratified situation is discussed and illustrated with data from a three-period crossover study and a repeated measurements study.

Matched case control studies in epidemiology also produce highly stratified data. In these studies, you match cases (persons with a disease or condition) to controls (persons without the disease or condition) on the basis of variables thought to be potential confounders such as age, sex, and location of residence. The use of conditional logistic regression for matched studies in epidemiological work is discussed and illustrated with two examples.

Finally, the use of exact logistic regression for the stratified setting is discussed for several examples.

10.2 Paired Observations from a Highly Stratified Cohort Study

Consider a randomized clinical trial where $h = 1, 2, \ldots, q$ centers are randomly selected, and, at each center, one randomly selected patient is placed on treatment and another randomly selected patient is placed on placebo. Interest lies in whether the patients improve; thus, improvement is the event of interest. Since there are only two observations per center, it is not possible to estimate a center effect (pair effect) without bias. As a general rule, you need each possible outcome to have five observations in each category of each categorical explanatory variable in the model for valid estimation to proceed.

Suppose $y_{hi} = 1$ if improvement occurs and $y_{hi} = 0$ if it does not ($i = 1$ for treatment and $i = 2$ for the placebo; $h = 1, 2, \ldots, q$). Suppose $x_{hi} = 1$ for treatment and $x_{hi} = 0$ for placebo, and $\mathbf{z}_{hi} = (z_{hi1}, z_{hi2}, \ldots, z_{hit})'$ represents the t explanatory variables.

The usual logistic likelihood for $\{y_{hi}\}$ is written

$$\Pr\{y_{hi}\} = \pi_{hi} = \frac{\exp\{\alpha_h + \beta x_{hi} + \boldsymbol{\gamma}'\mathbf{z}_{hi}\}}{1 + \exp\{\alpha_h + \beta x_{hi} + \boldsymbol{\gamma}'\mathbf{z}_{hi}\}}$$

where α_h is an intercept for the hth center, β is the treatment parameter, and $\boldsymbol{\gamma}' = (\gamma_1, \gamma_2, \ldots, \gamma_t)$ is the parameter vector for the covariates $\mathbf{z}$. Since there are only two observations per center, you can't estimate these parameters without bias. However, you can fit a model based on conditional probabilities that condition away the center effects, which results in a model that contains substantially fewer parameters. In this context, the α_h are known as *nuisance parameters*. It is useful to describe these data with a model that considers the probability of a pair's treatment patient improving and the pair's placebo patient not improving, compared to the probability that one of them improved.

You can write a conditional probability for $\{y_{hi}\}$ as the ratio of the joint probability of a pair's treatment patient improving and the pair's placebo patient not improving to the joint probability that either the treatment patient or the placebo patient improved.

$$\Pr\{y_{h1}=1, y_{h2}=0 \mid y_{h1}=1, y_{h2}=0 \text{ or } y_{h1}=0, y_{h2}=1\} =$$

$$\frac{\Pr\{y_{h1}=1\}\Pr\{y_{h2}=0\}}{\Pr\{y_{h1}=1\}\Pr\{y_{h2}=0\} + \Pr\{y_{h1}=0\}\Pr\{y_{h2}=1\}}$$

You write the probabilities in terms of the logistic model as

$$\Pr\{y_{h1}=1\}\Pr\{y_{h2}=0\} = \frac{\exp\{\alpha_h + \beta + \gamma' z_{h1}\}}{1 + \exp\{\alpha_h + \beta + \gamma' z_{h1}\}} \cdot \frac{1}{1 + \exp\{\alpha_h + \gamma' z_{h2}\}}$$

and

$$\Pr\{y_{h1}=1\}\Pr\{y_{h2}=0\} + \Pr\{y_{h1}=0\}\Pr\{y_{h2}=1\} =$$

$$\frac{\exp\{\alpha_h + \beta + \gamma' z_{h1}\}}{1 + \exp\{\alpha_h + \beta + \gamma' z_{h1}\}} \cdot \frac{1}{1 + \exp\{\alpha_h + \gamma' z_{h2}\}}$$

$$+ \frac{1}{1 + \exp\{\alpha_h + \beta + \gamma' z_{h1}\}} \cdot \frac{\exp\{\alpha_h + \gamma' z_{h2}\}}{1 + \exp\{\alpha_h + \gamma' z_{h2}\}}$$

Form their ratio to obtain

$$\frac{\exp\{\alpha_h + \beta + \gamma' z_{h1}\}}{\exp\{\alpha_h + \beta + \gamma' z_{h1}\} + \exp\{\alpha_h + \gamma' z_{h2}\}}$$

and, since the denominators cancel out, this expression reduces to

$$\frac{\exp\{\beta + \gamma'(z_{h1} - z_{h2})\}}{1 + \exp\{\beta + \gamma'(z_{h1} - z_{h2})\}}$$

which no longer contains the $\{\alpha_h\}$. Thus, by focusing on modeling a meaningful conditional probability, you develop a model with a reduced number of parameters that can be estimated without bias.

The conditional likelihood for the entire data is written

$$\prod_{h=1}^{q} \left\{ \frac{\exp\{\beta + \gamma'(z_{h1} - z_{h2})\}}{1 + \exp\{\beta + \gamma'(z_{h1} - z_{h2})\}} \right\}^{y_{h1}(1-y_{h2})} \left\{ \frac{1}{1 + \exp\{\beta + \gamma'(z_{h1} - z_{h2})\}} \right\}^{(1-y_{h1})y_{h2}}$$

This is the unconditional likelihood for the usual logistic model, except that the intercept is now β, the effect for treatment, and each observation represents a pair of observations from a center where the response is 1 if the pair represents the combination $\{y_{h1}=1$ and $y_{h2}=0\}$ and 0 if the pair has the combination $\{y_{h1}=0$ and $y_{h2}=1\}$. The explanatory variables are the differences in values of the explanatory variables for the treatment patient and the placebo patient. Since the likelihood is conditioned on the discordant pairs, the concordant pairs (the observations where $\{y_{h1}=1$ and $y_{h2}=1\}$ and $\{y_{h1}=0$ and $y_{h2}=0\}$) are uninformative and thus can be ignored.

Note that the ratio above can also be rewritten as

$$\frac{\exp\{\beta + \gamma' z_{h1}\}}{\exp\{\beta + \gamma' z_{h1}\} + \exp\{\gamma' z_{h2}\}}$$

and the corresponding likelihood for the entire data as

$$\prod_{h=1}^{q} \left\{\frac{\exp\{\beta + \gamma' z_{h1}\}}{\exp\{\beta + \gamma' z_{h1}\} + \exp\{\gamma' z_{h2}\}}\right\}^{y_{h1}(1-y_{h2})} \left\{\frac{\exp\{\gamma' z_{h2}\}}{\exp\{\beta + \gamma' z_{h1}\} + \exp\{\gamma' z_{h2}\}}\right\}^{(1-y_{h1})y_{h2}}$$

This is the same likelihood that applies to paired data in a simple case of the Cox regression model (proportional hazards model), which is used in the analysis of survival times.

If there are no covariates for such a study, so that the data represent a 2×2 table where the responses for treatment are cross-classified with the responses for placebo, then testing $\beta = 0$ is equivalent to McNemar's test. Also, it can be shown that e^{β} is estimated by n_{12}/n_{21}, where n_{12} and n_{21} are the off-diagonal counts from this table.

10.3 Clinical Trials Study Analysis

Researchers studying the effect of a new treatment on a skin condition collected information from 79 clinics. In each clinic, one patient received the treatment, and another patient received a placebo. Variables collected included age, sex, and an initial grade for the skin condition, which ranged from 1 to 4 for mild to severe. The response variable was whether the skin condition improved. Conditional logistic regression is suitable for the analysis of such data.

10.3.1 Analysis Using the LOGISTIC Procedure

You could proceed by taking differences of the observations as discussed earlier and fitting a standard logistic regression to them. However, you can perform conditional logistic regression for the original data by using the STRATA statement in PROC LOGISTIC.

The following DATA step creates the SAS data set TRIAL.

```
data trial;
   input center treatment $ sex $ age improve initial @@;
   datalines;
1  t f 27 0 1   1 p f 32 0 2  41 t f 13 1 2  41 p m 22 0 3
2  t f 41 1 3   2 p f 47 0 1  42 t m 31 1 1  42 p f 21 1 3
3  t m 19 1 4   3 p m 31 0 4  43 t f 19 1 3  43 p m 35 1 3
4  t m 55 1 1   4 p m 24 1 3  44 t m 31 1 3  44 p f 37 0 2
5  t f 51 1 4   5 p f 44 0 2  45 t f 44 0 1  45 p f 41 1 1
6  t m 23 0 1   6 p f 44 1 3  46 t m 41 1 2  46 p m 41 0 1
```

```
 7 t m 31 1 2   7 p f 39 0 2  47 t m 41 1 2  47 p f 21 0 4
 8 t m 22 0 1   8 p m 54 1 4  48 t f 51 1 2  48 p m 22 1 1
 9 t m 37 1 3   9 p m 63 0 2  49 t f 62 1 3  49 p f 32 0 3
10 t m 33 0 3  10 p f 43 0 3  50 t m 21 0 1  50 p m 34 0 1
11 t f 32 1 1  11 p m 33 0 3  51 t m 55 1 3  51 p f 35 1 2
12 t m 47 1 4  12 p m 24 0 4  52 t f 61 0 1  52 p m 19 0 1
13 t m 55 1 3  13 p f 38 1 1  53 t m 43 1 2  53 p m 31 0 2
14 t f 33 0 1  14 p f 28 1 2  54 t f 44 1 1  54 p f 41 1 1
15 t f 48 1 1  15 p f 42 0 1  55 t m 67 1 2  55 p m 41 0 1
16 t m 55 1 3  16 p m 52 0 1  56 t m 41 0 2  56 p m 21 1 4
17 t m 30 0 4  17 p m 48 1 4  57 t f 51 1 3  57 p m 51 0 2
18 t f 31 1 2  18 p m 27 1 3  58 t m 62 1 3  58 p m 54 1 3
19 t m 66 1 3  19 p f 54 0 1  59 t m 22 0 1  59 p f 22 0 1
20 t f 45 0 2  20 p f 66 1 2  60 t m 42 1 2  60 p f 29 1 2
21 t m 19 1 4  21 p f 20 1 4  61 t f 51 1 1  61 p f 31 0 1
22 t m 34 1 4  22 p f 31 0 1  62 t m 27 0 2  62 p m 32 1 2
23 t f 46 0 1  23 p m 30 1 2  63 t m 31 1 1  63 p f 21 0 1
24 t m 48 1 3  24 p f 62 0 4  64 t m 35 0 3  64 p m 33 1 3
25 t m 50 1 4  25 p m 45 1 4  65 t m 67 1 2  65 p m 19 0 1
26 t m 57 1 3  26 p f 43 0 3  66 t m 41 0 2  66 p m 62 1 4
27 t f 13 0 2  27 p m 22 1 3  67 t f 31 1 2  67 p m 45 1 3
28 t m 31 1 1  28 p f 21 0 1  68 t m 34 1 1  68 p f 54 0 1
29 t m 35 1 3  29 p m 35 1 3  69 t f 21 0 1  69 p m 34 1 4
30 t f 36 1 3  30 p f 37 0 3  70 t m 64 1 3  70 p m 51 0 1
31 t f 45 0 1  31 p f 41 1 1  71 t f 61 1 3  71 p m 34 1 3
32 t m 13 1 2  32 p m 42 0 1  72 t m 33 0 1  72 p f 43 0 1
33 t m 14 0 4  33 p f 22 1 2  73 t f 36 0 2  73 p m 37 0 3
34 t f 15 1 2  34 p m 24 0 1  74 t m 21 1 1  74 p m 55 0 1
35 t f 19 1 3  35 p f 31 0 1  75 t f 47 0 2  75 p f 42 1 3
36 t m 20 0 2  36 p m 32 1 3  76 t f 51 1 4  76 p m 44 0 2
37 t m 23 1 3  37 p f 35 0 1  77 t f 23 1 1  77 p m 41 1 3
38 t f 23 0 1  38 p m 21 1 1  78 t m 31 0 2  78 p f 23 1 4
39 t m 24 1 4  39 p m 30 1 3  79 t m 22 0 1  79 p m 19 1 4
40 t m 57 1 3  40 p f 43 1 3
;
```

First, consider the model where treatment is the only term. In this case, there is only one x_{hik}, treatment, and there are no z_{hi}. Table 10.1 displays the associated crosstabulation of pairs by treatment and response.

Table 10.1 Pairs Breakdown

Treatment	Improvement	
	No	Yes
No	7	34
Yes	20	18

McNemar's test statistic is computed as

$$\frac{(34-20)^2}{(34+20)} = 3.63$$

which is nearly significant. Also, $n_{12}/n_{21} = 1.7$ is an estimate of the odds ratio, which is also the exponentiated parameter for treatment in the conditional logistic regression model.

The following SAS statements produce the conditional logistic analysis with only treatment in the model. You place the stratification variable CENTER in the STRATA statement. SubidxSTRATA statementLOGISTIC procedureSubidxLOGISTIC procedureSTRATA statement

```
proc logistic data=trial;
   class treatment(ref="p") /param=ref;
   strata center;
   model improve(event="1")= treatment;
run;
```

Output 10.1 displays information about the analysis. The input data included 79 strata, of which 25 were uninformative (those in which each patient of the pair had the identical response).

Output 10.1 Model Information

Conditional Analysis

Model Information	
Data Set	WORK.TRIAL
Response Variable	improve
Number of Response Levels	2
Number of Strata	79
Number of Uninformative Strata	25
Frequency Uninformative	50
Model	binary logit
Optimization Technique	Newton-Raphson ridge

The response profiles shown in Output 10.2 indicate that improvement is the majority response and that the logistic regression focuses on the probability of improvement.

Output 10.2 Response Profiles

Response Profile		
Ordered Value	improve	Total Frequency
1	0	68
2	1	90

Probability modeled is improve=1.

Output 10.3 displays both the values of the classification variables and the reference level of placebo (p) for treatment.

Output 10.3 Classification Levels

Class Level Information		
Class	Value	Design Variables
treatment	p	0
	t	1

Output 10.4 provides summary information about the response patterns found in the strata; as previously discussed, 54 strata are informative since both response values are recorded. There are 25 uninformative strata, with 18 pairs recording improvement and 7 pairs recording no improvement.

Output 10.4 Strata Summary

Strata Summary				
Response Pattern	improve 0	improve 1	Number of Strata	Frequency
1	0	2	18	36
2	1	1	54	108
3	2	0	7	14

Output 10.5 contains the parameter estimates. Note that no intercept term is displayed; the intercepts correspond to the strata (centers) and they have been conditioned away.

Output 10.5 Treatment Effect Only Model

Analysis of Conditional Maximum Likelihood Estimates						
Parameter		DF	Estimate	Standard Error	Wald Chi-Square	Pr > ChiSq
treatment	t	1	0.5306	0.2818	3.5457	0.0597

Note that $e^\beta = e^{0.5306} = 1.70$, the same value for the odds ratio computed from Table 10.1. In addition, the Wald test takes the value 3.5457 with $p = 0.0597$, which is nearly significant. This value is also close to the value 3.63 computed for McNemar's test. As the sample size grows, the Wald statistic for the treatment effect and McNemar's test statistic become asymptotically equivalent.

The analysis continues with the consideration of other explanatory variables. The following PROC LOGISTIC invocation retains TREATMENT but considers other main effects through the forward selection process.

```
proc logistic data=trial;
    class sex (ref="f") treatment(ref="p") /param=ref;
    strata center;
    model improve(event="1") =  treatment initial sex age/
              selection=forward include=1 details;
run;
```

Only INITIAL was added to the model. Output 10.6 displays the residual score statistic ($Q_{RS} = 2.2008$ with 2 df and $p = 0.3327$), which is the joint test for AGE and SEX, indicating that they are unimportant. The relatively small size of n_{12} and n_{21} (20 and 34, respectively) limit the utility of a score test for the inclusion of additional terms such as pairwise interactions.

Output 10.6 Score Statistics

Conditional Analysis

Residual Chi-Square Test		
Chi-Square	DF	Pr > ChiSq
2.2008	2	0.3327

Analysis of Effects Eligible for Entry			
Effect	DF	Score Chi-Square	Pr > ChiSq
sex	1	0.9358	0.3334
age	1	1.2516	0.2632

If you examine the parameter estimates in Output 10.7, it appears that treatment becomes slightly more influential when initial score is added to the model ($p = 0.0413$).

Output 10.7 Maximum Likelihood Estimates

Analysis of Conditional Maximum Likelihood Estimates						
Parameter		DF	Estimate	Standard Error	Wald Chi-Square	Pr > ChiSq
treatment	t	1	0.7113	0.3487	4.1617	0.0413
initial		1	1.0774	0.3214	11.2395	0.0008

However, examine the statistics in the "Global Fit Statistics" table in Output 10.8. The likelihood ratio, score, and Wald statistics have fairly different values at 22.0627, 18.5853, and 12.9050, respectively, for 2 df. This suggests that asymptotic methods may not be appropriate for these data, and perhaps the number of off-diagonal pairs displayed in Table 10.1 is not high enough to support the analysis.

Output 10.8 Global Fit Statistics

Testing Global Null Hypothesis: BETA=0			
Test	Chi-Square	DF	Pr > ChiSq
Likelihood Ratio	22.0627	2	<.0001
Score	18.5853	2	<.0001
Wald	12.9050	2	0.0016

However, the exact methods discussed in Chapter 8 are also available for the stratified analysis. See Appendix B in this chapter for a discussion of the methodology involved; you are essentially conditioning away both nuisance parameters and stratum-specific intercepts in order to obtain an appropriate conditional likelihood function.

The specification of the exact conditional analysis in the LOGISTIC procedure is the same as previously discussed except for the addition of the STRATA statement. Exact analyses for both INITIAL and TREATMENT are requested by listing them in the EXACT statement. The ESTIMATE=BOTH option requests both parameter estimates and odds ratio estimates. The variable CENTER is specified in the STRATA statement, so this analysis will condition away the center-specific intercepts. The EXACTONLY option in the PROC LOGISTIC statement specifies that only the exact analysis be performed.

```
proc logistic data=trial exactonly;
   class treatment(ref="p") /param=ref;
   strata center;
   model improve(event="1") = treatment initial;
   exact treatment initial / estimate=both;
run;
```

Output 10.9 displays the results of the exact conditional analysis.

Output 10.9 Exact Tests

Exact Conditional Analysis

Exact Conditional Tests				
			p-Value	
Effect	Test	Statistic	Exact	Mid
treatment	Score	4.3986	0.0377	0.0305
	Probability	0.0145	0.0377	0.0305
initial	Score	15.7365	<.0001	<.0001
	Probability	0.000015	<.0001	<.0001

The *p*-values for both INITIAL and TREATMENT are lower than those reported in the asymptotic analysis. The exact score *p*-value is 0.0377 for treatment and < 0.0001 for initial grade.

Output 10.10 displays the exact parameter estimates, which are similar to but different from those from the asymptotic conditional analysis. Recall that the *p*-values reported in the "Exact Parameter Estimates" table are different from the ones reported in the "Exact Conditional Analysis" table, and the latter should generally be reported.

Output 10.10 Exact Parameter Estimates

Exact Parameter Estimates						
Parameter		Estimate	Standard Error	95% Confidence Limits		Two-sided p-Value
treatment	t	0.7034	0.3461	-0.005365	1.4836	0.0520
initial		1.0542	0.3171	0.4625	1.8221	<.0001

Output 10.11 displays the odds ratios based on the exact parameter estimates.

Output 10.11 Exact Odds Ratio

Exact Odds Ratios					
Parameter		Estimate	95% Confidence Limits		Two-sided p-Value
treatment	t	2.021	0.995	4.409	0.0520
initial		2.870	1.588	6.185	<.0001

The odds of improving for the patient receiving the treatment is $e^{0.7034} = 2.021$ times higher than for the patient receiving the placebo in each center. The odds of improvement also increase by a factor of 2.870 for each unit increase in the initial grade. Treatment has a nearly significant effect even after adjusting for the effect of initial grade. And the stratified analysis has taken center into account.

Finally, for illustration's sake, even though the sample sizes may not be sufficient, the final analysis of these data forces all the main effects in the model with the INCLUDE=4 option and makes all pairwise interaction terms available for consideration.

```
proc logistic data=trial;
   class sex (ref="f") treatment(ref="p") /param=ref;
   strata center;
   model improve(event="1") = initial age sex treatment
              sex*age sex*initial age*initial
              treatment*sex treatment*initial treatment*age /
              selection=forward include=4 details ;
run;
```

Output 10.12 shows both the residual score statistic ($Q_{RS} = 4.7214$ with 6 df and $p = 0.5800$) and the score statistics for the addition of the individual terms into the model. Since there are 20 strata with the less prevalent response (see cell n_{21} in Table 10.1), this model can support about $20/5 = 4$ terms. Thus, there are possibly too many terms to rely entirely on the residual score statistic to assess goodness of fit. However, considering both the residual test and the individual tests provides reasonable confidence that the model fits adequately. All of the individual tests have *p*-values greater than 0.08, and most of them have *p*-values greater than 0.5. This model doesn't require the addition of any interaction terms.

Output 10.12 Score Statistics

Conditional Analysis

Residual Chi-Square Test		
Chi-Square	DF	Pr > ChiSq
4.7214	6	0.5800

Analysis of Effects Eligible for Entry			
Effect	DF	Score Chi-Square	Pr > ChiSq
age*sex	1	0.6593	0.4168
initial*sex	1	0.1775	0.6736
initial*age	1	2.9195	0.0875
sex*treatment	1	0.2681	0.6046
initial*treatment	1	0.0121	0.9125
age*treatment	1	0.4336	0.5102

Output 10.13 and Output 10.14 display model fit statistics.

Output 10.13 Model Fit Statistics

Model Fit Statistics		
Criterion	Without Covariates	With Covariates
AIC	74.860	58.562
SC	74.860	70.813
-2 Log L	74.860	50.562

Output 10.14 Global Fit Statistics

Testing Global Null Hypothesis: BETA=0			
Test	Chi-Square	DF	Pr > ChiSq
Likelihood Ratio	24.2976	4	<.0001
Score	19.8658	4	0.0005
Wald	13.0100	4	0.0112

Output 10.15 contains the maximum likelihood estimates of the parameters. The treatment effect takes the value 0.7025, which is nearly significant with $p = 0.0511$. Neither age nor sex appear to be very influential but are left in the model as covariates. The effect for initial grade is highly significant ($p = 0.0011$).

Output 10.15 Maximum Likelihood Estimates

Analysis of Conditional Maximum Likelihood Estimates						
Parameter		DF	Estimate	Standard Error	Wald Chi-Square	Pr > ChiSq
initial		1	1.0915	0.3351	10.6106	0.0011
age		1	0.0248	0.0224	1.2253	0.2683
sex	m	1	0.5312	0.5545	0.9176	0.3381
treatment	t	1	0.7025	0.3601	3.8053	0.0511

Output 10.16 contains the odds ratios and their 95% confidence limits.

Output 10.16 Odds Ratios

Odds Ratio Estimates			
Effect	Point Estimate	95% Wald Confidence Limits	
initial	2.979	1.545	5.745
age	1.025	0.981	1.071
sex m vs f	1.701	0.574	5.043
treatment t vs p	2.019	0.997	4.089

However, as previously stated, the exact analysis may be most suitable for these data.

10.4 Crossover Design Studies

Conditional logistic regression is a useful technique in the analysis of the *crossover design study*, also called the *changeover study*. In these designs, often used in clinical trials, the study is divided into periods and patients receive a different treatment during each period. Thus, the patients act as their own controls. Interest lies in comparing the efficacy of the treatments, adjusting for period effects and carryover effects. The basic crossover design is a two-period design, but designs with three or more periods are also implemented. This section describes the use of conditional logistic regression for both two- and three- period designs.

10.4.1 Two-Period Crossover Design

A two-period crossover study can be considered another example of paired data. Table 10.2 contains data from a two-period crossover design clinical trial (Koch et al. 1977). Patients were stratified according to two age groups and then assigned to one of three treatment sequences. Responses were measured as favorable (F) or unfavorable (U); thus, FF indicates a favorable response in both Period 1 and Period 2.

Table 10.2 Two-Period Crossover Study

Age	Sequence	Response Profiles				Total
		FF	FU	UF	UU	
older	A:B	12	12	6	20	50
older	B:P	8	5	6	31	50
older	P:A	5	3	22	20	50
younger	B:A	19	3	25	3	50
younger	A:P	25	6	6	13	50
younger	P:B	13	5	21	11	50

Sequence A:B means that Drug A was administered during the first period and Drug B was administered during the second period. The value P indicates Placebo. There are six possible sequences over the two age groups; each sequence occurs for one set of 50 patients.

These data can be considered paired data in the sense that there is a response for both Period 1 and Period 2. One strategy for analyzing these data is to model the probability of improvement for each patient in the first period (and not the second) versus the probability of improvement in either the first or second period but not both. This can be expressed as the conditional probability

$$\frac{\Pr\{Period1=F\}\Pr\{Period2=U\}}{\Pr\{Period1=F\}\Pr\{Period2=U\} + \Pr\{Period1=U\}\Pr\{Period2=F\}}$$

Thus, the analysis strategy can proceed in the same manner as for the highly stratified paired data. In that example, the analysis adjusted out center-to-center variability (intercenter variability) and concentrated on intracenter variability. In this example, you are conditioning away, or adjusting out, patient-to-patient variability (interpatient variability) and concentrating on intrapatient information. This allows you to perform analyses that may not be possible with population-averaging methods (such as ordinary logistic regression) because of small sample size, although the resulting strategy may not be as efficient. These conditioning methods also lead to results with different interpretation; for example, the resulting odds ratios apply to each patient individually in the study rather than to patients on average.

The effects of interest are the period effect, effects for Drugs A and B, and a carryover effect for drugs A and B from Period 1 to Period 2. Table 10.3 and Table 10.4 display the explanatory variables that pertain to these effects for Period 1 and Period 2, using incremental effects parameterization.

Table 10.3 Period 1 Data

Age	Treatment	Period1	Period × Age	Drug A	Drug B	CarryA	CarryB
older	A	1	1	1	0	0	0
older	B	1	1	0	1	0	0
older	P	1	1	0	0	0	0
younger	B	1	0	0	1	0	0
younger	A	1	0	1	0	0	0
younger	P	1	0	0	0	0	0

Table 10.4 Period 2 Data

Age	Treatment	Period1	Period × Age	Drug A	Drug B	CarryA	CarryB
older	B	0	0	0	1	1	0
older	P	0	0	0	0	0	1
older	A	0	0	1	0	0	0
younger	A	0	0	1	0	0	1
younger	P	0	0	0	0	1	0
younger	B	0	0	0	1	0	0

Note that there are six response functions, logits based on FU versus UF, and thus six degrees of freedom with which to work. If you include the two effects for Drugs A and B, the age × period effect, and the period effect, then there are two degrees of freedom remaining. They can be used to explore the carryover effects or the age × drug effects. The two degree-of-freedom tests for both sets of effects are identical since both correspond to comparable simplifications of the saturated model.

The model employed includes the carryover effects. You can write this model as

$$\Pr\{FU|FU \text{ or } UF\} = \frac{\exp\{\beta + \tau'\mathbf{z}\}}{1 + \exp\{\beta + \tau'\mathbf{z}\}}$$

where $\mathbf{z}$ consists of the difference between the two periods for period × age, Drug A, Drug B, CarryA, and CarryB. The parameter β is the effect for period, τ_0 is the effect for period × age, τ_1 and τ_2 are the effects for Drug A and Drug B, respectively, and τ_3 and τ_4 are the effects for carryover for Drug A and carryover for Drug B, respectively.

The following DATA step inputs the cell counts of the table one response profile at a time.

```
data cross1 (drop=count);
   input age $ sequence $ time1 $ time2 $ count;
   do i=1 to count;
      output;
   end;
   datalines;
older AB F F 12
older AB F U 12
older AB U F 6
older AB U U 20
```

```
older   BP F F  8
older   BP F U  5
older   BP U F  6
older   BP U U 31
older   PA F F  5
older   PA F U  3
older   PA U F 22
older   PA U U 20
younger BA F F 19
younger BA F U  3
younger BA U F 25
younger BA U U  3
younger AP F F 25
younger AP F U  6
younger AP U F  6
younger AP U U 13
younger PB F F 13
younger PB F U  5
younger PB U F 21
younger PB U U 11
;
```

The DATA step that creates SAS data set CROSS2 creates observations for both periods, in addition to character variables for drug, carryover, and response based on the subject's values at each of the two periods.

```
data cross2; set cross1;
   subject=_n_;
   period=1;
      drug = substr(sequence, 1, 1);
      carry='none';
      response =time1;
      output;
   period=2;
      drug    =  substr(sequence, 2, 1);
      carry   =  substr(sequence, 1, 1);
      if carry='P' then carry='none';
      response =time2;
      output;
run;
```

The following PROC LOGISTIC statements request the desired analysis. The strata variable SUBJECT is specified in the STRATA statement; these are the effects that will be conditioned out. Variables DRUG, PERIOD, AGE, and CARRY are listed in the CLASS statement and reference parameterization is requested. Model effects included are period, drug, carryover, and the period × age interaction.

```
proc logistic data=cross2;
   class drug period age carry / param=ref;
   strata subject;
   model response = period drug period*age carry;
run;
```

Output 10.17 displays the fit statistics.

Output 10.17 Model Assessment Statistics

Model Fit Statistics		
Criterion	Without Covariates	With Covariates
AIC	166.355	129.579
SC	166.355	155.961
-2 Log L	166.355	117.579

Output 10.18 displays the global tests.

Output 10.18 Global Tests

Conditional Analysis

Testing Global Null Hypothesis: BETA=0			
Test	Chi-Square	DF	Pr > ChiSq
Likelihood Ratio	48.7761	6	<.0001
Score	43.6628	6	<.0001
Wald	32.8513	6	<.0001

The Type 3 analysis of effects results in Output 10.19 indicate that neither carryover effect is influential. There appears to be a period effect and a possible drug effect.

Output 10.19 Type 3 Results

Type 3 Analysis of Effects			
Effect	DF	Wald Chi-Square	Pr > ChiSq
period	1	4.1832	0.0408
drug	2	4.5691	0.1018
period*age	1	2.2056	0.1375
carry	2	0.2450	0.8847

Output 10.20 displays the parameter estimates. Drug A appears to drive the drug effect.

Output 10.20 Parameter Estimates

Analysis of Conditional Maximum Likelihood Estimates							
Parameter			DF	Estimate	Standard Error	Wald Chi-Square	Pr > ChiSq
period	1		1	-1.4370	0.7026	4.1832	0.0408
drug	A		1	1.2467	0.6807	3.3547	0.0670
drug	B		1	-0.00190	0.6412	0.0000	0.9976
period*age	1	older	1	0.6912	0.4654	2.2056	0.1375
carry	A		1	-0.1903	1.1125	0.0293	0.8642
carry	B		1	-0.5653	1.1556	0.2393	0.6247

Next, the reduced model that excludes the carryover effects is fit. Since the period × age effect is modestly suggestive, it is kept in the model. The following PROC LOGISTIC invocation fits this model. It also includes a test for whether Drug A and Drug B have similar effects by using the TEST statement. The ODDSRATIO statement is specified to produce odds ratios that compare each level of the drug effect to all other levels. When ODS Graphics is enabled, submitting an ODDSRATIO statement also produces the odds ratio plot.

```
ods graphics on;
proc logistic data=cross2;
   class drug period age carry / param=ref;
   strata subject;
   model response = period drug period*age;
   A_B: test drugA=drugB;
   oddsratio drug;
run;
ods graphics off;
```

Output 10.21 displays the model fit statistics for the reduced model. If you take the difference in $-2 \log L$ for the full model displayed in Output 10.17 and the reduced model, $117.826 - 117.579$, you obtain the log-likelihood ratio test for the carryover effects. Since $Q_L = 0.247$ with 2 df, this test is nonsignificant. (If you fit the model with age and drug interactions and perform a similar model reduction, this test would have the same value.)

Output 10.21 Model Assessment Statistics

Model Fit Statistics		
Criterion	Without Covariates	With Covariates
AIC	166.355	125.826
SC	166.355	143.413
-2 Log L	166.355	117.826

Output 10.22 displays the global tests.

Output 10.22 Global Tests

Conditional Analysis

Testing Global Null Hypothesis: BETA=0			
Test	Chi-Square	DF	Pr > ChiSq
Likelihood Ratio	48.5296	4	<.0001
Score	43.1385	4	<.0001
Wald	32.3405	4	<.0001

The maximum likelihood estimates are displayed in Output 10.23. The period effect remains clearly significant ($Q_W = 12.9534$, $p = 0.0003$). Drug A appears to be strongly significant relative to placebo, while Drug B appears to be nonsignificant.

Output 10.23 Maximum Likelihood Estimates

Analysis of Conditional Maximum Likelihood Estimates						
Parameter			DF	Estimate	Standard Error	Wald Chi-Square	Pr > ChiSq
period	1		1	-1.1905	0.3308	12.9534	0.0003
drug	A		1	1.3462	0.3289	16.7497	<.0001
drug	B		1	0.2662	0.3233	0.6777	0.4104
period*age	1	older	1	0.7102	0.4576	2.4088	0.1207

The period × age effect is still suggestive. Whether you remove this effect from the model depends on your approach to the analysis. If you think of the study as two separate studies of older and younger people, then you probably will want to keep this effect in the model. If your general structural purpose did not include the distinction of older and younger groups, then you will probably want to remove this effect.

Output 10.24 contains the results of the test that compares the Drug B effect and the Drug A effect. The test is clearly significant; the drugs have different effects.

Output 10.24 Drug A versus Drug B

Linear Hypotheses Testing Results			
Label	Wald Chi-Square	DF	Pr > ChiSq
A_B	10.9220	1	0.0010

Output 10.25 reports the odds ratios. For each patient, the odds of a favorable versus an unfavorable response is 3.84 higher for drug A than for the placebo. The odds of a favorable response is 2.95 higher for those receiving Drug A compared to Drug B. However, the odds of favorable response for Drug B are only 1.305 higher than the same odds for the placebo, and the corresponding Wald

confidence interval contains the value 1.

Output 10.25 Odds Ratios

Odds Ratio Estimates and Wald Confidence Intervals			
Label	Estimate	95% Confidence Limits	
drug A vs B	2.945	1.552	5.588
drug A vs P	3.843	2.017	7.322
drug B vs P	1.305	0.692	2.459

These results are illustrated with Output 10.26, which presents the odds ratio graphically, with the horizontal axis presented on a log 2 scale.

Output 10.26 Odds Ratios

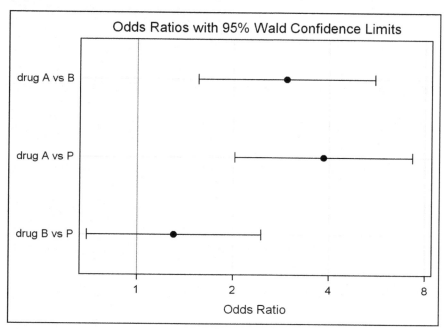

10.4.2 Three-Period Crossover Study

The three-period crossover study provides additional challenges. Consider the data from an exercise study in which participants with chronic respiratory conditions were exposed to low, medium, and high levels of air pollution while exercising on a stationary bike. The outcome was the level of respiratory distress as measured on a scale from 0 for none to 3 for severe. A dichotomous baseline reading of 0 for no distress and 1 for some distress was also recorded before each subject began pedaling. There was a two-week washout period between each of the sessions. As in the two-period crossover study, there is interest in examining carryover effects as well as period effects. The subjects were randomized to one of six sequences: HLM, HML, LHM, LMH, MHL, and MLH, where L, M, and H correspond to low, medium, and high amounts of air pollution. These data are loosely based on an example discussed in Tudor, Koch, and Catellier (2000), which is a useful discussion of biostatistical data from crossover studies.

The conditional analysis of these data provides a way to detect within-subject effects, namely the pollution effect, and also investigates the period and carryover effects. The response of interest is dichotomous—whether the subject had distress versus no distress—the 1s, 2s, and 3s versus the 0s. Table 10.5 contains the eight possible response profile frequencies by randomization sequence scheme and response profile.

Table 10.5 Response Profile Frequencies

Sequence	NNN	DNN	NDN	NND	DDN	DND	NDD	DDD	Total
HLM	0	0	2	0	2	9	0	11	24
HML	0	2	0	0	10	5	0	9	26
LHM	0	1	0	0	3	2	7	11	24
LMH	0	0	0	0	2	3	9	10	24
MHL	0	0	4	0	5	0	3	8	20
MLH	0	1	0	5	0	5	5	16	32
Total	0	4	6	5	22	24	24	65	150

Consider the possible outcome profiles similar to those discussed previously for the two-period crossover study. On page 309, the likelihood conditioned on the discordant pairs, or those response profiles where y_{h1} and y_{h2} were different. Cases where $y_{h1} = y_{h2}$ were considered uninformative. The sum $\sum_i^2 y_{hi}$ implies equal levels of y_{h1} and y_{h2} when it is 0 (0,0) and 2 (1,1) but not when the sum is 1. In conditional logistic regression, you are conditioning on the $\{\sum_i^r y_{hi}\}$, which are the sufficient statistics for the $\{\alpha_h\}$.

For the three-period case, $r = 3$ and eight possible profiles exist, two of which are uninformative, when $\{\sum_i^3 y_{hi}\} = 0$ or 3. When $\{\sum_i^3 y_{hi}\} = 1$ or 2, there are three possible patterns for (y_{h1}, y_{h2}, y_{h3}).

The contributions to the conditional likelihood are:

$$\frac{\Pr\{y_{hi} = 1, y_{hi'} = 0 \text{ for all } i' \neq i\}}{\Pr\{y_{h1} + y_{h2} + y_{h3} = 1\}} = \frac{\exp(\mathbf{x}'_{hi}\boldsymbol{\beta})}{\sum_{i'}^3 \exp(\mathbf{x}'_{hi'}\boldsymbol{\beta})} \text{ for } i = 1, 2, 3$$

and

$$\frac{\Pr\{y_{hi} = 0, y_{hi'} = 1 \text{ for all } i' \neq i\}}{\Pr\{y_{h1} + y_{h2} + y_{h3} = 2\}} = \frac{\exp(\sum_{i'=1}^3 \mathbf{x}'_{hi'}\boldsymbol{\beta} - \mathbf{x}'_{hi}\boldsymbol{\beta})}{\sum_{i'=1}^3 \exp(\sum_{i'=1}^3 \mathbf{x}'_{hi'}\boldsymbol{\beta} - \mathbf{x}'_{hi}\boldsymbol{\beta})}$$

This likelihood structure turns out to be the same as the trichotomous loglinear extension of logistic regression, and you can use unconditional maximum likelihood to estimate $\boldsymbol{\beta}$. The functions you would analyze are generalized logits, which can be fit by using the GLOGIT link in the LOGISTIC procedure. However, here the conditional logistic model is used to analyze these data.

For the exercise data, there is interest in evaluating whether pollution has an effect on respiratory distress and whether there are period effects and carryover effects. You could consider carryover effects from one level to another from Period 1 to Period 2, and then from Period 2 to Period 3. However, there are not enough degrees of freedom for these data to pursue such a strategy. Instead, the analysis focuses on whether there is a carryover effect from a medium-pollution period to the next period and a carryover effect from a high-pollution period to the next period. This is a reasonable strategy.

The next DATA step inputs the exercise data. There is one observation per subject per period. The variable SEQUENCE contains the sequence information, for example, observations with the value 'HML' received the sequence high in the first period, medium in the second period, and low in the third period. The variable ID is the subject ID within sequence group, and the variable PERIOD records the period number. The variable STRATA, a unique identifier for each subject, is created from SEQUENCE and ID.

The indicator variables HIGH and MEDIUM take the value 1 if the exposure is high or medium, respectively, for that period; the class variable EXPOSURE is created from them. Variables CARRYHIGH and CARRYMEDIUM are indicator variables for whether the previous period had high exposure or medium exposure; class variable CARRY is created from them. The variable BASELINE takes the value 1 for respiratory distress at the beginning of the study. The RESPONSE variable is defined as any distress.

```
data exercise;
input Sequence $ ID $ Period High Medium Baseline
      Response CarryHigh CarryMedium @@;
   strata=sequence||id;
   length exposure carry $ 10;
   If (High) then exposure='high';
   else if (Medium) then exposure='medium';
   else exposure='low';
   If (CarryHigh) then carry='high';
   else if (CarryMedium) then carry='medium';
   else carry='low';
   distress=(Response >0);
   datalines;
HML  1  1 1 0 0 3 0 0  HML  1  2 0 1 0 1 1 0  HML  1  3 0 0 0 0 0 1
HML  2  1 1 0 0 3 0 0  HML  2  2 0 1 0 2 1 0  HML  2  3 0 0 0 0 0 1
HML  3  1 1 0 1 3 0 0  HML  3  2 0 1 0 2 1 0  HML  3  3 0 0 0 0 0 1
HML  4  1 1 0 0 2 0 0  HML  4  2 0 1 0 0 1 0  HML  4  3 0 0 0 2 0 1
HML  5  1 1 0 0 3 0 0  HML  5  2 0 1 0 0 1 0  HML  5  3 0 0 0 1 0 1
HML  6  1 1 0 1 2 0 0  HML  6  2 0 1 0 1 1 0  HML  6  3 0 0 0 2 0 1
HML  7  1 1 0 0 3 0 0  HML  7  2 0 1 0 1 1 0  HML  7  3 0 0 0 2 0 1
HML  8  1 1 0 0 3 0 0  HML  8  2 0 1 0 2 1 0  HML  8  3 0 0 0 1 0 1

    ... more lines ...

LMH 22  1 0 0 0 0 0 0  LMH 22  2 0 1 0 2 0 0  LMH 22  3 1 0 0 2 0 1
LMH 23  1 0 0 0 0 0 0  LMH 23  2 0 1 0 1 0 0  LMH 23  3 1 0 0 2 0 1
LMH 24  1 0 0 1 0 0 0  LMH 24  2 0 1 0 2 0 0  LMH 24  3 1 0 0 2 0 1
;
```

The STRATA statement defines the strata. Variables PERIOD, CARRY, and EXPOSURE are listed in the CLASS statement as well as being specified as explanatory variables, along with BASELINE. The INCLUDE=2 option produces a score test for the joint effects of PERIOD and CARRY.

```
proc logistic data=exercise descending;
   class period carry exposure / param=ref order=data;
   strata strata;
   model distress = exposure baseline period carry  / include=2
      selection=forward details;
run;
```

Output 10.27 displays the Model Information. It shows that 65 of the 150 strata are uninformative. This corresponds to the DDD column in Table 10.5.

Output 10.27 Model Information

Conditional Analysis

Model Information	
Data Set	WORK.EXERCISE
Response Variable	distress
Number of Response Levels	2
Number of Strata	150
Number of Uninformative Strata	65
Frequency Uninformative	195
Model	binary logit
Optimization Technique	Newton-Raphson ridge

Output 10.28 contains the response profile and Output 10.29 contains the CLASS variable information.

Output 10.28 Response Profile

Response Profile		
Ordered Value	distress	Total Frequency
1	1	350
2	0	100

Probability modeled is distress=1.

Output 10.29 Class Level Information

Class Level Information			
Class	Value	Design Variables	
Period	1	1	0
	2	0	1
	3	0	0
carry	low	1	0
	high	0	1
	medium	0	0
exposure	high	1	0
	medium	0	1
	low	0	0

Output 10.30 displays the stratification summary.

Output 10.30 Strata Summary

Strata Summary				
Response Pattern	distress		Number of Strata	Frequency
	1	0		
1	1	2	15	45
2	2	1	70	210
3	3	0	65	195

The 70 strata with response pattern (2, 1) correspond to the response profiles with Ds in the first and second, first and third, or second and third periods from Table 10.5. The 15 strata with response pattern (1, 2) correspond to the response profiles with D in either the first, second, or third period. Together, the 70 and 15 strata comprise the informative strata that are the basis of the analysis.

Output 10.31 contains the residual score test.

Output 10.31 Residual Score Test

Residual Chi-Square Test		
Chi-Square	DF	Pr > ChiSq
0.6431	4	0.9582

The score test for including both CARRY and PERIOD in the model takes the value 0.6431 with $p = 0.9582$ for 4 df (two levels for each of these variables). Thus, you can conclude that carryover and period effects do not need to be further considered.

Output 10.32 contains the parameter estimates from this analysis.

Output 10.32 Parameter Estimates

Analysis of Conditional Maximum Likelihood Estimates						
Parameter		DF	Estimate	Standard Error	Wald Chi-Square	Pr > ChiSq
exposure	high	1	2.2527	0.3983	31.9938	<.0001
exposure	medium	1	0.6559	0.2547	6.6324	0.0100
Baseline		1	-0.4872	0.4457	1.1948	0.2744

Baseline appears to have a marginal influence.

The model with exposure and baseline is fit next; the forward selection method is used to include the exposure effect and then evaluate whether the baseline variable also enters the model.

```
proc logistic data=exercise descending;
   class exposure / param=ref order=data;
   strata strata;
   model distress = exposure  baseline
       / selection=forward include=1 details;
run;
```

Output 10.33 contains the score test, labeled "Residual Chi-Square Test," which has the value 1.2087 and indicates adequate goodness of fit with 1 df and $p = 0.2716$. Thus, the model that contains only treatment is reasonable.

Output 10.33 Score Test

Conditional Analysis

Residual Chi-Square Test		
Chi-Square	DF	Pr > ChiSq
1.2087	1	0.2716

Output 10.34 displays the parameter estimates for the final model.

Output 10.34 Final Model Parameter Estimates

Analysis of Conditional Maximum Likelihood Estimates						
Parameter		DF	Estimate	Standard Error	Wald Chi-Square	Pr > ChiSq
exposure	high	1	2.2635	0.3986	32.2410	<.0001
exposure	medium	1	0.6605	0.2526	6.8360	0.0089

With the other effects out of the model, the parameters for high and medium pollution levels are significant at the $\alpha = 0.05$ level of significance.

You might want to determine whether high and medium pollution are equivalent effects. You can submit another PROC LOGISTIC analysis with a CONTRAST statement to test this hypothesis.

```
proc logistic data=exercise descending;
   class exposure / param=ref order=data;
   strata strata;
   model distress = exposure ;
   contrast 'difference' exposure 1 -1 / estimate=parm;
   oddsratio exposure;
run;
```

Output 10.35 displays the test results for whether the effects of high pollution and medium pollution are equivalent.

Output 10.35 Test Results for Exposure

Conditional Analysis

Contrast Test Results			
Contrast	DF	Wald Chi-Square	Pr > ChiSq
difference	1	15.7264	<.0001

With a Wald chi-square statistic of 15.7264, 1 df, and $p < 0.0001$, this hypothesis is strongly rejected. High pollution has a much stronger effect on response than medium pollution.

Output 10.36 displays the odds ratios. The odds ratio listed for high level of pollution versus low level means that those subjects exposed to high levels of pollution have approximately ten times higher odds of experiencing respiratory distress than those subjects exposed to low levels of pollution. The 95% confidence limits for this odds ratio are (4.403, 21.006). Similarly, subjects exposed to medium levels of pollution levels have approximately two times higher odds of experiencing respiratory distress than those subjects exposed to low levels of pollution, with 95% confidence limits of (1.180, 3.176).

Output 10.36 Odds Ratios

Odds Ratio Estimates and Wald Confidence Intervals			
Label	Estimate	95% Confidence Limits	
exposure high vs medium	4.968	2.250	10.970
exposure high vs low	9.617	4.403	21.006
exposure medium vs low	1.936	1.180	3.176

10.5 General Conditional Logistic Regression

When you have paired data or responses that comprise profiles considered to come from a trinomial distribution, you can easily write down the possible response profiles and identify the informative and uninformative observations. In the case of the paired response, you can construct a likelihood that is similar to the unconditional likelihood for a dichotomous response; in the case of a trinomial outcome, you can create a model and likelihood that is based on generalized logits and is an extension of the loglinear model, which is discussed in Chapter 12, "Poisson Regression and Related Loglinear Models."

However, for more complicated data situations, equivalent unconditional strategies are not conveniently available. Consider the case of the diagnosis data displayed in Table 10.6 on page 323. Researchers studied subjects at two times under two conditions. You can treat these data as repeated measurements, which is done in Chapter 15, "Generalized Estimating Equations," or you can perform a conditional logistic regression, considering each patient to be a separate stratum. This is a reasonable strategy if you are only interested in within-subject effects since you are conditioning out subject to subject variability. In addition, the resulting odds ratios apply to subjects individually instead of on average. Such models are called subject-specific models versus population-averaged models, which are discussed in Chapter 15. If one purpose of your analysis is to produce a prediction model, such as one you might use in a clinical setting to determine treatment for patients, the subject-specific model may be appealing. Random effect models are other examples of subject-specific models.

If you consider the diagnosis data, you see that there are two possible outcomes at four different combinations of condition and time. Only two profiles are uninformative: the case where all of the responses are 'no' and the case where all of the responses are 'yes'. There are, however, fourteen other profiles (out of 2^4 total profiles): four in which only one 'yes' is recorded, six in which two 'yes's are recorded, and four profiles in which three 'yes's are recorded.

Consider the general model for stratified logistic regression:

$$\log\left\{\frac{\theta}{1-\theta}\right\} = \alpha_h + \mathbf{x}\boldsymbol{\beta}$$

The α_h are stratum-specific parameters for each stratum, $h = 1, \ldots, q$. In conditional inference, you treat the α_h as nuisance parameters and eliminate them from the likelihood function by conditioning on their sufficient statistic. Appendix B contains the methodological details.

10.5.1 Analyzing Diagnostic Data

MacMillan et al. (1981) analyze data from a one population observational study involving 793 subjects. For each subject, two diagnostic procedures (standard and test) were carried out at each of two times. The results of the four evaluations were classified as positive or negative. Since a dichotomous response was measured at $t = 4$ occasions, there are $r = 2^4 = 16$ response profiles. Table 10.6 displays the resulting data.

You can consider each of the subjects in this study to be a separate stratum, with four measurements in each stratum. By performing a conditional logistic regression, you are eliminating subject-to-subject variability. The effects of interest, time and treatment, are within-subject effects, which can be handled by conditional logistic regression. Note that other strategies would be required if between-subject effects were of interest, such as age, clinic, and sex.

Table 10.6 Diagnostic Test Results for 793 Subjects

Time 1		Time 2		No. of
Standard	Test	Standard	Test	Subjects
Negative	Negative	Negative	Negative	509
Negative	Negative	Negative	Positive	4
Negative	Negative	Positive	Negative	17
Negative	Negative	Positive	Positive	3
Negative	Positive	Negative	Negative	13
Negative	Positive	Negative	Positive	8
Negative	Positive	Positive	Negative	0
Negative	Positive	Positive	Positive	8
Positive	Negative	Negative	Negative	14
Positive	Negative	Negative	Positive	1
Positive	Negative	Positive	Negative	17
Positive	Negative	Positive	Positive	9
Positive	Positive	Negative	Negative	7
Positive	Positive	Negative	Positive	4
Positive	Positive	Positive	Negative	9
Positive	Positive	Positive	Positive	170

The following DATA step creates the SAS data set DIAGNOSIS.

```
data diagnosis;
   input std1 $ test1 $ std2 $ test2 $ count;
   do i=1 to count;
      output;
   end;
   datalines;
Neg Neg Neg Neg 509
Neg Neg Neg Pos   4
Neg Neg Pos Neg  17
Neg Neg Pos Pos   3
Neg Pos Neg Neg  13
```

```
Neg Pos Neg Pos   8
Neg Pos Pos Neg   0
Neg Pos Pos Pos   8
Pos Neg Neg Neg  14
Pos Neg Neg Pos   1
Pos Neg Pos Neg  17
Pos Neg Pos Pos   9
Pos Pos Neg Neg   7
Pos Pos Neg Pos   4
Pos Pos Pos Neg   9
Pos Pos Pos Pos 170
;
```

The next DATA step creates one record per measurement per subject as well as treatment and time variables. It creates a unique SUBJECT value for each subject in the study.

```
data diagnosis2;
   set diagnosis;
   drop std1 test1 std2 test2;
   subject=_n_;
   time=1; procedure='standard';
   response=std1; output;
   time=1; procedure='test';
   response=test1; output;
   time=2; procedure='standard';
   response=std2; output;
   time=2; procedure='test';
   response=test2; output;
run;
```

The following PROC LOGISTIC invocation requests the model including the variables TIME, TREATMENT, and their interaction. The variable SUBJECT is placed in the STRATA statement so that the estimation process conditions on subject. The REF=FIRST option is used in the CLASS statement to specify that the reference level is standard treatment at the first time.

```
proc logistic data=diagnosis2;
   class time (ref=first) procedure (ref=first) / param=ref;
   strata subject;
   model response(event="Neg")= time procedure
   time*procedure;
run;
```

Output 10.37 displays the strata summary. The majority of these strata are uninformative.

Output 10.37 Strata Summary

Conditional Analysis

Strata Summary				
Response Pattern	response		Number of Strata	Frequency
	Neg	Pos		
1	0	4	170	680
2	1	3	30	120
3	2	2	36	144
4	3	1	48	192
5	4	0	509	2036

A look at the table of parameter estimates in Output 10.38 indicates that the interaction is not important. The Wald statistic has the value 1.695 for 1 df and serves as a goodness-of-fit test for the main effects model.

Output 10.38 Parameter Estimates for Full Model

Analysis of Conditional Maximum Likelihood Estimates							
Parameter			DF	Estimate	Standard Error	Wald Chi-Square	Pr > ChiSq
time	2		1	-0.0625	0.2500	0.0625	0.8026
procedure	test		1	0.3848	0.2544	2.2881	0.1304
time*procedure	2	test	1	0.4726	0.3630	1.6952	0.1929

The model is refit with variables TIME and PROCEDURE. The options SELECTION=FORWARD, INCLUDE=2, and DETAILS are specified to obtain a score test to serve as a goodness-of-fit test for the model.

```
proc logistic data=diagnosis2;
   class time (ref=first) procedure (ref=first) / param=ref;
   strata subject;
   model response(event="Neg")= time procedure
         time*procedure
      /selection=forward include=2 details;
run;
```

Output 10.39 displays the score statistic based on the remaining influence of the time × procedure interaction. It takes the value 1.7002 with $p = 0.1923$, indicating an adequate model fit.

Output 10.39 Score Statistic

Conditional Analysis

Residual Chi-Square Test		
Chi-Square	DF	Pr > ChiSq
1.7002	1	0.1923

Output 10.40 contains the parameter estimates and odds ratios. The test procedure is highly significant, and the odds of a negative response are almost twice as much for the test procedure as for the standard procedure. The confidence limits for this odds ratio are (1.292, 2.653).

Output 10.40 Parameter Estimates for Main Effects Model

Analysis of Conditional Maximum Likelihood Estimates						
Parameter		DF	Estimate	Standard Error	Wald Chi-Square	Pr > ChiSq
time	2	1	0.1627	0.1807	0.8114	0.3677
procedure	test	1	0.6159	0.1836	11.2557	0.0008

Odds Ratio Estimates			
Effect	Point Estimate	95% Wald Confidence Limits	
time 2 vs 1	1.177	0.826	1.677
procedure test vs standard	1.851	1.292	2.653

10.6 Paired Observations in a Retrospective Matched Study

Epidemiological investigations often involve the use of retrospective, or case-control studies, where a person known to have the event of interest (case) is paired, or matched, with a person who doesn't have the event (control). The idea is to determine whether the exposure factor is associated with the event; this is presumably made less complicated by using matching to control for possible covariates.

- In a 1:1 matched study, the matched set consists of one case and one control from each stratum. This is the most common situation.

- In a 1:m matched study, the matched set consists of one case and m controls. Usually, m ranges between 2 and 5.

- In the m:n matched study, the matched set consists of n cases with m controls, where usually both m and n are between 1 and 5.

Then, data are collected to determine whether the case and control were exposed to certain risk factors, as measured by the explanatory variables. Through the use of a conditional likelihood, you can define a model that allows you to predict the odds for the event given the explanatory variables. This involves setting up the probabilities for having the exposure given the event and then using Bayes' theorem to determine a relevant conditional likelihood concerning the event. You derive the conditional likelihood by first focusing on the conditional probability of observing the explanatory variables given the outcome (event or not). The derivation of the likelihood in the matched pairs setting is discussed in Appendix A in this chapter. This likelihood is similar to that seen in the preceding sections for highly stratified data.

Note that the conditional likelihood for the matched pairs data is the unconditional likelihood for a logistic regression model where the response is always equal to 1, the covariate values are equal to the differences between the values for the case and the control, and there is no intercept. This means that you can use standard logistic regression software by configuring your data appropriately and eliminating the intercept term.

Through a similar process, you can show that the conditional likelihood for the 1:m matched setting is

$$\prod_{h=1}^{q}\left[1 + \sum_{i=1}^{m}\exp\left\{\boldsymbol{\gamma}'(\mathbf{z}_{hi} - \mathbf{z}_{h0})\right\}\right]^{-1}$$

where $i = 1, 2, \ldots, m$ indexes the controls and $i = 0$ corresponds to the case. However, this is not equivalent to any unconditional form, and the same goes for m:n matched data.

The following sections illustrate the use of PROC LOGISTIC in applications of conditional logistic regression for 1:1, 1:m, and m:n matching. See Breslow and Day (1980) and Collett (2003) for more detail on conditional logistic regression for matched studies.

10.6.1 1:1 Conditional Logistic Regression

Researchers studied women in a retirement community in the 1970s to determine if there was an association between the use of estrogen and the incidence of endometrial cancer (Mack et al. 1976).[1] Cases were matched to controls who were within a year of the same age, had the same marital status, and were living in the same community at the time of the diagnosis of the case. Information was also collected on obesity, hypertension, gallbladder disease history, and non-estrogen drug use. The data used here is a subset of the actual data. There are 63 matched pairs, with the variable CASE=1 indicating a case and CASE=0 indicating a control. The goal of the analysis is to determine whether the presence of endometrial disease is associated with any of the explanatory variables.

```
data match;
   input id case age est gall hyper nonest @@;
   datalines;
1  1 74 1 0 0  1    1 0 75 0 0 0  0
2  1 67 1 0 0  1    2 0 67 0 0 1  1
3  1 76 1 0 1  1    3 0 76 1 0 1  1
4  1 71 1 0 0  0    4 0 70 1 1 0  1
```

---

[1] Data provided by Norman Breslow.

```
 5 1 69 1 1 0   1   5 0 69 1 0 1   1
 6 1 70 1 0 1   1   6 0 71 0 0 0   0
 7 1 65 1 1 0   1   7 0 65 0 0 0   0
 8 1 68 1 1 1   1   8 0 68 0 0 1   1
 9 1 61 0 0 0   1   9 0 61 0 0 0   1
10 1 64 1 0 0   1  10 0 65 0 0 0   0
11 1 68 1 1 0   1  11 0 69 1 1 0   0
12 1 74 1 0 0   1  12 0 74 1 0 0   0
13 1 67 1 1 0   1  13 0 68 1 0 1   1
14 1 62 1 1 0   1  14 0 62 0 1 0   0
15 1 71 1 1 0   1  15 0 71 1 0 1   1

 ... more lines ...

61 1 67 1 1 0   1  61 0 67 1 1 0   1
62 1 74 1 0 1   1  62 0 75 0 0 0   1
63 1 68 1 1 0   1  63 0 69 1 0 0   1
;
```

The following PROC LOGISTIC invocation requests forward model selection.

```
proc logistic;
   strata id;
   model case (event="1") = gall est hyper age nonest /
   selection=forward details;
run;
```

Output 10.41 contains the response profiles. Since these are 1:1 case control data, there are 63 observations with 1 for variable CASE and 63 observations with 0 for variable CASE.

Output 10.41 Response Profile

Conditional Analysis

Ordered Value	case	Total Frequency
1	0	63
2	1	63

Probability modeled is case=1.

Output 10.42 displays the summary strata information. Each pair of subjects includes a case and a control.

Output 10.42 Strata Summary

Response Pattern	case 0	case 1	Number of Strata	Frequency
1	1	1	63	126

In the model selection process, only EST and GALL are entered into the model. Output 10.43 displays the residual score statistic, which has a value of 0.2077 with 3 df, indicating an adequate fit. Output 10.44 displays the score statistic for each variable's entry into the model; since all of the chi-square values are strongly nonsignificant, the model goodness of fit is supported.

Output 10.43 Residual Chi-Square

Residual Chi-Square Test		
Chi-Square	DF	Pr > ChiSq
0.2077	3	0.9763

Output 10.44 Model Selection Results

Analysis of Effects Eligible for Entry			
Effect	DF	Score Chi-Square	Pr > ChiSq
hyper	1	0.0186	0.8915
age	1	0.1432	0.7051
nonest	1	0.0370	0.8474

Output 10.45 displays the statistics that assess the model's explanatory capacity. The range of values indicate that exact analysis might be indicated.

Output 10.45 Explanatory Capacity

Testing Global Null Hypothesis: BETA=0			
Test	Chi-Square	DF	Pr > ChiSq
Likelihood Ratio	33.6457	2	<.0001
Score	27.0586	2	<.0001
Wald	15.3291	2	0.0005

Output 10.46 contains the parameter estimates for the model that contains the main effects EST and GALL. Output 10.47 contains the odds ratios.

Output 10.46 Parameter Estimates

Analysis of Conditional Maximum Likelihood Estimates					
Parameter	DF	Estimate	Standard Error	Wald Chi-Square	Pr > ChiSq
gall	1	1.6551	0.7980	4.3017	0.0381
est	1	2.7786	0.7605	13.3492	0.0003

Parameter estimates for both GALL and EST are significant. The odds ratio for GALL indicates that women with gallbladder disease history have 5.234 times higher odds of contracting endometrial cancer as women without it, adjusting for estrogen use and matched pairs. The odds ratio for EST indicates that women who used estrogen have 16.096 times higher odds for contracting endometrial cancer as women who don't use estrogen, adjusting for gallbladder disease history and matched pairs.

Output 10.47 Odds Ratios

Odds Ratio Estimates		
Effect	Point Estimate	95% Wald Confidence Limits
gall	5.234	1.095 25.006
est	16.096	3.626 71.457

An exact analysis is also performed. The following PROC LOGISTIC statements produce exact analyses for both GALL and EST.

```
proc logistic;
   strata id;
   model case (event="1") = gall est;
   exact gall est /estimate=both;
run;
```

Output 10.48 contains the exact analysis results.

Output 10.48 Exact Tests

Exact Conditional Analysis

Exact Conditional Tests				
			p-Value	
Effect	Test	Statistic	Exact	Mid
gall	Score	5.0341	0.0302	0.0202
	Probability	0.0200	0.0302	0.0202
est	Score	23.0075	<.0001	<.0001
	Probability	3.115E-7	<.0001	<.0001

The exact *p*-value for GALL is 0.0302, and the exact *p*-value for EST is near zero.

Output 10.49 and Output 10.50 display the exact parameter estimates and the odds ratios.

Output 10.49 Exact Parameter Estimates

Exact Parameter Estimates					
Parameter	Estimate	Standard Error	95% Confidence Limits		Two-sided p-Value
gall	1.6318	0.7826	0.0157	3.7162	0.0471
est	2.7124	0.7416	1.3086	4.8930	<.0001

Output 10.50 Exact Odds Ratios

Exact Odds Ratios				
Parameter	Estimate	95% Confidence Limits		Two-sided p-Value
gall	5.113	1.016	41.110	0.0471
est	15.066	3.701	133.346	<.0001

10.7 1:*m* Conditional Logistic Regression

Researchers in a midwestern county tracked flu cases that required hospitalization in residents aged 65 and older during a two-month period in one winter. They matched each case with two controls according to sex and age and also determined whether the cases and controls had a flu vaccine shot and whether they had lung disease. Vaccines were then verified by county health and individual medical practice records. Researchers were interested in whether vaccination had a protective influence on the odds of getting a severe case of flu.

This study is an example of a 1:2 matched study since two controls were chosen for each case. The following DATA step reads the data and computes the frequency of vaccine and lung disease for both cases and controls.

```
data matched;
   input id outcome lung vaccine @@;
   datalines;
1  1 0 0   1 0 1 0   1 0 0 0   2 1 0 0   2 0 0 0   2 0 1 0
3  1 0 1   3 0 0 1   3 0 0 0   4 1 1 0   4 0 0 0   4 0 1 0
5  1 1 0   5 0 0 1   5 0 0 1   6 1 0 0   6 0 0 0   6 0 0 1
7  1 0 0   7 0 0 0   7 0 0 1   8 1 1 1   8 0 0 0   8 0 0 1
9  1 0 0   9 0 0 1   9 0 0 0   10 1 0 0  10 0 1 0  10 0 0 0
11 1 1 0   11 0 0 1  11 0 0 0  12 1 1 1  12 0 0 1  12 0 0 0
13 1 0 0   13 0 0 1  13 0 1 0  14 1 0 0  14 0 0 0  14 0 0 1
```

```
15 1 1 0 15 0 0 0 15 0 0 1 16 1 0 1 16 0 0 1 16 0 0 1
17 1 0 0 17 0 1 0 17 0 0 0 18 1 1 0 18 0 0 1 18 0 0 1
19 1 1 0 19 0 0 1 19 0 0 1 20 1 0 0 20 0 0 0 20 0 0 0
21 1 0 0 21 0 0 1 21 0 0 1 22 1 0 1 22 0 0 0 22 0 1 0
23 1 1 1 23 0 0 0 23 0 0 0 24 1 0 0 24 0 0 1 24 0 0 1
25 1 1 0 25 0 1 0 25 0 0 0 26 1 1 1 26 0 0 0 26 0 0 0
27 1 1 0 27 0 0 1 27 0 0 0 28 1 0 1 28 0 1 0 28 0 0 0
29 1 0 0 29 0 0 0 29 0 1 1 30 1 0 0 30 0 0 0 30 0 0 0

  ... more lines ...

145 1 1 0 145 0 0 1 145 0 0 0 146 1 1 0 146 0 1 0 146 0 0 0
147 1 0 1 147 0 0 0 147 0 0 1 148 1 0 0 148 0 0 1 148 0 0 0
149 1 1 0 149 0 1 0 149 0 1 0 150 1 1 1 150 0 0 0 150 0 0 1
;
```

The following PROC FREQ statements request crosstabulations of vaccine by outcome status and lung disease by outcome status.

```
proc freq;
    tables outcome*lung outcome*vaccine /nocol nopct;
run;
```

Output 10.51 contains the frequencies of vaccine and lung disease for both cases and controls. In these data, 16% of the controls had lung disease, and 42% of the cases had lung disease. Also, 39% of the controls and 31% of the cases had been vaccinated.

Output 10.51 Frequencies of Vaccine and Lung Disease by Cases and Controls

Frequency Row Pct	Table of outcome by lung		
		lung	
outcome	0	1	Total
0	252 84.00	48 16.00	300
1	87 58.00	63 42.00	150
Total	339	111	450

Frequency Row Pct	Table of outcome by vaccine		
		vaccine	
outcome	0	1	Total
0	183 61.00	117 39.00	300
1	103 68.67	47 31.33	150
Total	286	164	450

The following statements request the conditional logistic regression analysis. The SELECTION=FORWARD option is specified to request forward selection model building.

```
proc logistic;
   class lung vaccine;
   strata id;
   model outcome(event="1") = lung vaccine lung*vaccine /
      selection=forward details;
run;
```

Output 10.52 displays the model building results.

Output 10.52 Model Building Results

Conditional Analysis

Residual Chi-Square Test		
Chi-Square	DF	Pr > ChiSq
3.2982	2	0.1922

Analysis of Effects Eligible for Entry			
Effect	DF	Score Chi-Square	Pr > ChiSq
vaccine	1	3.2529	0.0713

The variable LUNG is entered into the model, but the variable VACCINE and the interaction term are not. However, the p-value of 0.0713 for VACCINE is suggestive, so the model including LUNG and VACCINE is fit next. The interaction term is included to obtain the residual score test as a measure of goodness of fit.

```
proc logistic;
   class lung vaccine;
   strata id;
   model outcome(event="1") = lung vaccine lung*vaccine /
      selection=forward details include=2;
run;
```

Output 10.53 displays the residual score statistic. With a value of 0.0573 and $p = 0.8107$, this statistic supports goodness of fit.

Output 10.53 Residual Score Statistic

Conditional Analysis

Residual Chi-Square Test		
Chi-Square	DF	Pr > ChiSq
0.0573	1	0.8107

Output 10.54 includes the parameter estimates.

Output 10.54 Parameter Estimates

Analysis of Conditional Maximum Likelihood Estimates						
Parameter		DF	Estimate	Standard Error	Wald Chi-Square	Pr > ChiSq
lung	1	1	1.3053	0.2348	30.8967	<.0001
vaccine	1	1	-0.4008	0.2233	3.2223	0.0726

Odds Ratio Estimates		
Effect	Point Estimate	95% Wald Confidence Limits
lung 1 vs 0	3.689	2.328 5.845
vaccine 1 vs 0	0.670	0.432 1.038

The odds ratio for getting a case of flu resulting in hospitalization is $e^{-0.40078} = 0.67$ for those with vaccine versus those without vaccine. Thus, study participants with vaccination reduced their odds of getting hospitalizable flu by 33% compared to their nonvaccinated matched counterparts. This means that vaccination had a protective effect, controlling for lung disease status (and age and sex, via matching). The confidence limits for this odds ratio are (0.432, 1.038).

10.8 Exact Conditional Logistic Regression in the Stratified Setting

While conditional logistic regression often serves to counterbalance the small counts in a strata by conditioning away the strata effect, sometimes the data are so sparse that these methods also become inappropriate. Exact conditional logistic regression has already been applied in previous sections when fit statistics indicated that asymptotic-based analyses were questionable. However, in many situations, the sparseness of the data is so apparent that only exact methods should be considered.

In the asymptotic logistic regression setting, the methodologies for the unstratified and stratified analysis are different (the former is based on an unconditional likelihood and the latter is based on a conditional likelihood). In the exact setting, you use the same (conditioning) methodology. The only difference is that, in the unstratified case, you don't have stratification variables and you are conditioning away only explanatory variables; in the stratified case, you are conditioning away both stratification variables and explanatory variables.

The following example is from Luta et al. (1998), which describes methods for analyzing clustered binary data with exact methods. The data are from a cardiovascular study of eight animals who received various drug treatments. Researchers then arrested coronary flow, which led to the development of regional ischemia, and they recorded whether an adverse cardiovascular event occurred during an eight-minute interval. The heart was reperfused for 50 minutes to allow the heart to return to normal, and then another treatment was tested. Thus, there are up to five repeated

10.8. Exact Conditional Logistic Regression in the Stratified Setting

measurements on eight clusters, or animals. For various reasons, no animal received all of the five possible treatments. Because of the sequences of treatments used by the investigators, the investigation was not assumed to be a crossover study. Because of the reperfusion, the period and carryover effects were considered to be ignorable.

The data include relatively small counts so a reasonable strategy is exact stratified logistic regression, conditioning on the animals. The following DATA step inputs the data for this analysis. Only the observations that correspond to drug treatments are included; observations corresponding to the shunt treatment are eliminated (the shunt is simply the placement of the intracoronary artery catheter). The treatments are control (C) which is no drug, test drug and counteracting agent (DA), low-dose test drug (D1), and high-dose test drug (D2). For this analysis, the drug effect is assumed to be ordinal with equally spaced intervals; the variable ORDTREAT is coded as 1 for control to 4 for high-dose drug. The variable ANIMAL takes the values from 1 to 8, and the variable RESPONSE is 1 if an event was observed and 0 otherwise.

```
data cardio;
   input animal treatment $ response $ @@;
   if treatment='S' then delete;
   else if treatment='C'  then ordtreat=1;
   else if treatment='DA' then ordtreat=2;
   else if treatment='D1' then ordtreat=3;
   else if treatment='D2' then ordtreat=4;
   datalines;
1 S no   1 C  no   1 C  no   1 D2 yes 1 D1 yes
2 S no   2 D2 yes  2 C  no   2 D1 yes
3 S no   3 C  yes  3 D1 yes  3 DA no  3 C  no
4 S no   4 C  no   4 D1 yes  4 DA no  4 C  no
5 S yes  5 C  no   5 DA no   5 D1 no  5 C  no
6 S no   6 C  no   6 D1 yes  6 DA no  6 C  no
7 S no   7 C  no   7 D1 yes  7 DA no  7 C  no
8 S yes  8 C  yes  8 D1 yes
;
```

The following PROC LOGISTIC statements specify the exact analysis. The variable ANIMAL is placed in the STRATA statement, and the MODEL statement includes RESPONSE as the outcome variable and ORDTREAT as the explanatory variable. You then specify ORDTREAT in the EXACT statement, and ESTIMATE=BOTH requests that both parameter estimates and odds ratios be generated.

```
proc logistic data=cardio descending exactonly;
   strata animal;
   model response = ordtreat;
   exact ordtreat / estimate=both;
run;
```

When the EXACTONLY option is specified, PROC LOGISTIC prints the "Model Information" and "Response Profile" tables (not shown here) and then prints the results of the exact conditional analysis. Output 10.55 displays the exact tests for the treatment effect.

Output 10.55 Exact Tests

Exact Conditional Analysis

Exact Conditional Tests

Effect	Test	Statistic	p-Value Exact	p-Value Mid
ordtreat	Score	10.4411	0.0009	0.0005
	Probability	0.000723	0.0009	0.0005

Both the score and probability test have exact *p*-values of 0.0009, which is highly significant.

Output 10.56 displays the point estimate of the drug effect, 1.9421, and a 95% confidence interval (0.4824, 5.2932).

Output 10.56 Exact Parameter Estimate

Exact Parameter Estimates

Parameter	Estimate	Standard Error	95% Confidence Limits		Two-sided p-Value
ordtreat	1.9421	0.8932	0.4824	5.2932	0.0017

Output 10.57 displays the odds ratio, which takes the value 6.974. The odds of an adverse event were seven times higher with each unit change in treatment level.

Output 10.57 Exact Odds Ratio

Exact Odds Ratios

Parameter	Estimate	95% Confidence Limits		Two-sided p-Value
ordtreat	6.974	1.620	198.976	0.0017

The following PROC LOGISTIC statements request the conditional analysis. The options SELECTION=FORWARD and SLENTRY=.05 are used to produce the score statistic for the drug effect.

```
proc logistic data=cardio descending;
   strata animal;
   model response = ordtreat /selection=forward
      details slentry=.05;
run;
```

Output 10.58 displays the residual score statistic for treatment, which has the value 10.4411, and, with 1 df, a *p*-value of 0.0012.

Output 10.58 Residual Score Test

Conditional Analysis

Residual Chi-Square Test		
Chi-Square	DF	Pr > ChiSq
10.4411	1	0.0012

Output 10.59 contains the parameter estimate, which is 1.9421 with a p-value of 0.0297 for the Wald chi-square. The estimate is very close to the estimate in the exact analysis. Note that the asymptotic p-value is larger than the exact one, which is a bit unusual since most often you find that the exact p-value is larger than the asymptotic p-value.

Output 10.59 Parameter Estimate

Analysis of Conditional Maximum Likelihood Estimates					
Parameter	DF	Estimate	Standard Error	Wald Chi-Square	Pr > ChiSq
ordtreat	1	1.9421	0.8932	4.7275	0.0297

If you run an unstratified asymptotic analysis on these data (that is, use PROC LOGISTIC and regress RESPONSE on ANIMAL and ORDTREAT), you would get the following messages from PROC LOGISTIC:

```
Quasicomplete separation of data points detected.

WARNING: The maximum likelihood estimate may not exist.
WARNING: The LOGISTIC procedure continues in spite of the above
warning. Results shown are based on the last maximum likelihood
iteration. Validity of the model fit is questionable.
```

These messages are a signal that your data are probably not suitable for asymptotic analysis and that you should consider exact methods. In addition, in the table for global fit in the asymptotic analysis (not shown here), the score statistic has the chi-square value 19.1924 ($p = 0.0139$) and the Wald statistic has the value 6.4478 ($p = 0.5972$). As discussed previously, when these test statistics indicate very different results, you should consider whether your data are more suitable for exact analysis.

10.8.1 Printing More Digits

Occasionally, you may want to generate more digits for the p-values than are printed according to the default format in the LOGISTIC procedure. (PROC LOGISTIC computes the number of digits that machine accuracy allows.) You can do this easily with ODS. You output the table that contains the information, and then you print it using different formats. The table names are listed in the documentation for each procedure, and you can determine the variable names with PROC CONTENTS.

The following statements produce the exact results displayed in Output 10.55 with six digits for the p-values.

```
ods output ExactTests=ET;
proc logistic data=cardio descending;
   strata animal;
   model response = ordtreat;
   exact ordtreat / estimate=parm;
run;
proc print data=ET noobs label;
   format ExactPValue pvalue8.6 MidPValue pvalue8.6;
   var Effect--MidPValue;
run;
```

Output 10.60 contains the results with additional digits.

Output 10.60 Exact Tests Results

Effect	Test	Test Statistic	#=Degenerate	Exact p-Value	Mid p-Value
ordtreat	Score	10.4411		0.000868	0.000506
ordtreat	Probability	0.000723		0.000868	0.000506

These results make it easier to compare the results reported in the Luta et al. paper; see that paper for additional analyses performed on these data.

Appendix D displays how to update the default ODS template if you want to produce more digits for this table whenever you used PROC LOGISTIC.

10.9 Appendix A: Theory for the Case-Control Retrospective Setting

Suppose that you have q matched pairs, $h = 1, 2, \ldots, q$, and θ_{hi} is the probability of the ith subject in the hth matched pair having the event ($i = 1, 2$). Suppose that $\mathbf{z}_{hi}$ represents the set of explanatory variables for the ith subject in the hth matched pair.

The likelihood for the vector of explanatory variables being $\mathbf{z}_{h1}$ given that subject $h1$ is the case (e) and being $\mathbf{z}_{h2}$ given that subject $h2$ is the control ($\bar{e}$) is

$$\Pr\{\mathbf{z}_{h1}|e\}\Pr\{\mathbf{z}_{h2}|\bar{e}\}$$

The sum of this likelihood and that for its reverse counterpart, the likelihood for the vector of explanatory variables being $\mathbf{z}_{h1}$ given the control and being $\mathbf{z}_{h2}$ given the case, is

$$\Pr\{\mathbf{z}_{h1}|e\}\Pr\{\mathbf{z}_{h2}|\bar{e}\} + \Pr\{\mathbf{z}_{h1}|\bar{e}\}\Pr\{\mathbf{z}_{h2}|e\}$$

10.9. Appendix A: Theory for the Case-Control Retrospective Setting

and thus the conditional likelihood for a particular matched pair having the observed pairing of explanatory variables z_{h1} with the case e and the explanatory variables z_{h2} with the control $\bar{e}$ is

$$\frac{\Pr\{z_{h1}|e\}\Pr\{z_{h2}|\bar{e}\}}{\Pr\{z_{h1}|e\}\Pr\{z_{h2}|\bar{e}\} + \Pr\{z_{h1}|\bar{e}\}\Pr\{z_{h2}|e\}}$$

Applying Bayes' Theorem, $(P(A|B) = P(B|A)P(A)/P(B))$, to each of the six terms in the above expression, you can rewrite the preceding as

$$\frac{\Pr\{e|z_{h1}\}\Pr\{\bar{e}|z_{h2}\}}{\Pr\{e|z_{h1}\}\Pr\{\bar{e}|z_{h2}\} + \Pr\{\bar{e}|z_{h1}\}\Pr\{e|z_{h2}\}}$$

Thus, the conditional probabilities have been reversed so that they are the probabilities of the event given the explanatory variables.

If you assume a logistic model for θ_{hi}, the probability of the ith subject in the hth matched pair having the event, then you can make the appropriate substitutions into the conditional likelihood. The following is the logistic model for θ_{hi}.

$$\theta_{hi} = \frac{\exp\{\alpha_h + \gamma' z_{hi}\}}{1 + \exp\{\alpha_h + \gamma' z_{hi}\}}$$

where α_h is an effect for the hth stratum, or pair, the z_{hik} are the $k = 1, 2, \ldots, t$ explanatory variables for the ith subject in the hth matched pair, and the γ_k are the corresponding parameters.

Substituting θ_{hi} for $\Pr\{e|z_{hi}\}$ and $(1 - \theta_{hi})$ for $\Pr\{\bar{e}|z_{hi}\}$ produces

$$\frac{\exp\{\alpha_h + \gamma' z_{h1}\}}{\exp\{\alpha_h + \gamma' z_{h1}\} + \exp\{\alpha_h + \gamma' z_{h2}\}}$$

which is equivalent to

$$\frac{\exp\{\gamma'(z_{h1} - z_{h2})\}}{1 + \exp\{\gamma'(z_{h1} - z_{h2})\}}$$

Note that the α_h have dropped out and thus you have eliminated the stratum-specific parameters.

The conditional likelihood for the entire data is the product of the likelihoods for the individual strata.

$$\prod_{h=1}^{q} \frac{\exp\{\gamma'(z_{h1} - z_{h2})\}}{1 + \exp\{\gamma'(z_{h1} - z_{h2})\}}$$

For this conditional likelihood, matched pairs with $z_{h1k} = z_{h2k}$ for all k are uninformative (that is, their contribution to the likelihood is the constant 0.5), and so these matched pairs can be excluded from the analysis.

Through a similar process, you can show that the conditional likelihood for the 1:m matched setting is

$$\prod_{h=1}^{q} \left[1 + \sum_{i=1}^{m} \exp\{\gamma'(z_{hi} - z_{h0})\} \right]^{-1}$$

where $i = 1, 2, \ldots, m$ indexes the controls and $i = 0$ corresponds to the case. However, this is not equivalent to any unconditional form, so you have to use software designed for conditional logistic regression. g

10.10 Appendix B: Theory for General Conditional Logistic Regression

Consider the general model for stratified logistic regression:

$$\log\left\{\frac{\theta}{1-\theta}\right\} = \alpha_h + \mathbf{x}'\boldsymbol{\beta}$$

The α_h are stratum-specific parameters for each stratum, $h = 1, \ldots, q$. In conditional inference, you treat the α_h as nuisance parameters and eliminate them from the likelihood function by conditioning on their sufficient statistic.

The sufficient statistics for the a_h are as follows:

$$a_h = \sum_{i=1}^{n_h} y_{hi}$$

where $i = 1, 2, \ldots, n_h$ for subjects in stratum h. Recall the attention to the sum of the y_{hi}s in the previous discussions of the various profiles possible in the paired case and the three-period crossover. They were the sufficient statistics in those cases. The analysis, and the elimination of the αs in the likelihood, involved conditioning on those sufficient statistics.

Now, consider the model

$$\log\left\{\frac{\theta}{1-\theta}\right\} = \mathbf{J}\boldsymbol{\alpha} + \mathbf{X}\boldsymbol{\beta}$$

$\boldsymbol{\alpha}$ is the $q \times 1$ vector of stratum-specific intercepts

$\boldsymbol{\beta}$ is the $t \times 1$ vector of within-stratum parameters

$\mathbf{J}$ is block diagonal with $\mathbf{1}_{n_h}$ as diagonal blocks

With $\mathbf{T}$ as the vector of sufficient statistics for $\boldsymbol{\beta}$ with elements

$$T_k = \sum_{h=1}^{q} \sum_{i=1}^{n_h} x_{hik} y_{hi} \quad k = (1, \ldots, t)$$

The conditional probability density function of $\mathbf{T} = \mathbf{t}$ given $\mathbf{a} = \mathbf{a}_0$ is

$$f_{\boldsymbol{\beta}}(\mathbf{t}|\mathbf{a}_0) = \frac{C(\mathbf{a}_0, \mathbf{t}) \exp(\mathbf{t}'\boldsymbol{\beta})}{\sum_{\mathbf{u}} C(\mathbf{a}_0, \mathbf{u}) \exp(\mathbf{u}'\boldsymbol{\beta})}$$

where $C(\mathbf{a}_0, \mathbf{u})$ are the number of ys such that $\{\mathbf{J}'\mathbf{y} = \mathbf{a}_0, \mathbf{X}_1'\mathbf{y} = \mathbf{u}\}$ where $\mathbf{y} = (y_{11}, \ldots, y_{1n_1}, \ldots, y_{qn_q})'$.

You then use this conditional likelihood function with apply an algorithm such as Newton-Raphson to obtain maximum likelihood estimates.

See Mehta and Patel (1995) for more detail.

10.11 Appendix C: Theory for Exact Conditional Inference

Section 10.5 provides a brief overview of the methodological ideas behind conditional asymptotic inference.

Now, consider the model

$$\log\left\{\frac{\theta}{1-\theta}\right\} = \mathbf{J}\boldsymbol{\alpha} + \mathbf{X}\boldsymbol{\beta} = \mathbf{J}\boldsymbol{\alpha} + \mathbf{X}_1\boldsymbol{\beta}_1 + \mathbf{X}_2\boldsymbol{\beta}_2$$

for which you have partitioned the vector $\boldsymbol{\beta}$ into components $\boldsymbol{\beta}_1$ and $\boldsymbol{\beta}_2$. Consider $\boldsymbol{\beta}_2$ to be a vector of parameters of interest and $\boldsymbol{\beta}_1$ to be a vector of other nuisance parameters. Correspondingly, partition $\mathbf{X}$ into $\mathbf{X}_1$ and $\mathbf{X}_2$.

The sufficient statistics for $\boldsymbol{\beta}_1$ and $\boldsymbol{\beta}_2$ are $\mathbf{T}_1 = \mathbf{X}_1'\mathbf{y}$ and $\mathbf{T}_2 = \mathbf{X}_2'\mathbf{y}$.

If $\mathbf{a} = \mathbf{J}'\mathbf{y}$, $\mathbf{T}_1$, and $\mathbf{T}_2$ are the sufficient statistics corresponding to $\boldsymbol{\alpha}$, $\boldsymbol{\beta}_1$, and $\boldsymbol{\beta}_2$, then you can define the conditional probability density function $\mathbf{T}_2$ conditional on $\mathbf{T}_1 = \mathbf{t}_1$ and $\mathbf{a} = \mathbf{a}_0$ as

$$f_{\boldsymbol{\beta}_2}(\mathbf{t}_2|\mathbf{a}_0, \mathbf{t}_1) = \frac{C(\mathbf{a}_0, \mathbf{t}_1, \mathbf{t}_2) \exp(\mathbf{t}_2'\boldsymbol{\beta}_2)}{\sum_{\mathbf{u}_2} C(\mathbf{a}_0, \mathbf{t}_1, \mathbf{u}_2) \exp(\mathbf{u}_2'\boldsymbol{\beta}_2)}$$

where $C(\mathbf{a}_0, \mathbf{t}_1, \mathbf{t}_2)$ are the number of vectors $\mathbf{y}$ such that $\mathbf{J}'\mathbf{y} = \mathbf{a}_0$, $\mathbf{X}_1'\mathbf{y} = \mathbf{t}_1$, and $\mathbf{X}_2'\mathbf{y} = \mathbf{u}_2$.

The function $f_{\boldsymbol{\beta}_2}(\mathbf{t}_2|\mathbf{a}_0, \mathbf{t}_1)$ is also the conditional likelihood function for $\boldsymbol{\beta}_2$ given $\{\mathbf{a} = a_0, \mathbf{T}_1 = \mathbf{t}_1\}$.

You can maximize this likelihood to obtain MLEs and conditional tests in a similar fashion to the way you would proceed with the unconditional likelihood.

Conditional exact inference involves generating the conditional permutational distribution for $f_{\boldsymbol{\beta}_2}(\mathbf{t}_2|\mathbf{a}_0, \mathbf{t}_1)$ for the sufficient statistics of the parameter or parameters of interest. You could proceed by completely enumerating the joint distribution of $(\mathbf{a}_0, \mathbf{t}_1, \mathbf{t}_2)$ but that becomes computationally infeasible after a handful of observations. Hirji, Mehta, and Patel (1987) devised the multivariate shift algorithm, a network algorithm, which makes the creation of the exact joint distribution computationally possible. Refer to Derr (2009) for an overview of how the algorithm works for a simple data set.

You can test hypotheses $H_0: \boldsymbol{\beta}_2 = \mathbf{0}$ conditional on $\mathbf{a} = \mathbf{a}_0$ and $\mathbf{T}_1 = \mathbf{t}_1$ with the exact probability test or the exact conditional score test. Under H_0, the statistic for the exact probability test is

$$f_{\boldsymbol{\beta}_2=0}(\mathbf{t}_2|\mathbf{a}_0, \mathbf{t}_1) = \frac{C(\mathbf{a}_0, \mathbf{t}_1, \mathbf{t}_2)}{\sum_{u_2} C(\mathbf{a}_0, \mathbf{t}_1, \mathbf{u}_2)}$$

and the *p*-value is the probability of getting a more extreme statistic

$$p = \sum_{u \in \Re_p} f_{\boldsymbol{\beta}_2=0}(\mathbf{t}_2|\mathbf{a}_0, \mathbf{t}_1)$$

where $u \in \Re_p$ are the u such that $\mathbf{y}$ exist with $\mathbf{J}'\mathbf{y} = \mathbf{a}_0, \mathbf{X}_1'\mathbf{y} = \mathbf{t}_1, \mathbf{X}_2'\mathbf{y} = \mathbf{u}_2$ and $f_{\boldsymbol{\beta}_2=0}(\mathbf{u}_2|\mathbf{a}_0, \mathbf{t}_1) \leq f_{\boldsymbol{\beta}_2=0}(\mathbf{t}_2|\mathbf{a}_0, \mathbf{t}_1)$.

For the exact conditional score test, you define the conditional mean $\boldsymbol{\mu}_2$ and variance matrix $\mathbf{V}_2$ of $\mathbf{T}_2$ (conditional on $\mathbf{a} = \mathbf{a}_0, \mathbf{T}_1 = \mathbf{t}_1$) and compute the score statistic

$$s = (\mathbf{t}_2 - \boldsymbol{\mu}_2)'\mathbf{V}_2^{-1}(\mathbf{t}_2 - \boldsymbol{\mu}_2)$$

and compare it to the score for each member of the distribution

$$S = (\mathbf{T}_2 - \boldsymbol{\mu}_2)'\mathbf{V}_2^{-1}(\mathbf{T}_2 - \boldsymbol{\mu}_2)$$

The *p*-value is

$$p = \Pr(S \geq s) = \sum_{u \in \Re_s} f_{\boldsymbol{\beta}_2=0}(\mathbf{u}_2|\mathbf{a}_0, \mathbf{t}_1)$$

where $\mathbf{u} \in \Re_s$ are the $\mathbf{u}$ such that $\mathbf{y}$ exist with $\mathbf{J}'\mathbf{y} = \mathbf{a}_0, \mathbf{X}_1'\mathbf{y} = \mathbf{t}_1, \mathbf{X}_2'\mathbf{y} = \mathbf{u}$, and $S(u) \geq s$.

You obtain exact parameter estimates β_j by considering all the other parameters as nuisance parameters, forming the conditional pdf, and using Newton-Raphson to find the maximum exact conditional likelihood estimates. Likelihood ratio tests based on the conditional pdf are used to test $H_0: \beta_j = 0$.

Refer to Derr (2009) for more detail on the methods employed by the LOGISTIC procedure, including a basic illustration of how the network algorithm works. Refer to Mehta and Patel (1995) for a complete discussion of exact logistic regression methodology and numerous applications.

10.12 Appendix D: ODS Macro

The following code updates the default template for the exact tests output in the PROC LOGISTIC procedure to produce six decimal place for the exact *p*-values.

```
proc template;
   edit Stat.XCL.PValue;
   format=D8.6;
   end;
run;
proc logistic data=cardio descending;
   strata animal;
   model response = ordtreat;
   exact ordtreat / estimate=parm;
run;
```

The updated template resides in your SASUSER directory and will be found for subsequent invocations of the PROC LOGISTIC procedure. To delete it and revert back to the default templates, submit the statements

```
proc template;
   delete Stat.XCL.PValue;
run;
```

Note that the default templates shipped with SAS/STAT software are stored in the SASHELP directory and cannot be deleted. See the chapter "Using the Output Delivery System" in the *SAS/STAT User's Guide* for an overview of using ODS for the statistician.

Chapter 11

Quantal Response Data Analysis

Contents

11.1	Introduction	**345**
11.2	Estimating Tolerance Distributions	**346**
	11.2.1 Analyzing the Bacterial Challenge Data	348
11.3	Comparing Two Drugs	**354**
	11.3.1 Analysis of the Peptide Data	356
11.4	Analysis of Pain Study	**361**
11.5	Estimating Tolerance Distributions	**367**
Appendix A: SAS/IML Macro for Confidence Intervals of Ratios Using Fieller's Theorem		**371**

11.1 Introduction

Quantal response data analysis deals with subject response to a stimulus that occurs with greater and greater intensity. This covers a great deal of territory, from the evaluation of drugs at various dosages to the assessment of radiation impact to the study of growth and development in children. Often, the reagent is a new drug and the subjects are experimental animals. Other possible stimuli include radiation and environmental exposures, and other possible subjects include humans and bacteria. Researchers are interested in the tolerance of the subjects to the stimulus or drug, where tolerance is defined as the amount of the stimulus required to produce a response.

These methods were developed for *bioassay*, which is the process of determining the potency or strength of a reagent or stimuli based on the response it elicits in biological organisms. Researchers are also often interested in the relative potency of a new drug compared to a standard drug. In a direct assay, you steadily increase the doses until you generate the desired reaction. In an indirect assay, you observe the reaction of groups of subjects to specified sets of doses. The measured response to the drug in an indirect assay can be either quantitative or quantal. An example of a quantitative response is red blood cells per milliliter of blood, and an example of a quantal response is death or survival.

This chapter is concerned with quantal responses, which are analyzed with categorical data analysis strategies. See Tsutakawa (1982) for an overview of general bioassay methods, and see Finney (1978) and Govindarajulu (1988) for textbook discussion of these areas. See Bock and Jones (1968) and Bock (1975) for some statistical methodology related to child development and behavioral areas. See Landis and Koch (1979) for examples of categorical analysis of behavioral data.

346 Chapter 11: Quantal Response Data Analysis

This chapter provides examples of applying quantal response data analysis techniques to drug development and growth studies.

11.2 Estimating Tolerance Distributions

Table 11.1 displays data from an experiment in which animals were exposed to bacterial challenges after having one-quarter of their spleen removed (splenectomy). After 96 hours, their survival status was assessed. The stimulus is the bacterial challenge, and interest lies in assessing the tolerances of the animals' immune systems to the bacterial challenge after they have had partial splenectomies (Koch and Edwards 1985).

Table 11.1 Status 96 Hours after Bacterial Challenge

Bacterial Dose	Status	
	Dead	Alive
1.2×10^3	0	5
1.2×10^4	0	5
1.2×10^5	2	3
1.2×10^6	4	2
1.2×10^7	5	1
1.2×10^8	5	0

In bioassay analysis, you make the assumption that responses of subjects are determined through a tolerance distribution. This means that at certain levels of the dose (bacterial challenge in this case) the animals will die; that is, death will occur if dose exceeds the tolerance, and survival will occur when dose is below tolerance. Historically, the tolerances have been assumed to follow a normal distribution. This allows you to write the probability of death at a level x_i of the bacterial challenge as

$$p_i = \Phi\left(\frac{x_i - \mu}{\sigma}\right)$$

where Φ is the cumulative distribution function for the standard normal distribution with mean 0 and variance 1; the parameter μ is the mean (or median) of the tolerance distribution, and σ is the standard deviation.

If $\alpha = -\mu/\sigma$ and $\beta = 1/\sigma$, then

$$p_i = \Phi(\alpha + \beta x_i)$$

and

$$\Phi^{-1}(p_i) = \alpha + \beta x_i$$

The function $\Phi^{-1}(p_i)$ is called the *probit* (or *normit*), and its analysis is called probit analysis. Sometimes the value 5 is added to $\Phi^{-1}(p_i)$ in order to have positive values for all p_i.

11.2. Estimating Tolerance Distributions

Berkson (1951) pointed out that the logistic distribution also works well as a tolerance distribution, generating essentially the same results as the normal distribution. This is particularly true for values of p_i in the middle of the (0, 1) range and when the median μ of the tolerance distribution is the primary parameter. While sometimes a probit analysis of a data set is of more interest to researchers in some disciplines (for example, growth and development) because of the correspondence of its parameters to the mean and standard deviation of the underlying tolerance distribution, the focus in this chapter is on logistic analysis. Note that the measures discussed are also relevant to a model based on the probit.

If you assume the logistic distribution for the tolerances, then

$$p_i = \frac{\exp\{\alpha + \beta x_i\}}{1 + \exp\{\alpha + \beta x_i\}}$$

and

$$\log\left\{\frac{p_i}{1 - p_i}\right\} = \alpha + \beta x_i$$

The parameters α and β are estimated with maximum likelihood estimation. Usually, the log of the tolerances is most likely to have a logistic distribution, so frequently you work with the log of the drug or concentration under investigation as the x_i.

One parameter of interest for estimation is the median of the tolerance distribution, or the dose at which 50% of the subjects produce a response. When the response is death, this estimate is called the LD50, for lethal dose. Otherwise, this measure is called the ED50, for effective dose. If you are working with log dose levels, you compute the log LD50 and then exponentiate it if you are also interested in the actual LD50.

Suppose x_{50} represents the log LD50 and p_{50} represents the probability of response at the median of the tolerance distribution.

$$\log\left\{\frac{p_{50}}{1 - p_{50}}\right\} = \log\left\{\frac{.5}{.5}\right\} = 0$$

Thus, the logistic parameters $\hat{\alpha}$ and $\hat{\beta} x_{50}$ can be set to zero to obtain

$$\hat{x}_{50} = \frac{-\hat{\alpha}}{\hat{\beta}}$$

You can construct an approximate form of the variance of $\hat{x}_{50}$ based on linearized Taylor series for situations where β is clearly different from 0.

$$\text{var}\{\hat{x}_{50}\} = \{\hat{x}_{50}\}^2 \left\{\frac{V(\hat{\alpha})}{\hat{\alpha}^2} - \frac{2V(\hat{\alpha}, \hat{\beta})}{\hat{\alpha}\hat{\beta}} + \frac{V(\hat{\beta})}{\hat{\beta}^2}\right\}$$

where $V(\hat{\alpha})$, $V(\hat{\alpha}, \hat{\beta})$, and $V(\hat{\beta})$ represent the variance of $\hat{\alpha}$, the covariance of $\hat{\alpha}$ and $\hat{\beta}$, and the variance of $\hat{\beta}$, respectively.

This allows you to express the confidence interval for log LD50 as

$$\hat{x}_{50} \pm z_{1-\alpha/2} \sqrt{\mathrm{var}\{\hat{x}_{50}\}}$$

Fieller's theorem is also used to compute confidence intervals for these measures. This theorem is a general result that enables confidence intervals to be computed for the ratio of two normally distributed random variables. Refer to Read (1983) for a description of Fieller's formula, and refer to Collett (2003) for a discussion of how to apply it to LD50s. Zerbe (1978) describes a matrix implementation of Fieller's formula for use with the general linear model as illustrated in the following analysis. Zerbe's implementation is provided in a SAS/IML routine listed in the appendix and illustrated in this chapter.

In order to compute the LD50, the actual dosage at which 50% of the subjects die, you exponentiate $\hat{x}_{50}$ (and its confidence limits). Sometimes analysts work on the loglog scale for LD50 to produce more stable computations. In that case, you would use

$$\frac{\mathrm{var}(\hat{x}_{50})}{\hat{x}_{50}^2}$$

as the applicable variance for log $\hat{x}_{50}$, and you would double exponentiate the results to generate the estimate of the actual LD50 and its confidence interval.

The LOGISTIC and PROBIT procedures are used to fit these models. In the following section, a logistic model is fit to the data in Table 11.1, and the log LD50 is computed.

11.2.1 Analyzing the Bacterial Challenge Data

The following SAS statements input the data from Table 11.1 and compute two additional variables: LDOSE is the log dose (natural log), which results in more evenly spaced dose levels.

```
data bacteria;
   input dose status $ count @@;
   ldose=log(dose);
   datalines;
1200       dead    0   1200       alive 5
12000      dead    0   12000      alive 5
120000     dead    2   120000     alive 3
1200000    dead    4   1200000    alive 2
12000000   dead    5   12000000   alive 1
120000000  dead    5   120000000  alive 0
;

proc print;
run;
```

Output 11.1 displays these data.

Output 11.1 Data Listing

Obs	dose	status	count	ldose
1	1200	dead	0	7.0901
2	1200	alive	5	7.0901
3	12000	dead	0	9.3927
4	12000	alive	5	9.3927
5	120000	dead	2	11.6952
6	120000	alive	3	11.6952
7	1200000	dead	4	13.9978
8	1200000	alive	2	13.9978
9	12000000	dead	5	16.3004
10	12000000	alive	1	16.3004
11	120000000	dead	5	18.6030
12	120000000	alive	0	18.6030

In the following PROC LOGISTIC specification, LDOSE is listed in the MODEL statement, and so is its square, which is included to help assess goodness of fit. The SELECTION=FORWARD option is specified so that a residual score statistic for the quadratic term is computed. The COVB option requests that PROC LOGISTIC print the covariance matrix for the parameter estimates, quantities necessary to compute the confidence interval for the log LD50.

```
proc logistic data=bacteria descending;
   freq count;
   model status = ldose ldose*ldose / scale=none aggregate
         selection=forward include=1 details covb;
run;
```

Since the option INCLUDE=1 is specified, the first model fit includes the intercept and the LDOSE term. The residual score statistic for the quadratic term displayed in Output 11.2 is not significant with $Q_S = 0.2580$ and $p = 0.6115$, so clearly this term makes no contribution to the model. This result supports the satisfactory fit of the intercept and slope model; the residual score test serves as a goodness-of-fit test for this model.

Output 11.2 Residual Score Statistic

Residual Chi-Square Test		
Chi-Square	DF	Pr > ChiSq
0.2580	1	0.6115

The Pearson and deviance goodness-of-fit statistics also indicate that the model provides an adequate fit, as displayed in Output 11.3. However, note that the sampling requirements for these statistics are minimally met; certainly the expected values for all cell counts are not greater than 4 for several cells. In such cases, it is better to support assessment of fit with methods such as the residual score statistic for the addition of the quadratic term.

Output 11.3 Goodness-of-Fit Statistics

Deviance and Pearson Goodness-of-Fit Statistics				
Criterion	Value	DF	Value/DF	Pr > ChiSq
Deviance	1.7508	4	0.4377	0.7815
Pearson	1.3379	4	0.3345	0.8549

Number of unique profiles: 6

Output 11.4 contains the maximum likelihood estimates for α and β. The estimate $\hat{\beta} = 0.7071$ has $p = 0.0027$ for the test of its significance. The level of bacterial challenge has a significant effect on survival. The intercept $\hat{\alpha} = -9.2680$.

Output 11.4 Maximum Likelihood Estimates

Analysis of Maximum Likelihood Estimates					
Parameter	DF	Estimate	Standard Error	Wald Chi-Square	Pr > ChiSq
Intercept	1	-9.2680	3.1630	8.5857	0.0034
ldose	1	0.7071	0.2354	9.0223	0.0027

Output 11.5 contains the estimated covariance matrix for the parameter estimates. The variance of $\hat{\alpha}$ is 10.0046, the variance of $\hat{\beta}$ is 0.05542, and the covariance of $\hat{\alpha}$ and $\hat{\beta}$ is -0.7334. Taking the square root of the variances produces the standard errors displayed in Output 11.4.

Output 11.5 Estimated Covariance Matrix

Estimated Covariance Matrix		
Parameter	Intercept	ldose
Intercept	10.00458	-0.73338
ldose	-0.73338	0.055418

To compute the log LD50, use the estimated values of $\hat{\alpha}$ and $\hat{\beta}$.

$$\log \text{LD50} = \frac{-\hat{\alpha}}{\hat{\beta}} = \frac{9.2680}{0.7071} = 13.1070$$

Using the covariances from Output 11.5 in the formula for var$\{x_{50}\}$ on page 348 yields the value 0.6005. Thus, a confidence interval for the log LD50 is written

$$13.1070 \pm 1.96\sqrt{0.6005}$$

so that the confidence interval is (11.588, 14.626). To determine the LD50 on the actual dose scale, you exponentiate the LD50 for the log scale.

$$\text{actual LD50} = e^{13.1070} = 4.9238 \times 10^5$$

To determine its confidence interval, exponentiate both bounds of the confidence interval to obtain $(1.0780 \times 10^5, 2.2490 \times 10^6)$. This confidence interval describes the location of the median bacterial challenge for the death of animals with one-fourth of the spleen removed.

An alternative to using the LOGISTIC procedure for this analysis is to use the PROBIT procedure, which was specifically designed to perform quantal response data analysis. By default, it performs probit analysis, but it also provides logistic analysis. It computes the LD50 automatically, as well as computing the estimates for the dose values that yield user-defined response rates and the corresponding confidence intervals based on Fieller's theorem. Several graphs are also provided.

The following statements request the same analysis using the PROBIT procedure. First, new data set BACTERIA2 is created with counts of dead animals and total animals so the events/trials syntax can be used (PROC PROBIT doesn't include a FREQ statement).

```
data bacteria2;
   input dose dead total @@;
   ldose=log(dose);
   datalines;
1200        0   5
12000       0   5
120000      2   5
1200000     4   6
12000000    5   6
120000000   5   5
;
```

The PROC PROBIT statements follow. You can specify the LOG option in the PROC statement to request that the analysis be performed on the log dose scale. The DIST=LOGISTIC option requests logistic analysis instead of probit analysis, and the LACKFIT option requests goodness-of-fit tests. The INVERSECL option requests the LD50 and confidence limits as well as the dose required for the 0.25 and 0.75 response rates.

```
ods graphics on;
proc probit data=bacteria2 log plot=ippplot;
   model dead/total = dose / dist=logistic lackfit
           inversecl (prob=.25 .50 .75);
run;
ods graphics off;
```

Output 11.6 displays the model information.

Output 11.6 Probit Model Information

Model Information	
Data Set	WORK.BACTERIA2
Events Variable	dead
Trials Variable	total
Number of Observations	6
Number of Events	16
Number of Trials	32
Name of Distribution	Logistic
Log Likelihood	-10.76291779

Output 11.7 displays the goodness-of-fit tests, which take the same values as those from PROC LOGISTIC. However, note that no residual score test is available with the PROBIT procedure (and that test might be considered necessary with some of the small event counts).

Output 11.7 Goodness-of-Fit Statistics

Goodness-of-Fit Tests				
Statistic	Value	DF	Value/DF	Pr > ChiSq
Pearson Chi-Square	1.3379	4	0.3345	0.8549
L.R. Chi-Square	1.7508	4	0.4377	0.7815

Output 11.8 displays the parameter estimates.

Output 11.8 Probit Parameter Estimates

Analysis of Maximum Likelihood Parameter Estimates							
Parameter	DF	Estimate	Standard Error	95% Confidence Limits		Chi-Square	Pr > ChiSq
Intercept	1	-9.2680	3.1630	-15.4674	-3.0687	8.59	0.0034
Ln(dose)	1	0.7071	0.2354	0.2457	1.1685	9.02	0.0027

Output 11.9 displays the parameters for the tolerance distribution. The value for 'MU' is the estimated LD50, which is 13.107. The value for 'SIGMA' is $(1/\hat{\beta})$.

Output 11.9 LD50 Estimate

Probit Model in Terms of Tolerance Distribution	
MU	SIGMA
13.1070197	1.41421791

Output 11.10 contains the covariance matrix for these estimated parameters. The value of 0.600471 is var$\{x_{50}\}$.

Output 11.10 Covariance Estimate

Estimated Covariance Matrix for Tolerance Parameters		
	MU	SIGMA
MU	0.600471	-0.019844
SIGMA	-0.019844	0.221675

Output 11.11 displays the table of estimates for the doses that pertain to the specified probability levels (response rates). Note that the 95% confidence limits for the log LD50 are (11.0067, 15.0196), which are different from the ones on page 351 because they are based on Fieller's theorem, not the approximation based on linearized Taylor series.

Output 11.11 Log Dose Analysis

Probit Analysis on Ln(dose)			
Probability	Ln(dose)	95% Fiducial Limits	
0.25	11.5533	7.5414	13.0735
0.50	13.1070	11.0067	15.0196
0.75	14.6607	13.1430	18.2947

Output 11.12 contains the same information for the natural dose scale.

Output 11.12 Actual Dose Analysis

Probit Analysis on dose			
Probability	dose	95% Fiducial Limits	
0.25	104124	1885	476140
0.50	492387	60276	3333723
0.75	2328412	510434	88159975

The PROBIT procedure also provides predicted probability plots. The following code produces the inverse probability plot for log dose.

```
ods graphics on;
proc probit data=bacteria2 plot=ippplot;
   model dead/total = ldose / dist=logistic;
run;
ods graphics off;
```

Output 11.13 displays the resulting graph. The confidence limits are based on Fieller's theorem.

Output 11.13 Inverse Probability Plot

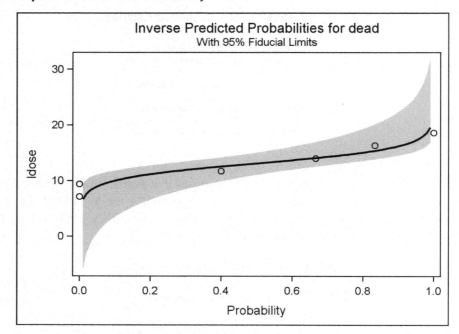

11.3 Comparing Two Drugs

Bioassay often involves the comparison of two drugs, usually a new drug versus a standard drug. Consider the data in Table 11.2. Researchers studied the effects of the peptides neurotensin and somatostatin in potentiating nonlethal doses of the barbiturate pentobarbital. Groups of mice were administered various dose levels of either neurotensin or somatostatin (Nemeroff et al. 1977; analyzed in Imrey, Koch, and Stokes 1982).

Many times, one drug acts as a dilution of another drug. If this is the case, then the dose response relationship is parallel on the logit scale. Assays that are designed for the dilution assumption are called *parallel lines assays*. The quantity that describes the relationship of such drugs to one another through the ratio of doses of the two drugs that produce the same response is called the *relative potency*.

11.3. Comparing Two Drugs

Table 11.2 N and S Comparison

Dose	Drug	Status Dead	Status Alive	Total
0.01	N	0	30	30
0.03	N	1	29	30
0.10	N	1	9	10
0.30	N	1	9	10
0.30	S	0	10	10
1.00	N	4	6	10
1.00	S	0	10	10
3.00	N	4	6	10
3.00	S	1	9	10
10.00	N	5	5	10
10.00	S	4	6	10
30.00	S	5	5	10
30.00	N	7	3	10
100.00	S	8	2	10

The dilution assumption for doses z_s of somatostatin and z_n of neurotensin can be stated as

$$z_s = \rho z_n$$

which means that the doses with comparable response for the two drugs are related by the constant ρ, the relative potency; that is, ρ units of somatostatin produce the same behavior as one unit of neurotensin. If x_n and x_s represent log doses, then the dilution assumption also implies that

$$x_s = \log \rho + x_n$$

Thus the logistic model structure for somatostatin is

$$p_s(x_{si}) = \{1 + \exp(-\alpha_s - \beta x_{si})\}^{-1}$$

where x_{si} denotes log dose levels of somatostatin. You can write the implied structure for log dose levels x_{ni} of neurotensin as

$$p_n(x_{ni}) = p_s(\log \rho + x_{ni}) = \{1 + \exp(-\alpha_s - \beta \log \rho - \beta x_{ni})\}^{-1}$$
$$= \{1 + \exp(-\alpha_n - \beta x_{ni})\}^{-1}$$

where $\alpha_n = \alpha_s + \beta \log \rho$.

By forming

$$\frac{p_n(x_{ni})}{1 - p_n(x_{ni})}$$

you obtain the results

$$\log\left\{\frac{p_n(x_{ni})}{1 - p_n(x_{ni})}\right\} = \{\alpha_s + \beta \log \rho\} + \beta x_{ni}$$
$$= \alpha_n + \beta x_{ni}$$

Similarly,

$$\log\left\{\frac{p_s(x_{si})}{1-p_s(x_{si})}\right\} = \alpha_s + \beta x_{si}$$

Thus, the dilution assumption can be tested by fitting a model with separate intercepts and slopes and then testing for a common slope.

The constant ρ is the relative potency, and since

$$\alpha_n = \alpha_s + \beta \log \rho$$

then

$$\rho = \exp\left\{\frac{\alpha_n - \alpha_s}{\beta}\right\}$$

This means that ρ units of somatostatin produce the same reaction as one unit of neurotensin.

Fieller's theorem can be used to produce confidence intervals for the relative potency.

11.3.1 Analysis of the Peptide Data

The following DATA step creates data set ASSAY for use with PROC LOGISTIC. Indicator variables INT_S and INT_N are created to form the intercepts for each drug, and LDOSE is dose on the log scale.

```
data assay;
   input drug $ dose status $ count;
   int_n=(drug='n');
   int_s=(drug='s');
   ldose=log(dose);
   datalines;
n  0.01    dead   0
n  0.01    alive  30
n   .03    dead   1
n   .03    alive  29
n   .10    dead   1
n   .10    alive  9
n   .30    dead   1
n   .30    alive  9
n  1.00    dead   4
n  1.00    alive  6
n  3.00    dead   4
n  3.00    alive  6
n 10.00    dead   5
n 10.00    alive  5
```

```
n  30.00  dead    7
n  30.00  alive   3
s   .30   dead    0
s   .30   alive  10
s  1.00   dead    0
s  1.00   alive  10
s  3.00   dead    1
s  3.00   alive   9
s 10.00   dead    4
s 10.00   alive   6
s 30.00   dead    5
s 30.00   alive   5
s 100.00  dead    8
s 100.00  alive   2
;
```

The following PROC LOGISTIC statements request the two intercepts and two slopes model. The NOINT option must be specified to suppress the default intercept. Crossing the intercept variables with LDOSE creates separate dose terms for each drug. Subsequently squaring these terms produces quadratic terms so that a test of quadratic terms can be performed to help assess goodness of fit; this is desirable since the cell counts in Table 11.2 are small.

The TEST statement requests a test for equality of the two slope parameters β_n and β_s. The coefficient names 'int_nldose' and 'int_sldose' pertain to the LDOSE*INT_N and LDOSE*INT_S terms in the MODEL statement; to determine the names to use in the TEST statement, you can create an OUTEST= data set and print it (not shown here).

```
proc logistic data=assay descending;
   freq count;
   model status = int_n int_s
              ldose*int_n ldose*int_s
              ldose*int_n*ldose*int_n
              ldose*int_s*ldose*int_s
              / noint
              scale=none aggregate
              include=4 selection=forward details;
   eq_slope: test int_nldose=int_sldose;
run;
```

Output 11.14 contains a listing of the response profile. The model is estimating the probability of death.

Output 11.14 Response Profile

Response Profile		
Ordered Value	status	Total Frequency
1	dead	41
2	alive	139

Probability modeled is status='dead'.

Output 11.15 displays the goodness-of-fit statistics. With values of 4.4144 and 3.6352 for Q_L and Q_P, respectively, these statistics support an adequate model fit.

Output 11.15 Goodness-of-Fit Statistics

Deviance and Pearson Goodness-of-Fit Statistics				
Criterion	Value	DF	Value/DF	Pr > ChiSq
Deviance	4.4144	10	0.4414	0.9267
Pearson	3.6352	10	0.3635	0.9623

Number of unique profiles: 14

Output 11.16 contains the results for the residual score test for the two quadratic terms. It is nonsignificant, and so are the individual tests. These results support the goodness of fit of the model.

Output 11.16 Tests for Quadratic Terms

Residual Chi-Square Test		
Chi-Square	DF	Pr > ChiSq
1.4817	2	0.4767

The parameter estimates are all significant, as seen in the "Analysis of Maximum Likelihood Estimates" table displayed in Output 11.17. However, if you examine the slope estimates (labeled 'int_n*ldose' and 'int_s*ldose') and their standard errors, you see that it is possible that these two slopes can be represented by one slope. The Wald statistic for the hypothesis test $H_0: \beta_n = \beta_s$ bears this out with a nonsignificant $p = 0.1490$, displayed in Output 11.18.

Output 11.17 Maximum Likelihood Estimates

Analysis of Maximum Likelihood Estimates					
Parameter	DF	Estimate	Standard Error	Wald Chi-Square	Pr > ChiSq
int_n	1	-1.1301	0.2948	14.6983	0.0001
int_s	1	-3.3782	0.8797	14.7479	0.0001
int_n*ldose	1	0.6199	0.1240	24.9907	<.0001
int_s*ldose	1	1.0615	0.2798	14.3914	0.0001

Output 11.18 Equal Slopes Hypothesis Test Results

	Linear Hypotheses Testing Results			
Label		Wald Chi-Square	DF	Pr > ChiSq
eq_slope		2.0820	1	0.1490

Thus, it appears that a parallel lines model fits these data, and the following PROC LOGISTIC statements request this model. The COVB option in the MODEL statement requests that the covariances of the parameters be printed, and the OUTEST=ESTIMATE and COVOUT options request that they be placed into a SAS data set for further processing. Without the specification of the COVOUT option, only the parameter estimates are placed in the OUTEST data set. For convenience, the _LINK_ and _LNLIKE_ variables placed in the OUTEST data set by default are dropped.

```
proc logistic data=assay descending outest=estimate
              (drop= intercept _link_ _lnlike_) covout;
   freq count;
   model status = int_n int_s ldose /
                  noint scale=none aggregate covb;
run;
```

Output 11.19 contains the goodness-of-fit statistics for this model, and they indicate that the model is adequate.

Output 11.19 Goodness-of-Fit Results

Deviance and Pearson Goodness-of-Fit Statistics				
Criterion	Value	DF	Value/DF	Pr > ChiSq
Deviance	6.8461	11	0.6224	0.8114
Pearson	5.6480	11	0.5135	0.8958

Number of unique profiles: 14

Output 11.20 contains the parameter estimates; all of them are clearly significant.

Output 11.20 Maximum Likelihood Estimates

Analysis of Maximum Likelihood Estimates					
Parameter	DF	Estimate	Standard Error	Wald Chi-Square	Pr > ChiSq
int_n	1	-1.1931	0.3158	14.2781	0.0002
int_s	1	-2.4476	0.4532	29.1632	<.0001
ldose	1	0.7234	0.1177	37.7681	<.0001

Output 11.21 contains the estimated covariance matrix.

Output 11.21 Covariance Matrix

Estimated Covariance Matrix			
Parameter	int_n	int_s	ldose
int_n	0.099702	0.025907	-0.00984
int_s	0.025907	0.20542	-0.03648
ldose	-0.00984	-0.03648	0.013856

The estimated log LD50s from this model are

$$\log \text{LD50}_n = \frac{-\hat{\alpha}_n}{\hat{\beta}} = \frac{1.1931}{0.7234} = 1.65$$

and

$$\log \text{LD50}_s = \frac{-\hat{\alpha}_s}{\hat{\beta}} = \frac{2.4476}{0.7234} = 3.38$$

The log relative potency is estimated as

$$\hat{\log \rho} = \frac{\hat{\alpha}_n - \hat{\alpha}_s}{\hat{\beta}} = \frac{-1.1931 - (-2.4476)}{0.7234} = 1.73$$

You can compute approximate confidence intervals for these quantities using the linearized Taylor series, as in the previous section for the log LD50, or you can produce confidence intervals based on Fieller's theorem.

The appendix to this chapter contains the SAS/IML routine that produces confidence intervals based on Fieller's theorem for ratios of estimates from a general linear model (Zerbe 1978). You specify coefficients for vectors **k** and **h** that premultiply the parameter vector to form the numerator and the denominator of the ratio of interest. For example, if $\boldsymbol{\beta} = \{\alpha_n, \alpha_s, \beta\}$, $\mathbf{k} = \{1, -1, 0\}$, and $\mathbf{h} = \{0, 0, 1\}$, then

$$\frac{\mathbf{k}'\boldsymbol{\beta}}{\mathbf{h}'\boldsymbol{\beta}} = \frac{\alpha_n - \alpha_s}{\beta}$$

which is the relative potency. Other choices of coefficients produce $\log \text{LD50}_n$ and $\log \text{LD50}_s$. See the appendix for the complete program listing to produce the results.

The ratio estimates for the log potency, $\log \text{LD50}_n$, and $\log \text{LD50}_s$ are displayed in Output 11.22. The lower and upper bounds of their confidence intervals appear under "l_bound" and "u_bound," respectively.

Output 11.22 Confidence Intervals Based on Fieller's Theorem

Confidence Intervals

95 % ci for ratio based on fieller

ratio	interval l_bound	u_bound
1.7341215	0.4262151	2.9994194

95 % ci for ratio based on fieller

ratio	interval l_bound	u_bound
1.6493371	0.8237277	2.6875216

95 % ci for ratio based on fieller

ratio	interval l_bound	u_bound
3.3834586	2.4863045	4.4505794

Table 11.3 contains these results as well as the exponentiated results for the natural dose. Thus, a dose of somatostatin must be 5.64 times higher than a dose of neurotensin to have the same effect, with a 95% confidence interval of (1.53, 20.07).

Table 11.3 Estimated Measures from Parallel Assay

Estimate	Value	95% Confidence Interval	Exponentiated Value	Exponentiated Confidence Interval
log(Potency)	1.73	(0.4262, 2.9994)	5.64	(1.53, 20.07)
log LD50$_n$	1.65	(0.8237, 2.6875)	5.21	(2.28, 14.69)
log LD50$_s$	3.38	(2.4863, 4.4506)	29.37	(12.02, 85.68)

11.4 Analysis of Pain Study

Researchers investigated a new drug for pain relief by studying its effect on groups of subjects with two different diagnoses. The drug was administered at five dosages, and the outcome measured was whether the subjects reported adverse effects. Table 11.4 contains the data. Interest lies in investigating the association of adverse effects with dose and diagnosis; in addition, there is interest in describing the influence of dose and diagnosis on reports of adverse effects with a statistical model.

Table 11.4 Pain Study

Dose	Diagnosis I		Diagnosis II	
	Adverse	Not	Adverse	Not
1	3	26	6	26
5	7	26	20	12
10	10	22	26	6
12	14	18	28	4
15	18	14	31	1

Unlike the previous bioassay analysis, this study does not compare the tolerance distributions of two drugs and is not strictly concerned with estimating the tolerance distribution for either drug. But even though the study does not completely fall into the usual realm of bioassay, it has a bioassay flavor. Its analysis also serves to illustrate the blend of hypothesis testing and model fitting that is often desired in a statistical analysis of categorical data.

Mantel-Haenzsel statistics are computed to determine if there is an association between adverse effects and dose, adverse effects and diagnosis, and adverse effects and dose, controlling for diagnosis. A logistic model is then fit to describe the influence of dose and diagnosis on adverse effects, and ED50s are estimated for both diagnosis groups.

The following DATA step statements input the data and create indicator variables to be used later for the PROC LOGISTIC runs.

```
data adverse;
    input diagnos $ dose status $ count @@;
    i_diagII=(diagnos='II');
    i_diagI= (diagnos='I');
datalines;
I    1    adverse  3 I     1 no 26
I    5    adverse  7 I     5 no 26
I   10    adverse 10 I    10 no 22
I   12    adverse 14 I    12 no 18
I   15    adverse 18 I    15 no 14
II   1    adverse  6 II    1 no 26
II   5    adverse 20 II    5 no 12
II  10    adverse 26 II   10 no  6
II  12    adverse 28 II   12 no  4
II  15    adverse 31 II   15 no  1
;

proc freq data=adverse;
    weight count;
    tables dose*status diagnos*status diagnos*dose*status /
        nopct nocol cmh;
run;
```

Output 11.23 contains the crosstabulation for DOSE × STATUS. There appears to be a positive association between dose level and proportion of adverse effects.

Output 11.23 Table of DOSE × STATUS

Confidence Intervals

Frequency
Row Pct

Table of dose by status

dose	status adverse	status no	Total
1	9 14.75	52 85.25	61
5	27 41.54	38 58.46	65
10	36 56.25	28 43.75	64
12	42 65.63	22 34.38	64
15	49 76.56	15 23.44	64
Total	163	155	318

Output 11.24 contains the Mantel-Haenszel statistics. Since the dose levels are numeric, the 1 df correlation statistic is appropriate. $Q_{CS} = 55.7982$, which is strongly significant. As the dose increases, the proportion of subjects who experienced adverse effects also increases.

Output 11.24 Mantel-Haenszel Statistics

Cochran-Mantel-Haenszel Statistics (Based on Table Scores)

Statistic	Alternative Hypothesis	DF	Value	Prob
1	Nonzero Correlation	1	55.7982	<.0001
2	Row Mean Scores Differ	4	57.1403	<.0001
3	General Association	4	57.1403	<.0001

Output 11.25 displays the crosstabulation for DIAGNOS × STATUS.

Output 11.25 DIAGNOS × STATUS Table

Frequency
Row Pct

Table of diagnos by status

diagnos	status adverse	status no	Total
I	52 32.91	106 67.09	158
II	111 69.38	49 30.63	160
Total	163	155	318

Output 11.26 contains the Mantel-Haenszel statistics. $Q_{MH} = 42.1732$ with 1 df, which is also strongly significant. Subjects with diagnosis II were more likely to experience adverse effects.

Output 11.26 Mantel-Haenszel Statistics

Cochran-Mantel-Haenszel Statistics (Based on Table Scores)				
Statistic	Alternative Hypothesis	DF	Value	Prob
1	Nonzero Correlation	1	42.1732	<.0001
2	Row Mean Scores Differ	1	42.1732	<.0001
3	General Association	1	42.1732	<.0001

Output 11.27 contains the extended Mantel-Haenszel statistics for the association of dose and status after adjusting for diagnosis. The correlation statistic is appropriate, and $Q_{CS} = 65.5570$ with 1 df, which is clearly significant.

Output 11.27 DIAGNOS*DRUG*STATUS

Cochran-Mantel-Haenszel Statistics (Based on Table Scores)				
Statistic	Alternative Hypothesis	DF	Value	Prob
1	Nonzero Correlation	1	65.5570	<.0001
2	Row Mean Scores Differ	4	67.4362	<.0001
3	General Association	4	67.4362	<.0001

The following PROC LOGISTIC statements fit a model that contains separate intercepts and slopes for the two diagnoses. First, the actual dose is used.

```
proc logistic data=adverse outest=estimate
              (drop= intercept _link_ _lnlike_) covout;
   freq count;
   model status = i_diagI i_diagII
                  dose*i_diagI dose*i_diagII /
             noint scale=none aggregate;
   eq_slope: test i_diagIdose=i_diagIIdose;
run;
```

Output 11.28 contains the response profiles and goodness-of-fit statistics. The model fit appears to be quite good.

11.4. Analysis of Pain Study

Output 11.28 Response Profiles and Goodness of Fit

Confidence Intervals

Response Profile		
Ordered Value	status	Total Frequency
1	adverse	163
2	no	155

Probability modeled is status='adverse'.

Deviance and Pearson Goodness-of-Fit Statistics				
Criterion	Value	DF	Value/DF	Pr > ChiSq
Deviance	2.7345	6	0.4557	0.8414
Pearson	2.7046	6	0.4508	0.8449

Number of unique profiles: 10

Output 11.29 contains the model parameters, and Output 11.30 contains the test for a common slope. The hypothesis of a common slope is rejected at the $\alpha = 0.05$ level of significance.

Output 11.29 Parameter Estimates

Analysis of Maximum Likelihood Estimates					
Parameter	DF	Estimate	Standard Error	Wald Chi-Square	Pr > ChiSq
i_diagI	1	-2.2735	0.4573	24.7197	<.0001
i_diagII	1	-1.4341	0.3742	14.6887	0.0001
i_diagI*dose	1	0.1654	0.0414	15.9478	<.0001
i_diagII*dose	1	0.3064	0.0486	39.8186	<.0001

Output 11.30 Hypothesis Test

Linear Hypotheses Testing Results			
Label	Wald Chi-Square	DF	Pr > ChiSq
eq_slope	4.8787	1	0.0272

Next, the model based on log doses is fit.

```
proc logistic data=adverse;
   freq count;
   model status = i_diagI i_diagII
                  dose*i_diagI dose*i_diagII /
                  noint scale=none aggregate;
   eq_slope: test i_diagIdose=i_diagIIdose;
run;
```

Output 11.31 contains the goodness-of-fit tests, which are not as supportive of this model as they are for the model based on actual dose; however, they are still entirely satisfactory.

Output 11.31 Goodness-of-Fit Tests

Confidence Intervals

Deviance and Pearson Goodness-of-Fit Statistics				
Criterion	Value	DF	Value/DF	Pr > ChiSq
Deviance	4.8774	6	0.8129	0.5596
Pearson	4.4884	6	0.7481	0.6109

Number of unique profiles: 10

Output 11.32 contains the results for the test that the slopes are equal.

Output 11.32 Hypothesis Test Results

Linear Hypotheses Testing Results			
Label	Wald Chi-Square	DF	Pr > ChiSq
eq_slope	2.4034	1	0.1211

With $p = 0.1211$, you would not usually reject the hypothesis that the slopes are equal.

Thus, both models fit the data, and one model offers the possibility of a parallel lines model. Frequently, you encounter different model choices in your analyses, and you need to make a decision about which model to present. Since this is not a true bioassay, in the sense of a study comparing two drugs, the fact that you can fit a model with a common slope has less motivation. Potency in this setting means only that the shape of the tolerance distribution of the analgesic is similar for the two diagnoses, which may not be as important as simply determining that the drug works differently for the two diagnoses.

The model with the actual dose is used, since it fits very well and since there is no *a priori* reason to use log doses. (One very good reason might be to compare results with other studies if they worked with dose on the log scale.) It's of interest to compute ED50s for both diagnoses, to describe the median impact on adverse effects for the two diagnoses. The SAS/IML routine is again used to compute the ED50s and to produce a confidence interval based on Fieller's formula.

The required coefficients are

```
k={ -1  0  0  0}`;
h={  0  0  1  0}`;
```

and

```
k={  0 -1  0  0}`;
h={  0  0  0  1}`;
```

Output 11.33 contains the results. You need 13.74 units of the analgesic to produce adverse effects in 50% of the subjects with Diagnosis I; you only need 4.68 units of the drug to produce adverse effects in 50% of the subjects with Diagnosis II. The respective confidence intervals are (11.5095, 18.2537) and (2.9651, 6.0377).

Output 11.33 ED50s

Confidence Intervals

95 % ci for ratio based on fieller

ratio	interval l_bound	u_bound
13.741832	11.509478	18.253683

95 % ci for ratio based on fieller

ratio	interval l_bound	u_bound
4.6799535	2.9651466	6.0377151

11.5 Estimating Tolerance Distributions

The previous example illustrates that quantal response data analysis techniques can be used for the analysis of data that are not strictly bioassay data but are concerned with the investigation of drug responses. These methods can also be extended to other application areas as well, such as child development. For example, concepts like ED50 can be applied to describe the median ages at which certain types of physical development occur.

The following example is loosely based on a study of secondary sexual characteristics in girls (Herman-Giddens et al. 1997). Practitioners saw secondary sexual characteristics in patients much younger than the standard pediatric textbooks suggested was the norm, and researchers wanted to determine if the age of puberty was indeed dropping. Subjects were girls who showed up at a network of practitioner offices throughout the United States; the study was observational but the girls' heights and weights were representative of their age groups when compared to a national sample of health outcomes data.

Table 11.5 displays data for the onset of breast development for a subset of the girls, ages 5–12.

Table 11.5 Girls with Development Characteristics

Age	Onset Number	Total Girls
5	5	209
6	8	126
7	21	136
8	54	143
9	72	115
10	90	112
11	121	126
12	90	91

Similar to bioassay, you can assume that the response (onset of breast development) is determined through a tolerance distribution, and that onset will occur for individuals at certain ages. Thus, you can determine the mean of the tolerance distribution, which is the average age for the onset of breast development with probit analysis, or the median age for the onset with logistic analysis.

The following statements create SAS data set DEVELOPMENT. Note that AGE is coded as age at the half-year point for ages 5–12.

```
data development;
  input age onset total @@;
datalines;
5.5      5 209
6.5      8 126
7.5     21 136
8.5     54 143
9.5     72 115
10.5    90 112
11.5   121 126
12.5    90  91
;
```

A probit analysis is requested with the PROBIT procedure. Since no DIST= option is specified in the MODEL statement, probit analysis is performed. The PLOTS=PREDPPLOT option in the PROC PROBIT statement requests the predicted probabilities plot when ODS Graphics is enabled. The LACKFIT option in the MODEL statement requests goodness-of-fit statistics.

```
ods graphics on;
proc probit order=data plots=predpplot;
   model onset/total= age / lackfit;
run;
ods graphics off;
```

Output 11.34 displays the model information. These data include a total of 461 events for 1058 girls.

11.5. Estimating Tolerance Distributions

Output 11.34 Probit Model Information

Confidence Intervals

Model Information	
Data Set	WORK.DEVELOPMENT
Events Variable	onset
Trials Variable	total
Number of Observations	8
Number of Events	461
Number of Trials	1058
Name of Distribution	Normal
Log Likelihood	-366.2110666

The goodness-of-fit statistics in Output 11.35 are reasonable, with p-values of 0.82 for both Q_P and Q_L.

Output 11.35 Goodness-of-Fit Statistics

Goodness-of-Fit Tests				
Statistic	Value	DF	Value/DF	Pr > ChiSq
Pearson Chi-Square	2.9603	6	0.4934	0.8138
L.R. Chi-Square	2.9181	6	0.4864	0.8191

Output 11.36 contains the parameter estimates.

Output 11.36 Probit Parameter Estimates

Analysis of Maximum Likelihood Parameter Estimates							
Parameter	DF	Estimate	Standard Error	95% Confidence Limits		Chi-Square	Pr > ChiSq
Intercept	1	-5.5836	0.2776	-6.1277	-5.0395	404.55	<.0001
age	1	0.6215	0.0310	0.5607	0.6823	401.44	<.0001

Output 11.37 contains the parameters of the tolerance distribution.

Output 11.37 Tolerance Parameter Estimates

Probit Model in Terms of Tolerance Distribution	
MU	SIGMA
8.98464317	1.6091091

Since the tolerance distribution is based on the probit, $\hat{\mu}$ represents the mean of the tolerance distribution; for these data, $\hat{\mu} = 8.98$ represents the average age of the subjects at which breast development occurred. Output 11.38 displays the covariance matrix of these estimated parameters, so the standard error of $\hat{\mu}$ is $\sqrt{0.0072} = 0.0849$.

Output 11.38 Estimated Covariances of Tolerance Parameters

Estimated Covariance Matrix for Tolerance Parameters		
	MU	SIGMA
MU	0.007187	0.000782
SIGMA	0.000782	0.006450

Output 11.39 contains the predicted probabilities plot.

Output 11.39 Predicted Probabilities

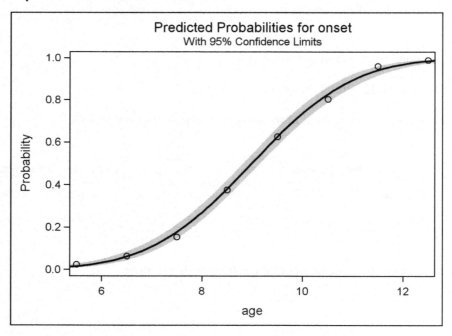

The analysis was rerun using the logistic distribution (DIST=LOGISTIC with the PROBIT procedure). The results are similar, as expected (The logistic fit was slightly better according to the

goodness-of-fit criteria). The following figure repeats the predicted probabilities plot for the probit analysis and also displays the same plot for the logistic analysis.

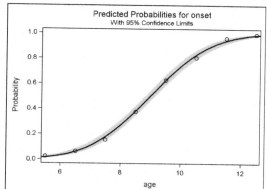

Figure 11.1 Probit Model

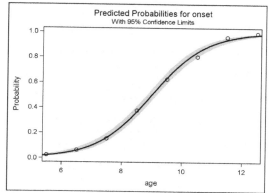

Figure 11.2 Logistic Model

You can see that the curves have somewhat different shapes, but they are very similar in their predictions.

Thus, quantal response data analysis methods can be applied to a variety of application areas, including growth and development. A related measure of interest is lifetime risk of an event, such as lifetime risk of a chronic disease such as osteoarthritis. See Murphy et al. (2008) for a study concerning this outcome conducted in Johnston County, North Carolina.

Appendix A: SAS/IML Macro for Confidence Intervals of Ratios Using Fieller's Theorem

The following SAS/IML code produces confidence intervals based on Fieller's theorem for ratios of estimates from a general linear model (Zerbe 1978). You specify coefficients for vectors **k** and **h** that premultiply the parameter vector to form the numerator and the denominator of the ratio of interest. For example, if $\beta = \{\alpha_n, \alpha_s, \beta\}$, $\mathbf{k} = \{1, -1, 0\}$, and $\mathbf{h} = \{0, 0, 1\}$, then

$$\frac{\mathbf{k}'\beta}{\mathbf{h}'\beta} = \frac{\alpha_n - \alpha_s}{\beta}$$

which is the log relative potency. Other choices of coefficients produce $\log LD50_n$ and $\log LD50_s$. You can produce the log LD50 for a single drug assay with the coefficients $\mathbf{k} = \{-1, 0\}$, and $\mathbf{h} = \{1, 0\}$.

The program inputs the covariance matrix for the parameters from an OUTEST= data set named ESTIMATE and applies the appropriate manipulations to produce the corresponding 95% confidence intervals for the ratios that are specified. The ratio estimates are displayed, and the lower and upper bounds of their Fieller-based confidence intervals appear under "l_bound" and "u_bound," respectively.

```
proc iml;
use estimate;
start fieller;
title 'Confidence Intervals';
use estimate;
read all into beta where (_type_='PARMS');
beta=beta`;
read all into cov where (_type_='COV');
ratio=(k`*beta) / (h`*beta);
a=(h`*beta)**2-(3.84)*(h`*cov*h);
b=2*(3.84*(k`*cov*h)-(k`*beta)*(h`*beta));
c=(k`*beta)**2 -(3.84)*(k`*cov*k);
disc=((b**2)-4*a*c);
if (disc<=0 | a<=0) then do;
print "confidence interval can't be computed", ratio;
stop; end;
sroot=sqrt(disc);
l_b=((-b)-sroot)/(2*a);
u_b=((-b)+sroot)/(2*a);
interval=l_b||u_b;
lname={"l_bound", "u_bound"};
print "95 % ci for ratio based on fieller", ratio interval[colname=lname];
finish fieller;
k={ -1  0  0  0}`;
h={  0  0  1  0}`;
run fieller;
k={  0 -1  0  0}`;
h={  0  0  0  1}`;
run fieller;
```

Chapter 12

Poisson Regression and Related Loglinear Models

Contents

12.1	Introduction	373
12.2	Methodology for Poisson Regression	374
12.3	Simple Poisson Counts Example	376
12.4	Poisson Regression for Incidence Densities	379
12.5	Overdispersion in Lower Respiratory Infection Example	383
12.6	Exact Poisson Regression	390
12.7	Loglinear Models	393
12.8	Analyzing Three-Way Cross-Classification Data with a Loglinear Model	395
	12.8.1 Analyzing Loglinear Models with the CATMOD Procedure	401
12.9	Correspondence between Logistic Models and Loglinear Models	405

12.1 Introduction

Categorical data often appear as discrete counts that are considered to be distributed as Poisson. Examples include colony counts for bacteria or viruses, accidents, equipment failures, insurance claims, and incidence of disease. Interest often lies in estimating a rate or incidence (bacteria counts per unit volume or cancer deaths per person-months of exposure to a carcinogen) and determining its relationship to a set of explanatory variables. Poisson regression became popularized as an analysis method in the 1970s and 1980s (Frome, Kutner, and Beauchamp 1973; Charnes, Frome, and Yu 1976; and Frome 1981), although Cochran pointed out the possibilities in a 1940 paper (Cochran 1940), along with the suggestion of the appropriateness of the loglinear model.

Poisson regression is a widely used modeling technique. Epidemiological applications include those where the events are occurrences of rare diseases (or experiences) for populations of different sizes; the enumeration of the number of events and determination of the population sizes are through possibly different data sources. Other epidemiological applications are occurrences of rare diseases (or experiences) for individuals with possibly differing amounts of exposure to risk; the explanatory variables are background covariables and risk factors.

Clinical trials provides another rich application area for Poisson regression. Events are occurrences of rare disorders for individuals with possibly different levels of exposure to risk; explanatory variables are treatment and background covariables. For example, the rare disorders in a vaccine study are the diseases to be prevented; in other studies, the events can correspond to an unfavorable side effect of a treatment or exacerbations of a disorder that treatment may reduce.

Some interesting uses of Poisson regression include a study homicide incidence study (Shahpar and Guohua 1999), a study of injuries incurred by electric utility workers (Loomis et al. 1999), and an evaluation of the risk of endometrial cancer as related to occupational physical activity (Moradi et al. 1998).

This chapter describes the methodology of Poisson regression in Section 12.2 and illustrates the use of the strategy in the next four sections. Section 12.5 describes the issue of overdispersion with Poisson data and discusses two techniques that adjust for it. Poisson regression is discussed in other chapters in this book as well. Chapter 15, "Generalized Estimating Equations," describes the analysis of Poisson-distributed correlated data with the GEE method in Section 15.9 and describes a GEE-based approach for managing overdispersion in Poisson regression for a univariate outcome in Section 15.13.

Loglinear model methodology is most appropriate when there is no clear distinction between response and explanatory variables, such as when all of the variables are observed simultaneously. The loglinear model point of view treats all variables as response variables, and the focus is on statistical independence and dependence. Loglinear modeling of categorical data is analogous to correlation analysis for normally distributed response variables and is useful in assessing patterns of statistical dependence among subsets of variables.

Although many investigators have made significant contributions, Goodman was a particularly influential researcher who popularized the method in the social sciences. Two of his key papers (Goodman 1968, 1970) summarize much of the earlier work. Bishop, Fienberg, and Holland (1975) first comprehensively described the methodology for the general statistical community. While loglinear models are used extensively in the social sciences, they have applicability to many areas. See Ahn et al. (2007) for the use of the loglinear model to study mutations along the HIV genome. Fienberg (2011) provides an historical perspective of the evolution of modeling methods for analyzing contingency tables, including the role of loglinear models.

There is a proportionality relationship between the likelihood for Poisson regression and the likelihood for the loglinear model, so the loglinear model can be fit using Poisson regression methods. Section 12.7 considers the loglinear model for three-way contingency tables, and Section 12.8 illustrates how to fit them using the GENMOD and CATMOD procedures. Section 12.9 describes the correspondence between logistic models and loglinear models.

In addition, the likelihoods for Poisson regression and the piecewise exponential model for analyzing time-to-event data are proportional, so the former can be used to fit the latter. This connection is illustrated in Chapter 13, "Categorized Time-to-Event Data."

12.2 Methodology for Poisson Regression

Suppose that a response variable Y is distributed as Poisson and has expected value μ. Recall that the variance of a Poisson variable is also μ. If you have a single explanatory variable x, you can write a regression model for μ as

$$g(\mu) = \alpha + x\beta$$

where g is a link function, in terms of a GLM (generalized linear model). Usually, g is taken to be the log function. If so, you have a loglinear model

$$\log(\mu) = \alpha + x\beta$$

You can rewrite this model as

$$\mu = e^\alpha e^{x\beta}$$

If you increase the explanatory variable x by one unit, it has a multiplicative effect of e^β on μ. Since this model is specified as a GLM, with a log link and a Poisson distribution, you can fit it with the GENMOD procedure and use the deviance and likelihood ratio tests to assess model fit and use Wald or score statistics to assess the model effects.

Frequently, discrete counts represent information collected over time (days, years) or in space (volume for bacteria counts), and interest lies in modeling rates. If the exposure time or volume is denoted as N, you write the rate as Y/N and write the expected value as μ/N. Modeling this rate with a loglinear model is written

$$\log \frac{\mu}{N} = \alpha + x\beta$$

which can be rearranged as

$$\log \mu = \alpha + x\beta + \log(N)$$

The term $\log(N)$ is called an *offset* and must be accounted for in the estimation process. Note that if you exponentiate both sides of this expression you obtain

$$\mu = \exp\{\alpha + x\beta + \log(N)\} = N e^\alpha e^{x\beta}$$

which indicates that the mean is proportional to N. Holding everything else constant, if you multiplied N by some number, you would be multiplying the expected mean by the same number.

More generally, when you have multiple explanatory variables, you can write the model in matrix terms as

$$\mu(\mathbf{x}) = \{N(\mathbf{x})\}\{g(\boldsymbol{\beta}|\mathbf{x})\}$$

where $\mu(\mathbf{x})$ is the expected value of the number of events $n(\mathbf{x})$, $\mathbf{x}$ is the vector of explanatory variables, $\mathbf{x} = (x_1, x_2, \ldots, x_t)'$, and $N(\mathbf{x})$ is the known total exposure to risk in the units in which the events occur (subject-days, for example). The rate for incidence is written as

$$\lambda(\mathbf{x}) = \mu(\mathbf{x})/N(\mathbf{x})$$

The loglinear model is written as

$$\log\left\{\frac{\mu(\mathbf{x})}{N(\mathbf{x})}\right\} = \mathbf{x}'\boldsymbol{\beta}$$

for counts $n(\mathbf{x})$ with independent Poisson distributions. An equivalent form is

$$\mu(\mathbf{x}) = \{N(\mathbf{x})\}\{\exp(\mathbf{x}'\boldsymbol{\beta})\}$$

If you have s independent groups referenced by $i = 1, 2, \ldots, s$, each with a vector $\mathbf{x}_i = (x_{i1}, x_{i2}, \ldots, x_{it})$ of t explanatory variables, you can write a likelihood function for the data as

$$\Phi(\mathbf{n}|\boldsymbol{\mu}) = \prod_{i=1}^{s} \mu_i^{n_i} \{\exp(-\mu_i)\}/n_i!$$

where $\mathbf{n} = (n_1, n_2, \ldots, n_s)'$ and $\boldsymbol{\mu} = (\mu_1, \mu_2, \ldots, \mu_s)'$.

The loglinear Poisson model is often written as

$$\log\{n_i\} = \log\{N_i\} + \mathbf{x}_i'\boldsymbol{\beta}$$

in the generalized linear models framework, where the quantity $\log\{N_i\}$ is the offset. For more information on Poisson regression, refer to Koch, Atkinson, and Stokes (1986).

12.3 Simple Poisson Counts Example

Table 12.1 presents salmonella counts for samples taken from two science labs (Margolin, Kaplan, and Zeiger 1981), reanalyzed in Koch, Atkinson, and Stokes (1986).

Table 12.1 Salmonella Counts in Two Science Labs

Science Lab	Counts
A	63 64 65 68 69 70 72 73
	75 80 82 83 83 84 84 85 90 91
B	168 171 174 175 185 189 190
	191 195 197 198 198 203 205 205
	207 210 214 216 218

Bacteria counts can be considered to be distributed according to the Poisson distribution. A simple Poisson regression model investigates whether the distributional assumption is correct and determines whether the counts for one science lab are different from those from the other science lab. You perform such an analysis with the GENMOD procedure.

The following SAS statements create data set SALMONELLA. Variable LAB indicates the A or B lab, and the COUNTS variable contains the bacteria counts for the samples taken (18 samples in lab A and 20 samples in lab B).

```
data salmonella;
   input lab $ counts @@;
datalines;
A 63 A 64 A 65 A 68 A 69 A 70 A 72 A 73
A 75 A 80 A 82 A 83 A 83 A 84 A 84 A 85 A 90 A 91
```

```
B 168 B 171 B 174 B 175 B 185 B 189 B 190
B 191 B 195 B 197 B 198 B 198 B 203 B 205 B 205
B 207 B 210 B 214 B 216 B 218
;
```

The next set of statements requests Poisson regression with the GENMOD procedure. The variable COUNTS is specified as the response variable, and the variable LAB is specified as the explanatory variable. The DIST=POISSON option and the LINK=LOG option request Poisson regression. The TYPE3 option requests the Type 3 analysis of effects.

```
proc genmod;
   class lab;
   model counts = lab / dist=poisson link=log type3;
run;
```

Output 12.1 displays the model information for this analysis, and Output 12.2 displays the CLASS variable levels for the explanatory variable.

Output 12.1 Model Information

Model Information	
Data Set	WORK.SALMONELLA
Distribution	Poisson
Link Function	Log
Dependent Variable	counts

Output 12.2 Class Information

Class Level Information		
Class	Levels	Values
lab	2	A B

Output 12.3 contains the table with the goodness-of-fit statistics. Note that the value of the deviance/df is 1.1207, and Q_p/df is 1.1113. Since these ratios are close to 1, there is good evidence that the data are distributed as Poisson, since these test statistics should be roughly equal to the corresponding degrees of freedom.

Output 12.3 Assessment of Fit

Criteria For Assessing Goodness Of Fit			
Criterion	DF	Value	Value/DF
Deviance	36	40.3451	1.1207
Scaled Deviance	36	40.3451	1.1207
Pearson Chi-Square	36	40.0077	1.1113
Scaled Pearson X2	36	40.0077	1.1113
Log Likelihood		21324.9700	
Full Log Likelihood		-146.8471	
AIC (smaller is better)		297.6942	
AICC (smaller is better)		298.0370	
BIC (smaller is better)		300.9693	

The Type 3 analysis is displayed in Output 12.4. The variable LAB is highly significant.

Output 12.4 Type 3 Analysis

LR Statistics For Type 3 Analysis			
Source	DF	Chi-Square	Pr > ChiSq
lab	1	1007.19	<.0001

The parameter estimates are presented in Output 12.5.

Output 12.5 Type 3 Analysis

Analysis Of Maximum Likelihood Parameter Estimates						Wald 95% Confidence Limits		Wald Chi-Square	Pr > ChiSq
Parameter			DF	Estimate	Standard Error				
Intercept			1	5.2753	0.0160	5.2440	5.3067	108783	<.0001
lab	A		1	-0.9351	0.0313	-0.9965	-0.8738	892.34	<.0001
lab	B		0	0.0000	0.0000	0.0000	0.0000	.	.
Scale			0	1.0000	0.0000	1.0000	1.0000		

Note: The scale parameter was held fixed.

The model-predicted log mean count is 5.2753 for lab B and $5.2753 - 0.9351 = 4.3402$ for lab A, corresponding to mean count estimates of 195.45 and 76.72, respectively.

12.4 Poisson Regression for Incidence Densities

Most of the time, Poisson regression is performed when you have counts and some measure of exposure. The next example concerns counts of melanoma cases as well as information on exposure. Thus, you are interested in fitting a model to the log rate, or incidence densities, of melanoma exposure. This requires adding an offset variable in the model.

Consider Table 12.2. The counts n_{hi} are the number of new melanoma cases reported in 1969–1971 for white males in two areas (Gail 1978; Koch, Imrey et al. 1985). The totals N_{hi} are the sizes of the estimated populations at risk; they may represent counts of people or counts of exposure units. Researchers were interested in whether the rates n_{hi}/N_{hi}, which are incidence densities, varied across age groups or region ($h = 1$ for the northern region, $h = 2$ for the southern region; $i = 1, 2, 3, 4, 5, 6$ for ascending age groups).

Table 12.2 New Melanoma Cases Among White Males: 1969–1971

Region	Age Group	Cases	Total
Northern	< 35	61	2880262
Northern	35–44	76	564535
Northern	45–54	98	592983
Northern	55–64	104	450740
Northern	65–74	63	270908
Northern	> 75	80	161850
Southern	< 35	64	1074246
Southern	35–44	75	220407
Southern	45–54	68	198119
Southern	55–64	63	134084
Southern	65–74	45	70708
Southern	> 75	27	34233

For this application of Poisson regression, the model of interest includes incremental effects for age levels and region. The following DATA step inputs the melanoma data and computes the variable LTOTAL, which is the log of the estimated exposure.

```
data melanoma;
   input age $ region $ cases total;
   ltotal=log(total);
   datalines;
35-44  south  75   220407
45-54  south  68   198119
55-64  south  63   134084
65-74  south  45    70708
75+    south  27    34233
<35    south  64  1074246
35-44  north  76   564535
45-54  north  98   592983
55-64  north 104   450740
65-74  north  63   270908
```

```
75+    north 80   161850
<35    north 61   2880262
;
```

The next statements invoke the GENMOD procedure.

```
proc genmod data=melanoma;
   class region (ref='north') age (ref='<35') / param=ref;
   model cases = age region
       / dist=poisson link=log offset=ltotal;
run;
```

The REF= option specifies that subjects from the North who are less than 35 years old are the reference group. The PARAM=REF option specifies incremental effects parameterization.

The MODEL statement specifies that a main effects model be fit. The response variable is CASES, and the explanatory variables are AGE and REGION. The DIST=POISSON option specifies the Poisson distribution, and the LINK=LOG option specifies that the log be the link function. The variable LTOTAL is to be treated as the offset. If you look at the preceding DATA step, you see that LTOTAL is the log of TOTAL. Thus, you are fitting a loglinear model to the ratio of cancer incidence to exposure.

Output 12.6 contains model specification information, including the fact that the model includes variable LTOTAL as an offset.

Output 12.6 Model Information

Model Information	
Data Set	WORK.MELANOMA
Distribution	Poisson
Link Function	Log
Dependent Variable	cases
Offset Variable	ltotal

Output 12.7 contains information about the CLASS variables. It confirms that the reference level for the parameterization corresponds to persons from the northern region who are younger than 35.

12.4. Poisson Regression for Incidence Densities

Output 12.7 Class Information

Class	Value	Design Variables				
region	north	0				
	south	1				
age	35-44	1	0	0	0	0
	45-54	0	1	0	0	0
	55-64	0	0	1	0	0
	65-74	0	0	0	1	0
	75+	0	0	0	0	1
	<35	0	0	0	0	0

Output 12.8 contains information on assessment of fit. Since $Q_P = 6.1151$ and the deviance has the value 6.2149, each with 5 df for their approximately chi-square distributions, the fit is satisfactory.

Output 12.8 Assessment of Fit

Criteria For Assessing Goodness Of Fit			
Criterion	DF	Value	Value/DF
Deviance	5	6.2149	1.2430
Scaled Deviance	5	6.2149	1.2430
Pearson Chi-Square	5	6.1151	1.2230
Scaled Pearson X2	5	6.1151	1.2230
Log Likelihood		2694.9262	
Full Log Likelihood		-39.2199	
AIC (smaller is better)		92.4398	
AICC (smaller is better)		120.4398	
BIC (smaller is better)		95.8342	

Output 12.9 contains the table of estimated model parameters. The log incidence density increases with increasing age level, and it also increases for the southern region.

Output 12.9 Estimated Model Parameters

Analysis Of Maximum Likelihood Parameter Estimates								
Parameter		DF	Estimate	Standard Error	Wald 95% Confidence Limits		Wald Chi-Square	Pr > ChiSq
Intercept		1	-10.6583	0.0952	-10.8449	-10.4718	12538.4	<.0001
age	35-44	1	1.7974	0.1209	1.5604	2.0344	220.92	<.0001
age	45-54	1	1.9131	0.1184	1.6810	2.1452	260.90	<.0001
age	55-64	1	2.2418	0.1183	2.0099	2.4737	358.89	<.0001
age	65-74	1	2.3657	0.1315	2.1080	2.6235	323.56	<.0001
age	75+	1	2.9447	0.1320	2.6859	3.2035	497.30	<.0001
region	south	1	0.8195	0.0710	0.6803	0.9587	133.11	<.0001
Scale		0	1.0000	0.0000	1.0000	1.0000		

Note: The scale parameter was held fixed.

You can exponentiate these parameters to express incidence density ratios in a manner similar to exponentiating parameters in logistic regression to obtain odds ratios. For example, exponentiating the parameter estimate for the increment for ages 45–54, $e^{1.9131} = 6.774$, gives you the ratio of the incidence of melanoma for those aged 45–54 relative to those less than 35. Similarly, $e^{0.8195} = 2.269$ is the ratio of the incidence of melanoma for those from the southern region relative to those from the northern region.

By submitting the ESTIMATE statements in the GENMOD procedure, you can produce these estimates as well as their confidence intervals.

```
proc genmod data=melanoma;
   class region (ref='north') age (ref='<35') / param=ref;
   model cases = age region
     / dist=poisson link=log offset=ltotal;
   estimate '45-54 vs. < 35' age 0 1 0 0/ exp;
   estimate 'South vs. North' region 1 / exp;
run;
```

Output 12.10 displays the results of this analysis. The confidence interval for the IDR for ages 45–54 relative to those less than 35 is (5.3707, 8.5440), and the confidence interval for the IDR for southern region relative to northern region is (1.9744, 2.6083).

Output 12.10 Estimated Incidence Density Ratios

Label	Mean			L'Beta Estimate	Standard Error	Alpha	L'Beta	
	Mean Estimate	Confidence Limits					Confidence Limits	
45-54 vs. < 35	6.7740	5.3707	8.5440	1.9131	0.1184	0.05	1.6810	2.1452
Exp(45-54 vs. < 35)				6.7740	0.8023	0.05	5.3707	8.5440
South vs. North	2.2693	1.9744	2.6083	0.8195	0.0710	0.05	0.6803	0.9587
Exp(South vs. North)				2.2693	0.1612	0.05	1.9744	2.6083

Contrast Estimate Results

Label	Chi-Square	Pr > ChiSq
45-54 vs. < 35	260.90	<.0001
Exp(45-54 vs. < 35)		
South vs. North	133.11	<.0001
Exp(South vs. North)		

Note that other methods for modeling incidence densities are available; see Saville, LaVange, and Koch (2011) for a discussion of nonparametric methods used to estimate incidence densities for multiple time intervals.

12.5 Overdispersion in Lower Respiratory Infection Example

Researchers studying the incidence of lower respiratory illness in infants took repeated observations of the children over one year. They studied 284 children and examined them every two weeks. Explanatory variables evaluated included passive smoking (one or more smokers in the household), socioeconomic status, and crowding. Refer to LaVange et al. (1994) for more information on the study and a discussion of the analysis of incidence densities. One outcome of interest was the total number of times, or counts, of lower respiratory infection recorded for the year. The strategy was to model these counts with Poisson regression. However, it is reasonable to expect that the children experiencing colds are more likely to have other infections; therefore, there may be some additional variance, or overdispersion, in these data.

Section 8.2.7 mentions overdispersion in the case of logistic regression. Overdispersion occurs when the observed variance is larger than the nominal variance for a particular distribution. It occurs with some regularity in the analysis of proportions and discrete counts. This is not surprising for the assumed distributions (binomial and Poisson, respectively) because the respective variances are fixed by a single parameter, the mean. When present, overdispersion can have a major impact on inference, so it needs to be taken into account. Note that underdispersion also occurs. Refer to McCullagh and Nelder (1989) and Dean (1998) for more detail on overdispersion.

One way to manage the overdispersion is to assume a more flexible distribution, such as the negative binomial in the case of overdispersed Poisson data. You can also adjust the covariance matrix of a Poisson-based analysis with a scaling factor. You expect that the goodness-of-fit chi-squares have values close to their degrees of freedom with this distribution; an indication of overdispersion is when their ratio is greater than 1. One way to manage this is to allow the variance function to have a multiplicative factor, that is, the variance is assumed to be $\phi\mu$ instead of μ. The chi-square statistic value divided by its degrees of freedom is used as the scaling factor ϕ. The covariance matrix is pre-multiplied by the scaling factor, and the scaled deviance and the log likelihoods are divided by ϕ, as is the profile likelihood function used in computing the confidence limits. Note that when there are indications of overdispersion, you also have to consider other causes besides overdispersion such as outliers and a misspecified model.

The following DATA step inputs the data into a SAS data set named LRI.[1]

```
data lri;
    input id count risk passive crowding ses agegroup race @@;
    logrisk =log(risk/52);
    datalines;
1  0 42 1 0 2 2 0    96 1 41 1 0 1 2 0    191 0 44 1 0 0 2 0
2  0 43 1 0 0 2 0    97 1 26 1 1 2 2 0    192 0 45 0 0 0 2 1
3  0 41 1 0 1 2 0    98 0 36 0 0 0 2 0    193 0 42 0 0 0 2 0
4  1 36 0 1 0 2 0    99 0 34 0 0 0 2 0    194 1 31 0 0 0 2 1
5  1 31 0 0 0 2 0   100 1  3 1 1 2 3 1    195 0 35 0 0 0 2 0
6  0 43 1 0 0 2 0   101 0 45 1 0 0 2 0    196 1 35 1 0 0 2 0
7  0 45 0 0 0 2 0   102 0 38 0 0 1 2 0    197 1 27 1 0 1 2 0
8  0 42 0 0 0 2 1   103 0 41 1 1 1 2 1    198 1 33 0 0 0 2 0
9  0 45 0 0 0 2 1   104 1 37 0 1 0 2 0    199 0 39 1 0 1 2 0
10 0 35 1 1 0 2 0   105 0 40 0 0 0 2 0    200 3 40 0 1 2 2 0
11 0 43 0 0 0 2 0   106 1 35 1 0 0 2 0    201 4 26 1 0 1 2 0
12 2 38 0 0 0 2 0   107 0 28 0 1 2 2 0    202 0 14 1 1 1 1 1
13 0 41 0 0 0 2 0   108 3 33 0 1 2 2 0    203 0 39 0 1 1 2 0
14 0 12 1 1 0 1 0   109 0 38 0 0 0 2 0    204 0  4 1 1 1 3 0

    ... more lines ...

90 1 38 1 1 1 2 1   185 0 43 0 0 0 2 0    280 0 31 0 0 0 2 0
91 0 32 1 1 1 2 0   186 0 42 0 0 0 2 0    281 0 18 0 0 0 2 0
92 1  3 1 0 1 3 1   187 0 42 0 0 0 2 0    282 1 32 1 0 2 2 0
93 0 26 1 0 0 2 1   188 0 38 0 0 0 2 0    283 0 22 1 1 2 2 1
94 0 35 1 0 0 2 0   189 0 36 1 0 0 2 0    284 0 35 0 0 0 2 1
95 3 37 1 0 0 2 0   190 0 39 0 1 0 2 0
;
```

The next set of statements requests Poisson regression for these data, including all of the explanatory variables. Variables PASSIVE and CROWDING are kept as (0, 1) variables and not included in the CLASS statement. Incremental effects parameterization is requested with the PARAM=REF option in the CLASS statement.

---

[1] Data provided by Lisa LaVange.

12.5. Overdispersion in Lower Respiratory Infection Example

```
proc genmod data=lri;
   class ses race agegroup / param=ref;
   model count = passive crowding ses race agegroup /
      dist=poisson offset=logrisk type3;
run;
```

Output 12.11 contains the general model information.

Output 12.11 Model Information

Model Information	
Data Set	WORK.LRI
Distribution	Poisson
Link Function	Log
Dependent Variable	count
Offset Variable	logrisk

Output 12.12 contains the goodness-of-fit statistics, along with the ratios of their values to their degrees of freedom. With values of 1.4788 for the deviance/df and 1.7951 for Pearson/df, there is evidence of overdispersion. The model-based estimates of standard errors may not be appropriate, and therefore any inference is questionable. (When such ratios are close to 1, you conclude that little evidence of over- or underdispersion exists.)

Output 12.12 Goodness-of-Fit Statistics

Criteria For Assessing Goodness Of Fit			
Criterion	DF	Value	Value/DF
Deviance	276	408.1549	1.4788
Scaled Deviance	276	408.1549	1.4788
Pearson Chi-Square	276	495.4493	1.7951
Scaled Pearson X2	276	495.4493	1.7951
Log Likelihood		-260.4117	
Full Log Likelihood		-337.5776	
AIC (smaller is better)		691.1551	
AICC (smaller is better)		691.6788	
BIC (smaller is better)		720.3469	

The model is refit with a scaling factor specified to adjust for the overdispersion. This is requested with the SCALE=PEARSON option, which computes a scaling factor that is the Pearson Q statistic divided by its degrees of freedom.

```
proc genmod data=lri;
    class ses race agegroup / param=ref;
    model count = passive crowding ses race agegroup /
        dist=poisson offset=logrisk type3 scale=pearson;
run;
```

Output 12.13 displays the goodness-of-fit statistics. Note that the scaled deviance and the scaled Pearson chi-square have different values because they have been divided by the scaling factor. The scaled Pearson chi-square is now 1 because the scaling factor requested was the Pearson chi-square value divided by the df.

Output 12.13 Assessment of Fit

Criteria For Assessing Goodness Of Fit			
Criterion	DF	Value	Value/DF
Deviance	276	408.1549	1.4788
Scaled Deviance	276	227.3709	0.8238
Pearson Chi-Square	276	495.4493	1.7951
Scaled Pearson X2	276	276.0000	1.0000
Log Likelihood		-145.0676	
Full Log Likelihood		-337.5776	
AIC (smaller is better)		691.1551	
AICC (smaller is better)		691.6788	
BIC (smaller is better)		720.3469	

Output 12.14 contains the results of the Type 3 analysis. Note that this table also includes F statistics; the chi-square approximation to the likelihood ratio test may have a less clear basis in this situation. Refer to *SAS/STAT User's Guide* for more detail about their computation.

Output 12.14 Type 3 Analysis

LR Statistics For Type 3 Analysis						
Source	Num DF	Den DF	F Value	Pr > F	Chi-Square	Pr > ChiSq
passive	1	276	3.89	0.0494	3.89	0.0484
crowding	1	276	5.86	0.0162	5.86	0.0155
ses	2	276	1.22	0.2966	2.44	0.2950
race	1	276	0.38	0.5408	0.38	0.5403
agegroup	2	276	1.07	0.3443	2.14	0.3429

Both passive smoking and crowding are strongly significant. Socioeconomic status and race do not appear to be influential, and neither does age group.

Finally, Output 12.15 contains the parameter estimates. The standard errors are adjusted due to the scaling factor, and they are larger than the standard errors for the unadjusted model, which are displayed in Output 12.16.

Output 12.15 Estimated Parameters for Adjusted Model

Analysis Of Maximum Likelihood Parameter Estimates								
Parameter		DF	Estimate	Standard Error	Wald 95% Confidence Limits		Wald Chi-Square	Pr > ChiSq
Intercept		1	0.6047	0.7304	-0.8269	2.0362	0.69	0.4077
passive		1	0.4310	0.2214	-0.0029	0.8649	3.79	0.0515
crowding		1	0.5199	0.2166	0.0953	0.9444	5.76	0.0164
ses	0	1	-0.3970	0.2886	-0.9627	0.1687	1.89	0.1690
ses	1	1	-0.0681	0.2627	-0.5830	0.4469	0.07	0.7956
race	0	1	0.1402	0.2309	-0.3123	0.5928	0.37	0.5436
agegroup	1	1	-0.4792	0.9043	-2.2516	1.2931	0.28	0.5962
agegroup	2	1	-0.9919	0.6858	-2.3361	0.3522	2.09	0.1481
Scale		0	1.3398	0.0000	1.3398	1.3398		

Note: The scale parameter was estimated by the square root of Pearson's Chi-Square/DOF.

Output 12.16 Estimated Parameters for Unadjusted Model

Analysis Of Maximum Likelihood Parameter Estimates								
Parameter		DF	Estimate	Standard Error	Wald 95% Confidence Limits		Wald Chi-Square	Pr > ChiSq
Intercept		1	0.6047	0.5452	-0.4638	1.6732	1.23	0.2673
passive		1	0.4310	0.1652	0.1072	0.7548	6.81	0.0091
crowding		1	0.5199	0.1617	0.2030	0.8367	10.34	0.0013
ses	0	1	-0.3970	0.2154	-0.8192	0.0252	3.40	0.0653
ses	1	1	-0.0681	0.1961	-0.4524	0.3163	0.12	0.7285
race	0	1	0.1402	0.1723	-0.1975	0.4780	0.66	0.4158
agegroup	1	1	-0.4792	0.6749	-1.8020	0.8436	0.50	0.4777
agegroup	2	1	-0.9919	0.5119	-1.9951	0.0113	3.76	0.0526
Scale		0	1.0000	0.0000	1.0000	1.0000		

Note: The scale parameter was held fixed.

A better way to account for the overdispersion might be to assume that the counts are distributed as negative binomial instead of Poisson. The negative binomial distribution allows for larger variance than the Poisson distribution does because it employs an additional parameter. The negative binomial variance equals $\mu + k\mu^2$, where k is the negative binomial dispersion parameter, thus allowing for larger variance than the Poisson distribution (when $k = 0$, the two distributions take the same form.) See Keene, Jones, and Lane (2007) for an application of negative binomial regression for clinical trials data.

The following SAS statements request negative binomial regression for these data by including the DIST=NB option in the MODEL statement. All other options remain the same.

```
proc genmod data=lri;
    class ses id race agegroup / param=ref;
    model count = passive crowding ses race agegroup /
        dist=nb offset=logrisk type3;
run;
```

Output 12.17 contains the resulting fit statistic. The values of 0.9310 and 1.08 for the statistic/DF for the deviance and Pearson chi-square, respectively, are not indicative of over- or underdispersion.

Output 12.17 Fit Statistics

Criteria For Assessing Goodness Of Fit			
Criterion	DF	Value	Value/DF
Deviance	276	256.9686	0.9310
Scaled Deviance	276	256.9686	0.9310
Pearson Chi-Square	276	298.2408	1.0806
Scaled Pearson X2	276	298.2408	1.0806
Log Likelihood		-242.2932	
Full Log Likelihood		-319.4590	
AIC (smaller is better)		656.9181	
AICC (smaller is better)		657.5750	
BIC (smaller is better)		689.7589	

Output 12.18 contains the Type 3 analysis, the results of which are similar to those for the scaled analysis.

12.5. Overdispersion in Lower Respiratory Infection Example

Output 12.18 Type 3 Analysis

LR Statistics For Type 3 Analysis			
Source	DF	Chi-Square	Pr > ChiSq
passive	1	4.43	0.0353
crowding	1	5.83	0.0158
ses	2	2.39	0.3034
race	1	0.26	0.6112
agegroup	2	2.92	0.2328

Output 12.19 contains the parameter estimates for the negative binomial regression. These estimates are also similar to those for the scaled analysis. However, note that the p-value for passive smoking is 0.0346 in this analysis compared to a p-value of 0.0515 for the scaled analysis. In general, the negative binomial regression approach would be suggested since it identifies the appropriate distributional assumption for the data rather than applying a fix to an inappropriate analysis.

Output 12.19 Estimated Parameters for Negative Binomial Regression

Analysis Of Maximum Likelihood Parameter Estimates								
Parameter		DF	Estimate	Standard Error	Wald 95% Confidence Limits		Wald Chi-Square	Pr > ChiSq
Intercept		1	0.6751	0.6333	-0.5661	1.9163	1.14	0.2864
passive		1	0.4530	0.2144	0.0329	0.8732	4.47	0.0346
crowding		1	0.5017	0.2061	0.0978	0.9057	5.93	0.0149
ses	0	1	-0.3987	0.2933	-0.9736	0.1762	1.85	0.1740
ses	1	1	-0.0857	0.2775	-0.6296	0.4582	0.10	0.7574
race	0	1	0.1178	0.2320	-0.3368	0.5725	0.26	0.6115
agegroup	1	1	-0.5652	0.8082	-2.1494	1.0189	0.49	0.4843
agegroup	2	1	-1.0131	0.6006	-2.1902	0.1641	2.85	0.0917
Dispersion		1	0.9760	0.2593	0.5798	1.6430		

Note: The negative binomial dispersion parameter was estimated by maximum likelihood.

See Section 15.13 in Chapter 15 for another method to adjust for overdispersion in these data.

12.6 Exact Poisson Regression

Exact Poisson regression is a useful strategy when you have small numbers of events because it does not depend on asymptotic results. Instead it relies on estimation of the model parameters by using the conditional distributions of the sufficient statistics of the parameters. The exact analysis is also available through the GENMOD procedure.

Table 12.3 displays the vaccine data discussed in Chapter 2 for both the United States and Latin America. There is interest in evaluating the effects of the vaccine with adjustment for the different regions. Since the event counts are small (1, 3), exact Poisson regression is indicated.

Table 12.3 Medical Events for Rotavirus Vaccine Study

Region	Vaccine		Placebo	
	Events	Person Years	Events	Person Years
United States	3	7500	58	7250
Latin America	1	1250	10	1250

The following SAS statements create the SAS data set ROTAVIRUS:

```
data rotavirus;
    input region $ treatment $ counts years_risk @@ ;
    log_risk=log(years_risk);
datalines;
US      Vaccine  3 7500 US     Placebo 58 7250
LA      Vaccine  1 1250 LA     Placebo 10 1250
;
run;
```

The following PROC GENMOD statements specify Poisson regression and exact Poisson regression. The ESTIMATE statement is used to produce the estimated incidence density ratio (IDR) for the standard analysis, comparing Vaccine and Placebo, and the ESTIMATE=ODDS and CLTYPE=EXACT options are used to produce the same quantities in the exact analysis (CLTYPE=EXACT is the default), even though IDRs are being estimated, not odds ratios.

```
proc genmod;
    class region  treatment/ param=ref;
    model counts = treatment region / dist=poisson offset= log_risk type3;
    estimate 'treatment' treatment 1 /exp;
    exact treatment / estimate=odds cltype=exact;
run;
```

Output 12.20 contains the criteria for evaluating goodness of fit. With p-values of 0.2979 for the deviance and 0.3431 for the Pearson chi-square (both 1 df), the model fit appears appropriate, although the number of events is too small for formal use of these statistics.

Output 12.20 Model Fit

Criteria For Assessing Goodness Of Fit			
Criterion	DF	Value	Value/DF
Deviance	1	0.2979	0.2979
Scaled Deviance	1	0.2979	0.2979
Pearson Chi-Square	1	0.3431	0.3431
Scaled Pearson X2	1	0.3431	0.3431
Log Likelihood		189.6784	
Full Log Likelihood		-7.6740	
AIC (smaller is better)		21.3481	
AICC (smaller is better)		.	
BIC (smaller is better)		19.5069	

Output 12.21 contains the maximum likelihood parameter estimates.

Output 12.21 Parameter Estimates

Analysis Of Maximum Likelihood Parameter Estimates								
Parameter		DF	Estimate	Standard Error	Wald 95% Confidence Limits		Wald Chi-Square	Pr > ChiSq
Intercept		1	-4.7886	0.3028	-5.3820	-4.1951	250.11	<.0001
treatment	Vaccine	1	-2.8620	0.5145	-3.8704	-1.8536	30.94	<.0001
region	US	1	-0.0467	0.3276	-0.6888	0.5953	0.02	0.8865
Scale		0	1.0000	0.0000	1.0000	1.0000		

Note: The scale parameter was held fixed.

Region is clearly not an influential effect, but it is considered a study design factor that should be maintained regardless of its *p*-value.

Output 12.22 contains the estimate of the incidence density ratio for vaccine compared to placebo. The IDR takes the value of 0.057, which is $e^{-2.862}$ with a confidence interval of (0.0209, 0.1567) based on the Wald chi-square statistic.

Output 12.22 ESTIMATE Statement Results

		Contrast Estimate Results						
		Mean					L'Beta	
Label	Mean Estimate	Confidence Limits		L'Beta Estimate	Standard Error	Alpha	Confidence Limits	
treatment	0.0572	0.0209	0.1567	-2.8620	0.5145	0.05	-3.8704	-1.8536
Exp(treatment)				0.0572	0.0294	0.05	0.0209	0.1567

Contrast Estimate Results		
Label	Chi-Square	Pr > ChiSq
treatment	30.94	<.0001
Exp(treatment)		

Output 12.24 contains the exact analysis results for treatment. Clearly, it is significant.

Output 12.23 Exact Tests

Exact Conditional Analysis

			p-Value	
Effect	Test	Statistic	Exact	Mid
treatment	Score	58.7561	<.0001	<.0001
	Probability	8.62E-17	<.0001	<.0001

Exact Conditional Tests header spans above.

Output 12.24 displays the estimate of IDR, which is 0.057, with an exact confidence interval of (0.015, 0.153), which is a bit different from the asymptotic results (0.0209, 0.1567). The percent rate reduction of events due to the vaccine is 94%.

Output 12.24 Exact IDR Estimates and Confidence Intervals

Exact Odds Ratios						
Parameter		Estimate	95% Confidence Limits		Two-sided p-Value	Type
treatment	Vaccine	0.057	0.015	0.153	<.0001	Exact

Output 12.24 displays the estimate of IDR, which is 0.057, with an exact confidence interval of (0.015, 0.153), which is a bit different from the asymptotic results (0.0209, 0.1567). The percent rate reduction of events due to the vaccine is 94%.

Note that these methods can also be used to generate the exact confidence interval for the IDR for the U.S. data analyzed in Section 2.8 in Chapter 2. The following statements create the data set ROTAVIRUS2 for just the U.S.

```
data rotavirus2;
   input region $ treatment $ counts years_risk @@ ;
   log_risk=log(years_risk);
   datalines;
US     Vaccine   3 7500 US     Placebo 58 7250
;
```

The GENMOD invocation produces a saturated model, since you are effectively modeling two counts with two parameters, and thus the usual goodness-of-fit statistics have the value 0.

```
proc genmod  order=data;
   class treatment/ param=ref;
   model counts = treatment / dist=poisson offset= log_risk type3;
   estimate 'treatment' treatment 1 /exp;
   exact treatment / estimate=odds cltype=exact;
run;
```

However, the exponentiated parameter for TREATMENT is the IDR, and when you request the exact analysis with the ESTIMATE=ODDS option, you also generate the exact confidence interval, as displayed in Output 12.25.

Output 12.25 Exact IDR Estimates and Confidence Intervals

Exact Conditional Analysis

Exact Odds Ratios						
Parameter		Estimate	95% Confidence Limits		Two-sided p-Value	Type
treatment	Vaccine	0.050	0.010	0.154	<.0001	Exact

The estimated IDR is 0.05, with (0.01002, 0.15355) as the confidence interval, which are very similar to the results produced in Section 2.8.

12.7 Loglinear Models

Suppose $\mathbf{n} = (n_1, n_2, \ldots, n_s)'$ denotes a vector of independent Poisson variables, and suppose $\boldsymbol{\mu} = (\mu_1, \mu_2, \ldots, \mu_s)'$ denotes the corresponding vector of expected values. Suppose variation among the elements of $\boldsymbol{\mu}$ can be described with the loglinear model

$$\boldsymbol{\mu} = \exp(\mathbf{X}_p \boldsymbol{\beta}_p)$$

where $\mathbf{X}_p = [\mathbf{1}, \mathbf{X}]$ is an $(s \times (t+1))$ matrix of known coefficients with full rank $(t+1) \leq s$ and $\boldsymbol{\beta}_p = [\beta_0, \boldsymbol{\beta}']'$ is a $(t+1)$ vector of unknown coefficients. The likelihood function for $\mathbf{n}$ is

$$\Phi(\mathbf{n}|\boldsymbol{\mu}) = \prod_{i=1}^{s} \mu_i^{n_i} \{\exp(-\mu_i)\}/n_i! = \{\mu_+^{n_+} \{\exp(-\mu_+)\}/n_+!\} \bullet \{n_+!/ \prod_{i=1}^{s} n_i!\}\{\prod_{i=1}^{s} \pi_i^{n_i}\}$$

where $n_+ = \sum_{i=1}^{s} n_i$, $\mu_+ = \sum_{i=1}^{s} \mu_i$, and $\pi_i = (\mu_i/\mu_+)$ for $i = 1, 2, \ldots, s$. Thus, the likelihood for $\mathbf{n}$ can be expressed as the product of a Poisson likelihood for n_+ and a multinomial likelihood $\phi(\mathbf{n}|n_+, \boldsymbol{\pi})$ with $\boldsymbol{\pi} = (\pi_1, \pi_2, \ldots, \pi_s)'$ for the conditional distribution of $\mathbf{n}$ given n_+. Since

$$\boldsymbol{\mu} = \exp(\mathbf{X}_p \boldsymbol{\beta}_p) = \exp(\beta_0) \exp(\mathbf{X}\boldsymbol{\beta})$$

it follows that

$$(\pi_1, \pi_2, \ldots, \pi_s)' = \boldsymbol{\pi} = \frac{\exp(\mathbf{X}\boldsymbol{\beta})}{\mathbf{1}'_s \exp(\mathbf{X}\boldsymbol{\beta})}$$

where $\mathbf{1}_s$ is the $(s \times 1)$ vectors of 1s. The structure shown here for $\boldsymbol{\pi}$ corresponds to the loglinear model for counts $\mathbf{n}$ with a multinomial distribution, either on the basis of conditioning independent Poisson counts on their sum or through simple random sampling (with replacement).

Since the maximization of the Poisson likelihood

$$\phi(\mathbf{n}|\boldsymbol{\mu}) = \phi(\mathbf{n}|\beta_0, \boldsymbol{\beta})$$
$$= \phi(n_+|\beta_0, \mathbf{1}'_t \exp(\mathbf{X}\boldsymbol{\beta}))\phi(\mathbf{n}|n_+, \boldsymbol{\beta})$$

relative to $\boldsymbol{\beta}_p = (\beta_0, \boldsymbol{\beta}')'$ correspondingly involves maximization of the multinomial likelihood $\phi(\mathbf{n}|n_+, \boldsymbol{\beta})$ relative to $\boldsymbol{\beta}$ as a by-product, the maximum likelihood estimates of $\boldsymbol{\beta}$ in the Poisson regression model

$$\boldsymbol{\mu} = \exp(\beta_0) \exp(\mathbf{X}\boldsymbol{\beta})$$

are also the maximum likelihood estimates of $\boldsymbol{\beta}$ in the multinomial loglinear model

$$\boldsymbol{\pi} = \frac{\exp(\mathbf{X}, \boldsymbol{\beta})}{(\mathbf{1}'_t \exp(\mathbf{X}\boldsymbol{\beta}))}$$

Also, the maximum likelihood estimate for the covariance matrix of $\hat{\boldsymbol{\beta}}$ is the same in both situations. A convenient consequence of these considerations is that $\hat{\boldsymbol{\beta}}$ for the multinomial loglinear model can have convenient computation through the use of Poisson regression via the GENMOD procedure to determine the maximum likelihood estimator

$$\hat{\boldsymbol{\beta}}_p = (\hat{\beta}_0, \hat{\boldsymbol{\beta}}')'$$

for $\boldsymbol{\beta}_p$ followed by removal of $\hat{\beta}_0$. In other words, the estimator $\hat{\beta}_0$ for the intercept in Poisson regression is ignored when this method is used to obtain $\hat{\boldsymbol{\beta}}$ for the multinomial loglinear model. A further point of interest is that the Poisson loglinear model is strictly loglinear since

$$\log(\boldsymbol{\mu}) = \mathbf{X}_p \boldsymbol{\beta}_p$$

whereas the multinomial loglinear model corresponds to

$$\log(\boldsymbol{\pi}) = \mathbf{X}\boldsymbol{\beta} - \{\log(\mathbf{1}'_s \exp(\mathbf{X}\boldsymbol{\beta}))\}\mathbf{1}_s$$

and so is loglinear with a constraint to ensure $\mathbf{1}'_s \boldsymbol{\pi} = \sum_{i=1}^{s} \pi_i = 1$.

12.8 Analyzing Three-Way Cross-Classification Data with a Loglinear Model

Table 12.4 displays the three-way cross-classification of quality of management, supervisor's job satisfaction, and worker's job satisfaction for a random sample of 715 workers selected from Danish industry (Andersen 1991, p. 155). Quality of management was categorized from an external evaluation of each factory, while the job satisfaction ratings were based on questionnaires completed by each worker and his or her supervisor. Since all three variables are response variables, the use of loglinear models to investigate the patterns of association among management quality, supervisor's job satisfaction, and worker's job satisfaction seems appropriate.

Table 12.4 Job Satisfaction Data*

Quality of Management	Supervisor's Job Satisfaction	Worker's Job Satisfaction		Total
		Low	High	
Bad	Low	103	87	190
	High	32	42	74
Good	Low	59	109	168
	High	78	205	283

*Reprinted by permission of Springer-Verlag.

Suppose X, Y, and Z denote quality of management, supervisor's job satisfaction, and worker's job satisfaction, respectively, and π_{ijk}s denotes the corresponding multinomial cell probabilities for $i = 1, 2$, $j = 1, 2$, and $k = 1, 2$. The saturated loglinear model is often written as follows:

$$\log(m_{ijk}) = \mu + \lambda_i^X + \lambda_j^Y + \lambda_k^Z + \lambda_{ij}^{XY} + \lambda_{ik}^{XZ} + \lambda_{jk}^{YZ} + \lambda_{ijk}^{XYZ}$$

for which the λ_i^X, λ_j^Y, and λ_k^Z correspond to the first order marginal distribution of X, Y, and Z; the λ_{ij}^{XY}, λ_{ik}^{XZ}, and λ_{jk}^{YZ} correspond to the associations (or interactions) between X and Y, X and Z, and Y and Z in their pairwise bivariate distribution; λ_{ijk}^{XYZ} corresponds to the second order association (or interaction) among X, Y, and Z in their joint distribution. When $\lambda_{ijk}^{XYZ} = 0$,

$$\frac{m_{111} m_{121}}{m_{211} m_{221}} = \frac{m_{112} m_{122}}{m_{212} m_{222}}$$

applies and corresponds to homogeneity of the odds ratios for the association between each pair of variables (for example, X and Y) across the categories of the third (for example, Z). If additionally $\lambda_{ij}^{XY} = 0$, then

$$\frac{m_{111} m_{221}}{m_{211} m_{121}} = \frac{m_{112} m_{222}}{m_{212} m_{122}} = 1$$

and so X and Y are conditionally independent for each category of Z.

If $\lambda_{ij}^{XY} = \lambda_{ik}^{XZ} = \lambda_{jk}^{YZ} = 0$ as well as $\lambda_{ijk}^{XYZ} = 0$, then you can verify that

$$m_{ijk} = \frac{m_{i++}m_{+j+}m_{++k}}{n^2}$$

where $m_{i++} = \sum_{j=1}^{2} \sum_{k=1}^{2} m_{ijk}$, $m_{+j+} = \sum_{i=1}^{2} \sum_{k=1}^{2} m_{ijk}$, and $m_{++k} = \sum_{i=1}^{2} \sum_{j=1}^{2} m_{ijk}$, and so independence applies to the distributions of X, Y, and Z. See Bishop, Fienberg, and Holland (1975), Roy and Mitra (1956), and Roy and Kastenbaum (1956) for primary sources for loglinear theory.

Alternatively, the model can be written as

$$\log\{m_{ijk}\} = \beta_0 + \beta_1 u_1 + \beta_2 u_2 + +\beta_3 u_3 + \beta_4 u_4 + \beta_5 u_5 + \beta_6 u_6 + \beta_7 u_7$$

where $u_1 = 1$ if $X = 1$ and $u_1 = -1$ if $X = 2$; $u_2 = 1$ if $Y = 1$ and $u_2 = -1$ if $Y = 2$; and $u_3 = 1$ if $Z = 1$ and $u_3 = -1$ if $Z = 2$. Also, $u_4 = u_1 u_2$, $u_5 = u_1 u_3$, $u_6 = u_2 u_3$, and $u_7 = u_1 u_2 u_3$.

The parameters correspond to loglinear model notation above as $\beta_0 = \mu$, and $\beta_1 = \lambda_i^X$, $\beta_2 = \lambda_1^Y$, $\beta_3 = \lambda_1^Z$, $\beta_4 = \lambda_{11}^{XY}$, $\beta_5 = \lambda_{11}^{XZ}$, $\beta_6 = \lambda_{11}^{YZ}$, and $\beta_7 = \lambda_{111}^{XYZ}$.

Also, conventions such as

$$\lambda_1^X + \lambda_2^X = 0, \lambda_1^Y + \lambda_2^Y = 0, \lambda_1^Z + \lambda_2^Z = 0$$

apply, as well as the sums of λ_{ij}^{XY}, λ_{ik}^{XZ}, λ_{jk}^{YZ}, and λ_{ijk}^{XYZ} over i, j, or k all equaling 0.

As previously mentioned, you fit the loglinear model with the GENMOD procedure. The following SAS statements input the job satisfaction data into data set SATISFY.

```
data satisfy;
    input management $ supervisor $ worker $ count;
    datalines;
Bad  Low   Low   103
Bad  Low   High   87
Bad  High  Low    32
Bad  High  High   42
Good Low   Low    59
Good Low   High  109
Good High  Low    78
Good High  High  205
;
```

The following SAS statements request a loglinear model analysis. You request the usual Poisson regression with the DIST=POISSON and LOG=LINK options, and you specify TYPE3 to obtain

12.8. Analyzing Three-Way Cross-Classification Data with a Loglinear Model

the tests for the effects (The WALD option is specified to produce Wald statistics for later comparison with the CATMOD procedure, but you would typically use the default Type 3 score statistics.) The saturated model is requested. The PARAM=EFFECT requests deviation from the mean parameterization, which is generally used for loglinear models.

```
proc genmod order=data;
   class management supervisor worker /param=effect;
   model count= management|supervisor|worker /
       link=log dist=poisson type3 wald;
run;
```

Output 12.26 displays the class levels for the response variables. Their cross-classification represents a single multinomial sample with eight categories of response. Since the model is saturated, the likelihood ratio test of fit is equal to zero.

Output 12.26 Class Levels

Class Level Information		
Class	Value	Design Variables
management	Bad	1
	Good	-1
supervisor	Low	1
	High	-1
worker	Low	1
	High	-1

Output 12.27 contains the Wald test of the three-factor interaction, which is nonsignificant ($Q_W = 0.06$, 1 df, $p = 0.7990$).

Output 12.27 Type 3 Wald Tests: Saturated Model

Wald Statistics For Type 3 Analysis			
Source	DF	Chi-Square	Pr > ChiSq
management	1	38.30	<.0001
supervisor	1	8.10	0.0044
managemen*supervisor	1	65.66	<.0001
worker	1	23.59	<.0001
management*worker	1	18.17	<.0001
supervisor*worker	1	5.24	0.0221
manage*superv*worker	1	0.06	0.7990

The second model fit includes only the main effects and two-factor interactions.

```
proc genmod order=data;
   class management supervisor worker /param=effect;
   model count= management|supervisor|worker@2 /
       link=log dist=poisson type3 wald;
run;
```

The deviance statistic (G^2 in loglinear terminology) in Output 12.28 is a likelihood ratio test that compares this model to the saturated model and thus tests the null hypothesis of no three-factor interaction. In this example, the G^2 statistic of 0.06 is the same as the Wald statistic for the three-way interaction from the saturated model. Although the two statistics are asymptotically equivalent, they are not identical in general.

Output 12.28 Fit Statistics for No Three-Factor Interaction Model

Criteria For Assessing Goodness Of Fit			
Criterion	DF	Value	Value/DF
Deviance	1	0.0649	0.0649
Scaled Deviance	1	0.0649	0.0649
Pearson Chi-Square	1	0.0649	0.0649
Scaled Pearson X2	1	0.0649	0.0649
Log Likelihood		2601.7363	
Full Log Likelihood		-24.7703	
AIC (smaller is better)		63.5406	
AICC (smaller is better)		.	
BIC (smaller is better)		64.0967	

Output 12.29 Wald Tests for No Three-Factor Interaction Model

Wald Statistics For Type 3 Analysis			
Source	DF	Chi-Square	Pr > ChiSq
management	1	38.37	<.0001
supervisor	1	8.32	0.0039
managemen*supervisor	1	67.06	<.0001
worker	1	25.96	<.0001
management*worker	1	19.57	<.0001
supervisor*worker	1	5.33	0.0210

The Wald tests of the two-factor interactions and main effects are all significant. This indicates that a more parsimonious model for the data may not be justified. However, you may wish to fit each of

12.8. Analyzing Three-Way Cross-Classification Data with a Loglinear Model

the three models containing only two of the two-factor interactions and compare these models to the model with no three-factor interaction using likelihood ratio tests. The SAS statements to build these models are as follows:

```
proc genmod order=data;
   class management supervisor worker /param=effect;
   model count= management|supervisor management|worker /
         link=log dist=poisson type3 wald;
run;

proc genmod order=data;
   class management supervisor worker /param=effect;
   model count= management|supervisor supervisor|worker /
         link=log dist=poisson type3 wald;
run;

proc genmod order=data;
   class management supervisor worker /param=effect;
   model count= management|worker supervisor|worker /
         link=log dist=poisson type3 wald;
run;
```

The corresponding likelihood ratio statistics for goodness of fit (output not shown) are $G^2 = 5.39$, 19.71, and 71.90, all with 2 df. The 1 df likelihood ratio statistics comparing each of these three models to the model with no three-factor interaction are $5.39 - 0.06 = 5.33$, $19.71 - 0.06 = 19.65$, and $71.90 - 0.06 = 71.84$, respectively. Relative to the chi-square distribution with 1 df, all indicate a significant lack of fit.

The model with no three-factor interaction provides a good fit to the observed data. Thus, no pair of variables is conditionally independent. In this model, the conditional odds ratios between any two variables are identical at each level of the third variable. For example, the odds ratio for the association between the employee's job satisfaction and the supervisor's job satisfaction is the same at each level of management quality. You can compute the estimated odds ratios from the parameter estimates (Output 12.30).

From the model with no three-factor interaction, the log odds of low job satisfaction for employees, at fixed levels of management quality and supervisor's job satisfaction, are

$$\begin{aligned} \log(m_{ij1}/m_{ij2}) &= \log(m_{ij1}) - \log(m_{ij2}) \\ &= \lambda_1^Z + \lambda_{i1}^{XZ} + \lambda_{j1}^{YZ} - (\lambda_2^Z + \lambda_{i2}^{XZ} + \lambda_{j2}^{YZ}) \\ &= 2\lambda_1^Z + 2\lambda_{i1}^{XZ} + 2\lambda_{j1}^{YZ} \end{aligned}$$

since $\lambda_1^Z + \lambda_2^Z = 0$, $\lambda_{i1}^{XZ} + \lambda_{i2}^{XZ} = 0$, and $\lambda_{j1}^{YZ} + \lambda_{j2}^{YZ} = 0$. Thus, at a fixed level of management quality, the log of the odds ratio at low and high levels of supervisor satisfaction is

$$\begin{aligned}\log(m_{i11}/m_{i12}) - \log(m_{i21}/m_{i22}) &= (2\lambda_1^Z + 2\lambda_{i1}^{XZ} + 2\lambda_{11}^{YZ}) - (2\lambda_1^Z + 2\lambda_{i1}^{XZ} + 2\lambda_{21}^{YZ}) \\ &= 2\lambda_{11}^{YZ} - 2\lambda_{21}^{YZ} \\ &= 4\lambda_{11}^{YZ}\end{aligned}$$

Output 12.30 contains the parameter estimates.

Output 12.30 Parameter Estimates from Model with No Three-Factor Interaction

Analysis Of Maximum Likelihood Parameter Estimates						Wald 95% Confidence Limits		Wald Chi-Square
Parameter			DF	Estimate	Standard Error			
Intercept			1	4.3448	0.0432	4.2602	4.4295	10128.3
management	Bad		1	-0.2672	0.0431	-0.3518	-0.1827	38.37
supervisor	Low		1	0.1243	0.0431	0.0398	0.2087	8.32
managemen*supervisor	Bad	Low	1	0.3491	0.0426	0.2655	0.4326	67.06
worker	Low		1	-0.2065	0.0405	-0.2860	-0.1271	25.96
management*worker	Bad	Low	1	0.1870	0.0423	0.1041	0.2698	19.57
supervisor*worker	Low	Low	1	0.0962	0.0417	0.0145	0.1778	5.33
Scale			0	1.0000	0.0000	1.0000	1.0000	

Analysis Of Maximum Likelihood Parameter Estimates			
Parameter			Pr > ChiSq
Intercept			<.0001
management	Bad		<.0001
supervisor	Low		0.0039
managemen*supervisor	Bad	Low	<.0001
worker	Low		<.0001
management*worker	Bad	Low	<.0001
supervisor*worker	Low	Low	0.0210
Scale			

Note: The scale parameter was held fixed.

Since the estimate of λ_{11}^{YZ} from Output 12.30 is 0.0962, the odds of low worker job satisfaction are estimated to be $\exp(4 \times 0.0962) = 1.47$ times higher when the supervisor's job satisfaction is low than when the supervisor's job satisfaction is high. Note that this estimate of the odds ratio is the same for factories with bad and good management quality. Using the observed counts from

Table 12.4, the observed odds ratios are

$$\frac{103 \times 42}{87 \times 32} = 1.55$$

in factories where the external evaluation of management quality was bad and

$$\frac{59 \times 205}{109 \times 78} = 1.42$$

in factories where the quality of management was good.

You can estimate additional odds ratios using the parameter estimates listed in Output 12.30. For a fixed level of supervisor job satisfaction, the odds of low worker satisfaction are estimated to be $\exp(4 \times 0.1870) = 2.1$ times higher when the quality of management is bad than when the management quality is good. This value is in between the corresponding observed odds ratios of

$$\frac{103 \times 109}{87 \times 59} = 2.19$$

when supervisor job satisfaction is low and

$$\frac{32 \times 205}{42 \times 78} = 2.00$$

when supervisor job satisfaction is high. Similarly, for a fixed level of worker job satisfaction, the odds of low supervisor job satisfaction are estimated to be $\exp(4 \times 0.3491) = 4.0$ times higher when the quality of management is bad than when the management quality is good. This value is in between the corresponding observed odds ratios of

$$\frac{103 \times 78}{32 \times 59} = 4.26$$

when worker job satisfaction is low and

$$\frac{87 \times 205}{42 \times 109} = 3.90$$

when worker job satisfaction is high.

These results show that the odds of low worker job satisfaction are somewhat more affected by the quality of management than by the supervisor's job satisfaction. In addition, bad management has a greater effect on the job satisfaction of supervisors than on worker job satisfaction.

12.8.1 Analyzing Loglinear Models with the CATMOD Procedure

You can also fit the loglinear model with the CATMOD procedure. While this procedure was designed to perform weighted least squares analysis of categorical response data, it also provides a convenient way to fit loglinear models with maximum likelihood estimation. Chapter 14, "Weighted Least Squares," contains a full discussion about using PROC CATMOD, ; the discussion in this section is limited to fitting the loglinear model.

The following statements fit the model with no three-way interaction. The COUNT variable is listed in the WEIGHT statement, which works much like a FREQ statement in other SAS/STAT procedures. The response variables MANAGEMENT, SUPERVISOR, and WORKER are crossed on the left-hand side of the MODEL statement, and the keyword _RESPONSE_ is used as the "explanatory variable." The specification of the model terms is done with the LOGLIN statement, and it is the same specification as used with PROC GENMOD in its MODEL statement. An advantage of using PROC CATMOD is that it produces predicted cell counts with the P=FREQ option in the MODEL statement.

```
proc catmod order=data data=satisfy;
   weight count;
   model management*supervisor*worker=_response_ / p=freq;
   loglin management|supervisor|worker@2;
run;
```

Output 12.31 displays the response profiles. The data come from a single multinomial population based on the cross-classification of the dichotomous variables MANAGEMENT, SUPERVISOR, and WORKER.

Output 12.31 Response Profile

Response Profiles			
Response	management	supervisor	worker
1	Bad	Low	Low
2	Bad	Low	High
3	Bad	High	Low
4	Bad	High	High
5	Good	Low	Low
6	Good	Low	High
7	Good	High	Low
8	Good	High	High

The results displayed in Output 12.32 contains the Wald statistics for model terms as well as an overall likelihood ratio test for the model compared to the saturated model. These results are the same as displayed in Output 12.29 and Output 12.30, respectively.

Output 12.32 Analysis of Variance Tests

Maximum Likelihood Analysis of Variance			
Source	DF	Chi-Square	Pr > ChiSq
management	1	38.37	<.0001
supervisor	1	8.32	0.0039
management*supervisor	1	67.06	<.0001
worker	1	25.96	<.0001
management*worker	1	19.57	<.0001
supervisor*worker	1	5.33	0.0210
Likelihood Ratio	1	0.06	0.7989

The CATMOD procedure also produces parameter estimates; unlike the GENMOD procedure, it does not produce an intercept since it is explicitly fitting a loglinear model. These estimates are displayed in Output 12.33.

Output 12.33 Parameter Estimates

Analysis of Maximum Likelihood Estimates					
Parameter		Estimate	Standard Error	Chi-Square	Pr > ChiSq
management	Bad	-0.2672	0.0431	38.37	<.0001
supervisor	Low	0.1243	0.0431	8.32	0.0039
management*supervisor	Bad Low	0.3491	0.0426	67.06	<.0001
worker	Low	-0.2065	0.0405	25.96	<.0001
management*worker	Bad Low	0.1870	0.0423	19.57	<.0001
supervisor*worker	Low Low	0.0962	0.0417	5.33	0.0210

Output 12.34 displays the predicted cell counts for this model. Instead of using the parameter estimates from Output 12.33, you could compute the estimated odds ratios using the predicted cell frequencies.

Output 12.34 Predicted Cell Counts

			Maximum Likelihood Predicted Values for Frequencies				
			Observed		Predicted		
management	supervisor	worker	Frequency	Standard Error	Frequency	Standard Error	Residual
Bad	Low	Low	103	9.389475	102.2639	8.904231	0.736105
Bad	Low	High	87	8.741509	87.73611	8.283433	-0.73611
Bad	High	Low	32	5.528818	32.73611	4.783695	-0.73611
Bad	High	High	42	6.287517	41.26389	5.525364	0.736105
Good	Low	Low	59	7.357409	59.73611	6.811457	-0.73611
Good	Low	High	109	9.611619	108.2639	9.138839	0.736105
Good	High	Low	78	8.336121	77.26389	7.782186	0.736105
Good	High	High	205	12.0923	205.7361	11.75535	-0.73611

The CATMOD procedure also offers the iterative proportional fitting (IPF) method for analyzing hierarchical loglinear models. If the contingency table is large in terms of number of cells in the multi-way cross-classification, possibly prohibitively large for maximum likelihood or weighted least squares estimation to proceed, you can still apply IPF to obtain the goodness-of-fit and the predicted cell frequencies and compute likelihood ratio statistics for model comparison (IPF does not estimate the parameters or covariance matrices).

The IPF algorithm is illustrated for the no three-way interaction model for the job satisfaction data. You simply add the ML=IPF option to the MODEL statement as shown.

```
proc catmod order=data;
   weight count;
   model management*supervisor*worker=_response_
      / ml=ipf;
   loglin management|supervisor|worker@2;
run;
```

The likelihood ratio statistic in Output 12.35 takes the value 0.06, the same value as computed in previous analyses.

Output 12.35 Analysis of Variance Table

Maximum Likelihood Analysis of Variance			
Source	DF	Chi-Square	Pr > ChiSq
Likelihood Ratio	1	0.06	0.7989

The predicted cell frequencies shown in Output 12.36 are very similar to the ones reported in Output 12.34. They would be the same with enough iterations.

Output 12.36 Predicted Cell Frequencies

			\multicolumn{2}{c|}{Observed}	\multicolumn{2}{c|}{Predicted}	
management	supervisor	worker	Frequency	Standard Error	Frequency	Standard Error	Residual
Bad	Low	Low	103	9.389475	102.2641	.	0.735932
Bad	Low	High	87	8.741509	87.73572	.	-0.73572
Bad	High	Low	32	5.528818	32.73639	.	-0.73639
Bad	High	High	42	6.287517	41.26388	.	0.736121
Good	Low	Low	59	7.357409	59.73593	.	-0.73593
Good	Low	High	109	9.611619	108.2643	.	0.735716
Good	High	Low	78	8.336121	77.26361	.	0.736385
Good	High	High	205	12.0923	205.7361	.	-0.73612

12.9 Correspondence between Logistic Models and Loglinear Models

Now suppose that the data displayed in Table 12.4 had instead been obtained from the four subpopulations defined by the cross-classification of quality of management and supervisor job satisfaction. In this case, you would be interested in modeling the probability of worker job satisfaction as a function of management quality and supervisor's job satisfaction. In practice, for situations where all of the variables are technically response variables, interest often focuses on modeling one of the variables as a function of the remaining ones.

The following PROC LOGISTIC statements model the logit of the probability of low worker job satisfaction as a function of management quality and supervisor job satisfaction.

```
proc logistic data=satisfy order=data;
   class management supervisor /param=effect;
   freq count;
   model worker=management supervisor
      / link=logit;
run;
```

Output 12.37 displays the resulting analysis of variance table. For comparison, examine Output 12.38 for the corresponding analysis of variance table from the loglinear model with no three-factor interaction.

Output 12.37 Analysis of Variance Table from Logistic Model

Type 3 Analysis of Effects			
Effect	DF	Wald Chi-Square	Pr > ChiSq
management	1	19.5629	<.0001
supervisor	1	5.3263	0.0210

Output 12.38 Analysis of Variance Table from Loglinear Model

Wald Statistics For Type 3 Analysis			
Source	DF	Chi-Square	Pr > ChiSq
management	1	38.37	<.0001
supervisor	1	8.32	0.0039
managemen*supervisor	1	67.06	<.0001
worker	1	25.96	<.0001
management*worker	1	19.57	<.0001
supervisor*worker	1	5.33	0.0210

The likelihood ratio goodness-of-fit statistics for the two models are identical (not shown here). In addition, the logistic model Wald chi-square statistics for the management and supervisor main effects are identical to the loglinear model Wald statistics for the management × worker and supervisor × worker interactions.

Output 12.39 displays the parameter estimates from the logistic model. The logistic model INTERCEPT chi-square statistic is identical to the loglinear model Wald statistic for the worker main effect. In addition, the log odds of low worker satisfaction are estimated to be exp(2 × 0.3739) = 2.1 times higher when the quality of management is bad than when the management quality is good and exp(2 × 0.1924) = 1.47 times higher when the supervisor's job satisfaction is low than when the supervisor's job satisfaction is high. These estimates are the same as those computed in Section 12.8.

Output 12.39 Parameter Estimates from Logistic Model

Analysis of Maximum Likelihood Estimates						
Parameter		DF	Estimate	Standard Error	Wald Chi-Square	Pr > ChiSq
Intercept		1	-0.4130	0.0811	25.9525	<.0001
management	Bad	1	0.3739	0.0845	19.5629	<.0001
supervisor	High	1	-0.1923	0.0833	5.3263	0.0210

12.9. Correspondence between Logistic Models and Loglinear Models

In summary, the cross-classification of the explanatory variables is fixed for logistic models, and the effects of the factors are specified explicitly in the MODEL statement. The loglinear model counterpart has the effects of factors specified through interactions with the response. In addition, the cross-classification of the explanatory variables is incorporated as a further component of the structure of the model.

The general result is that you can always rewrite a logistic analysis with one response variable as a loglinear model.

Chapter 13

Categorized Time-to-Event Data

Contents

13.1	Introduction	**409**
13.2	Life Table Estimation of Survival Rates	**410**
	13.2.1 Computing Survival Estimates with the LIFETEST Procedure	412
13.3	Mantel-Cox Test	**416**
13.4	Piecewise Exponential Models	**419**
	13.4.1 An Application of the Proportional Hazards Piecewise Exponential Model	421
	13.4.2 Using PROC LOGISTIC to Fit the Piecewise Exponential Model	424

13.1 Introduction

Categorical data often are generated from studies that have time from treatment or exposure until some event as their outcome. Such data are known as *time-to-event* data. The event may be death, the recurrence of some condition, or the emergence of a developmental characteristic. Often, the outcome is the actual lifetime (or waiting time), which is the response analyzed in typical survival analyses. However, due to resource constraints or the need to perform a diagnostic procedure, you sometimes can determine only the interval of time during which an event occurs. Examples include examining dental patients for caries at six-month periods, evaluating animals every four hours after their exposure to bacteria, and examining patients every six weeks for the recurrence of a medical condition for which they've been treated. Such data are often referred to as *grouped survival data* or as *categorized survival data*.

Since the study is conducted over a period of time, some subjects may leave before the study ends. This is called *withdrawal*. There may be protocol violations, subjects may join the study in progress and not complete the desired number of evaluations, or the subjects may drop out for other reasons. Thus, not only is status determined for each interval between successive evaluations, but the number of withdrawals for that interval is also determined. Most analysis strategies assume that withdrawal is independent of the condition being studied and that multiple withdrawals occur uniformly throughout the interval.

Frequently, interest lies in computing the survival rates. Section 13.2 discusses life table methods for computing these results. In addition, you generally want to compare survival rates for treatment groups and determine whether there is a treatment effect. Section 13.3 discusses the Mantel-Cox test, one strategy for addressing this question. It is similar to the logrank test used in traditional survival analysis. In addition to hypothesis testing, you may be interested in describing the variation in survival rates. Section 13.4 discusses the piecewise exponential model, one that is commonly used to model grouped survival data, as well as how to implement it using a Poisson regression strategy.

For an overview of grouped survival data analysis, refer to Deddens and Koch (1988).

13.2 Life Table Estimation of Survival Rates

The prevention of recurrences of medical conditions is the goal of many drugs under development. See Elashoff and Koch (1991) and Koch et al. (1984) for statistical methodology used in the analysis of clinical trials evaluating such treatments, which is described below.

Consider Table 13.1. Investigators were interested in comparing an active and control treatment to prevent the recurrence of a medical condition that had been healed. They applied a diagnostic procedure at the end of the first, second, and third years to determine whether there was a recurrence (based on Johnson and Koch 1978).

Table 13.1 Recurrences of Medical Condition

Treatment	Withdrawals			Recurrences			No Recurrence	Total
	Year 1	Year 2	Year 3	Year 1	Year 2	Year 3		
Control	9	7	6	15	13	7	17	74
Active	9	3	4	12	7	10	45	90

The survival rate, or the waiting time rate, is a key measure in the analysis of time-to-event data. It is written

$$S(y) = 1 - F(y) = \Pr\{Y \geq y\}$$

where Y denotes the continuous lifetime of a subject and $F(y) = \Pr\{Y \leq y\}$ is the cumulative probability distribution function. The exact form of $S(y)$ depends on the nature of $F(y)$, the probability distribution of Y. Common choices are the Weibull distribution and the exponential distribution, with the latter being a special case of the former.

One way of estimating survival rates is with the *life table*, or *actuarial*, method. Table 13.2 displays the life table format for the data displayed in Table 13.1. You determine the number of subjects at risk for each interval (the sum of those with no recurrence, those with recurrences, and those who withdrew). By knowing the number who survived all three intervals with no recurrence, you can determine the number with no recurrence for each interval.

13.2. Life Table Estimation of Survival Rates

Table 13.2 Life Table Format for Medical Condition Data

Interval	Controls			
	No Recurrences	Recurrences	Withdrawals	At Risk
0–1 Years	50	15	9	74
1–2 Years	30	13	7	50
2–3 Years	17	7	6	30
Interval	Active			
	No Recurrences	Recurrences	Withdrawals	At Risk
0–1 Years	69	12	9	90
1–2 Years	59	7	3	69
2–3 Years	45	10	4	59

Define n_{ijk} to be the number of patients in the ith group with the jth status for the kth time interval, where $j = 0$ corresponds to no recurrence during the time interval and $j = 1, 2$ corresponds to those with recurrence and those withdrawn during the kth interval, respectively; $i = 1, 2$ for the control and active groups and $k = 1, 2, \ldots, t$. The n_{i0k} are determined from

$$n_{i0k} = \sum_{j=1}^{2} \sum_{g=k+1}^{t} n_{ijg} + n_{i0t}$$

The life table estimates for the probability of surviving at least k intervals are computed as

$$G_{ik} = \prod_{g=1}^{k} \frac{n_{i0g} + 0.5 n_{i2g}}{n_{i0g} + n_{i1g} + 0.5 n_{i2g}} = \prod_{g=1}^{k} p_{ig}$$

where p_{ig} denotes the estimated conditional probability for surviving the gth interval given that survival of all preceding intervals has occurred.

The standard error of G_{ik} is estimated as

$$\text{s.e.}(G_{ik}) = G_{ik} \left\{ \sum_{g=1}^{k} \frac{(1 - p_{ig})}{(n_{i0g} + n_{i1g} + 0.5 n_{i2g}) p_{ig}} \right\}^{1/2}$$

$$= G_{ik} \left\{ \sum_{g=1}^{k} \frac{(1 - p_{ig})}{(n_{i0g} + 0.5 n_{i2g})} \right\}^{1/2}$$

where $(n_{i0g} + n_{i1g} + 0.5 n_{i2g})$ is the effective number at risk during the gth interval. Since

$$p_{ig} = \frac{n_{i0g} + 0.5 n_{i2g}}{n_{i0g} + n_{i1g} + 0.5 n_{i2g}}$$

then

$$1 - p_{ig} = \frac{n_{i1g}}{n_{i0g} + n_{i1g} + 0.5 n_{i2g}}$$

The quantity $0.5 \times n_{i2g}$ is used in the numerator and denominator of p_{ig} since uniform withdrawals throughout the interval are assumed. The average exposure to risk for the withdrawing subjects is assumed to be one-half the interval, and withdrawals are assumed not to have the event at the time of withdrawal.

For the active treatment, the life table estimates of surviving the kth interval are

$$G_{21} = \frac{69 + 0.5(9)}{69 + 12 + 0.5(9)} = 0.8596$$

$$G_{22} = 0.8596 \times \frac{59 + 0.5(3)}{59 + 7 + 0.5(3)} = 0.7705$$

$$G_{23} = 0.7705 \times \frac{45 + 0.5(4)}{45 + 10 + 0.5(4)} = 0.6353$$

Their standard errors are computed as follows:

$$\text{s.e.}(G_{21}) = 0.8596 \times \left\{ \frac{12/85.5}{69 + 0.5(9)} \right\}^{1/2} = 0.0376$$

$$\text{s.e.}(G_{22}) = 0.7705 \times \left\{ \frac{12/85.5}{69 + 0.5(9)} + \frac{7/67.5}{59 + 0.5(3)} \right\}^{1/2} = 0.0464$$

$$\text{s.e.}(G_{23}) = 0.6352 \times \left\{ \frac{12/85.5}{69 + 0.5(9)} + \frac{7/67.5}{59 + 0.5(3)} + \frac{10/57}{45 + 0.5(4)} \right\}^{1/2} = 0.0545$$

Table 13.3 contains the estimated survival rates and their standard errors for both active treatment and controls. The estimated survival rates for the active treatment are higher than for the controls for each of the intervals.

Table 13.3 Life Table Estimates for Medical Condition Data

	Estimated Survival Rates	Standard Errors
Controls		
0–1 Years	0.7842	0.0493
1–2 Years	0.5649	0.0627
2–3 Years	0.4185	0.0665
Active		
0–1 Years	0.8596	0.0376
1–2 Years	0.7705	0.0464
2–3 Years	0.6353	0.0545

13.2.1 Computing Survival Estimates with the LIFETEST Procedure

The LIFETEST procedure provides life table analysis for categorized time-to-event data. You need to supply the survival time interval, a censoring indicator, and variables for any stratification (treatment). A subject who withdrew during the study would be censored at the interval during which he or she withdrew. Subjects who completed the study with no events are considered to be censored at the end of the study.

The following SAS statements create the data set MEDICAL. There are two observations for each interval: one for the subjects who withdrew (censored) and one for the subjects who had a recurrence. An additional observation is added for the "fourth" interval, which is simply the number of subjects who didn't have a recurrence before the study ended. The variable INTERVAL takes the value 0, 1, 2, or 3, and the variable CENSOR takes the value 1 for censored subjects and 0 otherwise. The variable TREATMENT reflects whether the subjects received the active treatment or acted as controls, and the variable COUNT contains the frequency counts.

```
data medical;
   input interval treatment $ censor count @@;
datalines;
0 control 1 9 0 control 0 15
1 control 1 7 1 control 0 13
2 control 1 6 2 control 0  7
3 control 1 17
0 active 1 9 0 active 0 12
1 active 1 3 1 active 0  7
2 active 1 4 2 active 0 10
3 active 1 45
;
```

The next set of SAS statements requests the computations. By default, PROC LIFETEST calculates survival estimates according to the Kaplan-Meier method, so you need to specify the METHOD=LT option in the PROC LIFETEST statement to specify the life table method. The TIME statement includes the interval variable crossed by the censoring variable, with the value indicating censoring in parentheses (1 in this case). The ODS OUTPUT statements create data sets containing the life table estimates of interest.

```
ods graphics on;
proc lifetest data=medical method=lt plots=(s,ls)
        intervals=0 to 3 by 1;
   freq count;
   strata treatment;
   time interval*censor(1);
   ods output Lifetest.Stratum1.LifetableEstimates=my
       (keep=STRATUM treatment LowerTime UpperTime Survival StdErr);
   ods output Lifetest.Stratum2.LifetableEstimates=my2
       (keep=STRATUM treatment LowerTime UpperTime Survival StdErr);
run;
ods graphics off;

data all;
   set my my2;
run;

proc print data=all noobs;
run;
```

Output 13.1 contains summary information. Ninety subjects received the active treatment, and twenty-nine subjects had recurrences; seventy-four subjects acted as controls and thirty-five subjects had recurrences.

Output 13.1 Censoring Information

Summary of the Number of Censored and Uncensored Values					
Stratum	treatment	Total	Failed	Censored	Percent Censored
1	active	90	29	61	67.78
2	control	74	35	39	52.70
Total		164	64	100	60.98

Output 13.2 contains the survival estimates for active treatment and control. The column labeled "Survival" displays these estimates, and the column labeled "StdErr" displays the corresponding standard errors.

Output 13.2 Survival Estimates

STRATUM	treatment	LowerTime	UpperTime	Survival	StdErr
1	active	0	1	1.0000	0
1	active	1	2	1.0000	0
1	active	2	3	0.8596	0.0376
1	active	3	4	0.7705	0.0464
1	active	4	.	0.6353	0.0545
2	control	0	1	1.0000	0
2	control	1	2	1.0000	0
2	control	2	3	0.7842	0.0493
2	control	3	4	0.5649	0.0627
2	control	4	.	0.4185	0.0665

The results in Output 13.2 are the same as those displayed in Table 13.3. Output 13.3 displays a plot of the survival curves for active treatment and control.

Output 13.3 Survival Plot

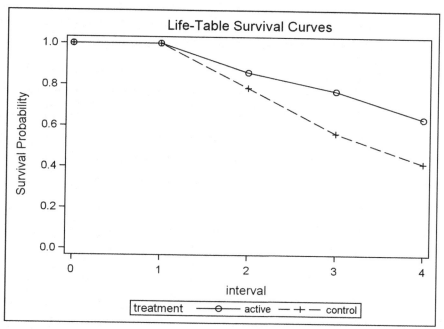

You generally are interested in comparing these curves to determine which treatment had the more favorable outcome. The LIFETEST procedure offers several different tests to assess whether there is equality over strata. These tests address withdrawals by grouping them with the no recurrences.

Output 13.4 displays the results of the logrank, Wilcoxon, and likelihood ratio tests for the homogeneity of the survival curves.

Output 13.4 Homogeneity of Survival Curves

Test of Equality over Strata			
Test	Chi-Square	DF	Pr > Chi-Square
Log-Rank	5.8836	1	0.0153
Wilcoxon	5.3880	1	0.0203
-2Log(LR)	6.0888	1	0.0136

The logrank and Wilcoxon tests are nonparametric tests that require no assumptions for the event time distribution. The Wilcoxon test is more powerful when there is a tendency for one group to have fewer early events than the other as well as longer survival times. The logrank test is more powerful when the two groups have similar rates of early events with one having more long-term survivors than the other. In this case, the logrank test might be preferred, but both tests have small p-values; the logrank test statistic has the value 5.8836 with $p = 0.0153$, and the Wilcoxon test statistic takes the value 5.3880 with $p = 0.0203$.

The likelihood ratio test is more powerful than the logrank test with the assumption that the event time distribution is exponential. This assumption can be evaluated by examining a plot of the negative log of the event times distribution. A linear trend through the origin would indicate consistency with exponential event times. Output 13.5 displays this plot for these data. There is the suggestion of linearity, which would make the likelihood ratio test potentially of more interest. However, there is uncertainty as to whether the exponential distribution applies.

Output 13.5 Negative Log Survival Plot

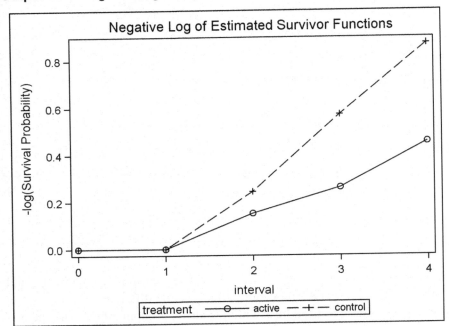

13.3 Mantel-Cox Test

An alternative to the tests of homogeneity discussed in the previous section is the Mantel-Cox test, which tests the null hypothesis that the survival rates are the same. Mantel (1966) and later Cox (1972) suggested an extension of the Mantel-Haenszel methodology that applies to survival data. You restructure the usual frequency table format of the data to a set of 2×2 tables, each with a life table format, and perform the Mantel-Haenszel computations on that set of tables.

The tables are generated by regarding treatment as the row variable, the numbers recurred and not recurred as the column variable, and the intervals as the strata. You are thus proceeding as though the time interval results are uncorrelated; methodological results for survival analysis establish that you can consider the respective time intervals to be essentially uncorrelated risk sets for survival information. The Mantel-Cox test for grouped data is equivalent to the logrank test for comparing survival curves for ungrouped data (refer to Koch, Sen, and Amara 1985). Withdrawals are handled by either grouping them with the no recurrences or eliminating them entirely (which is usually preferable when withdrawals have unknown status at the time of withdrawal, and it avoids the assumption of withdrawals not having the event through the end of the interval).

Table 13.4 contains the life table format for the study of the medical condition recurrence with the data grouped together by intervals and with the withdrawals excluded.

13.3. Mantel-Cox Test

Table 13.4 Medical Condition Data

Years	Treatment	Recurrences	No Recurrences
0–1	Control	15	50
	Active	12	69
1–2	Control	13	30
	Active	7	59
2–3	Control	7	17
	Active	10	45

The following DATA step inputs these data, and the PROC FREQ statements specify that the MH test be computed. Recall that for sets of 2 × 2 tables, all scores are equivalent, so no scores need to be specified.

```
data clinical;
    input time $ treatment $ status $ count @@;
    datalines;
0-1 control recur 15 0-1 control not 50
0-1 active  recur 12 0-1 active  not 69
1-2 control recur 13 1-2 control not 30
1-2 active  recur  7 1-2 active  not 59
2-3 control recur  7 2-3 control not 17
2-3 active  recur 10 2-3 active  not 45
;

proc freq order=data;
    weight count;
    tables time*treatment*status / cmh;
run;
```

Output 13.6 contains the PROC FREQ output (the individual printed tables are not displayed). $Q_{MC} = 8.0294$ with 1 df ($p = 0.0046$). There is a significant treatment effect on survival. The p-value for this test is much lower than those seen in the tests produced by PROC LIFETEST. However, those tests had withdrawals grouped with no recurrences, and this analysis excluded them.

Output 13.6 Results for Mantel-Cox Test

Summary Statistics for treatment by status
Controlling for time

Cochran-Mantel-Haenszel Statistics (Based on Table Scores)				
Statistic	Alternative Hypothesis	DF	Value	Prob
1	Nonzero Correlation	1	8.0294	0.0046
2	Row Mean Scores Differ	1	8.0294	0.0046
3	General Association	1	8.0294	0.0046

Note that if you did include the withdrawals with the no recurrences and recomputed the Mantel-Cox test, the results would be identical to those displayed in Output 13.4 for the logrank test.

418 Chapter 13: Categorized Time-to-Event Data

You can also apply the Mantel-Cox test when you have additional explanatory variables. Table 13.5 contains data from a study on gastrointestinal patients being treated for ulcers. Investigators compared an active treatment to a placebo in three medical centers.

Table 13.5 Healing for Gastrointestinal Patients

Center	Treatment	Healed at Two Weeks	Healed at Four Weeks	Not Healed at Four Weeks	Total
1	A	15	17	2	34
1	P	15	17	7	39
2	A	17	17	10	44
2	P	12	13	15	40
3	A	7	17	16	40
3	P	3	17	18	38

Table 13.6 contains the life table format for the same data.

Table 13.6 Healing for Gastrointestinal Patients

Center	Weeks	Treatment	Number Healed	Number Not Healed	Total
1	0–2	A	15	19	34
		P	15	24	39
	2–4	A	17	2	19
		P	17	7	24
2	0–2	A	17	27	44
		P	12	28	40
	2–4	A	17	10	27
		P	13	15	28
3	0–2	A	7	33	40
		P	3	35	38
	2–4	A	17	16	33
		P	17	18	35

The following DATA step inputs these data, and the PROC FREQ statements specify that the MH test be computed. For these data, both TIME and CENTER are used as stratification variables.

```
data duodenal;
   input center time $ treatment $ status $ count @@;
   datalines;
1 0-2 A healed 15 1 0-2 A not 19
1 0-2 P healed 15 1 0-2 P not 24
1 2-4 A healed 17 1 2-4 A not  2
1 2-4 P healed 17 1 2-4 P not  7
2 0-2 A healed 17 2 0-2 A not 27
2 0-2 P healed 12 2 0-2 P not 28
2 2-4 A healed 17 2 2-4 A not 10
2 2-4 P healed 13 2 2-4 P not 15
3 0-2 A healed  7 3 0-2 A not 33
```

```
3 0-2 P healed   3  3 0-2 P not 35
3 2-4 A healed 17  3 2-4 A not 16
3 2-4 P healed 17  3 2-4 P not 18
;

proc freq;
   weight count;
   tables center*time*treatment*status / cmh;
run;
```

Output 13.7 contains the results.

Output 13.7 Results for Mantel-Cox Test

Summary Statistics for treatment by status
Controlling for center and time

Cochran-Mantel-Haenszel Statistics (Based on Table Scores)				
Statistic	Alternative Hypothesis	DF	Value	Prob
1	Nonzero Correlation	1	4.2527	0.0392
2	Row Mean Scores Differ	1	4.2527	0.0392
3	General Association	1	4.2527	0.0392

The null hypothesis is that within each center, the distribution of time to healing is the same for placebo and active treatment. $Q_{MC} = 4.2527$ with 1 df ($p = 0.0392$), so there is a significant effect of active treatment on time to healing after adjusting for center.

13.4 Piecewise Exponential Models

Statistical models can extend the analysis of grouped survival data by providing a description of the pattern of event rates. They can describe this pattern over time as well as describe the variation due to the influence of treatment and other explanatory variables. One particularly useful model is the piecewise exponential model.

Consider Table 13.7, which contains information pertaining to the experience of patients undergoing treatment for duodenal ulcers (based on Johnson and Koch 1978). One of two types of surgeries was randomly assigned: vagotomy and drainage or antrectomy, or vagotomy and hemigastrectomy. The patients were evaluated at 6 months, 24 months, and 60 months. Death and recurrence are considered failure events, and reoperation and loss to follow-up are considered withdrawal events.

Table 13.7 Comparison of Two Surgeries for Duodenal Ulcer

Operation	Time (months)	Death or Recurrence	Reoperation or Lost	Satisfactory	Exposure (months)
V + D/A	0–6	23	15	630	3894
	7–24	32	20	578	10872
	25–60	45	71	462	18720
V + H	0–6	9	5	329	2016
	7–24	5	17	307	5724
	25–60	10	24	273	10440

In this study there are two treatment groups with $i = 1$ for V and D/A, $i = 2$ for V + H and three time intervals with $k = 1$ for 0–6 months, $k = 2$ for 7–24 months, and $k = 3$ for 25–60 months.

If you can make the following assumptions, then you can fit the piecewise exponential model to these data.

- The withdrawals are uniformly distributed during the time intervals in which they occur. They are unrelated to treatment failures and do not have the event at the time of withdrawal.

- The within-interval probabilities of the treatment failures are small and have independent exponential distributions.

The piecewise exponential likelihood is written

$$\Phi_{PE} = \prod_{i=1}^{2} \prod_{k=1}^{3} \lambda_{ik}^{n_{i1k}} \left\{ \exp[-\lambda_{ik} N_{ik}] \right\}$$

where n_{i1k} is the number of failures for the ith group during the kth interval, N_{ik} is the total person-months of exposure, and λ_{ik} is the hazard parameter. The piecewise exponential model assumes that there are independent exponential distributions with hazard parameters λ_{ik} for the respective time periods.

The N_{ik} are computed as

$$N_{ik} = a_k(n_{i0k} + 0.5 n_{i1k} + 0.5 n_{i2k})$$

where $a_k = 6, 18, 36$ is the length of the kth interval, n_{i0k} is the number of patients completing the kth interval without failure or withdrawal, and n_{i2k} denotes the number of withdrawals. The quantity n_{i1k} is the number of failures during the interval.

If you think of the number of deaths n_{i1k}, conditional on the exposures N_{ik}, as having independent Poisson distributions, then you can write a Poisson likelihood for these data.

$$\Phi_{PO} = \prod_{i=1}^{2} \prod_{k=1}^{3} (N_{ik} \lambda_{ik})^{n_{i1k}} \left\{ \frac{\exp[-N_{ik} \lambda_{ik}]}{n_{i1k}!} \right\}$$

$$= \Phi_{PE} \left\{ \prod_{i=1}^{2} \prod_{k=1}^{3} \frac{N_{ik}^{n_{i1k}}}{n_{i1k}!} \right\}$$

Since these likelihoods are proportional, whatever maximizes Φ_{PO} also maximizes Φ_{PE}. Thus, you can still assume the piecewise exponential model but obtain the estimates from Poisson regression computations, which are more accessible, regardless of whether you want to make the conditional arguments necessary to assume a Poisson distribution.

The relationship of the failure events to the explanatory variables is specified through models for the λ_{ik}. One class of models has the structure

$$\lambda_{ik} = \exp(\mathbf{x}'_{ik}\boldsymbol{\beta})$$

A useful subset of these models has the specification

$$\lambda_{ik} = \exp(\alpha + \eta_k + \mathbf{x}'_i\boldsymbol{\beta})$$

This latter model has the proportional hazards structure, where $\{\alpha + \eta_k\}$ is the constant value of the log hazard function within the kth interval when $\mathbf{x}_i = 0$. The parameter vector $\boldsymbol{\beta}$ relates the hazard function for the ith population to the explanatory variables $\mathbf{x}_i$.

Those readers familiar with survival analysis may recognize the general form of the proportional hazards model as

$$h(y, \mathbf{x}) = h_0(y)\{\exp(\mathbf{x}'\boldsymbol{\beta})\}$$

where y denotes continuous time and $h_0(y)$ is the hazard function for the reference population. In reference to this general form, $\exp(\alpha + \eta_k)$ corresponds to $h_0(y)$ for y in the kth interval.

13.4.1 An Application of the Proportional Hazards Piecewise Exponential Model

Since the GENMOD procedure fits Poisson regression models, you also use it to fit piecewise exponential models. The DATA step inputs the duodenal ulcer data and computes the variable NMONTHS as the log of MONTHS. The following PROC GENMOD statements request that the main effects model consisting of time and treatment be fit. The variable TREATMENT has the value 'vda' for V and D/A and the value 'vh' for V + H. Note that the value for 0–6 months for the variable TIME is '_0-6' so that it will sort last and thus become the reference value in the PROC GENMOD parameterization.

```
data vda;
    input treatment $ time $ failure months;
    nmonths=log(months);
    datalines;
vda   _0-6    23    3894
vda   7-24    32   10872
vda  25-60    45   18720
vh    _0-6     9    2016
vh    7-24     5    5724
vh   25-60    10   10440
;
```

```
proc genmod data=vda;
   class treatment time;
   model failure = time treatment
       / dist=poisson link=log offset=nmonths;
run;
```

Both TIME and TREATMENT are defined as CLASS variables. The LINK=LOG option is specified so that the model is in loglinear form, and the OFFSET=NMONTHS is specified since the quantity n_{ik}/N_{ik} is being modeled.

Information about the model specification and the sort levels of the CLASS variables are displayed in Output 13.8.

Output 13.8 Model Information

Model Information	
Data Set	WORK.VDA
Distribution	Poisson
Link Function	Log
Dependent Variable	failure
Offset Variable	nmonths

Class Level Information		
Class	Levels	Values
treatment	2	vda vh
time	3	25-60 7-24 _0-6

Statistics for assessing fit are displayed in Output 13.9. Q_P and the deviance both indicate an adequate fit, with values of 2.6730 and 2.5529, respectively, and 2 df for their approximately chi-square distributions.

Output 13.9 Goodness-of-Fit Criteria

Criteria For Assessing Goodness Of Fit			
Criterion	DF	Value	Value/DF
Deviance	2	2.5529	1.2764
Scaled Deviance	2	2.5529	1.2764
Pearson Chi-Square	2	2.6730	1.3365
Scaled Pearson X2	2	2.6730	1.3365
Log Likelihood		279.8914	
Full Log Likelihood		-15.0909	
AIC (smaller is better)		38.1819	
AICC (smaller is better)		78.1819	
BIC (smaller is better)		37.3489	

The "Analysis of Parameter Estimates" table in Output 13.10 includes an intercept parameter, incremental effects for 7–24 months and 25–60 months, and an incremental effect for the V + D/A treatment. The 0–6 months time interval and the V + H treatment are the reference cell.

Output 13.10 Parameter Estimates

Analysis Of Maximum Likelihood Parameter Estimates					Wald 95% Confidence Limits		Wald Chi-Square	Pr > ChiSq
Parameter		DF	Estimate	Standard Error				
Intercept		1	-5.8164	0.2556	-6.3174	-5.3153	517.66	<.0001
time	25-60	1	-1.0429	0.2223	-1.4787	-0.6071	22.00	<.0001
time	7-24	1	-0.8847	0.2414	-1.3579	-0.4116	13.43	0.0002
time	_0-6	0	0.0000	0.0000	0.0000	0.0000	.	.
treatment	vda	1	0.8071	0.2273	0.3616	1.2527	12.61	0.0004
treatment	vh	0	0.0000	0.0000	0.0000	0.0000	.	.
Scale		0	1.0000	0.0000	1.0000	1.0000		

Note: The scale parameter was held fixed.

All of these effects are significant. Table 13.8 contains the parameter interpretations.

Table 13.8 Parameter Interpretations

GENMOD Parameter	Model Parameter	Value	Interpretation
INTERCEPT	α	−5.8164	log incidence density for V + H, 0–6 months (reference)
TIME 25-60	η_2	−1.0429	increment for 25–60 interval
TIME 7-24	η_1	−0.8847	increment for 7–24 interval
TREAT vda	β	0.8071	increment for treatment V + D/A

Log incidence density decreases with the 7–24 interval and further decreases with the 25–60 interval. The V + D/A treatment increases log incidence density. What this means is that the failure rate is highest for the first interval and lower for the other two intervals; the failure rate is higher for V + D/A than for V + H. Table 13.9 displays the estimated failure rates (incidence densities) per person-month for each group and interval.

The survival rates can be calculated as follows:

$$\Pr\{\text{survival for } k \text{ intervals}\} = \Pr\{\text{survival for } k-1 \text{ intervals}\} \times e^{-\hat{\lambda}_{ik} a_k}$$

where a_k is the length of the kth interval in months, $k = 1, 2, \ldots, t$, and $\hat{\lambda}_{ik}$ is the predicted hazard parameter from the model for the ith group during the k interval.

Table 13.9 contains the survival estimates for each interval for each treatment group.

Table 13.9 Model-Estimated Failure Rates

Group	Interval	Failure Rate Formula	Estimated Failure Rate	Estimated Survival Rate
V + H	0–6	$e^{\hat{\alpha}}$	0.002978	0.9823
V + H	7–24	$e^{\hat{\alpha}+\hat{\eta}_1}$	0.001230	0.9608
V + H	25–60	$e^{\hat{\alpha}+\hat{\eta}_2}$	0.001050	0.9252
V + D/A	0–6	$e^{\hat{\alpha}+\hat{\beta}}$	0.006676	0.9607
V + D/A	7–24	$e^{\hat{\alpha}+\hat{\beta}+\hat{\eta}_1}$	0.002756	0.9142
V + D/A	25–60	$e^{\hat{\alpha}+\hat{\beta}+\hat{\eta}_2}$	0.002353	0.8399

13.4.2 Using PROC LOGISTIC to Fit the Piecewise Exponential Model

When the incidence rates $\{\lambda_{ik}\}$ are small (less than 0.05) and the exposures N_{ik} are very large, then you can approximate Poisson regression with logistic regression (Vine et al. 1990). Thus, you can take advantage of the features of the LOGISTIC procedure, such as its model-building facilities, to fit models such as the piecewise exponential model. You can facilitate the approximation by rescaling the exposure factor by multiplying it by a number such as 10,000; the only adjustment you need to make after parameter estimation is to add the log of the multiplier you choose to the resulting intercept estimate.

Using PROC LOGISTIC to Fit the Piecewise Exponential Model

The following SAS statements fit a piecewise exponential model to the duodenal ulcer data using the LOGISTIC procedure. In the DATA step, the variable SMONTHS is the exposure in months multiplied by a factor of 100,000.

The events/trials syntax is employed in the MODEL statement, with FAILURE in the numerator and SMONTHS in the denominator. The SELECTION=FORWARD option specifies forward model selection, and INCLUDE=2 specifies that the first two variables listed in the MODEL statement, TIME and TREATMENT, be forced into the first model so that a score test is produced for the contribution of the remaining variables to the model (interactions). This serves as a goodness-of-fit test.

```
data vda;
    input treatment time $ failure months;
    smonths=100000*months;
    datalines;
1   _0-6    23   3894
1   7-24    32   10872
1   25-60   45   18720
0   _0-6    9    2016
0   7-24    5    5724
0   25-60   10   10440
;

proc logistic;
    class time/param=ref;
    model failure/smonths = time treatment time*treatment /
                    scale=none include=2 selection=forward;
run;
```

Output 13.11 contains the resulting statistics for explanatory variable contribution and the score statistic (Residual Chi-Square) for the contribution of variables not in the model. Q_S has a value of 2.6730 and is nonsignificant with 2 df and $p = 0.2628$. Also, Q_S is identical to the Pearson chi-square goodness-of-fit statistic in Output 13.9 and Output 13.12. The model fits adequately and the proportional hazards assumption is reasonable (no time × treatment interaction).

Output 13.11 Score Statistic

Testing Global Null Hypothesis: BETA=0			
Test	Chi-Square	DF	Pr > ChiSq
Likelihood Ratio	34.7700	3	<.0001
Score	39.0554	3	<.0001
Wald	36.2836	3	<.0001

Residual Chi-Square Test		
Chi-Square	DF	Pr > ChiSq
2.6730	2	0.2628

Output 13.12 displays the goodness-of-fit statistics, which are adequate.

Output 13.12 Goodness-of-Fit Statistics

Deviance and Pearson Goodness-of-Fit Statistics				
Criterion	Value	DF	Value/DF	Pr > ChiSq
Deviance	2.5529	2	1.2764	0.2790
Pearson	2.6730	2	1.3365	0.2628

Number of events/trials observations: 6

The results are very similar to what was obtained with the PROC GENMOD analysis.

The resulting parameter estimates are displayed in Output 13.13. The parameter estimates for the time and treatment effects are very close to the estimates resulting from the PROC GENMOD analysis in the previous section, and if you add log(100000) to the intercept, -17.3293, you obtain -5.8164, which is identical to the intercept estimate obtained from PROC GENMOD. This approximation is usually very good.

Output 13.13 Parameter Estimates

Analysis of Maximum Likelihood Estimates						
Parameter		DF	Estimate	Standard Error	Wald Chi-Square	Pr > ChiSq
Intercept		1	-17.3293	0.2556	4595.1888	<.0001
time	25-60	1	-1.0430	0.2223	22.0023	<.0001
time	7-24	1	-0.8848	0.2414	13.4324	0.0002
treatment		1	0.8071	0.2273	12.6068	0.0004

At this point, you would proceed to produce survival rates and survival estimates as computed in the previous section.

Chapter 14

Weighted Least Squares

Contents

14.1	Introduction	**427**
14.2	Weighted Least Squares Methodology	**428**
	14.2.1 Weighted Least Squares Framework	429
	14.2.2 Weighted Least Squares Estimation	430
	14.2.3 Model Parameterization	432
14.3	Using PROC CATMOD for Weighted Least Squares Analysis	**432**
14.4	Obstetrical Pain Data: Advanced Modeling of Means	**439**
	14.4.1 Performing the Analysis with PROC CATMOD	441
14.5	Analysis of Survey Sample Data	**449**
	14.5.1 HANES Data	449
	14.5.2 Direct Input of Response Functions	450
	14.5.3 The FACTOR Statement	451
	14.5.4 Preliminary Analysis	452
	14.5.5 Inputting the Model Matrix Directly	455
14.6	Modeling Rank Measures of Association Statistics	**457**
14.7	Repeated Measurements Analysis	**463**
	14.7.1 WLS Methodology for Repeated Measurements	465
	14.7.2 One Population, Dichotomous Response	466
	14.7.3 Two Populations, Polytomous Response	471
	14.7.4 One Population Regression Analysis of Logits	478
	Appendix A: Statistical Methodology for Weighted Least Squares	**482**
	Appendix B: CONTRAST statements for Obstetrical Pain	**485**

14.1 Introduction

Previous chapters discussed statistical modeling of categorical data with logistic regression. Maximum likelihood (ML) estimation was used to estimate parameters for models based on logits and cumulative logits. Logistic regression is suitable for many situations, particularly for dichotomous response outcomes. However, there are situations where modeling techniques other than logistic regression are of interest. You may be interested in modeling functions besides logits, such as mean scores, proportions, or more complicated functions of the responses. In addition, the analysis framework may dictate a different modeling approach, such as in the case of repeated measurements studies.

Weighted least squares (WLS) estimation provides a methodology for modeling a wide range of categorical data outcomes. This chapter focuses on the application of weighted least squares for the modeling of mean scores and proportions in the stratified simple random sampling framework, as well as for the modeling of estimates produced by more complex sampling mechanisms, such as those required for complex sample surveys. The methodology is explained in the context of a basic example.

The CATMOD procedure is a general tool for modeling categorical data. It uses weighted least squares estimation to model a variety of response functions. PROC CATMOD also uses maximum likelihood estimation for logistic regression when the response functions are generalized logits. This chapter discusses the use of PROC CATMOD for numerous applications of weighted least squares analyses, including repeated measurements analysis.

You should be familiar with the material in Chapter 8, "Logistic Regression I: Dichotomous Response," and Chapter 9, "Logistic Regression II: Polytomous Response," before proceeding with this chapter.

14.2 Weighted Least Squares Methodology

To motivate the discussion of weighted least squares methodology, consider the following example. Epidemiologists investigating air pollution effects conducted a study of childhood respiratory disease (Stokes 1986). Investigators visited groups of children at two times and recorded whether they were exhibiting symptoms of colds. The children were recorded as having no periods with a cold, one period with a cold, or two periods with a cold. Investigators were interested in determining whether sex or residence affected the distribution of colds. These data are displayed in Table 14.1.

Table 14.1 Colds in Children

Sex	Residence	Periods with Colds			Total
		0	1	2	
Female	Rural	45	64	71	180
Female	Urban	80	104	116	300
Male	Rural	84	124	82	290
Male	Urban	106	117	87	310

As previously discussed, statistical modeling addresses the question of how a response outcome is distributed across the various levels of the explanatory variables. In the standard linear model, this is done by fitting a model to the response mean. In logistic regression, the function that is modeled is the logit or cumulative logit. For these data, a response measure of interest is the mean number of periods with colds. However, because there are a small, discrete number of response values, it is unlikely that the normality or variance homogeneity assumptions usually required for the standard linear model are met. Weighted least squares methodology provides a useful strategy for analyzing these data.

14.2.1 Weighted Least Squares Framework

Underlying most types of weighted least squares methods for categorical data analysis is a contingency table. The general idea is to model the distribution of the response variable, represented in the columns of the table, across the levels of the explanatory variables, represented by the rows of the table. These rows are determined by the cross-classification of the levels, or values, of the explanatory variables. The contingency table for the colds data displayed in Table 14.1 has four rows and three columns. There are four rows since there are four combinations of sex and residence; there are three columns because the response variable has three possible outcomes: 0, 1, and 2.

The general contingency table is displayed in Table 14.2, where s represents the number of rows (groups) in the table and r represents the number of responses. The rows of the table are also referred to as subpopulations.

Table 14.2 General Contingency Table

Group	Response 1	2	...	r	Total
1	n_{11}	n_{12}	...	n_{1r}	n_{1+}
2	n_{21}	n_{22}	...	n_{2r}	n_{2+}
...	...	...	...	...	...
s	n_{s1}	n_{s2}	...	n_{sr}	n_{s+}

The proportion of subjects in each group who have each response is written as

$$p_{ij} = n_{ij}/n_{i+}$$

where n_{ij} is the number of subjects in the ith group who have the jth response. For example, $p_{11} = 45/180$ in Table 14.1. You can put the proportions for one group together in a proportion vector that describes the response distribution for that group. For the colds data, it looks like the following:

$$\mathbf{p}_i = (p_{i1}, p_{i2}, p_{i3})'$$

You can then form a proportion vector for each group in the contingency table. The proportions for each group add up to 1. All the functions that can be modeled with weighted least squares methodology are generated from these proportion vectors.

The rows of the contingency table are considered to be simple random samples from the multinomial distribution; since the rows are independent, the entire table is distributed as product multinomial. You can write the estimated covariance matrix for the proportions in the ith row as

$$\mathbf{V}_i = \frac{1}{n_{i+}} \begin{bmatrix} p_{i1}(1-p_{i1}) & -p_{i1}p_{i2} & \cdots & -p_{i1}p_{ir} \\ -p_{i2}p_{i1} & p_{i2}(1-p_{i2}) & \cdots & -p_{i2}p_{ir} \\ \vdots & \vdots & \vdots & \vdots \\ -p_{ir}p_{i1} & -p_{ir}p_{i2} & \cdots & p_{ir}(1-p_{ir}) \end{bmatrix}$$

So, then you can write the estimated covariance matrix for the entire table as

$$\mathbf{V_p} = \begin{bmatrix} \mathbf{V}_1 & 0 & \cdots & 0 \\ 0 & \mathbf{V}_2 & \cdots & 0 \\ \vdots & \vdots & \vdots & \vdots \\ 0 & 0 & \cdots & \mathbf{V}_s \end{bmatrix}$$

where $\mathbf{V}_i$ is the estimated covariance matrix for the ith row.

14.2.2 Weighted Least Squares Estimation

Once the proportion vector and its estimated covariance matrix are computed, the modeling phase begins with the choice of a response function. You can model any of the following: the proportions themselves; mean scores, which are simple linear functions of the proportions; logits, which are constructed by taking a linear function (difference) of the log proportions; and a number of more complicated functions that are created by combinations of various transformations of the proportions, such as the kappa statistic for observer agreement (see Landis and Koch 1977) or rank measures of association (see Koch and Edwards 1988).

For the colds data, the response function is the mean number of periods with a cold. You construct these means from the proportions of responses in each row of Table 14.1 and then apply a statistical model that determines the effect of sex and residence on their distribution. Table 14.3 displays the row proportions.

Table 14.3 Colds in Children

Sex	Residence	Periods with Colds			Total
		0	1	2	
Female	Rural	0.25	0.36	0.39	1.00
Female	Urban	0.27	0.35	0.39	1.00
Male	Rural	0.29	0.43	0.28	1.00
Male	Urban	0.34	0.38	0.28	1.00

For example, to compute the mean number of periods with colds for females in a rural residence, you would perform the following computation:

$$\begin{aligned} \text{mean colds} &= 0*p_{11} + 1*p_{12} + 2*p_{13} \\ &= 0*(0.25) + 1*(0.36) + 2*(0.39) \\ &= 1.14 \end{aligned}$$

In matrix terms, you have multiplied the proportion vector by a linear transformation matrix $\mathbf{A}$.

$$\mathbf{A}\mathbf{p}_1 = \begin{bmatrix} 0 & 1 & 2 \end{bmatrix} \begin{bmatrix} 0.25 \\ 0.36 \\ 0.39 \end{bmatrix} = 1.14$$

Means are generated for each sex × residence group to produce a total of four functions for the table. The ith function is denoted $F(\mathbf{p}_i)$. The following expression shows how you generate a function vector by applying a linear transformation matrix to the total proportion vector $\mathbf{p} = (\mathbf{p}_1', \mathbf{p}_2', \mathbf{p}_3', \mathbf{p}_4')'$ to produce the four means of interest.

$$F(\mathbf{p}) = \mathbf{A}\mathbf{p} = \begin{bmatrix} 0 & 1 & 2 & 0 & 0 & 0 & 0 & 0 & 0 & 0 & 0 & 0 \\ 0 & 0 & 0 & 0 & 1 & 2 & 0 & 0 & 0 & 0 & 0 & 0 \\ 0 & 0 & 0 & 0 & 0 & 0 & 0 & 1 & 2 & 0 & 0 & 0 \\ 0 & 0 & 0 & 0 & 0 & 0 & 0 & 0 & 0 & 0 & 1 & 2 \end{bmatrix} \mathbf{p} = \begin{bmatrix} 1.14 \\ 1.12 \\ 0.99 \\ 0.94 \end{bmatrix}$$

If the groups have sufficient sample size (usually $n_{i+} \geq 25$), then the variation among the response functions can be investigated by fitting linear regression models with weighted least squares:

$$E_A\{F(\mathbf{p})\} = F(\boldsymbol{\pi}) = \mathbf{X}\boldsymbol{\beta}$$

E_A denotes asymptotic expectation, and $\boldsymbol{\pi} = E\{\mathbf{p}\}$ denotes the vector of population probabilities for all the populations together. The vector $\boldsymbol{\beta}$ contains the parameters that describe the variation among the response functions, and $\mathbf{X}$ is the model specification matrix. The equations for WLS estimation are similar to those for least squares estimation:

$$\mathbf{b} = (\mathbf{X}'\mathbf{V}_F^{-1}\mathbf{X})^{-1}\mathbf{X}'\mathbf{V}_F^{-1}\mathbf{F}$$

$\mathbf{V}_F$ is the estimated covariance matrix for the vector of response functions and is usually nonsingular when the sample sizes n_{i+} are sufficiently large (for example, $n_{i+} \geq 25$ and at least two $n_{ij} \geq 1$ in each row). This is the weight matrix component of weighted least squares estimation. Its form depends on the nature of the response functions. In the case of the colds data, where the response functions are means computed as $\mathbf{A}\mathbf{p}$, the estimated covariance matrix is computed as

$$\mathbf{V}_F = \mathbf{A}\mathbf{V}_p\mathbf{A}'$$

The estimated covariance matrix for $\mathbf{b}$ is written

$$\mathbf{V}(\mathbf{b}) = (\mathbf{X}'\mathbf{V}_F^{-1}\mathbf{X})^{-1}$$

Model adequacy is assessed with Wald goodness-of-fit statistics, which are computed as

$$Q_W = (\mathbf{F} - \mathbf{X}\mathbf{b})'\mathbf{V}_F^{-1}(\mathbf{F} - \mathbf{X}\mathbf{b})$$

For an applicable model, Q_W is distributed as chi-square for moderately large sample sizes (for example, all $n_{i+} \geq 25$), and its degrees of freedom are equal to the difference between the number of rows of $F(\mathbf{p})$ and the number of parameters. If only one response function is created per row of the contingency table, then this is the number of table rows minus the number of estimated parameters.

You can address questions about the parameters with the use of hypothesis tests. Each hypothesis is written in the form

$$H_0: \mathbf{C}\boldsymbol{\beta} = \mathbf{0}$$

and can investigate whether specified linear combinations of the parameters are equal to zero. The test statistic employed is a Wald statistic that is expressed as

$$Q_C = (\mathbf{C}\mathbf{b})'[\mathbf{C}(\mathbf{X}'\mathbf{V}_F^{-1}\mathbf{X})^{-1}\mathbf{C}']^{-1}(\mathbf{C}\mathbf{b})$$

Under H_0, Q_C is distributed as chi-square with degrees of freedom equal to the number of linearly independent rows in **C**.

You can also generate predicted values $\hat{\mathbf{F}} = \mathbf{Xb}$ of the response functions and their estimated covariance matrix $\mathbf{V}_{\hat{\mathbf{F}}} = \mathbf{XV(b)X'}$. See Appendix A in this chapter for more statistical theory concerning weighted least squares estimation.

14.2.3 Model Parameterization

The preliminary model of interest for a WLS analysis is often the *saturated* model, in which all the variation is explained by the parameters. In a saturated model, there are as many parameters in the model as there are response functions. For these data, the saturated model is written

$$E \begin{bmatrix} F(\mathbf{p}_1) \\ F(\mathbf{p}_2) \\ F(\mathbf{p}_3) \\ F(\mathbf{p}_4) \end{bmatrix} = \begin{bmatrix} \alpha + \beta_1 + \beta_2 + \beta_3 \\ \alpha + \beta_1 - \beta_2 - \beta_3 \\ \alpha - \beta_1 + \beta_2 - \beta_3 \\ \alpha - \beta_1 - \beta_2 + \beta_3 \end{bmatrix} = \begin{bmatrix} 1 & 1 & 1 & 1 \\ 1 & 1 & -1 & -1 \\ 1 & -1 & 1 & -1 \\ 1 & -1 & -1 & 1 \end{bmatrix} \begin{bmatrix} \alpha \\ \beta_1 \\ \beta_2 \\ \beta_3 \end{bmatrix}$$

which is the differential effect parameterization described in Section 8.3.3 in Chapter 8. Here, α is a centered intercept, β_1 is the differential effect for sex, β_2 is the differential effect for residence, and β_3 represents their interaction. The intercept is the mean number of colds averaged over all the groups. This full rank parameterization is the default for the CATMOD procedure.

14.3 Using PROC CATMOD for Weighted Least Squares Analysis

Since the CATMOD procedure is very general, it offers great flexibility in its input. Standard uses that take advantage of defaults may require no more than three or four statements. More statements are required if you take advantage of the facilities for repeated measurements analysis or loglinear model analysis. And the input can be quite rich if you choose to create your own response functions through the specification of the appropriate matrix operations or create your own parameterization by directly inputting your model matrix.

The analysis for the colds data requires minimal input. You need to specify the input data set, the WEIGHT variable if the data are in count form, the response function, and the desired model in a MODEL statement. The MODEL statement is the only required statement for PROC CATMOD.

First, a SAS data set is created for the colds data.

```
data colds;
   input sex $ residence $ periods count @@;
   datalines;
female rural 0 45 female rural 1 64   female rural 2 71
female urban 0 80 female urban 1 104 female urban 2 116
male   rural 0 84 male   rural 1 124 male   rural 2 82
male   urban 0 106 male  urban 1 117 male   urban 2 87
;
```

14.3. Using PROC CATMOD for Weighted Least Squares Analysis

The following set of SAS statements requests that a weighted least squares analysis be performed for the mean response, using the saturated model.

```
proc catmod;
   weight count;
   response means;
   model periods = sex residence sex*residence /freq prob design;
run;
```

The WEIGHT statement works the same as it does for the FREQ procedure; the WEIGHT variable contains the count of observations that have the values listed in the data line. As with PROC FREQ, you can supply input data in raw form–one observation per data line–or in count form. The RESPONSE statement specifies the response functions. If you leave out this statement, PROC CATMOD models generalized logits with maximum likelihood estimation. Specifying the MEANS keyword requests that mean response functions be constructed for each subpopulation; the default estimation method for functions other than generalized logits is weighted least squares.

The MODEL statement requests that PROC CATMOD fit a model that includes both the main effects for sex and residence and their interaction. The effects specification is similar to that used in the GLM procedure. The effects for sex and residence each have 1 df, and their interaction also has 1 df. Since the model also includes an intercept by default, this model is saturated. There are four parameters for the four response functions.

PROC CATMOD uses the explanatory variables listed in the right-hand side of the MODEL statement to determine the rows of the underlying contingency table. Since the variable SEX has two levels and the variable RESIDENCE has two levels, PROC CATMOD forms a contingency table that has four rows. The columns of the underlying contingency table are determined by the number of values for the response variable on the left-hand side of the MODEL statement. Since there can be 0, 1, or 2 periods with colds, there are three columns in this table.

The FREQ and PROB options in the MODEL statement cause the frequencies and proportions from the underlying contingency table to be printed, and the DESIGN option specifies that the response functions and model matrix are to be printed.

Output 14.1 displays the population and response profiles, which represent the rows and columns of the underlying table, respectively. Output 14.2 displays the underlying frequency table and the corresponding table of proportions. PROC CATMOD labels each group or subpopulation "Sample n"; you often need to refer back to the "Population Profiles" table to interpret other parts of the PROC CATMOD output. You should always check the population and response profiles to ensure that you have defined the underlying frequency table as you intended.

Output 14.1 Population and Response Profiles

Population Profiles			
Sample	sex	residence	Sample Size
1	female	rural	180
2	female	urban	300
3	male	rural	290
4	male	urban	310

Response Profiles	
Response	periods
1	0
2	1
3	2

Output 14.2 Table Frequencies and Proportions

Response Frequencies			
	Response Number		
Sample	1	2	3
1	45	64	71
2	80	104	116
3	84	124	82
4	106	117	87

Response Probabilities			
	Response Number		
Sample	1	2	3
1	0.25000	0.35556	0.39444
2	0.26667	0.34667	0.38667
3	0.28966	0.42759	0.28276
4	0.34194	0.37742	0.28065

The table of response function values and the model matrix (labeled 'Design Matrix') are displayed in Output 14.3. The response functions are the mean number of periods with colds for each of the populations.

14.3. Using PROC CATMOD for Weighted Least Squares Analysis

Output 14.3 Observed Response Functions and Model Matrix

Response Functions and Design Matrix					
Sample	Response Function	Design Matrix			
		1	2	3	4
1	1.14444	1	1	1	1
2	1.12000	1	1	-1	-1
3	0.99310	1	-1	1	-1
4	0.93871	1	-1	-1	1

Model-fitting results are displayed in Output 14.4 in a table labeled "Analysis of Variance" for its similarity in function to an ANOVA table.

Output 14.4 ANOVA Table

Analysis of Variance			
Source	DF	Chi-Square	Pr > ChiSq
Intercept	1	1841.13	<.0001
sex	1	11.57	0.0007
residence	1	0.65	0.4202
sex*residence	1	0.09	0.7594
Residual	0	.	.

The effects listed in the right-hand side of the MODEL statement are listed under "Source." Unless otherwise specified, an intercept is included in the model. If there is one response function per subpopulation, the intercept has 1 df. The statistics printed under "Chi-Square" are Wald statistics. Also provided are the degrees of freedom for each effect and corresponding p-value.

The last row contains information labeled "Residual." Normally, this line contains a chi-square value that serves as a goodness-of-fit test for the specified model. However, in this case, the model uses four parameters to fit four response functions. The fit must necessarily be perfect, and thus the model explains all the variation among the response functions. The degrees of freedom are zero since the degrees of freedom for Q_W are equal to the difference in the number of response functions and the number of parameters. SAS prints out missing values under "Chi-Square" and "Pr > ChiSq" for zero degrees of freedom.

Since the model fits, it is appropriate to examine the chi-square statistics for the individual effects. With a chi-square value of 0.09 and $p = 0.7594$, the SEX*RESIDENCE interaction is clearly nonsignificant. SEX appears to be a strong effect and RESIDENCE a negligible effect, but these are better assessed in the context of the main effects model that remains after the interaction term is deleted, since the estimation of these main effects is better in the absence of the interaction.

Chapter 14: Weighted Least Squares

The following statements request the main effects model and produce the analysis of variance table displayed in Output 14.5.

```
proc catmod;
   weight count;
   response means;
   model periods = sex residence;
run;
```

Output 14.5 Main Effects ANOVA

Analysis of Variance			
Source	DF	Chi-Square	Pr > ChiSq
Intercept	1	1882.77	<.0001
sex	1	12.08	0.0005
residence	1	0.76	0.3839
Residual	1	0.09	0.7594

Look at the goodness-of-fit statistic $Q_W = 0.09$ with 1 df and $p = 0.7594$. The main effects model adequately fits the data. The smaller the goodness-of-fit chi-square value, and correspondingly the larger the p value, the better the fit. This is different from the model F statistic in the usual linear model setting, where the F value is high for a model that fits the data well in the sense of explaining a large amount of the variation. Strictly speaking, using the usual significance level of $\alpha = 0.05$, any p-value greater than 0.05 supports an adequate model fit. However, many analysts are more comfortable with goodness-of-fit p-values that are greater than 0.15.

The effect for sex is highly significant, $p < 0.001$. However, the effect for residence remains nonsignificant when the interaction is removed from the model, $p = 0.3839$. These results suggest that a model with a single main effect for SEX is appropriate.

Consider the following statements to perform this task. The MODEL statement contains the response variable PERIODS and a single explanatory variable, SEX. This should produce the desired model. However, recall that the variables listed in the right-hand side of the MODEL statement are also used to determine the underlying contingency table structure. This table has its rows determined by both SEX and RESIDENCE. If RESIDENCE is *not* included in the MODEL statement, as shown in the following statements, then PROC CATMOD would create two groups based on SEX instead of four groups based on SEX and RESIDENCE.

```
proc catmod;
   weight count;
   response means;
   model periods = sex;
run;
```

However, you need to maintain the sampling structure of the underlying table. The solution is the addition of the POPULATION statement. When a POPULATION statement is included, the variables listed in it determine the populations, not the variables listed in the MODEL statement. So, you can let the right-hand variables on the MODEL statement determine the populations so long as all the necessary variables are included; if not, you need to use a POPULATION statement. Some analysts use the POPULATION statement for all PROC CATMOD invocations as a precautionary measure.

The following statements request the single main effect model.

```
proc catmod;
   population sex residence;
   weight count;
   response means;
   model periods = sex;
run;
```

The table of population profiles in Output 14.6 is identical to the tables produced by previous invocations of PROC CATMOD without the POPULATION statement when the model included both SEX and RESIDENCE as explanatory variables.

Output 14.6 POPULATION Statement Results

Population Profiles			
Sample	sex	residence	Sample Size
1	female	rural	180
2	female	urban	300
3	male	rural	290
4	male	urban	310

The analysis of variance table in Output 14.7 includes only one main effect, SEX. The residual goodness-of-fit $Q_W = 0.85$, with 2 df and $p = 0.6531$, indicates an adequate fit.

Output 14.7 Single Main Effect ANOVA Table

Analysis of Variance			
Source	DF	Chi-Square	Pr > ChiSq
Intercept	1	1899.55	<.0001
sex	1	11.53	0.0007
Residual	2	0.85	0.6531

Compare this analysis of variance table with that displayed in Output 14.5.

Note that Q_W for the reduced model (0.85) is the sum of Q_W for the two effects model ($Q_W = 0.09$) plus the value of the Wald statistic for the effect for residence (0.76). This is a property of weighted least squares. When you delete a term from a model, the residual chi-square for the goodness of fit for the new model is equal to the old model's residual chi-square value plus the chi-square value for the particular effect. This is also true for maximum likelihood estimation when likelihood ratio tests are used for goodness of fit and for particular effects, but not when the

Wald statistic is used with maximum likelihood estimation. Similarly, note that the $Q_W = 0.09$ for the two main effects model of Output 14.5 is equal to the chi-square for the interaction term in the saturated model (Output 14.4).

When an effect is deleted, any variation attributed to that effect is put into the residual variation, which is the variation that the model does not explain; this variation is essentially random for well-fitting models. If the residual variation is low, the residual chi-square will be small, indicating that the model explains the variation in the response fairly well. If the residual variation is high, the residual chi-square will be large, with a correspondingly low *p*-value, indicating that the residual variation is significantly different from zero. The implication is that the model lacks necessary terms.

Finally, note that the degrees of freedom for the goodness of fit for the reduced model are increased by the number of degrees of freedom for the deleted effect, in this case from 1 to 2, since residence had one degree of freedom.

PROC CATMOD also prints out a table that contains the parameter estimates. Since the model fits, it is appropriate to examine this table, displayed in Output 14.8.

Output 14.8 Single Main Effect Model

Analysis of Weighted Least Squares Estimates					
Parameter		Estimate	Standard Error	Chi-Square	Pr > ChiSq
Intercept		1.0477	0.0240	1899.55	<.0001
sex	female	0.0816	0.0240	11.53	0.0007

Listed under "Parameter" are the parameters estimated for the model. Since sex is represented by one parameter, only one estimate is listed. Since females are listed first under SEX in the population profile of Output 14.6, the effect for sex is the differential effect for females. If an effect has more than one parameter, each of them is listed, in addition to the associated standard error, Wald statistic, and *p*-value. Since sex is represented by only one parameter, the chi-square value listed in the table of WLS estimates is identical to that listed in the analysis of variance table. This won't happen for those effects that are comprised of more than one parameter, since the effect test listed in the analysis of variance table is the test of whether all the effect parameters are jointly zero, and the chi-square tests listed in the parameter estimates table are always one degree of freedom tests for each of the individual parameters.

To summarize, the model that most effectively describes these data is a single main effect model where sex is the main effect. Its goodness of fit is satisfactory, and the model is parsimonious in the sense of not including factors with essentially no association with the response. Girls reported more colds than boys; the model-predicted mean number of periods with colds for girls is

$$\bar{F}_{\text{girls}} = \hat{\alpha} + \hat{\beta}_1 = 1.0477 + 0.0816 = 1.1293$$

and the model-predicted mean number of periods with colds for boys is

$$\bar{F}_{\text{boys}} = \hat{\alpha} - \hat{\beta}_1 = 1.0477 - 0.0816 = 0.9661$$

14.4 Obstetrical Pain Data: Advanced Modeling of Means

Table 14.4 displays data from a multicenter randomized study of obstetrical-related pain for women who had recently delivered a baby (Koch et al. 1985). Investigators were interested in comparing four treatments: placebo, Drug a, Drug b, and a combined treatment of Drug a and Drug b. Each patient was classified as initially having some pain or a lot of pain. Then, a randomly assigned treatment was administered at the beginning of the study period and again at 4 hours. Each patient was observed at hourly intervals for 8 hours, and pain status was recorded as either little or no pain or as some or more pain. The response measure of interest is the average proportion of hours for which the patient reported little or no pain.

The patients for each center × initial status × treatment group can be considered to be representative of some corresponding large target population in a manner that is consistent with stratified simple random sampling. Each patient's responses can also be assumed to be independent of other patient responses. Thus, the data in Table 14.4 are distributed as product multinomial. There are many small cell frequencies (less than 5) in this table. This means that the asymptotic requirements that are necessary for modeling functions such as multiple cell proportions or generalized logits are not met. However, the average proportion of hours with little or no pain is a reasonable response measure, and there is sufficient sample size for modeling means.

Table 14.4 Number of Hours with Little or No Pain for Women Who Recently Delivered a Baby

Center	Initial Pain Status	Treatment	Hours with Little or No Pain									Total
			0	1	2	3	4	5	6	7	8	
1	lot	placebo	6	1	2	2	2	3	7	3	0	26
1	lot	a	6	3	1	2	4	4	7	1	0	28
1	lot	b	3	1	0	4	2	3	11	4	0	28
1	lot	ba	0	0	0	1	1	7	9	6	2	26
1	some	placebo	1	0	3	0	2	2	4	4	2	18
1	some	a	2	1	0	2	1	2	4	5	1	18
1	some	b	0	0	0	1	0	3	7	6	2	19
1	some	ba	0	0	0	0	1	3	5	4	6	19
2	lot	placebo	7	2	3	2	3	2	3	2	2	26
2	lot	a	3	1	0	0	3	2	9	7	1	26
2	lot	b	0	0	0	1	1	5	8	7	4	26
2	lot	ba	0	1	0	0	1	2	8	9	5	26
2	some	placebo	2	0	2	1	3	1	2	5	4	20
2	some	a	0	0	1	1	1	8	1	7		19
2	some	b	0	2	0	1	0	1	4	6	6	20
2	some	ba	0	0	0	1	3	0	4	7	5	20
3	lot	placebo	6	0	2	2	2	6	1	2	1	22
3	lot	a	4	2	1	5	1	1	3	2	3	22
3	lot	b	5	0	2	3	1	0	2	6	7	26
3	lot	ba	3	2	1	0	0	2	5	9	4	26
3	some	placebo	5	0	0	1	3	1	4	4	5	23
3	some	a	1	0	0	1	3	5	3	3	6	22
3	some	b	3	0	1	1	0	0	3	7	11	26
3	some	ba	0	0	0	1	1	4	2	4	13	25
4	lot	placebo	4	0	1	3	2	1	1	2	2	16
4	lot	a	0	1	3	1	1	6	1	3	6	22
4	lot	b	0	0	0	0	2	7	2	2	9	22
4	lot	ba	1	0	3	0	1	2	3	4	8	22
4	some	placebo	1	0	1	1	4	1	1	0	10	19
4	some	a	0	0	0	1	0	2	2	1	13	19
4	some	b	0	0	0	1	1	1	1	5	11	20
4	some	ba	1	0	0	0	0	2	2	2	14	21

The proportion vector for each group i is written

$$\mathbf{p}_i = (p_{i0}, p_{i1}, p_{i2}, p_{i3}, p_{i4}, p_{i5}, p_{i6}, p_{i7}, p_{i8})'$$

where p_{ij} is the proportion of patients with j hours of pain for the ith group, and $i = 1, \ldots, 32$ is the group corresponding to the ith row of Table 14.4. You compute the average proportion response function by applying the following matrix operation to the proportion vector for each group:

$$F_i = F(\mathbf{p}_i) = A\mathbf{p}_i = \left[\tfrac{0}{8}, \tfrac{1}{8}, \tfrac{2}{8}, \tfrac{3}{8}, \tfrac{4}{8}, \tfrac{5}{8}, \tfrac{6}{8}, \tfrac{7}{8}, \tfrac{8}{8} \right] \mathbf{p}_i$$

A useful preliminary model is one that includes effects for center, initial pain, treatment, and initial pain × treatment interaction; initial pain and treatment are believed to be similar across centers, so their interactions with the center effects are not included.

14.4.1 Performing the Analysis with PROC CATMOD

The following DATA step creates the SAS data set PAIN. The raw data contains an observation for each line of Table 14.4, although they are in a different order. The ARRAY and OUTPUT statements in the DATA step modify the input data by creating an individual observation for each different response value, or number of hours with little or no pain. It creates the variables NO_HOURS and COUNT. The CATMOD procedure requires that each response value be represented on a different observation. This data set contains 288 observations (9 observations for each of the 32 original data lines). The values for the variable INITIAL are 'some' for some pain and 'lot' for a lot of pain.

```
data pain (drop=h0-h8);
   input center initial $ treat $ h0-h8;
   array hours h0-h8;
   do i=1 to 9;
      no_hours=i-1; count=hours(i); output;
   end;
   datalines;
1 some placebo  1 0 3 0 2 2 4 4 2
1 some treat_a  2 1 0 2 1 2 4 5 1
1 some treat_b  0 0 0 1 0 3 7 6 2
1 some treat_ba 0 0 0 0 1 3 5 4 6
1 lot  placebo  6 1 2 2 2 3 7 3 0
1 lot  treat_a  6 3 1 2 4 4 7 1 0
1 lot  treat_b  3 1 0 4 2 3 11 4 0
1 lot  treat_ba 0 0 0 1 1 7 9 6 2
2 some placebo  2 0 2 1 3 1 2 5 4
2 some treat_a  0 0 0 1 1 1 8 1 7
2 some treat_b  0 2 0 1 0 1 4 6 6

   ... more lines ...

4 some treat_b  0 0 0 1 1 1 1 5 11
4 some treat_ba 1 0 0 0 0 2 2 2 14
4 lot  placebo  4 0 1 3 2 1 1 2 2
4 lot  treat_a  0 1 3 1 1 6 1 3 6
4 lot  treat_b  0 0 0 0 2 7 2 2 9
4 lot  treat_ba 1 0 3 0 1 2 3 4 8
;

proc print data=pain(obs=9);
run;
```

Output 14.9 displays the observations for the group from Center 1 who had some initial pain and received the placebo.

Output 14.9 Partial Listing of Data Set PAIN

Obs	center	initial	treat	i	no_hours	count
1	1	some	placebo	1	0	1
2	1	some	placebo	2	1	0
3	1	some	placebo	3	2	3
4	1	some	placebo	4	3	0
5	1	some	placebo	5	4	2
6	1	some	placebo	6	5	2
7	1	some	placebo	7	6	4
8	1	some	placebo	8	7	4
9	1	some	placebo	9	8	2

The following SAS statements invoke the CATMOD procedure and fit the preliminary model. Note that the RESPONSE statement includes the coefficients required to compute the average proportions per group. Using the MEANS keyword in the RESPONSE statement would compute the mean number of hours with little or no pain, not the average proportion of hours with little or no pain, which is desired here. Actually the results will be the same; the decision is whether you want the parameter estimates to apply to proportions or means.

```
proc catmod;
   weight count;
   response 0 .125 .25 .375 .5 .625 .75 .875 1;
   model no_hours =  center initial treat
                     treat*initial;
run;
```

The population profiles and the response profiles are displayed in Output 14.10 and Output 14.11, respectively.

Output 14.10 Population Profiles

Population Profiles			
Sample	center	initial	treat	Sample Size
1	1	lot	placebo	26
2	1	lot	treat_a	28
3	1	lot	treat_b	28
4	1	lot	treat_ba	26
5	1	some	placebo	18
6	1	some	treat_a	18
7	1	some	treat_b	19
8	1	some	treat_ba	19
9	2	lot	placebo	26
10	2	lot	treat_a	26
11	2	lot	treat_b	26
12	2	lot	treat_ba	26
13	2	some	placebo	20
14	2	some	treat_a	19
15	2	some	treat_b	20
16	2	some	treat_ba	20
17	3	lot	placebo	22
18	3	lot	treat_a	22
19	3	lot	treat_b	26
20	3	lot	treat_ba	26
21	3	some	placebo	23
22	3	some	treat_a	22
23	3	some	treat_b	26
24	3	some	treat_ba	25
25	4	lot	placebo	16
26	4	lot	treat_a	22
27	4	lot	treat_b	22
28	4	lot	treat_ba	22
29	4	some	placebo	19
30	4	some	treat_a	19
31	4	some	treat_b	20
32	4	some	treat_ba	21

Output 14.11 Response Profiles

Response Profiles	
Response	no_hours
1	0
2	1
3	2
4	3
5	4
6	5
7	6
8	7
9	8

The goodness-of-fit statistic for this preliminary model is $Q_W = 26.90$ with 21 df, as displayed in Output 14.12. With $p = 0.1743$, this indicates that the model fits the data adequately. All the constituent effects are highly significant, $p < 0.01$.

Output 14.12 Preliminary ANOVA Table

Analysis of Variance			
Source	DF	Chi-Square	Pr > ChiSq
Intercept	1	5271.98	<.0001
center	3	29.02	<.0001
initial	1	62.65	<.0001
treat	3	92.15	<.0001
initial*treat	3	12.63	0.0055
Residual	21	26.90	0.1743

It is useful to examine further the interaction between the treatments and initial pain status. The significant interaction means that some of the treatment effects depend on the level of initial pain status.

Some questions of interest are:

- Which treatment effects depend on the level of initial pain status? Where exactly is the interaction occurring?
- In which levels of initial pain do treatments differ?

You can address these questions with the same model, but it can be easier to address them with a differently parameterized model. If you nest the effects of treatment within levels of initial pain, then these questions can be addressed with the use of contrasts. By using the nested effects model, you trade the 3 df for TREAT and 3 df for TREAT*INITIAL for 6 df for the nested effect TREAT(INITIAL). Accordingly, the TREAT(INITIAL) effect is associated with six parameters, three of which pertain to the effects of treatment within some initial pain and three of which pertain to the effects of treatment within a lot of initial pain.

The following PROC CATMOD statements fit the nested model. The difference between specifying the nested effect TREAT(INITIAL) and the nested-by-values effects TREAT(INITIAL=some) and TREAT(INITIAL=lot) is that the former yields the 6 df test in the ANOVA table that tests whether the six parameters for treatment effects within both some and a lot of initial pain levels are essentially zero; the latter results in two separate 3 df tests in the ANOVA table, one for whether the three treatment parameters for some pain are essentially zero and one for whether the three treatment parameters for a lot of pain are essentially zero.

```
proc catmod;
   weight count;
   response 0 .125 .25 .375 .5 .625 .75 .875 1;
   model no_hours = center initial
                    treat(initial);
run;
```

Submitting these statements produces the results contained in Output 14.13.

Output 14.13 Nested Value ANOVA Table

Analysis of Variance			
Source	DF	Chi-Square	Pr > ChiSq
Intercept	1	5271.98	<.0001
center	3	29.02	<.0001
initial	1	62.65	<.0001
treat(initial)	6	102.70	<.0001
Residual	21	26.90	0.1743

The model goodness-of-fit test is the same, $Q_W = 26.90$ with 21 df, since no model reduction was performed. The model simply redistributes the variation over different degrees of freedom. Geometrically, you can think of the model space as being spanned by a different, but equivalent, vector set. The tests for CENTER effect and INITIAL effect also remain the same. However, now there is the 6 df nested effect TREAT(INITIAL) in place of a TREAT effect and a TREAT*INITIAL interaction.

Output 14.14 displays the parameter estimates.

Output 14.14 Parameter Estimates

Analysis of Weighted Least Squares Estimates					
Parameter		Estimate	Standard Error	Chi-Square	Pr > ChiSq
Intercept		0.6991	0.00963	5271.98	<.0001
center	1	-0.0484	0.0145	11.24	0.0008
	2	0.0187	0.0145	1.66	0.1982
	3	-0.0415	0.0176	5.56	0.0184
initial	lot	-0.0753	0.00951	62.65	<.0001
treat(initial)	placebo lot	-0.1739	0.0283	37.81	<.0001
	treat_a lot	-0.0644	0.0255	6.39	0.0115
	treat_b lot	0.0952	0.0206	21.45	<.0001
	placebo some	-0.1159	0.0284	16.68	<.0001
	treat_a some	0.00740	0.0217	0.12	0.7331
	treat_b some	0.0347	0.0206	2.84	0.0921

Consider testing to see whether the effect for treatment a is the same as the effect for placebo for patients with some pain. The appropriate hypothesis is stated

$$H_0: \beta_9 - \beta_8 = 0$$

Since the implicit effect for treatment ba is written in terms of the other treatment parameters,

$$\beta_{\text{treatment ba}} = -\beta_8 - \beta_9 - \beta_{10}$$

the hypothesis test to see whether the effect for treatment ba is the same as the effect for placebo is written

$$H_0: -2\beta_8 - \beta_9 - \beta_{10} = 0$$

The hypotheses of interest and their corresponding contrasts and coefficients are displayed in Table 14.5. The coefficients are required in the CONTRAST statement in PROC CATMOD to perform a particular contrast test.

Table 14.5 Hypothesis Tests

Hypothesis	Initial Pain	Contrast	Coefficients					
			β_5	β_6	β_7	β_8	β_9	β_{10}
treatment a vs. placebo	a lot	$-\beta_5 + \beta_6$	-1	1	0	0	0	0
treatment b vs. placebo	a lot	$-\beta_5 + \beta_7$	-1	0	1	0	0	0
treatment ba vs. placebo	a lot	$-2\beta_5 - \beta_6 - \beta_7$	-2	-1	-1	0	0	0
treatment ba vs. a	a lot	$-\beta_5 - 2\beta_6 - \beta_7$	-1	-2	-1	0	0	0
treatment ba vs. b	a lot	$-\beta_5 - \beta_6 - 2\beta_7$	-1	-1	-2	0	0	0
treatment a vs. placebo	some	$-\beta_8 + \beta_9$	0	0	0	-1	1	0
treatment b vs. placebo	some	$-\beta_8 + \beta_{10}$	0	0	0	-1	0	1
treatment ba vs. placebo	some	$-2\beta_8 - \beta_9 - \beta_{10}$	0	0	0	-2	-1	-1
treatment ba vs. a	some	$-\beta_8 - 2\beta_9 - \beta_{10}$	0	0	0	-1	-2	-1
treatment ba vs. b	some	$-\beta_8 - \beta_9 - 2\beta_{10}$	0	0	0	-1	-1	-2

In Chapter 8, the CONTRAST statements in PROC LOGISTIC and PROC GENMOD were discussed. These statements serve the same purpose as the CONTRAST statement in PROC CATMOD: testing linear combinations of the parameters.

You list a character string that labels the contrast, list the effect whose parameters you are interested in, and then supply a coefficient for each of the effect parameters that PROC CATMOD estimates. Remember that since PROC CATMOD uses full rank parameterization, it produces $g - 1$ estimated parameters for an effect that has g levels.

The following CONTRAST statements request that the CATMOD procedure perform the appropriate tests. The statements can be submitted interactively, following the previous nested model invocation, or in batch, included at the end of the nested model invocation. Since you are only interested in the parameters corresponding to the TREAT(INITIAL) effect, you list that effect in the CONTRAST statement and then specify the appropriate six coefficients that pertain to the contrast involving the parameters $\beta_5 - \beta_{10}$.

```
contrast 'lot: a-placebo'   treat(initial) -1  1  0  0  0  0 ;
contrast 'lot: b-placebo'   treat(initial) -1  0  1  0  0  0 ;
contrast 'lot: ba-placebo'  treat(initial) -2 -1 -1  0  0  0 ;
contrast 'lot: ba-a'        treat(initial) -1 -2 -1  0  0  0 ;
contrast 'lot: ba-b'        treat(initial) -1 -1 -2  0  0  0 ;
contrast 'some:a-placebo'   treat(initial)  0  0  0 -1  1  0 ;
contrast 'some:b-placebo'   treat(initial)  0  0  0 -1  0  1 ;
contrast 'some:ba-placebo'  treat(initial)  0  0  0 -2 -1 -1 ;
contrast 'some:ba-a'        treat(initial)  0  0  0 -1 -2 -1 ;
contrast 'some:ba-b'        treat(initial)  0  0  0 -1 -1 -2 ;
run;
```

These statements produce the "Analysis of Contrasts" table shown in Output 14.15. The table includes the results for all the individual hypothesis tests.

Output 14.15 Contrast Results

Analysis of Contrasts			
Contrast	DF	Chi-Square	Pr > ChiSq
lot: a-placebo	1	5.59	0.0180
lot: b-placebo	1	42.81	<.0001
lot: ba-placebo	1	61.48	<.0001
lot: ba-a	1	32.06	<.0001
lot: ba-b	1	2.59	0.1076
some:a-placebo	1	8.19	0.0042
some:b-placebo	1	12.83	0.0003
some:ba-placebo	1	21.45	<.0001
some:ba-a	1	4.37	0.0365
some:ba-b	1	1.67	0.1964

Most of these contrasts are significant using the $\alpha = 0.05$ criterion; however, it appears that the difference between the ba treatment and the b treatment is marginal for both some initial pain and a lot of initial pain.

Additional contrasts of interest are the individual components of the treatment × initial pain status interaction as well as the individual components of the overall treatment effect. Output 14.16 displays these results (the statements are listed in Appendix B).

Output 14.16 Contrast Results

Analysis of Contrasts			
Contrast	DF	Chi-Square	Pr > ChiSq
interact:a-placebo	1	0.05	0.8266
interact:b-placebo	1	4.05	0.0441
interact:ba-placebo	1	4.89	0.0271
interact:ba-a	1	8.61	0.0033
interact:ba-b	1	0.04	0.8344
average:a-placebo	1	13.51	0.0002
average:b-placebo	1	50.93	<.0001
average:ba-placebo	1	77.60	<.0001
average:ba-a	1	31.42	<.0001
average:ba-b	1	4.22	0.0399
interaction	3	12.63	0.0055
treatment effect	3	92.15	<.0001

These results indicate that the interaction component corresponding to the comparison of treatment a and placebo is nonsignificant; therefore, treatment a has similar effects for an initial pain status of some or a lot. Treatments b and ba do appear to have different effects at the different levels of initial pain when compared to placebo; however, these effects may be quite similar. The interaction component corresponding to the comparison of treatment a and treatment ba is significant; their difference depends on the level of initial pain. This is not the case for treatment b compared to treatment ba; their interaction component is nonsignificant.

All of the components of the treatment effect–those tests labeled average–are significant.

The 'interaction' and the 'treatment effect' are testing the same thing as the TREAT effect and the TREAT*INITIAL effects listed in the ANOVA table for the preliminary model. Compare the Q_W values, 12.63 and 92.15, to those listed in that table. They are identical.

These results address the pertinent questions of this analysis. Other approaches may include fitting a reduced model that incorporates the results of these hypothesis tests, which is not pursued here.

14.5 Analysis of Survey Sample Data

In addition to analyzing data based on an underlying contingency table, the CATMOD procedure provides a convenient way to analyze data that come in the form of a function vector and covariance matrix. Often, such data come from complex surveys, and the covariance matrix has been computed using other software that takes the sampling design into account. If the number of response function estimates and the corresponding covariance matrix is large, then software designed for survey data analysis may be more appropriate.

14.5.1 HANES Data

The following data are from the Health and Nutrition Examination Survey (HANES) that was conducted in the United States from 1971 to 1974. This survey obtained various information concerning health from over 10,000 households in the United States. One of the measures constructed for analysis of these data was a well-being index, a composite index comprised from the answers to a questionnaire on general well-being. Table 14.6 contains the well-being ordered categorical estimates and standard errors for a cross-classification based on sex and age. The covariance matrix was computed using other software that used balanced repeated replication and took into account the sampling framework of the survey (Koch and Stokes 1979).

Table 14.6 Well-Being Index

Sex	Age	Estimate	S.E.
Male	25–34	7.937	0.086
Male	35–44	7.925	0.108
Male	45–54	7.828	0.102
Male	55–64	7.737	0.116
Male	65–74	8.168	0.120
Female	25–34	7.250	0.105
Female	35–44	7.190	0.153
Female	45–54	7.360	0.103
Female	55–64	7.319	0.152
Female	65–74	7.552	0.139

14.5.2 Direct Input of Response Functions

In the typical PROC CATMOD analysis, you input a contingency table or raw data and specify the response functions, and PROC CATMOD computes the appropriate covariance matrix based on the product multinomial distribution. In this case, your data are the vector of response functions and its covariance matrix. Thus, in order to describe the variation of these functions across various groups, you need to inform the procedure of the structure of your underlying cross-classification. Ordinarily, you would do this with the explanatory variables that define the contingency table. For this case, you need to rely on the FACTOR statement to express the cross-classification relationships.

The following DATA step inputs the response functions and their covariance matrix into the SAS data set WBEING. The first two data lines are the well-being estimates, listed in the same order as they appear in Table 14.6. The subsequent data lines contain the 10×10 covariance matrix that corresponds to the estimates; each row takes two lines. The variable _TYPE_ identifies whether each data line corresponds to parameter estimates or covariance estimates. The value 'parms' identifies the lines with parameter estimates, and the value 'cov' identifies the lines with the covariance estimates. The variable _NAME_ identifies the name of the variable that has its covariance elements stored in that data line. Note that the diagonal element in the ith row of the covariance matrix is the variance for the ith well-being estimate. (The square root of the element is the standard error for the estimate.)

```
data wbeing(type=est);
   input   b1-b5    _type_ $  _name_ $  b6-b10 #2;
   datalines;
   7.93726  7.92509   7.82815   7.73696   8.16791  parms     .
   7.24978  7.18991   7.35960   7.31937   7.55184
   0.00739  0.00019   0.00146  -0.00082   0.00076  cov       b1
   0.00189  0.00118   0.00140  -0.00140   0.00039
   0.00019  0.01172   0.00183   0.00029   0.00083  cov       b2
  -0.00123 -0.00629  -0.00088  -0.00232   0.00034
   0.00146  0.00183   0.01050  -0.00173   0.00011  cov       b3
   0.00434 -0.00059  -0.00055   0.00023  -0.00013
  -0.00082  0.00029  -0.00173   0.01335   0.00140  cov       b4
```

```
    0.00158   0.00212   0.00211   0.00066   0.00240
    0.00076   0.00083   0.00011   0.00140   0.01430   cov   b5
   -0.00050  -0.00098   0.00239  -0.00010   0.00213
    0.00189  -0.00123   0.00434   0.00158  -0.00050   cov   b6
    0.01110   0.00101   0.00177  -0.00018  -0.00082
    0.00118  -0.00629  -0.00059   0.00212  -0.00098   cov   b7
    0.00101   0.02342   0.00144   0.00369   0.00253
    0.00140  -0.00088  -0.00055   0.00211   0.00239   cov   b8
    0.00177   0.00144   0.01060   0.00157   0.00226
   -0.00140  -0.00232   0.00023   0.00066  -0.00010   cov   b9
   -0.00018   0.00369   0.00157   0.02298   0.00918
    0.00039   0.00034  -0.00013   0.00240   0.00213   cov   b10
   -0.00082   0.00253   0.00226   0.00918   0.01921
;
```

14.5.3 The FACTOR Statement

The FACTOR statement is where you define the cross-classification structure of the estimates. You need to specify names for the CATMOD procedure to use internally that correspond to grouping variables. You specify the number of levels for each and whether their values are character. This is done in the first part of the FACTOR statement. The internal variable SEX has two levels, and its values are character, as denoted by the dollar sign; the internal variable AGE has five levels, and its values are also character. The values for these internal variables are listed under the PROFILE option after a slash (/) in the FACTOR statement. The values are listed according to the order of the estimates; thus, they are listed in the same order as they appear in Table 14.6.

Since SEX and AGE are internal variables, not part of the input data set, you cannot refer to them in the MODEL statement. Thus, you use the keyword _RESPONSE_ to specify the desired model effects. In the following statements, the saturated model is assigned to the keyword _RESPONSE_ in the FACTOR statement. This keyword is later used on the right-hand side of the MODEL statement. The _RESPONSE_ construction is also used to perform repeated measurements analyses and loglinear model analyses with the CATMOD procedure. Since the response functions are input directly, the keyword _F_ is used to represent them on the left-hand side of the MODEL statement.

The following PROC CATMOD statements invoke the procedure and specify that a saturated model be fit to the data. The keyword READ in the RESPONSE statement tells PROC CATMOD that the response functions and covariance matrix are to be directly input. The variables B1–B10 listed after the keyword READ specify that ten response functions are involved and thus that the covariance matrix is 10×10.

```
proc catmod data=wbeing;
   response read b1-b10;
   factors sex $ 2, age $ 5 / _response_=sex|age
                              profile=(male      '25-34',
                                       male      '35-44',
                                       male      '45-54',
                                       male      '55-64',
                                       male      '65-74',
                                       female    '25-34',
```

```
                                            female    '35-44',
                                            female    '45-54',
                                            female    '55-64',
                                            female    '65-74');
   model _f_ = _response_ / design;
run;
```

14.5.4 Preliminary Analysis

Since populations and responses are not determined by input data variables, the population profiles and response profiles are not printed as usual at the beginning of the PROC CATMOD output. Instead, the first table displayed contains the response functions and the model matrix, as shown in Output 14.17.

Output 14.17 Directly Input Response Functions

			\multicolumn{10}{c	}{Response Functions and Design Matrix}

| Sample | Function Number | Response Function | \multicolumn{10}{c|}{Design Matrix} |
			1	2	3	4	5	6	7	8	9	10
1	1	7.93726	1	1	1	0	0	0	1	0	0	0
	2	7.92509	1	1	0	1	0	0	0	1	0	0
	3	7.82815	1	1	0	0	1	0	0	0	1	0
	4	7.73696	1	1	0	0	0	1	0	0	0	1
	5	8.16791	1	1	-1	-1	-1	-1	-1	-1	-1	-1
	6	7.24978	1	-1	1	0	0	0	-1	0	0	0
	7	7.18991	1	-1	0	1	0	0	0	-1	0	0
	8	7.35960	1	-1	0	0	1	0	0	0	-1	0
	9	7.31937	1	-1	0	0	0	1	0	0	0	-1
	10	7.55184	1	-1	-1	-1	-1	-1	1	1	1	1

The analysis of variance table is shown in Output 14.18. The internal variables SEX and AGE are listed under "Source" just as if they were explanatory variables in the input data set. The SEX*AGE interaction is clearly nonsignificant, with $p = 0.5713$. Thus, the additive model with effects SEX and AGE has an adequate goodness of fit with $Q_W = 2.92$ and 4 df.

Output 14.18 Saturated Model

Analysis of Variance			
Source	DF	Chi-Square	Pr > ChiSq
Intercept	1	27117.73	<.0001
sex	1	47.07	<.0001
age	4	10.87	0.0281
sex*age	4	2.92	0.5713
Residual	0	.	.

The additive model is fit next, with contrasts requested to determine whether any of the parameters for age are essentially the same. The following statements fit the additive model.

```
proc catmod data=wbeing;
   response read b1-b10;
   factors sex $ 2, age $ 5 /
      _response_ = sex age
      profile = (male '25-34' ,
                 male '35-44',
                 male '45-54' ,
                 male '55-64',
                 male '65-74' ,
                 female '25-34',
                 female '35-44',
                 female '45-54',
                 female '55-64' ,
                 female '65-74');
   model _f_ = _response_;
```

The contrasts are set up to compare the first parameter for the age effect with each of the others. Recall that the implicit parameter for the last level of the age effect (ages 65–74) is the negative of the sum of the other parameters. Since the response functions are input directly, coefficients must be supplied for all the effects, including the intercept. Thus, the ALL_PARMS keyword is required. When you specify this keyword, you must supply coefficients for all the model parameters. Here, 0s are supplied on all contrasts for the intercept term and the sex effect term, and the final four coefficients apply to the age effect.

```
   contrast '25-34 vs. 35-44' all_parms 0 0 1 -1 0  0;
   contrast '25-34 vs. 45-54' all_parms 0 0 1  0 -1 0;
   contrast '25-34 vs. 55-64' all_parms 0 0 1  0  0 -1;
   contrast '25-34 vs. 65-74' all_parms 0 0 2  1  1  1;
run;
```

When these statements are submitted, they produce the results displayed in Output 14.19 and Output 14.20. There are six parameters: one for the intercept, one for the sex effect, and four for the age effect. None of the age effect parameters listed appears to be of much importance. However, there does appear to be suggestive variation among age groups, with the *p*-value for the age effect at 0.0561.

Output 14.19 Additive Model

Analysis of Variance			
Source	DF	Chi-Square	Pr > ChiSq
Intercept	1	28089.07	<.0001
sex	1	65.84	<.0001
age	4	9.21	0.0561
Residual	4	2.92	0.5713

Analysis of Weighted Least Squares Estimates					
Effect	Parameter	Estimate	Standard Error	Chi-Square	Pr > ChiSq
Intercept	1	7.6319	0.0455	28089.07	<.0001
sex	2	0.2900	0.0357	65.84	<.0001
age	3	-0.00780	0.0645	0.01	0.9037
	4	-0.0465	0.0636	0.54	0.4642
	5	-0.0343	0.0557	0.38	0.5387
	6	-0.1098	0.0764	2.07	0.1506

The contrasts indicate that the first four age groups act essentially the same and that the oldest age group is responsible for the age effect, $p = 0.0744$; note that its estimate is $-\{-0.008 - 0.046 - 0.034 - 0.110\} = 0.198$.

Output 14.20 Contrasts for Age Effect

Analysis of Contrasts			
Contrast	DF	Chi-Square	Pr > ChiSq
25-34 vs. 35-44	1	0.16	0.6937
25-34 vs. 45-54	1	0.12	0.7288
25-34 vs. 55-64	1	0.72	0.3954
25-34 vs. 65-74	1	3.18	0.0744

One more contrast is specified to test the joint hypothesis that the lower four age groups are essentially the same. The following CONTRAST statement is submitted; these three sets of coefficients, separated by commas, result in a 3 df test.

```
contrast '25-64 the same'    all_parms  0 0 1 -1  0  0,
                             all_parms  0 0 1  0 -1  0,
                             all_parms  0 0 1  0  0 -1;
run;
```

The result of this test is nonsignificant, $p = 0.8678$, as shown in Output 14.21.

Output 14.21 Joint Test for Ages 25–64

Analysis of Contrasts			
Contrast	DF	Chi-Square	Pr > ChiSq
25-64 the same	3	0.72	0.8678

14.5.5 Inputting the Model Matrix Directly

These results suggest that an appropriate model for these data is one that includes the sex effect and an effect for the oldest age group. You do this with the CATMOD procedure by specifying your model matrix directly.

The following MODEL statement specifies a direct model matrix. You write the coefficients for the model matrix row-wise, separating each row with a comma. The entire matrix is enclosed by parentheses. This is similar to how you would input a matrix in the SAS/IML matrix programming language.

```
model _f_ = ( 1 0 0 ,
              1 0 0 ,
              1 0 0 ,
              1 0 0 ,
              1 0 1 ,
              1 1 0 ,
              1 1 0 ,
              1 1 0 ,
              1 1 0 ,
              1 1 1 );
```

This matrix represents an incremental effects model. The first column of 1s is for the intercept, the second column is for the incremental effect of sex, where the reference level is for males and the incremental effect is for females, and the third column is an incremental effect for age, where the increment is for those aged 65–74.

Model matrices can be inputted directly for all applications of the CATMOD procedure, if desired.

The following PROC CATMOD statements fit this model. After the model matrix is listed a set of labels for the effects; the numbers correspond to the columns of the model matrix. Without the _RESPONSE_ keyword in the MODEL statement, the CATMOD procedure has no way of knowing how to divide the model variability into various components. You can request that various column parameters be tested jointly, or singly, as specified here. Refer to the CATMOD procedure chapter in the *SAS/STAT User's Guide* for more detail. If you don't specify information concerning the columns, PROC CATMOD performs a joint test for the significance of the model beyond an overall mean, labeling this effect MODEL|MEAN in the ANOVA table.

```
proc catmod data=wbeing;
   response read b1-b10;
   factors sex $ 2, age $  5 /
```

```
            _response_ = sex age
              profile = (male '25-34',
                        male '35-44',
                        male '45-54',
                        male '55-64',
                        male '65-74',
                        female    '25-34',
                        female    '35-44',
                        female    '45-54',
                        female    '55-64',
                        female    '65-74');
    model _f_ = ( 1 0 0 ,
                  1 0 0 ,
                  1 0 0 ,
                  1 0 0 ,
                  1 0 1 ,
                  1 1 0 ,
                  1 1 0 ,
                  1 1 0 ,
                  1 1 0 ,
                  1 1 1 ) (1='Intercept', 2='Sex', 3='65-74')
                         / pred;
```

The resulting tables for the analysis of variance and the parameter estimates are displayed in Output 14.22. The model fits very well, with a Q_W of 3.64 and 7 df, which results in $p = 0.8198$. The effects are listed as specified in the MODEL statement, and each is significant.

Output 14.22 Reduced Model

Analysis of Variance			
Source	DF	Chi-Square	Pr > ChiSq
Intercept	1	27165.20	<.0001
Sex	1	72.64	<.0001
65-74	1	8.49	0.0036
Residual	7	3.64	0.8198

Analysis of Weighted Least Squares Estimates					
Effect	Parameter	Estimate	Standard Error	Chi-Square	Pr > ChiSq
Model	1	7.8680	0.0477	27165.20	<.0001
	2	-0.5601	0.0657	72.64	<.0001
	3	0.2607	0.0895	8.49	0.0036

Finally, the predicted values are displayed in Output 14.23. Fitting this model has resulted in estimates of the standard error that are on the order of half as large as the standard errors for the original data.

Output 14.23 Reduced Model

Predicted Values for Response Functions					
	Observed		Predicted		
Function Number	Function	Standard Error	Function	Standard Error	Residual
1	7.93726	0.085965	7.867955	0.047737	0.069305
2	7.92509	0.108259	7.867955	0.047737	0.057135
3	7.82815	0.10247	7.867955	0.047737	-0.0398
4	7.73696	0.115542	7.867955	0.047737	-0.13099
5	8.16791	0.119583	8.128703	0.095929	0.039207
6	7.24978	0.105357	7.30786	0.060831	-0.05808
7	7.18991	0.153036	7.30786	0.060831	-0.11795
8	7.3596	0.102956	7.30786	0.060831	0.05174
9	7.31937	0.151592	7.30786	0.060831	0.01151
10	7.55184	0.1386	7.568608	0.098147	-0.01677

14.6 Modeling Rank Measures of Association Statistics

Many studies include outcomes that are ordinal in nature. When the treatment is either dichotomous or ordinal, you can model rank measures of correlation using WLS methods and use that framework to investigate treatment effects and interactions. Such an analysis can complement statistical models such as the proportional odds model. See Carr, Hafner, and Koch (1989) for an example of such an analysis applied to Goodman-Kruskal rank correlation coefficients, also known as gamma coefficients.

The Mann-Whitney rank measure of association statistics are useful statistics for assessing the association between an ordinal outcome and a dichotomous explanatory variable. Consider the following data in Table 14.7 from a randomized clinical trial of chronic pain. Investigators compared a new treatment with a placebo and assessed the response for a particular condition. Patients were obtained from investigators at two centers whose design included stratification relative to four diagnostic classes.

458   Chapter 14: Weighted Least Squares

Table 14.7 Chronic Pain Clinical Trial

Center	Diagnostic Class	Treatment	Excellent	Good	Moderate	Fair	Poor
I	A	Active	1	3	2	5	1
I	A	Placebo	2	4	3	4	3
I	B	Active	3	10	1	4	2
I	B	Placebo	2	4	1	5	2
I	C	Active	6	1	1	1	0
I	C	Placebo	0	5	1	1	3
I	D	Active	3	5	1	6	1
I	D	Placebo	3	3	2	4	5
II	A	Active	0	4	3	1	8
II	A	Placebo	0	3	3	0	5
II	B	Active	2	3	3	0	2
II	B	Placebo	1	8	0	0	5
II	C	Active	2	2	1	0	1
II	C	Placebo	1	1	0	1	1
II	D	Active	0	1	2	2	3
II	D	Placebo	1	1	1	0	7

You may be interested in computing the Mann-Whitney rank measure of association as a way of assessing the extent to which patients with active treatments are more likely to have better response status than those with placebo. You may then be interested in seeing whether diagnostic class and center influence this association through model-fitting. You proceed by computing the Mann-Whitney statistics and their standard errors and using these estimates as input to the CATMOD procedure.

Recall that you can compute the Mann-Whitney measures g_h and their standard errors s_h as functions of the Somers' D measures, which are produced by the FREQ procedure:

$$g_h = \frac{\{\text{Somers'}\,DC|R + 1\}}{2} \text{ and } s_h = \frac{SE(\text{Somers'}\,DC|R)}{2}$$

The following SAS statements input the chronic pain data.

```
data cpain;
   input  center $ diagnosis $ treat $ status $ count @@;
   datalines;
I   A active excellent  1 I   A active   good  3 I A active moderate 2
I   A active fair    5 I   A active   poor  1
I   A placebo excellent 2 I   A placebo good  4 I A placebo moderate 3
I   A placebo fair  4 I   A placebo poor  3
I   B active excellent  3 I   B active   good 10 I B active moderate 1
I   B active fair    4 I   B active   poor  2
I   B placebo excellent 2 I   B placebo good  4 I B placebo moderate 1
I   B placebo fair 5 I   B placebo poor  2
I   C active excellent   6 I   C active   good  1 I C active moderate 1
```

14.6. Modeling Rank Measures of Association Statistics

```
    I   C  active  fair       1  I    C  active   poor    0
    I   C  placebo excellent  0  I    C  placebo  good    5  I  C  placebo  moderate  1
    I   C  placebo fair       1  I    C  placebo  poor    3
    I   D  active  excellent  3  I    D  active   good    5  I  D  active   moderate  1
    I   D  active  fair       6  I    D  active   poor    1
    I   D  placebo excellent  3  I    D  placebo  good    3  I  D  placebo  moderate  2
    I   D  placebo fair       4  I    D  placebo  poor    5
    II  A  active  excellent  0  II   A  active   good    4  II A  active   moderate  3
    II  A  active  fair       1  II   A  active   poor    8
    II  A  placebo excellent  0  II   A  placebo  good    3  II A  placebo  moderate  3
    II  A  placebo fair       0  II   A  placebo  poor    5
    II  B  active  excellent  2  II   B  active   good    3  II B  active   moderate  3
    II  B  active  fair       0  II   B  active   poor    2
    II  B  placebo excellent  1  II   B  placebo  good    8  II B  placebo  moderate  0
    II  B  placebo fair       0  II   B  placebo  poor    5
    II  C  active  excellent  2  II   C  active   good    2  II C  active   moderate  1
    II  C  active  fair       0  II   C  active   poor    1
    II  C  placebo excellent  1  II   C  placebo  good    1  II C  placebo  moderate  0
    II  C  placebo fair       1  II   C  placebo  poor    1
    II  D  active  excellent  0  II   D  active   good    1  II D  active   moderate  2
    II  D  active  fair       2  II   D  active   poor    3
    II  D  placebo excellent  1  II   D  placebo  good    1  II D  placebo  moderate  1
    II  D  placebo fair       0  II   D  placebo  poor    7
    ;
```

The following statements produce measures of association for the eight 2 × 5 tables formed for the cross-classification of center and diagnostic class.

```
proc freq data=cpain order=data;
   weight count;
   tables center*diagnosis*treat*status/ measures;
run;
```

Output 14.24 displays the table for Center I and Diagnostic Class A. Output 14.25 displays the measures of association for that table.

Output 14.24 Frequency Counts

Frequency Percent Row Pct Col Pct	Table 1 of treat by status					
	Controlling for center=I diagnosis=A					
	status					
treat	excellen	good	moderate	fair	poor	Total
active	1 3.57 8.33 33.33	3 10.71 25.00 42.86	2 7.14 16.67 40.00	5 17.86 41.67 55.56	1 3.57 8.33 25.00	12 42.86
placebo	2 7.14 12.50 66.67	4 14.29 25.00 57.14	3 10.71 18.75 60.00	4 14.29 25.00 44.44	3 10.71 18.75 75.00	16 57.14
Total	3 10.71	7 25.00	5 17.86	9 32.14	4 14.29	28 100.00

Output 14.25 Measures of Association

Statistics for Table 1 of treat by status
Controlling for center=I diagnosis=A

Statistic	Value	ASE
Gamma	-0.0201	0.2727
Kendall's Tau-b	-0.0125	0.1687
Stuart's Tau-c	-0.0153	0.2073
Somers' D C|R	-0.0156	0.2116
Somers' D R|C	-0.0099	0.1346
Pearson Correlation	-0.0166	0.1853
Spearman Correlation	-0.0138	0.1870
Lambda Asymmetric C|R	0.0000	0.1489
Lambda Asymmetric R|C	0.0833	0.2394
Lambda Symmetric	0.0323	0.1596
Uncertainty Coefficient C|R	0.0147	0.0256
Uncertainty Coefficient R|C	0.0332	0.0576
Uncertainty Coefficient Symmetric	0.0204	0.0355

Since the fifth and sixth subtables in Table 14.7 have zero columns, statistics requested by the MEASURES option are not computed for them in a multiway table specification in PROC FREQ. However, zero columns present no problem for Somers' D, and the statistics are produced when the tables are specified with a simple two-way specification, as follows.

```
proc sort data=cpain; by diagnosis;
proc freq data=cpain order=data; by diagnosis;
   weight count;
   where (center='II' and (diagnosis='A' or diagnosis='B'));
   tables treat*status/ measures;
run;
```

Table 14.8 displays Somers' D values and asymptotic standard errors produced by the FREQ procedure and the calculated values of g_h and s_h. These estimates can be generated by using the OUTPUT statement in PROC FREQ and some basic PROC IML programming statements as demonstrated in Section 4.3.3.

14.6. Modeling Rank Measures of Association Statistics

Table 14.8 Mann-Whitney Statistics

Center	Diagnosis	Somers'D	ASE	g_h	s_h
I	A	−0.0156	0.2116	0.4922	0.1058
I	B	0.1893	0.1923	0.5946	0.0959
I	C	0.6778	0.1834	0.8389	0.0917
I	D	0.2022	0.1915	0.6011	0.0959
II	A	−0.0625	0.2100	0.4688	0.1049
II	B	0.0571	0.2305	0.5286	0.1153
II	C	0.2083	0.3622	0.6042	0.1811
II	D	0.2000	0.2514	0.6000	0.1257

To produce the appropriate input for PROC CATMOD, you compute the variances by squaring the s_h and then create a data set that contains the estimates g_h and the covariance matrix. The following DATA step creates the data set MANNWHITNEY.

```
data MannWhitney;
   input b1-b8 _type_ $ _name_ $8.;
   datalines;
.4922  .5946  .8389  .6011  .4688  .5286  .6042  .6000  parms
.0112  .0000  .0000  .0000  .0000  .0000  .0000  .0000  cov b1
.0000  .0092  .0000  .0000  .0000  .0000  .0000  .0000  cov b2
.0000  .0000  .0084  .0000  .0000  .0000  .0000  .0000  cov b3
.0000  .0000  .0000  .0092  .0000  .0000  .0000  .0000  cov b4
.0000  .0000  .0000  .0000  .0110  .0000  .0000  .0000  cov b5
.0000  .0000  .0000  .0000  .0000  .0133  .0000  .0000  cov b6
.0000  .0000  .0000  .0000  .0000  .0000  .0328  .0000  cov b7
.0000  .0000  .0000  .0000  .0000  .0000  .0000  .0158  cov b8
;
```

This data set is then input into the CATMOD procedure. Instead of generating functions from an underlying contingency table, the CATMOD procedure directly models the input functions and uses the input covariance matrix as the weights. You define the profiles for each function with the PROFILE option in the FACTORS statement. You also define your factors, or explanatory variable structure, along with the number of levels. Then you describe the effects you want to include in your model with the _RESPONSE_ option.

```
proc catmod data=MannWhitney;
   response read b1-b8;
   factors center $ 2 , diagnosis $ 4 /
     _response_ = center diagnosis
   profile = (I    A,
              I    B,
              I    C,
              I    D,
              II   A,
              II   B,
              II   C,
              II   D);
   model _f_ = _response_ / cov;
run;
```

The ANOVA table results are displayed in Output 14.26. The residual Wald test is a test of the diagnostic class and investigator interaction on the treatment effect, which is nonsignificant with a *p*-value of 0.8158. Neither diagnostic class nor center appear to explain significant variation, with diagnostic class appearing to be modestly influential with a *p*-value of 0.0745.

Output 14.26 ANOVA Table

Analysis of Variance			
Source	DF	Chi-Square	Pr > ChiSq
Intercept	1	229.83	<.0001
center	1	0.62	0.4327
diagnosis	3	6.92	0.0745
Residual	3	0.94	0.8158

By submitting another MODEL statement that specifies the vector of 1s as the model matrix, you can obtain a test of the hypothesis that the measures have the same value for each diagnostic class and investigator combination through the residual Wald test.

```
model _f_ =( 1,
             1,
             1,
             1,
             1,
             1,
             1,
             1 );
```

This is the seven degree of freedom test that is labeled 'Residual' in the "Analysis of Variance" table. Note that there are no degrees of freedom left over for the "Model|Mean" source of variation, which is why the redundant or restricted parameter message appears. Output 14.27 contains the results.

Output 14.27 ANOVA Table

Analysis of Variance			
Source	DF	Chi-Square	Pr > ChiSq
Model|Mean	0*	.	.
Residual	7	9.78	0.2013

Note: Effects marked with '*' contain one or more redundant or restricted parameters.

Analysis of Weighted Least Squares Estimates					
Effect	Parameter	Estimate	Standard Error	Chi-Square	Pr > ChiSq
Model	1	0.6017	0.0382	248.23	<.0001

The p-value of 0.2013 suggests that this hypothesis is compatible with these data. The estimate of the common Mann-Whitney rank measure for the eight strata is 0.6017 with standard error 0.0382. It is interpretable as an estimated probability of 0.6017 for a randomly selected patient with active treatment having better responses than a patient with placebo. A two-sided 95% confidence interval is about (0.527, 0.677) which, by excluding 0.50, indicates a significant treatment difference with $p < 0.05$.

See Kawaguchi et al. (2011) for additional analysis of these data.

14.7 Repeated Measurements Analysis

Many types of studies have research designs that involve multiple measurements of a response variable. Longitudinal studies, in which repeated measures are obtained over time from each subject, are one important and commonly used type of repeated measures study. In other applications, the response from each experimental unit is measured under multiple conditions rather than at multiple time points. In some settings in which repeated measures data are obtained, the independent experimental units are not individual subjects. For example, in a toxicological study the experimental units might be litters; responses are then obtained from the multiple newborns in each litter. In a genetic study, experimental units might be defined by families; responses are then obtained from the members of each family.

There are two main difficulties in the analysis of data from repeated measures studies. First, the analysis is complicated by the dependence among repeated observations made on the same experimental unit. Second, the investigator often cannot control the circumstances for obtaining measurements, so that the data may be unbalanced or partially incomplete. For example, in a longitudinal study the response from a subject may be missing at one or more of the time points due to factors that are unrelated to the outcome of interest. In toxicology or genetic studies, litter or family sizes are variable rather than fixed; hence, the number of repeated measures is not constant across experimental units.

While many approaches to the analysis of repeated measures data have been studied, most are restricted to the setting in which the response variable is normally distributed and the data are balanced and complete. Although the development of methods for the analysis of repeated measures categorical data has received substantially less attention in the past, this has recently become an important and active area of research. Still, the methodology is not nearly as well developed as the methodology for continuous, normally distributed outcomes.

SAS provides several useful methodologies for analyzing repeated measures categorical data. These methodologies are applicable when a univariate response variable is measured repeatedly for each independent experimental unit.

One of these approaches, based on Mantel-Haenszel (MH) test statistics, is described as an advanced topic in Chapter 6, "Sets of $s \times r$ Tables." The MH methodology is useful for testing the null hypothesis of no association between the response variable and the repeated time points or conditions within each subject (that is, interchangeability). Although these randomization model methods require minimal assumptions and the sample size requirements are less stringent than for other methods, they have important limitations. First, the MH methods are restricted to the analysis of data from a single sample; thus, the effects of additional factors (for example, treatment group) cannot be incorporated. In addition, the methods are oriented primarily to hypothesis testing rather than to parameter estimation.

Another approach is to model categorical repeated measurements in terms of a parsimonious number of parameters. Chapter 8, "Logistic Regression I: Dichotomous Response," introduces statistical modeling of categorical data using maximum likelihood to estimate parameters of models for logits, and earlier sections of this chapter describe weighted least squares (WLS) methodology for modeling a wide range of types of categorical data outcomes. However, both of these approaches focus on statistical modeling of the relationship between a single dependent categorical variable and one or more explanatory variables. When you model repeated measurements data, you are dealing with multiple dependent variables that reflect different times or conditions under which the outcome of interest was measured.

This section describes methods for analyzing repeated measurements data with weighted least squares methods. The WLS techniques are a direct extension of the general approach introduced and described earlier in this chapter. The WLS methodology is an extremely versatile modeling approach that can be used efficiently for parameter estimation and hypothesis testing. However, the price of this versatility is that large sample sizes are required.

While such methods are still useful in analyzing repeated measurements data, the generalized estimating equations (GEE) technique has become popular for the analysis of repeated categorical measures and clustered data. GEE methods handle continuous explanatory variables, missing data, and time-dependent covariates. Chapter 15 discusses the GEE methodology and its application through a series of practical examples.

Many repeated measurements analyses are now undertaken with the GEE strategy. However, the weighted least squares approach is still reasonable for data that meet the sample size requirements and include a minimum number of discrete explanatory variables, complete data, and limited time-dependent explanatory variables. For these situations, weighted least squares offers full efficiency, provides well-defined goodness-of-fit statistics, accounts for all degrees of freedom, and is asymptotically equivalent to maximum likelihood methods. You do lose these properties with the GEE approach, but you gain greater scope in your analysis. In some sense, you can consider the analogy to the classical MANOVA model in the continuous response setting. The weighted least squares and MANOVA methods have very desirable properties but limited scope; GEE methods, like mixed model methods, extend the possible scope of analyses with some reasonable choice of assumptions. Choosing one method or another depends on the data at hand and your analysis objectives.

14.7.1 WLS Methodology for Repeated Measurements

Previously, this chapter considered situations in which there is a single outcome (dependent) variable, so that the response profiles are defined by the r possible levels of the dependent variable. In each group, at most $r - 1$ linearly independent response functions can be analyzed. Thus, in the applications in which the response variable is dichotomous, there is one response function per group. In other applications, the response is polytomous but a single response function, such as a mean score, is computed for each group. In both of these situations, there are s independent response functions (one for each row of the table) and their estimated covariance matrix $\mathbf{V_F}$ is diagonal.

However, the methodology can also be used when there are multiple response functions per group. In these situations, the response functions from the same group are correlated and their covariance matrix $\mathbf{V_F}$ is block diagonal. Since the usual covariance structure based on the multinomial distribution accounts for correlated proportions, it is a natural candidate for handling the correlation structure of repeated measurements.

In repeated measures applications, interest generally focuses on the analysis of the marginal distributions of the response at each time point, that is, regardless of the responses at the other time points. Thus, there are multiple response functions per group, and the correlation structure induced by the repeated measures must be taken into consideration. In the general situation in which a c-category response variable is measured at t time points, the cross-classification of the possible outcomes results in $r = c^t$ response profiles. You generally consider $t(c - 1)$ correlated marginal proportions, generalized logits, or cumulative logits, or t correlated mean scores (if the response is ordinal), in the analysis.

Provided that the appropriate covariance matrix is computed for these correlated response functions, the WLS computations are no different from those described previously. Koch and Reinfurt (1971) and Koch et al. (1977) first described the application of WLS to repeated measures categorical data. Further work is described in Stanish, Gillings, and Koch (1978), Koch et al. (1985), and Koch et al. (1989). Stanish (1986), Landis et al. (1988), Agresti (1988, 1989), and Davis (1992) further developed this methodology and also illustrated various aspects of the use of the CATMOD procedure in analyzing categorical repeated measures.

The following sections illustrate several basic types of WLS analyses of repeated measurements data when the outcome is categorical. The examples progress in difficulty and gradually introduce more sophisticated analyses.

14.7.2 One Population, Dichotomous Response

Grizzle, Starmer, and Koch (1969) analyze data in which 46 subjects were treated with three Drugs (A, B, and C). The response to each drug was recorded as favorable or unfavorable. The null hypothesis of interest is that the marginal probability of a favorable response is the same for all three drugs, that is, the hypothesis of marginal homogeneity (see page 157 in Chapter 6). Since the same 46 subjects were used in testing each of the three drugs, the estimates for the three marginal probabilities are correlated. In Section 6.4, this null hypothesis was tested using the Mantel-Haenszel general association statistic. This analysis concluded that there was a statistically significant difference among the three marginal probabilities.

Table 14.9 displays the data from Section 6.4 in the general WLS framework. There is one subpopulation (since there is a single group of subjects) and $r = 2^3 = 8$ response profiles, which correspond to the possible combinations of favorable and unfavorable response for the three drugs. For example, there are 6 subjects who had a favorable response to all three drugs (FFF) and 16 subjects who responded favorably to Drugs A and B and unfavorably to Drug C (FFU). In the current notation, $s = 1$, $c = 2$, $t = 3$, and $r = 2^3 = 8$.

Based on the underlying multinomial distribution of the cell counts, computation of response functions of interest and subsequent analysis using the WLS approach follows the same principles described earlier. However, the eight response profiles are not defined by the eight levels of a single response but rather by the response combinations resulting from the measurement of three dichotomous variables. From the proportions of these eight profiles, you can construct three correlated marginal proportions that correspond to those subjects who responded favorably to Drug A, Drug B, and Drug C, respectively.

Table 14.9 Drug Response Data

	F=favorable, U=unfavorable								
Drug A response	F	F	F	F	U	U	U	U	
Drug B response	F	F	U	U	F	F	U	U	
Drug C response	F	U	F	U	F	U	F	U	Total
Number of subjects	6	16	2	4	2	4	6	6	46

Suppose that p_i denotes the observed proportion of subjects in the ith response profile (ordered from left to right as displayed in Table 14.9), and let $\mathbf{p} = (p_1, \ldots, p_8)'$. For example, $p_1 = \Pr\{FFF\}$ is the probability of a favorable response to all three drugs. Now let p_A, p_B, and p_C denote the marginal proportions with a favorable response to Drugs A, B, and C, respectively. For example, $p_A = \Pr\{FFF \text{ or } FFU \text{ or } FUF \text{ or } FUU\}$. The vector of response functions $\mathbf{F}(\mathbf{p}) = (p_A, p_B, p_C)'$ can be computed by the linear transformation $\mathbf{F}(\mathbf{p}) = \mathbf{A}\mathbf{p}$, where

$$\mathbf{A} = \begin{bmatrix} 1 & 1 & 1 & 1 & 0 & 0 & 0 & 0 \\ 1 & 1 & 0 & 0 & 1 & 1 & 0 & 0 \\ 1 & 0 & 1 & 0 & 1 & 0 & 1 & 0 \end{bmatrix}$$

The first row of $\mathbf{A}$ sums p_1, p_2, p_3, and p_4 to compute the proportion of subjects with a favorable response to Drug A. Similarly, the second row of $\mathbf{A}$ sums p_1, p_2, p_5, and p_6 to yield the proportion with a favorable response to Drug B. Finally, the corresponding proportion for Drug C is computed by summing p_1, p_3, p_5, and p_7. The hypothesis of marginal homogeneity specifies that the marginal proportions with a favorable response to Drugs A, B, and C are equal. This hypothesis can be tested by fitting a model of the form $\mathbf{F}(\boldsymbol{\pi}) = \mathbf{X}\boldsymbol{\beta}$, where $\boldsymbol{\pi}$ is the vector of population probabilities estimated by $\mathbf{p}$, $\mathbf{X}$ is a known model matrix, and $\boldsymbol{\beta}$ is a vector of unknown parameters. If the drug effect is significant, then the hypothesis of marginal homogeneity can be rejected.

This analysis is performed with the CATMOD procedure. The following statements create the SAS data set DRUG. The variables DRUGA, DRUGB, and DRUGC contain the responses for Drugs A, B, and C, respectively.

```
data drug;
   input druga $ drugb $ drugc $ count;
   datalines;
F F F  6
F F U 16
F U F  2
F U U  4
U F F  2
U F U  4
U U F  6
U U U  6
;
```

The next group of statements requests a repeated measurements analysis that tests the hypothesis of marginal homogeneity.

```
proc catmod;
   weight count;
   response marginals;
   model druga*drugb*drugc=_response_ / oneway cov;
   repeated drug 3 / _response_=drug;
run;
```

A major difference between this PROC CATMOD invocation and those discussed in previous sections is the syntax of the MODEL statement. One function of the MODEL statement is to specify the underlying $s \times r$ contingency table; that is, it defines the r response profiles by the values of the response variable and the s population profiles by the cross-classification of the levels of the explanatory variables. The fundamental distinction of repeated measures analyses is that there are now multiple response variables and they determine both the response functions and the variation to be modeled.

The response variables are crossed (separated by asterisks) on the left-hand side of the MODEL statement, and the r response profiles are defined by the cross-classification of their levels.

```
model druga*drugb*drugc=_response_ / oneway design cov;
```

The response profiles are ordered so that the rightmost variable on the left-hand side of the MODEL statement varies fastest and the leftmost variable varies slowest. In this example, the Drug C response changes from favorable to unfavorable most rapidly, followed by Drug B, with Drug A changing the slowest. Look ahead to Output 14.29 to see these response profiles listed in the resulting PROC CATMOD output. Since the MARGINALS option is specified in the RESPONSE statement, the marginal proportions for Drug A, Drug B, and Drug C are computed as the three response functions, as seen in Output 14.30.

Since the right-hand side of the MODEL statement does not include any explanatory variables, the data are correctly structured as a single population with $r = 8$ response profiles. The keyword _RESPONSE_ specifies that the variation among the dependent variables is to be modeled; by default, PROC CATMOD builds a full factorial _RESPONSE_ effect with respect to the repeated measurement factors. In this case, there is only one repeated factor, drug, so the full factorial includes only the drug main effect.

However, you can specify a different model matrix in the REPEATED statement, which is usually used in repeated measurements analysis. The general purpose of the REPEATED statement is to specify how to incorporate repeated measurement factors into the model. You can specify a name for each repeated measurement factor in the REPEATED statement, and you can specify the type (numeric or character), number of levels, and the label or value of each level. You can also define the model matrix in terms of the repeated measurement factors.

```
repeated drug 3 / _response_=drug;
```

In this example, the REPEATED statement specifies that there is a single repeated measurement factor that has three levels (drugs A, B, C). Although it is convenient to name this factor DRUG, any valid SAS variable name can be used, with the restriction that it cannot be the same as the name of an existing variable in the data set. If there is only one repeated measurements factor and the number of levels is omitted, then the CATMOD procedure assumes that the number of levels is equal to the number of response functions per group. So, in this case, the number 3 could have been omitted from the REPEATED statement.

The _RESPONSE_= option in the REPEATED statement specifies the effects to be included in the model matrix as a result of using the _RESPONSE_ keyword in the MODEL statement. The variables named in the effects must be listed in the REPEATED statement. If this option is omitted, then PROC CATMOD builds a full factorial _RESPONSE_ effect with respect to the repeated measurement factors. In this example, the _RESPONSE_ option specifies that the model matrix include a DRUG main effect. Since there is only one repeated measurement factor, you could replace the preceding REPEATED statement with

```
repeated drug;
```

The ONEWAY option in the MODEL statement prints one-way marginal frequency distributions for each response variable in the MODEL statement. This is very useful in verifying that your model is set up as intended. The COV option in the MODEL statement prints the covariance matrix of the vector of response functions $\mathbf{F}(\mathbf{p})$.

Output 14.28 displays the one-way frequency distributions of the variables DRUGA, DRUGB, and DRUGC; they are useful for checking that the response functions are defined as desired. The variables DRUGA, DRUGB, and DRUGC have two levels, so the marginal proportion of subjects with the first level (F) is computed for each variable.

Output 14.28 One-Way Frequency Distributions

Variable	Value	Frequency
druga	F	28
	U	18
drugb	F	28
	U	18
drugc	F	16
	U	30

Output 14.29 displays the population and response profiles.

Output 14.29 Population and Response Profiles

Population Profiles

Sample	Sample Size
1	46

Response Profiles

Response	druga	drugb	drugc
1	F	F	F
2	F	F	U
3	F	U	F
4	F	U	U
5	U	F	F
6	U	F	U
7	U	U	F
8	U	U	U

Output 14.30 displays the vector of response functions, its covariance matrix, and the model matrix. Compare these three response functions with the one-way distributions in Output 14.28 and verify that they are equal to the marginal proportions with a favorable response to drugs A, B, and C, respectively; for example, $28/(28+18) = 0.6087$ for Drugs A and B, $16/(16+30) = 0.34783$ for Drug C. The covariance matrix $\mathbf{V_F} = \mathbf{A V_p A'}$ of the response function vector $\mathbf{F}$ is printed because the COV option was specified in the MODEL statement. While $\mathbf{V_p}$ is the 8×8 covariance matrix of the proportions in the eight response categories, $\mathbf{V_F}$ is the 3×3 covariance matrix of $\mathbf{F}$. Note that the off-diagonal elements of $\mathbf{V_F}$ are nonzero, since the three marginal proportions are correlated. The model matrix has three columns, and the corresponding parameters are an overall intercept, an

effect for Drug A, and an effect for Drug B.

Output 14.30 Response Functions and Model Matrix

Response Functions and Covariance Matrix					
Sample	Function Number	Response Function	Covariance Matrix		
			1	2	3
1	1	0.60870	0.0051779	0.0023424	-0.000822
	2	0.60870	0.0023424	0.0051779	-0.000822
	3	0.34783	-0.000822	-0.000822	0.0049314

Design Matrix				
Sample	Function Number	Design Matrix		
		1	2	3
1	1	1	1	0
	2	1	0	1
	3	1	-1	-1

Output 14.31 displays the analysis of variance (ANOVA) table. The source of variation labeled "drug" tests the null hypothesis that the probability of a favorable response is the same for all three drugs. Since the observed value of the 2 df test statistic is 6.58, the hypothesis of marginal homogeneity is rejected at the 0.05 level of significance ($p = 0.0372$).

Output 14.31 ANOVA Table

Analysis of Variance			
Source	DF	Chi-Square	Pr > ChiSq
Intercept	1	146.84	<.0001
drug	2	6.58	0.0372
Residual	0	.	.

From inspection of the marginal proportions of favorable response, it is clear that Drug C is inferior to Drugs A and B. You can test the equality of Drugs A and C using a contrast statement. Since β_2 and β_3 are the parameters for Drugs A and B (corresponding to the second and third columns of the model matrix in Output 14.30), the null hypothesis is

$$H_0: \beta_2 = -\beta_2 - \beta_3$$

or, equivalently,

$$H_0: 2\beta_2 + \beta_3 = 0$$

The corresponding CONTRAST statement is

```
contrast 'A versus C' _response_ 2 1;
```

Note that the keyword _RESPONSE_ is specified in the CONTRAST statement. You could also test this hypothesis using the ALL_PARMS keyword; the CONTRAST statement would be

```
contrast 'A versus C' all_parms 0 2 1;
```

The results in Output 14.32 indicate a significant difference between Drugs A and C ($Q_W = 5.79$, 1 df, $p = 0.0161$).

Output 14.32 Contrast Results

Analysis of Contrasts			
Contrast	DF	Chi-Square	Pr > ChiSq
A versus C	1	5.79	0.0161
A versus C	1	5.79	0.0161

14.7.3 Two Populations, Polytomous Response

The previous section describes repeated measures analyses when the response variable is dichotomous. In these situations, there is a single response function at each time point (level of the repeated measurement factor). This section describes the application of the WLS methodology when the response variable has more than two levels and the repeated measurement factor isn't time.

Table 14.10 displays unaided distance vision data from 30–39 year-old employees of United Kingdom Royal Ordnance factories during the years 1943–1946 (Kendall and Stuart 1961, pp. 564 and 586). Vision was graded in both the right eye and the left eye on a four-point ordinal categorical scale where 1 represents the highest grade and 4 represents the lowest grade. Interest focuses on determining whether the marginal vision grade distributions are the same in the right eye as in the left eye, whether the marginal distributions differ between females and males, and whether differences between right eye and left eye vision are the same for females and males.

472 Chapter 14: Weighted Least Squares

Table 14.10 Unaided Distance Vision Data[*]

Gender	Right Eye Grade	Left Eye Grade				Total
		1	2	3	4	
Female	1	1520	266	124	66	1976
	2	234	1512	432	78	2256
	3	117	362	1772	205	2456
	4	36	82	179	492	789
	Total	1907	2222	2507	841	7477
Male	1	821	112	85	35	1053
	2	116	494	145	27	782
	3	72	151	583	87	893
	4	43	34	106	331	514
	Total	1952	791	919	480	3242

[*]Reprinted by permission of Edward Arnold.

In this example, there are two populations (females, males). Two measurements of an ordered four-category response variable were obtained from each subject. Thus, there are $r = 4^2 = 16$ response profiles defined by the possible combinations of right-eye and left-eye vision grade. The between subjects variation is due to differences between females and males and the within subjects variation is due to differences between the right eye and the left eye.

The following SAS statements read in the counts displayed in Table 14.10 and create the SAS data set VISION.

```
data vision;
   input gender $ right left count;
   datalines;
F 1 1 1520
F 1 2  266
F 1 3  124
F 1 4   66
F 2 1  234
F 2 2 1512
F 2 3  432
F 2 4   78
F 3 1  117
F 3 2  362
F 3 3 1772
F 3 4  205
F 4 1   36
F 4 2   82
F 4 3  179
F 4 4  492
M 1 1  821
M 1 2  112
M 1 3   85
M 1 4   35
M 2 1  116
M 2 2  494
M 2 3  145
```

```
M 2 4    27
M 3 1    72
M 3 2   151
M 3 3   583
M 3 4    87
M 4 1    43
M 4 2    34
M 4 3   106
M 4 4   331
;
```

Since there are two populations, the null hypothesis of marginal homogeneity can be tested separately for females and males. The marginal distribution of vision grade in each eye involves three linearly independent proportions, since the proportions in the four categories sum to one. Thus, the null hypothesis of marginal homogeneity has 3 df for each gender. The following statements produce the analysis.

```
proc catmod;
   weight count;
   response marginals;
   model right*left=gender _response_(gender='F')
                          _response_(gender='M') / design;
   repeated eye 2;
run;
```

The RESPONSE statement computes six correlated marginal proportions in each of the two populations. The first three response functions in each population are the proportions of subjects with right-eye vision grades of 1, 2, and 3, while the next three are the proportions with left-eye vision grades of 1, 2, and 3. For example, the response function for sample 1 (females), function number 1 (right-eye vision grade of 1) is the marginal proportion of subjects in this category:

$$\frac{\text{number of females with right-eye grade 1}}{\text{total number of females}} = \frac{1976}{7477} = 0.26428$$

In this example, you must specify that the repeated measures factor labeled EYE has two levels. If this specification is omitted, PROC CATMOD constructs a model matrix to test the 5 df null hypothesis that the six response functions from each population are equal. However, it is not necessary to include the option _RESPONSE_=EYE in the REPEATED statement, since there is only one repeated measures factor and the default factorial _RESPONSE_ effect is desired.

Output 14.33 displays the population and response profiles, and Output 14.34 displays the response functions and model matrix.

Output 14.33 Population and Response Profiles

Population Profiles		
Sample	gender	Sample Size
1	F	7477
2	M	3242

Output 14.33 *continued*

Response Profiles		
Response	right	left
1	1	1
2	1	2
3	1	3
4	1	4
5	2	1
6	2	2
7	2	3
8	2	4
9	3	1
10	3	2
11	3	3
12	3	4
13	4	1
14	4	2
15	4	3
16	4	4

Output 14.34 Response Functions and Model Matrix

			Response Functions and Design Matrix											
			Design Matrix											
Sample	Function Number	Response Function	1	2	3	4	5	6	7	8	9	10	11	12
1	1	0.26428	1	0	0	1	0	0	1	0	0	0	0	0
	2	0.30173	0	1	0	0	1	0	0	1	0	0	0	0
	3	0.32847	0	0	1	0	0	1	0	0	1	0	0	0
	4	0.25505	1	0	0	1	0	0	-1	0	0	0	0	0
	5	0.29718	0	1	0	0	1	0	0	-1	0	0	0	0
	6	0.33529	0	0	1	0	0	1	0	0	-1	0	0	0
2	1	0.32480	1	0	0	-1	0	0	0	0	0	1	0	0
	2	0.24121	0	1	0	0	-1	0	0	0	0	0	1	0
	3	0.27545	0	0	1	0	0	-1	0	0	0	0	0	1
	4	0.32449	1	0	0	-1	0	0	0	0	0	-1	0	0
	5	0.24399	0	1	0	0	-1	0	0	0	0	0	-1	0
	6	0.28347	0	0	1	0	0	-1	0	0	0	0	0	-1

The first three parameters, which correspond to the first three columns of the model matrix, are overall intercepts for the probability of vision grades 1, 2, and 3. Recall that with a dichotomous response, the first column of the model matrix is an overall intercept for the probability of the first level of response. Likewise, with a polytomous response with r levels, there are $r - 1$ columns in the model matrix that corresponds to overall intercepts for the probability of the first $r - 1$ levels of response, respectively. The next three parameters compare females to males at vision grades 1, 2, and 3, respectively. Parameters 7–9 (10–12) compare the right eye to the left eye at grades 1, 2, and 3 for females (males).

Output 14.35 displays the resulting ANOVA table. The test of marginal homogeneity is clearly significant in females ($Q_W = 11.98$, 3 df, $p = 0.0075$), but the differences between the right- and left-eye vision grade distributions in males are not statistically significant ($Q_W = 3.68$, 3 df, $p = 0.2984$).

Output 14.35 ANOVA Table for Gender-Specific Tests of Marginal Homoge

Analysis of Variance			
Source	DF	Chi-Square	Pr > ChiSq
Intercept	3	71753.50	<.0001
gender	3	142.07	<.0001
eye(gender=F)	3	11.98	0.0075
eye(gender=M)	3	3.68	0.2984
Residual	0	.	.

If the differences between right-eye and left-eye vision are the same for females and males, there is no interaction between gender and eye. This hypothesis is tested using the following CONTRAST statement to compare parameters within the EYE(GENDER=F) and EYE(GENDER=M) effects.

```
contrast 'Interaction' all_parms 0 0 0 0 0 0 1 0 0 -1  0  0,
                       all_parms 0 0 0 0 0 0 0 1 0  0 -1  0,
                       all_parms 0 0 0 0 0 0 0 0 1  0  0 -1;
run;
```

The results in Output 14.36 indicate that there is evidence of interaction ($Q_W = 8.27$, 3 df, $p = 0.0407$).

Output 14.36 Test of Interaction

Analysis of Contrasts			
Contrast	DF	Chi-Square	Pr > ChiSq
Interaction	3	8.27	0.0407

You could also test the hypothesis of no interaction between gender and eye by fitting the following model.

```
model right*left=gender|_response_;
repeated eye 2;
run;
```

and looking at the GENDER*EYE effect in the ANOVA table. Although this model would provide a more straightforward test of no interaction, it would not provide tests of marginal homogeneity in females and males.

Since vision grade is an ordinal dependent variable, an alternative approach is to assign scores to its four levels and test the hypothesis that the average vision scores in the right and left eyes are the same. Using the scores 1, 2, 3, and 4 (the actual vision grades recorded), you can test the hypothesis of homogeneity for females and males by requesting that mean scores be computed as follows.

```
proc catmod;
   weight count;
```

```
      response means;
      model right*left=gender _response_(gender='F')
            _response_(gender='M') / noprofile design;
      repeated eye;
   run;
```

You do not need to specify the number of levels of the repeated measures factor in the REPEATED statement since there are only two response functions per group and, by default, the model matrix will be constructed to test their equality.

Output 14.37 displays the response functions and model matrix. Response function 1 in sample 1 is the average right-eye vision grade for females. This is computed as follows:

$$\frac{1 \times 1976 + 2 \times 2256 + 3 \times 2456 + 4 \times 789}{7477} = 2.27524$$

The model matrix now includes an overall intercept, a gender effect, and two eye effects (one for females and one for males).

Output 14.37 Response Functions and Model Matrix for Mean Score Model

Response Functions and Design Matrix						
Sample	Function Number	Response Function	\multicolumn{4}{c}{Design Matrix}
			1	2	3	4
1	1	2.27524	1	1	1	0
	2	2.30520	1	1	-1	0
2	1	2.26774	1	-1	0	1
	2	2.25509	1	-1	0	-1

Output 14.38 displays the resulting ANOVA table. The test of homogeneity is again clearly significant in females ($Q_W = 11.97$, 1 df, $p = 0.0005$), and the difference between the right- and left- average vision scores in males is not statistically significant ($Q_W = 0.73$, 1 df, $p = 0.3916$).

Output 14.38 ANOVA Table for Mean Score Model

Analysis of Variance			
Source	DF	Chi-Square	Pr > ChiSq
Intercept	1	50866.50	<.0001
gender	1	2.04	0.1534
eye(gender=F)	1	11.97	0.0005
eye(gender=M)	1	0.73	0.3916
Residual	0	.	.

The following CONTRAST statement tests the null hypothesis that the mean score differences between right eye and left eye are equal for females and males.

```
contrast 'Interaction' all_parms 0 0 1 -1;
run;
```

The results in Output 14.39 again indicate that there is evidence of interaction ($Q_W = 6.20$, 1 df, $p = 0.0128$).

Output 14.39 Test of Interaction for Mean Score Model

Analysis of Contrasts			
Contrast	DF	Chi-Square	Pr > ChiSq
Interaction	1	6.20	0.0128

Note that the values of the test statistics are affected only by the spacing between scores, not by their values. Thus, the same test statistics would have been obtained using any set of equally spaced scores, for example, vision scores of (1 3 5 7) instead of (1 2 3 4). If it is not reasonable to assume that the vision grades levels are equally spaced, you may redefine the values of the RIGHT and LEFT variables to a different set of scores in a DATA step prior to invoking PROC CATMOD.

14.7.4 One Population Regression Analysis of Logits

In a longitudinal study of the health effects of air pollution (Ware, Lipsitz, and Speizer 1988), children were examined annually at ages 9, 10, 11, and 12. At each examination, the response measured was the presence of wheezing. Two questions of interest are:

- Does the prevalence of wheezing change with age?
- Is there a quantifiable trend in the age-specific prevalence rates?

Table 14.11, from Agresti (2002, p. 478), displays data from 1,019 children included in this study. In this single population example, the cross-classification of a dichotomous outcome at four time points defines $r = 2^4 = 16$ response profiles.

Table 14.11 Breath Test Results at Four Ages for 1019 Children*

Wheeze				No. of
Age 9	Age 10	Age 11	Age 12	Children
Present	Present	Present	Present	94
Present	Present	Present	Absent	30
Present	Present	Absent	Present	15
Present	Present	Absent	Absent	28
Present	Absent	Present	Present	14
Present	Absent	Present	Absent	9
Present	Absent	Absent	Present	12
Present	Absent	Absent	Absent	63
Absent	Present	Present	Present	19
Absent	Present	Present	Absent	15
Absent	Present	Absent	Present	10
Absent	Present	Absent	Absent	44
Absent	Absent	Present	Present	17
Absent	Absent	Present	Absent	42
Absent	Absent	Absent	Present	35
Absent	Absent	Absent	Absent	572

*Reprinted by permission of John Wiley & Sons, Inc. Copyright ©John Wiley & Sons.

The following SAS statements read the observed counts for each of the 16 response profiles.

```
data wheeze;
   input wheeze9 $ wheeze10 $ wheeze11 $ wheeze12 $ count;
   datalines;
Present Present Present Present  94
Present Present Present Absent   30
Present Present Absent  Present  15
Present Present Absent  Absent   28
Present Absent  Present Present  14
Present Absent  Present Absent    9
Present Absent  Absent  Present  12
Present Absent  Absent  Absent   63
Absent  Present Present Present  19
Absent  Present Present Absent   15
Absent  Present Absent  Present  10
Absent  Present Absent  Absent   44
Absent  Absent  Present Present  17
Absent  Absent  Present Absent   42
Absent  Absent  Absent  Present  35
Absent  Absent  Absent  Absent  572
;

proc catmod order=data;
   weight count;
   response marginals;
   model wheeze9*wheeze10*wheeze11*wheeze12=_response_ / oneway;
   repeated age;
run;
```

Chapter 8 describes logistic models for dichotomous response variables. As an alternative to modeling the probability of wheezing as a linear function of age (not shown here), you could choose to model the marginal logit of the probability of wheezing. In this case, the logarithm of the odds is modeled as a linear function of age. Even if there are no substantive grounds for preferring a logit analysis over the analysis on the proportion scale, you may decide to consider both types of models and select the model that provides the simplest interpretation.

Since it is not possible to analyze repeated measurements using the LOGISTIC procedure, maximum likelihood parameter estimates cannot be obtained. However, you can use PROC CATMOD to estimate model parameters using weighted least squares.

Suppose L_x denotes the observed log odds of wheezing at age x, for $x = 9, 10, 11, 12$, respectively, that is,

$$L_x = \log\left(\frac{p_x}{1 - p_x}\right)$$

where p_x denotes the marginal probability of wheezing at age x. The following statements fit the regression model

$$L_x = \alpha + \beta x$$

```
proc catmod order=data;
    weight count;
    response logits;
    model wheeze9*wheeze10*wheeze11*wheeze12=(1  9,
                                              1 10,
                                              1 11,
                                              1 12)
                              (1='Intercept',
                               2='Linear Age') / noprofile design;
run;
```

Output 14.40 displays the marginal logit response functions and the model matrix. The ANOVA table in Output 14.41 indicates that the regression model for marginal logits also provides a good fit to the observed data ($Q_W = 0.67$, 2 df, $p = 0.7167$) and that the linear effect of age is clearly significant ($Q_W = 11.77$, 1 df, $p = 0.0006$).

Output 14.40 Response Functions and Model Matrix

Response Functions and Design Matrix				
			Design Matrix	
Sample	Function Number	Response Function	1	2
1	1	-1.04566	1	9
	2	-1.09730	1	10
	3	-1.17737	1	11
	4	-1.31308	1	12

Output 14.41 ANOVA Table

Analysis of Variance			
Source	DF	Chi-Square	Pr > ChiSq
Intercept	1	0.76	0.3824
Linear Age	1	11.77	0.0006
Residual	2	0.67	0.7167

The model for predicting the log odds of wheezing (Output 14.42) is

$$\text{logit}[\Pr\{\text{wheezing}\}] = -0.2367 - 0.0879 \times \text{age in years}$$

The parameter estimates are interpreted in the same manner as was described in Chapter 8. For example, $e^{-0.0879} = 0.916$ is the extent to which the odds of wheezing decrease for each one-year increase in age.

Output 14.42 Parameter Estimates

Analysis of Weighted Least Squares Estimates					
Effect	Parameter	Estimate	Standard Error	Chi-Square	Pr > ChiSq
Model	1	-0.2367	0.2710	0.76	0.3824
	2	-0.0879	0.0256	11.77	0.0006

The logit function is the default response function for the CATMOD procedure, and maximum likelihood is the default estimation method. However, Output 14.42 displays weighted least squares parameter estimates. In a repeated measures analysis, the specification

```
response logits;
```

analyzes marginal logits using weighted least squares. If the RESPONSE statement is omitted in this example, 15 generalized logits would be computed, comparing each of the first 15 response profiles with the last one. Since the model matrix has only four rows, an error message would then be printed.

Appendix A: Statistical Methodology for Weighted Least Squares

Consider the general contingency table displayed in Table 14.12, where s represents the number of rows (groups) in the table, and r represents the number of responses.

Table 14.12 General Contingency Table

Group	Response 1	2	...	r	Total
1	n_{11}	n_{12}	...	n_{1r}	n_{1+}
2	n_{21}	n_{22}	...	n_{2r}	n_{2+}
...	...	...	...	...	...
s	n_{s1}	n_{s2}	...	n_{sr}	n_{s+}

The proportion of subjects in the ith group who have the jth response is written

$$p_{ij} = n_{ij}/n_{i+}$$

Suppose $\mathbf{n}'_i = (n_{i1}, n_{i2}, \ldots, n_{ir})$ represents the vector of responses for the ith subpopulation. If $\mathbf{n}' = (\mathbf{n}'_1, \mathbf{n}'_2, \ldots, \mathbf{n}'_s)$, then $\mathbf{n}$ follows the product multinomial distribution, given that each group has an independent sample. You can write the likelihood of $\mathbf{n}$ as

$$\Pr\{\mathbf{n}\} = \prod_{i=1}^{s} n_{i+}! \prod_{j=1}^{r} \pi_{ij}^{n_{ij}}/n_{ij}!$$

where π_{ij} is the probability that a randomly selected subject from the ith group has the jth response profile. The π_{ij} satisfy the natural restrictions

$$\sum_{j=1}^{r} \pi_{ij} = 1 \text{ for } i = 1, 2, \ldots, s$$

Suppose $\mathbf{p}_i = \mathbf{n}_i/n_{i+}$ is the $r \times 1$ vector of observed proportions associated with the ith group and suppose $\mathbf{p}' = (\mathbf{p}'_1, \mathbf{p}'_2, \ldots, \mathbf{p}'_s)$ is the $(sr \times 1)$ compound vector of proportions.

A consistent estimator of the covariance matrix for the proportions in the ith row is

$$\mathbf{V}(\mathbf{p}_i) = \frac{1}{n_i} \begin{bmatrix} p_{i1}(1-p_{i1}) & -p_{i1}p_{i2} & \cdots & -p_{i1}p_{ir} \\ -p_{i2}p_{i1} & p_{i2}(1-p_{i2}) & \cdots & -p_{i2}p_{ir} \\ \vdots & \vdots & \vdots & \vdots \\ -p_{ir}p_{i1} & -p_{ir}p_{i2} & \cdots & p_{ir}(1-p_{ir}) \end{bmatrix}$$

and the estimated covariance matrix for the vector $\mathbf{p}$ is

$$\mathbf{V_p} = \begin{bmatrix} \mathbf{V}_1 & 0 & \cdots & 0 \\ 0 & \mathbf{V}_2 & \cdots & 0 \\ \vdots & \vdots & \vdots & \vdots \\ 0 & 0 & \cdots & \mathbf{V}_s \end{bmatrix}$$

where $\mathbf{V}_i$ is the estimated covariance matrix for $\mathbf{p}_i$.

Suppose $\mathbf{F}_1(\mathbf{p}), \mathbf{F}_2(\mathbf{p}), \ldots, \mathbf{F}_u(\mathbf{p})$ is a set of u functions of $\mathbf{p}$. Each of the functions is required to have continuous partial derivatives through order two, and $\mathbf{F}$ must have a nonsingular covariance matrix, which can be written

$$\mathbf{V_F}(\pi) = [\mathbf{H}(\pi)][\mathbf{V}(\pi)][\mathbf{H}(\pi)]'$$

where $\mathbf{H}(\pi) = [\partial \mathbf{F}/\partial \mathbf{z} | \mathbf{z} = \pi]$ is the first derivative matrix of $\mathbf{F}(\mathbf{z})$.

$\mathbf{F}$ is a consistent estimator of $\mathbf{F}(\pi)$, so you can investigate the variation among the elements of $\mathbf{F}(\pi)$ with the linear model

$$E_A\{\mathbf{F}(\mathbf{p})\} = \mathbf{F}(\pi) = \mathbf{X}\boldsymbol{\beta}$$

where $\mathbf{X}$ is a known model matrix with rank $t \leq u$, $\boldsymbol{\beta}$ is a $t \times 1$ vector of unknown parameters, and E_A means asymptotic expectation.

The goodness of fit of the model is assessed with

$$Q(\mathbf{X}, \mathbf{F}) = (\mathbf{WF})'[\mathbf{WV_F W}']^{-1}\mathbf{WF}$$

where $\mathbf{W}$ is any full rank $[(u-t) \times u]$ matrix orthogonal to $\mathbf{X}$. When the model applies, the quantity $Q(\mathbf{X}, \mathbf{F})$ is approximately distributed as chi-square with $(u-t)$ degrees of freedom when the sample sizes n_{i+} are large enough so that the elements of $\mathbf{F}$ have an approximate multivariate normal distribution. Such statistics are known as Wald statistics (Wald 1943).

The following statistic

$$Q_W = (\mathbf{F} - \mathbf{Xb})'\mathbf{V_F}^{-1}(\mathbf{F} - \mathbf{Xb})$$

is identical to $Q(\mathbf{X}, \mathbf{F})$ and is obtained by using weighted least squares to produce an estimate for $\boldsymbol{\beta}$,

$$\mathbf{b} = (\mathbf{X}'\mathbf{V_F}^{-1}\mathbf{X})^{-1}\mathbf{X}'\mathbf{V_F}^{-1}\mathbf{F}$$

which is the minimum modified chi-square estimator (Neyman 1949).

A consistent estimator for the covariance matrix of $\mathbf{b}$ is given by

$$\mathbf{V}(\mathbf{b}) = (\mathbf{X}'\mathbf{V_F}^{-1}\mathbf{X})^{-1}$$

If the model adequately characterizes the data as indicated by the goodness-of-fit criterion, then linear hypotheses of the form $\mathbf{C}\boldsymbol{\beta} = \mathbf{0}$, where $\mathbf{C}$ is a known $c \times t$ matrix of constants of rank c, can be tested with the Wald statistic

$$Q_C = (\mathbf{Cb})'[\mathbf{C}(\mathbf{X}'\mathbf{V_F}^{-1}\mathbf{X})^{-1}\mathbf{C}']^{-1}(\mathbf{Cb})$$

Q_C is distributed as chi-square with degrees of freedom equal to c under the hypothesis.

Predicted values for $\mathbf{F}(\pi)$ can be calculated from

$$\hat{\mathbf{F}} = \mathbf{Xb} = \mathbf{X}(\mathbf{X}'\mathbf{V_F}^{-1}\mathbf{X})^{-1}\mathbf{X}'\mathbf{V_F}^{-1}\mathbf{F}$$

and consistent estimators for the variances of $\hat{\mathbf{F}}$ can be obtained from the diagonal elements of

$$\mathbf{V}_{\hat{\mathbf{F}}} = \mathbf{X}(\mathbf{X}'\mathbf{V}_{\mathbf{F}}^{-1}\mathbf{X})^{-1}\mathbf{X}'$$

While the functions $\mathbf{F}(\mathbf{p})$ can take on a wide range of forms, a few functions are commonly used. In particular, you can fit a strictly linear model

$$\mathbf{F}(\mathbf{p}) = \mathbf{A}\mathbf{p}$$

where $\mathbf{A}$ is a matrix of known constants. The covariance matrix of $\mathbf{F}$ is written

$$\mathbf{V}_{\mathbf{F}} = \mathbf{A}\mathbf{V}_{\mathbf{p}}\mathbf{A}'$$

Another common model is loglinear:

$$\mathbf{F}(\mathbf{p}) = \mathbf{A}\log\mathbf{p}$$

where log transforms a vector to the corresponding vector of natural logarithms and $\mathbf{A}$ is orthogonal to $\mathbf{1}$ (vector of 1s), that is, $\mathbf{A}\mathbf{1} = \mathbf{0}$. In this case,

$$\mathbf{V}_{\mathbf{F}} = \mathbf{A}\mathbf{D}_{\mathbf{p}}^{-1}\mathbf{A}'$$

where $\mathbf{D}_{\mathbf{p}}$ is a diagonal matrix with the elements of $\mathbf{p}$ on the diagonal.

Many other useful functions can be generated as a sequence of linear, logarithmic, and exponential operations on the vector $\mathbf{p}$.

- linear transformations: $\mathbf{F}_1(\mathbf{p}) = \mathbf{A}_1\mathbf{p} = \mathbf{a}_1$
- logarithmic: $\mathbf{F}_2(\mathbf{p}) = \log(\mathbf{p}) = \mathbf{a}_2$
- exponential: $\mathbf{F}_3(\mathbf{p}) = \exp(\mathbf{p}) = \mathbf{a}_3$

The corresponding $\mathbf{H}_k$ matrix operators needed to produce the covariance matrix for $\mathbf{F}$ are

- $\mathbf{H}_1 = \mathbf{A}_1$
- $\mathbf{H}_2 = \mathbf{D}_{\mathbf{p}}^{-1}$
- $\mathbf{H}_3 = \mathbf{D}_{\mathbf{a}_3}$

$\mathbf{V}_{\mathbf{F}}$ is estimated by $\mathbf{V}_{\mathbf{F}} = [\mathbf{H}(\mathbf{p})]\mathbf{V}_{\mathbf{p}}[\mathbf{H}(\mathbf{p})]'$ where $\mathbf{H}(\mathbf{p})$ is a product of the first derivative matrices $\mathbf{H}_k(\mathbf{p})$ where k indicates the ith operation in accordance with the chain rule.

Appendix B: CONTRAST Statements for Obstetrical Pain Example

```
proc catmod;
   weight count;
   response 0 .125 .25 .375 .5 .625 .75 .875 1;
   model no_hours = center initial
                    treat(initial)/design ;
contrast 'interact:a-placebo'   treat(initial) -1  1  0  1 -1  0 ;
contrast 'interact:b-placebo'   treat(initial) -1  0  1  1  0 -1 ;
contrast 'interact:ba-placebo'  treat(initial) -2 -1 -1  2  1  1 ;
contrast 'interact:ba-a'        treat(initial) -1 -2 -1  1  2  1 ;
contrast 'interact:ba-b'        treat(initial) -1 -1 -2  1  1  2 ;
contrast 'average:a-placebo'    treat(initial) -1  1  0 -1  1  0 ;
contrast 'average:b-placebo'    treat(initial) -1  0  1 -1  0  1 ;
contrast 'average:ba-placebo'   treat(initial) -2 -1 -1 -2 -1 -1 ;
contrast 'average:ba-a'         treat(initial) -1 -2 -1 -1 -2 -1 ;
contrast 'average:ba-b'         treat(initial) -1 -1 -2 -1 -1 -2 ;
contrast 'interaction'          treat(initial) -1  1  0  1 -1  0 ,
                                treat(initial) -1  0  1  1  0 -1 ,
                                treat(initial) -2 -1 -1  2  1  1 ;
contrast 'treatment effect'     treat(initial) -1  1  0 -1  1  0 ,
                                treat(initial) -1  0  1 -1  0  1 ,
                                treat(initial) -2 -1 -1 -2 -1 -1 ;
run;
```

Chapter 15

Generalized Estimating Equations

Contents

15.1	Introduction	**487**
15.2	Methodology	**488**
	15.2.1 Motivation	488
	15.2.2 Generalized Linear Models	489
	15.2.3 Generalized Estimating Equations Methodology	490
15.3	Summary of the GEE Methodology	**494**
	15.3.1 Marginal Model	496
15.4	Passive Smoking Example	**497**
15.5	Using a Modified Wald Statistic to Assess Model Effects	**505**
15.6	Crossover Example	**506**
15.7	Respiratory Data	**514**
15.8	Diagnostic Data	**522**
15.9	Using GEE for Count Data	**528**
15.10	Fitting the Proportional Odds Model	**533**
15.11	GEE Analyses for Data with Missing Values	**537**
	15.11.1 Crossover Study with Missing Data	538
15.12	Alternating Logistic Regression	**542**
	15.12.1 Respiratory Data	544
15.13	Using GEE to Account for Overdispersion: Univariate Outcome	**549**
	Appendix A: Steps to Find the GEE Solution	**554**
15.14	Appendix B: Macro for Adjusted Wald Statistic	**555**

15.1 Introduction

The weighted least squares methodology described in Chapter 14 is a useful approach to the analysis of repeated binary and ordered categorical outcome variables. However, it can only accommodate categorical explanatory variables and can't easily handle missing values. In addition, the WLS methodology requires sufficient sample size for the marginal response functions for each assessment in each subpopulation to have an approximately multivariate normal distribution. This requirement can be very restrictive.

The generalized estimating equation (GEE) approach (Liang and Zeger 1986) is an extension of generalized linear models that provides a semiparametric approach to longitudinal data analysis with univariate outcomes for which the quasi-likelihood formulation is sensible, for example, normal, Poisson, binomial, and gamma response variables. This approach encompasses a broad range of data situations, including missing observations, continuous explanatory variables, and time-dependent explanatory variables.

The scope for the GEE strategy is useful for many situations, including the following:

- a two-period crossover study in which researchers study the effects of two treatments and a placebo

- a longitudinal study on the efficacy of a new drug designed to prevent fractures in the elderly. The outcome of interest is the number of fractures that occur.

- an observational study on diagnostic procedures in which subjects had assessments for test and standard procedures at two different times

In this chapter, the generalized estimating equations approach for the analysis of repeated measurements is discussed and illustrated with a series of examples using the GENMOD procedure. In addition, the use of GEE methods for the analysis of a univariate response outcome with an overdispersed distribution is also discussed.

15.2 Methodology

15.2.1 Motivation

Correlated data come from many sources: longitudinal studies on health care outcomes, crossover studies concerned with drug comparisons, split plot experiments in agriculture, and clinical trials investigating new treatments with baseline and follow-up visits. You may have multiple measurements taken at the same time, such as in a psychometric study. You may also have clusters of correlated measurements: one example results from group randomization, such as randomizations of litters of animals to experimental conditions. Another example is sample selection of physician practices and the assessment of all of the patients in each practice, or cluster. Often, particularly with longitudinal studies, missing data are common.

An important consideration in each of these situations is how to account for the correlated measurements in the analysis. Within-subject factors (visit, time) are likely to have correlated measurements, while between-subject factors (age, gender) are likely to have independent measurements. The correlation must be taken into account, or you may produce incorrect standard errors. In the presence of positive correlations, you would underestimate the standard errors of the between-subject effects and overestimate the standard errors of the within-subject effects, resulting in inefficient estimation.

As discussed in Chapter 14, weighted least squares provides a reasonable strategy for repeated categorical outcomes when you have all of the following:

- complete data
- small number of discrete explanatory variables
- samples that are large enough to support approximately normal distributions

However, when you have continuous explanatory variables, a larger number of categorical variables, missing response values, and/or time-dependent covariates, the WLS approach does not apply. The GEE strategy, however, can handle these situations.

When you have continuous outcomes, the general linear multivariate model for normally distributed errors is often appropriate. It requires complete data for all outcomes and requires the covariates to be measured at the cluster level. If you can assume that the covariances have a spherical structure (compound symmetry), then repeated measures ANOVA applies for univariate tests about the within-subject effects. However, if you have time-dependent covariates, missing data, or nonnormality, that approach may not be adequate. You might consider the mixed model, which handles these issues, but that requires certain covariance matrix assumptions. If these are not met, the GEE method provides an alternative strategy.

GEEs were introduced by Liang and Zeger (1986) as a way of handling correlated data that, except for the correlation among responses, can be modeled with a generalized linear model (GLM). GEEs are ideal for discrete response data such as binary outcomes and Poisson counts. They work for longitudinal studies data and cluster sampling data. You model the correlated data using the same link functions and linear predictor setup as you do in the GLM for the independent case. The difference is that you account for the structure of the covariances of the response outcomes through its specification in the estimating process. This is much like the specification of covariance structure in mixed model analysis, but there is robustness to it. See Liang and Zeger (1986), Zeger and Liang (1986), Wei and Stram (1988), Stram, Wei, and Ware (1988), Moulton and Zeger (1989), and Zhao and Prentice (1990) for more detail.

The focus of this chapter is the analysis of categorical repeated measurements; however, as mentioned above, the GEE methodology also applies to continuous outcomes and often is used as an adjunct to other types of analyses.

15.2.2 Generalized Linear Models

The GEE method is an extension of generalized linear models (GLM), which are an extension of traditional linear models (Nelder and Wedderburn 1972). The GLM relates a mean response to a vector of explanatory variables through a link function

$$g(E(y_i)) = g(\mu_i) = \mathbf{x_i}'\boldsymbol{\beta}$$

where y_i is a response variable ($i = 1, \ldots, n$), $\mu_i = E(y_i)$, g is a link function, x_i is a vector of independent variables, and $\boldsymbol{\beta}$ is a vector of regression parameters to be estimated.

Additionally:

- The variance of y_i is $v_i = v_i(\mu_i)$ and is a specified function of its mean μ_i.

- The y_i are from the exponential family. This includes the binomial, Poisson, normal, gamma, and inverse Gaussian distributions. When you assume the normal distribution and specify the identity link function $g(\mu_i) = \mu_i$, you are fitting the same model as the general linear model.

For logistic regression, the link and variance functions are

$$g(\mu) = \log\left\{\frac{\mu}{1-\mu}\right\} \text{ and } v(\mu) = \mu(1-\mu)$$

For Poisson regression, the link and variance functions are

$$g(\mu) = \log(\mu) \text{ and } v(\mu) = \mu$$

You obtain the maximum likelihood estimator $\hat{\beta}$ of the $p \times 1$ parameter vector β by solving the estimating equations, which are the score equations shown below. These estimators also maximize the log likelihood. The estimating equations are written as

$$\sum_{i=1}^{n} \frac{\partial \mu_i'}{\partial \beta} v_i^{-1}(y_i - \mu_i(\beta)) = 0$$

Generally, these estimating equations are a set of nonlinear equations with no closed form solution, so you must solve them iteratively. The Newton-Raphson or Fisher scoring methods are often used; the fitting algorithm in the GENMOD procedure begins with a few Fisher scoring steps and then switches to a ridge-stabilized Newton-Raphson method.

15.2.3 Generalized Estimating Equations Methodology

Generalized estimating equations are an extension of GLMs to accommodate correlated data; they are an extension of quasi-score equations. The GEE methodology models a known function of the marginal expectation of the dependent variable as a linear function of one or more explanatory variables. With quasi-likelihood methods, you pursue statistical models by making assumptions about the link function and the relationship between the first two moments, but without fully specifying the complete distribution of the response. With GEEs, you describe the random component of the model for each marginal response with a common link and variance function, similar to what you do with a GLM model. However, unlike GLMs, you have to account for the covariance structure of the correlated measures, although there is robustness to how this is done.

The GEE methodology provides consistent estimators of the regression coefficients and their variances under weak assumptions about the actual correlation among a subject's observations. This approach avoids the need for multivariate distributions by assuming only a functional form for the marginal distribution at each time point or condition. The covariance structure across time or conditions is managed as a nuisance parameter. The method relies on the independence across subjects to consistently estimate the variance of the proposed estimators even when the specified working correlation structure is incorrect. Zeger (1988), Zeger, Liang, and Albert (1988), and Liang, Zeger, and Qaqish (1992) provide further detail on the GEE methodology.

15.2.3.1 Data Structure

Suppose repeated measurements are obtained at t_i time points, $1 \leq t_i \leq t$, from each of n subjects. (Note that if the number and spacing of the repeated measurements are fixed and do not vary among subjects, t is equal to the total number of distinct measurement times.) Although this notation is most natural for longitudinal studies, it also applies to the general case of correlated responses. For example, t might instead denote the number of conditions under which dependent measurements are obtained, or there might be n clusters with at most t experimental units per cluster.

Now, suppose y_{ij} denotes the response from subject i at time or condition j, for $i = 1, \ldots, n$ and $j = 1, \ldots, t_i$. These y_{ij} may be binary outcomes or Poisson counts, for example. Also, suppose $x_{ij} = (x_{ij1}, \ldots, x_{ijp})'$ denotes a $p \times 1$ vector of explanatory variables (covariates) associated with y_{ij}. If all covariates are time independent, then $x_{i1} = x_{i2} = \cdots = x_{it}$. Note that y_{ij} and x_{ij} are missing if observations are not obtained at time j.

15.2.3.2 Generalized Estimating Equations

Assume that you have chosen a model that relates a marginal mean μ_{ij} to the linear predictor $x'_{ij}\beta$ through a link function. The generalized estimating equations for estimating β, an extension of the GLM estimating equation, are

$$\sum_{i=1}^{n} \frac{\partial \mu'_i}{\partial \beta} \mathbf{V}_i^{-1} (\mathbf{Y}_i - \mu_i(\beta)) = \mathbf{0}$$

where $\mu_i = \mu_i(\beta)$ is the corresponding vector of means $\mu_i = (\mu_{i1}, \ldots, \mu_{it_i})'$, $\mathbf{Y}_i = (y_{i1}, y_{i2}, \ldots, y_{it_i})$, and $\mathbf{V}_i$ is an estimator of the covariance matrix of $\mathbf{Y}_i$. These equations are similar to the GLM estimating equations except that, since you have multiple outcomes, they include a vector of means instead of a single mean and a covariance matrix instead of a scalar variance. The covariance matrix of $\mathbf{Y}_i$ is specified as the estimator

$$\mathbf{V}_i = \phi \mathbf{A}_i^{\frac{1}{2}} \mathbf{R}_i(\alpha) \mathbf{A}_i^{\frac{1}{2}}$$

where $\mathbf{A}_i$ is a $t_i \times t_i$ diagonal matrix with $v(\mu_{ij})$ as the jth diagonal element. Note that $\mathbf{V}_i$ can be different from subject to subject, but generally you use a specification that approximates the average dependence among repeated observations over time. Note that the GEE facilities in the GENMOD procedure only allow you to specify the same form of $\mathbf{V}_i$ for all subjects.

$\mathbf{R}_i(\alpha)$ is the *working correlation matrix*. The (j, j') element of $\mathbf{R}_i(\alpha)$ is the known, hypothesized, or estimated correlation between y_{ij} and $y_{ij'}$. This working correlation matrix may depend on a vector of unknown parameters α, which is the same for all subjects. You specify that $\mathbf{R}_i(\alpha)$ is known except for a fixed number of parameters α that must be estimated from the data.

15.2.3.3 Choosing the Working Correlation Matrix

Several possibilities for the working correlation structure have been suggested (Liang and Zeger 1986). First, when the number of subjects is large relative to the number of observations per subject, the influence of correlation is often small enough that the GLM regression coefficients are nearly efficient. The correlations among repeated measures, however, may have a substantial effect on the estimated variances of the regression coefficients and hence must be taken into account to make correct inferences.

The following are some choices for **R** with matrix formulations for $t = 4$.

Independence: $\mathbf{R} = \mathbf{R}_0 = \mathbf{I}$.

$$\mathbf{R} = \begin{bmatrix} 1 & 0 & 0 & 0 \\ 0 & 1 & 0 & 0 \\ 0 & 0 & 1 & 0 \\ 0 & 0 & 0 & 1 \end{bmatrix}$$

The independence model adopts the working specification that repeated observations for a subject are independent. In this case, solving the GEE is the same as fitting the usual regression models for independent data, and the resulting parameter estimates are the same. However, their standard errors are different. You are choosing not to specify the correlation explicitly, but the GEE method still accounts for that correlation by operating at the cluster level. However, the estimation is done with estimation of β only at each step, and not α, so it doesn't improve the precision of the parameter estimates with additional iterations. In this case, the GEE simplifies to the GLM estimating equations.

Fixed: $\mathbf{R} = \mathbf{R}_0$.

Fixed correlation matrices arise when you have determined the form from a previous analysis. You simply input your covariance matrix directly.

Exchangeable:

$$\text{Corr}(y_{ij}, y_{i,j'}) = \begin{Bmatrix} 1 & j = j' \\ \alpha & j \neq j' \end{Bmatrix}$$

$$\mathbf{R} = \begin{bmatrix} 1 & \alpha & \alpha & \alpha \\ \alpha & 1 & \alpha & \alpha \\ \alpha & \alpha & 1 & \alpha \\ \alpha & \alpha & \alpha & 1 \end{bmatrix}$$

The *exchangeable* working correlation specification makes constant the correlations between any two measurements within a subject, that is, $R_{jj'} = \alpha$, for $j \neq j'$. This is the correlation structure assumed in a random effects model with a random intercept and is also known as *compound symmetry* in the repeated measures ANOVA literature. Although the specification of constant correlation between any two repeated measurements may not be justified in a longitudinal study, it is often reasonable in situations in which the repeated measures are not obtained over time. It is probably reasonable when there are a few repeated measurements. An arbitrary number of observations per subject is permissible with both the independence and exchangeable working correlation structures. This structure is commonly used and is relatively easy to explain to investigators. The exchangeable structure is also appropriate when cluster sampling is involved, such as studies in which physician practices are selected as clusters and measurements are obtained for the patients in those practices.

Unstructured:

$$\text{Corr}(y_{ij}, y_{i,j'}) = \begin{cases} 1 & j = j' \\ \alpha_{jj'} & j \neq j' \end{cases}$$

$$\mathbf{R} = \begin{bmatrix} 1 & \alpha_{21} & \alpha_{31} & \alpha_{41} \\ \alpha_{21} & 1 & \alpha_{32} & \alpha_{42} \\ \alpha_{31} & \alpha_{32} & 1 & \alpha_{43} \\ \alpha_{41} & \alpha_{42} & \alpha_{43} & 1 \end{bmatrix}$$

When the correlation matrix is completely unspecified, there are $t(t-1)/2$ parameters to be estimated. This provides the most efficient estimator for $\boldsymbol{\beta}$ but is useful only when there are relatively few observation times or conditions. In addition, when there are missing data and/or varying numbers of observations per subject, estimation of the complete correlation structure may result in a nonpositive definite matrix and parameter estimation may not proceed.

m-dependent:

$$\text{Corr}(y_{ij}, y_{i,j+s}) = \begin{cases} 1 & s = 0 \\ \alpha_s & s = 1, 2, \ldots, m \\ 0 & s > m \end{cases}$$

$$\mathbf{R} = \begin{bmatrix} 1 & \alpha_1 & \alpha_2 & 0 \\ \alpha_1 & 1 & \alpha_1 & \alpha_2 \\ \alpha_2 & \alpha_1 & 1 & \alpha_1 \\ 0 & \alpha_2 & \alpha_1 & 1 \end{bmatrix}$$

With the m-dependent structure, the correlations depend on the distances between measures; eventually, they diminish to zero for $s > m$.

Auto-regressive (AR-1):

$$\text{Corr}(y_{ij}, y_{i,j+s}) = \alpha^s \quad s = 0, 1, 2, \ldots, t_i - j$$

$$\mathbf{R} = \begin{bmatrix} 1 & \alpha & \alpha^2 & \alpha^3 \\ \alpha & 1 & \alpha & \alpha^2 \\ \alpha^2 & \alpha & 1 & \alpha \\ \alpha^3 & \alpha^2 & \alpha & 1 \end{bmatrix}$$

With an auto-regressive correlation structure, the correlations also depend on the distance between the measures; they diminish with increasing distance.

See the PROC GENMOD documentation for specific estimators of the $\mathbf{R}_i(\boldsymbol{\alpha})$ parameters for each of the working correlation matrix types; they involve using the current value of $\boldsymbol{\beta}$ to compute functions of the Pearson residual

$$r_{ij} = \frac{y_{ij} - \hat{\mu}_{ij}}{\sqrt{v(\hat{\mu}_{ij})}}$$

$\mathbf{R}$ is called a working correlation matrix because, for nonnormal data, the actual values may depend on the mean value and on $\mathbf{x}'_i \boldsymbol{\beta}$. See Appendix A at the end of this chapter for more detail on the steps in the GEE solution.

15.2.3.4 Estimating the Covariance of the Parameter Estimates

The model-based estimator of the covariance matrix for $\hat{\beta}$ is the inverse of the observed information matrix

$$\Sigma_m(\hat{\beta}) = I_0^{-1}$$

where

$$I_0 = \sum_{i=1}^{n} \frac{\partial \mu_i}{\partial \beta}' V_i^{-1} \frac{\partial \mu_i}{\partial \beta}$$

This is a consistent estimator if the model and working correlation matrix are correctly specified. Its use may be preferable in those situations where you have only a moderate (rather than clearly large) number of clusters (Albert and McShane 1995).

The empirical sandwich (robust) estimator of $\text{Cov}(\hat{\beta})$ is given by

$$\Sigma_e = I_0^{-1} I_1 I_0^{-1} = V_{\hat{\beta}}$$

where

$$I_1 = \sum_{i=1}^{n} \frac{\partial \mu_i}{\partial \beta}' V_i^{-1} \text{Cov}(Y_i) V_i^{-1} \frac{\partial \mu_i}{\partial \beta}$$

$\text{Cov}(Y_i)$ is estimated by

$$(Y_i - \mu_i(\hat{\beta}))(Y_i - \mu_i(\hat{\beta}))'$$

This is a consistent estimator even when $\text{Var}(y_{ij}) \neq v(\mu_{ij})$, or when $\mathbf{R}_i(\alpha)$ is not the correlation matrix of $\mathbf{Y}_i$, or when the true correlation varies across clusters. You lose efficiency with the misspecification, but if the working correlation structure is approximately correct, the asymptotic efficiency is expected to be relatively high.

You can test linear hypotheses of the form $H_0: \mathbf{C}\beta = \mathbf{0}$, where $\mathbf{C}$ is a known $c \times p$ matrix of constants of rank c, with the Wald statistic

$$Q_C = (\mathbf{C}\hat{\beta})'[\mathbf{C} V_{\hat{\beta}} \mathbf{C}']^{-1}(\mathbf{C}\hat{\beta})$$

The statistic Q_C is approximately distributed as chi-square under H_0 with degrees of freedom equal to c.

15.3 Summary of the GEE Methodology

The GEE method is a practical strategy for the analysis of repeated measurements, particularly categorical repeated measurements. It provides a way to handle continuous explanatory variables, a moderate number of explanatory categorical variables, and time-dependent explanatory variables. It handles missing values, that is, the number of measurements in each cluster can vary from 1 to t.

The following are the important properties of the GEE method:

15.3. Summary of the GEE Methodology

- GEEs reduce to GLM estimating equations for $t = 1$.

- GEEs are the maximum likelihood score equations for multivariate Gaussian data when you specify unstructured correlation.

- The regression parameter estimates are consistent as the number of clusters become large, even if you have misspecified the working correlation matrix, as long as the model for the mean is correct.

- The empirical sandwich estimator of the covariance matrix of $\hat{\beta}$ is also consistent relative to the number of clusters becoming large, even if you have misspecified the working correlation matrix, as long as the model for the mean is correct.

While the GEE method handles missing values, it is important to note that the method requires the missing data to be missing completely at random (MCAR), which roughly means that the missing values may depend only on the explanatory variables that appear in the model. This requirement is more restrictive than the missing at random (MAR) assumption, which is the assumption for likelihood-based inference.

The GEE method depends on asymptotic theory; the number of clusters needs to be large for the method to produce consistent estimates. That means that the sample size needs to be large enough to support the properties of consistency and approximate normality for the estimates from the method. The number of clusters determines adequate sample size, not the number of measurements per cluster or the total number of measurements. The desired number of clusters depends on other factors: if you have a very small number of continuous or dichotomous explanatory variables, 25 clusters may be minimally enough so that you aren't badly misled by your results. If you have 5–12 explanatory variables, you need at least 100 clusters. If you want to be reasonably confident, you probably need 200 clusters. Note that if the correlations are relatively small, you may be able to handle more time-dependent explanatory variables within a subject than if you have a high degree of correlation.

The Z statistics and Wald statistics (with the former being the square root of the latter) used in the GENMOD procedure to assess parameter significance and Type 3 contrasts require around 200 clusters to reasonably support confidence concerning assessments of statistical significance at the 0.05 confidence level or smaller; the score statistics produced in the Type 3 analyses of the model effects procedure have similar properties (Boos 1992, Rotnitzky and Jewell 1990) although they are often more conservative in the presence of small numbers of clusters. As the number of degrees of freedom of the contrast for the hypothesis test approaches the number of clusters, these tests are likely to become less reliable. Several simulation studies (for example, Hendricks et al. 1996) show that the Type I errors associated with the robust variance estimators can be inflated. Researchers are investigating adjustments to the Wald statistic based on the number of clusters in order to produce statistics with better properties for moderate sample sizes. Shah, Holt, and Folsom (1977) discuss such strategies in the context of sample survey data analysis. See Section 15.5 for an example of the use of one of these adjusted Wald statistics and the availability of a SAS macro to compute it.

GEE methods are robust to an assigned correlation structure; you can misspecify that correlation structure and still obtain consistent parameter estimates. However, note that the closer the working correlation matrix is to the true structure, the more efficient your estimates will be. You can compare this property to the mixed model, which heavily leverages the correlation assumption; this means that if you have misspecified the correlation structure, you may obtain biased estimates.

The previous discussion did not include goodness-of-fit criteria for the GEE model. Since the GEE method is quasi-likelihood based, the usual fit statistics for maximum likelihood estimation do not apply. Pan (2001) describes a quasi-likelihood model assessment criterion that is a modification of the Akaike information criterion (AIC) often used in model assessment for maximum likelihood methods.

The quasi-likelihood is defined as

$$Q(\hat{\boldsymbol{\beta}}(R), \phi) = \sum_{i=1}^{n} \sum_{j=1}^{t_i} Q(\hat{\boldsymbol{\beta}}(R), \phi; (Y_{ij}, \mathbf{X}_{ij}))$$

under the independence working correlation assumption. The $\hat{\boldsymbol{\beta}}(R)$ are the parameter estimates obtained from the GEEs with the working correlation of interest, and ϕ is any dispersion parameter ($\phi = 1$ for the binomial and Poisson distributions).

QIC is defined as

$$\text{QIC} = -2Q(\hat{\boldsymbol{\beta}}(R), \phi) + 2\text{trace}(\hat{\Omega}_I \mathbf{V}_{\hat{\boldsymbol{\beta}}})$$

where $\mathbf{V}_{\hat{\boldsymbol{\beta}}}$ is the robust covariance estimate and $\hat{\Omega}_I$ is the inverse of the model-based covariance estimate under the independent working correlation assumption, evaluated at $\hat{\boldsymbol{\beta}}(R)$.

You can use QIC to compare the fit of competing models, similar to the way in which the AIC criterion is used to compare models. You can use QIC to select both regression models and working correlation models. The GENMOD procedure also produces an approximation to QIC, which is QIC_U; it should only be used to select regression models (Pan 2001).

Other techniques are available to address GEE model fit as well. Barnhart and Williamson (1998) describe an empirical procedure for GEE fit based on the Hosmer and Lemeshow approach for logistic regression (1989). Preisser and Qaqish (1996) describe diagnostics for GEE that are extensions of Cook's D and DBETA for linear regression. The GENMOD procedure produces these diagnostics but they are beyond the realm of this discussion.

15.3.1 Marginal Model

The robustness of the GEE method is due to the fact that the GEE method produces a *marginal model*. It models a known function of the marginal expectation of the dependent variable as a linear function of the explanatory variables. The resulting parameter estimates are population-averaged, or estimates "on the average." You can also think of the GEE model as a variational model in which you use estimation to describe the variation among a set of population parameters (Koch, Gillings, and Stokes 1980). You are relying on the independence across clusters to consistently estimate the variance; the covariance matrix parameters are effectively managed as nuisance parameters.

Compare the marginal model to the subject-specific model fit with the conditional logistic regression method described in Chapter 10 or with mixed models. In those analyses, you characterize behavior as a process for individuals. The predictions you produce are individual-based, rather than predictions that apply on average. Your choice of strategy often depends on the goals of your analysis—whether you want to make population statements about your results, on average, or whether you want to produce a model that permits individual prediction. Note that in the standard linear model there is no distinction between the marginal and subject-specific models. Refer to Diggle, Liang, and Zeger (1994) and Zeger, Liang, and Albert (1988) for more discussion of marginal models in longitudinal data analysis.

15.4 Passive Smoking Example

The data in Table 15.1 are from a hypothetical study of the effects of air pollution on children. Researchers followed 25 children and recorded whether they were exhibiting wheezing symptoms during the periods of evaluation at ages 8, 9, 10, and 11. The response is recorded as 1 for symptoms and 0 for no symptoms. Explanatory variables included age, city, and a passive smoking index with values 0, 1, and 2 that reflected the degree of smoking in the home.

Table 15.1 Pollution Study Data

ID	City	Age 8 Smoke	Age 8 Symp	Age 9 Smoke	Age 9 Symp	Age 10 Smoke	Age 10 Symp	Age 11 Smoke	Age 11 Symp
1	steelcity	0	1	0	1	0	1	0	0
2	steelcity	2	1	2	1	2	1	1	0
3	steelcity	2	1	2	0	1	0	0	0
4	greenhills	0	0	1	1	1	1	0	0
5	steelcity	0	0	1	0	1	0	1	0
6	greenhills	0	1	0	0	0	0	0	1
7	steelcity	1	1	1	1	0	1	0	0
8	greenhills	1	0	1	0	1	0	2	0
9	greenhills	2	1	2	0	1	1	1	0
10	steelcity	0	0	0	0	0	0	1	0
11	steelcity	1	1	0	0	0	0	0	1
12	greenhills	0	0	0	0	0	0	0	0
13	steelcity	2	1	2	1	1	0	0	1
14	greenhills	0	1	0	1	0	0	0	0
15	steelcity	2	0	0	0	0	0	2	1
16	greenhills	1	0	1	0	0	0	1	0
17	greenhills	0	0	0	1	0	1	1	1
18	steelcity	1	1	2	1	0	0	1	0
19	steelcity	2	1	1	0	0	1	0	0
20	greenhills	0	0	0	1	0	1	0	0
21	steelcity	1	0	1	0	1	0	2	1
22	greenhills	0	1	0	1	0	0	0	0
23	steelcity	1	1	1	0	0	1	0	0
24	greenhills	1	0	1	1	1	1	2	1
25	greenhills	0	1	0	0	0	0	0	0

Note that age and the passive smoking index are time-dependent explanatory variables; their values depend on the period of measurement. This example provides a basic introduction to fitting GEE models with the GENMOD procedure. The dichotomous outcome is modeled with a logistic regression analysis; while four response times may be pushing the limits for the exchangeable structure, the small number of clusters makes a single-parameter covariance structure a more reasonable choice. Since there are only 25 experimental units, or clusters, only a few explanatory variables can be included in the analysis.

In fact, the small number of clusters requires some justification for performing a GEE analysis at all. If the proportion of children with symptoms is very low or very high (close to 0 or 1), then it would be difficult to do. However, the following table produced with the FREQ procedure shows that the proportions with symptoms are generally between 0.4 and 0.6 by age group with the exception of the 11-year-olds, and their proportion with symptoms is still reasonable at 0.28.

Output 15.1 Proportions with Symptoms by Age Group

Frequency Row Pct	Table of age by symptom		
		symptom	
	age	0	1	Total
	8	11 44.00	14 56.00	25
	9	14 56.00	11 44.00	25
	10	15 60.00	10 40.00	25
	11	18 72.00	7 28.00	25
	Total	58	42	100

The following DATA step inputs the measurements into the SAS data set named CHILDREN. Note that the data are stored with all of a particular child's measurements on a single data line. However, the GENMOD procedure requires that each repeated measure be managed as a separate observation. So, the DO loop included in the DATA step statements inputs each measure, age, and the passive smoking index and outputs them, along with the variable CITY, to the CHILDREN data set. You often need to rearrange data in this manner (sometimes called rolling out the data) when you are dealing with repeated measurements.

```
data children;
   input id city$ @@;
   do i=1 to 4;
      input age smoke symptom @@;
      output;
   end;
   datalines;
1 steelcity   8 0 1   9 0 1   10 0 1   11 0 0
2 steelcity   8 2 1   9 2 1   10 2 1   11 1 0
3 steelcity   8 2 1   9 2 0   10 1 0   11 0 0
4 greenhills  8 0 0   9 1 1   10 1 1   11 0 0
5 steelcity   8 0 0   9 1 0   10 1 0   11 1 0
```

```
 6 greenhills 8 0 1   9 0 0   10 0 0   11 0 1
 7 steelcity  8 1 1   9 1 1   10 0 1   11 0 0
 8 greenhills 8 1 0   9 1 0   10 1 0   11 2 0
 9 greenhills 8 2 1   9 2 0   10 1 1   11 1 0
10 steelcity  8 0 0   9 0 0   10 0 0   11 1 0
11 steelcity  8 1 1   9 0 0   10 0 0   11 0 1
12 greenhills 8 0 0   9 0 0   10 0 0   11 0 0
13 steelcity  8 2 1   9 2 1   10 1 0   11 0 1
14 greenhills 8 0 1   9 0 1   10 0 0   11 0 0
15 steelcity  8 2 0   9 0 0   10 0 0   11 2 1
16 greenhills 8 1 0   9 1 0   10 0 0   11 1 0
17 greenhills 8 0 0   9 0 1   10 0 1   11 1 1
18 steelcity  8 1 1   9 2 1   10 0 0   11 1 0
19 steelcity  8 2 1   9 1 0   10 0 1   11 0 0
20 greenhills 8 0 0   9 0 1   10 0 1   11 0 0
21 steelcity  8 1 0   9 1 0   10 1 0   11 2 1
22 greenhills 8 0 1   9 0 1   10 0 0   11 0 0
23 steelcity  8 1 1   9 1 0   10 0 1   11 0 0
24 greenhills 8 1 0   9 1 1   10 1 1   11 2 1
25 greenhills 8 0 1   9 0 0   10 0 0   11 0 0
;
```

The PROC GENMOD invocation includes the usual MODEL statement as well as the REPEATED statement. You use the MODEL statement to request the logit link function, binomial distribution, and a Type 3 analysis by specifying LINK=LOGIT, DIST=BIN, and TYPE3, respectively. So far, this specification is the same as for any logistic regression using PROC GENMOD. The DESCENDING option in the PROC statement specifies that the model is based on the probability of the largest value of the response variable, which is 1.

```
proc genmod data=children descending;
   class id city;
   model symptom = city age smoke  /
                   link=logit dist=bin type3;
   repeated subject=id / type=exch covb corrw;
run;
```

You request a GEE analysis with the REPEATED statement. The SUBJECT=ID identifies the clustering variable. The SUBJECT= variable must be listed in the CLASS statement and needs to have a unique value for each cluster. Specifying TYPE=EXCH requests the exchangeable working correlation structure. The COVB option requests that the parameter estimate covariance matrix be printed, and the CORRW option specifies that the final working correlation matrix be printed.

Output 15.2 displays the "Model Information" table, which provides information about the model specifications, including the specified distribution and link function. In addition, the table describes on which level of the outcome variable the model is based.

Output 15.2 Basic Model Information

Model Information	
Data Set	WORK.CHILDREN
Distribution	Binomial
Link Function	Logit
Dependent Variable	symptom

PROC GENMOD is modeling the probability that symptom='1'.

Output 15.3 displays the class levels and response profiles, respectively.

Output 15.3 Class Levels and Response Profiles

Class Level Information		
Class	Levels	Values
id	25	1 2 3 4 5 6 7 8 9 10 11 12 13 14 15 16 17 18 19 20 21 22 23 24 25
city	2	greenhil steelcit

Response Profile		
Ordered Value	symptom	Total Frequency
1	1	42
2	0	58

Output 15.4 displays information concerning the parameters, including which parameter pertains to which level of the CLASS variables.

Output 15.4 Information About Parameters

Parameter Information		
Parameter	Effect	city
Prm1	Intercept	
Prm2	city	greenhil
Prm3	city	steelcit
Prm4	age	
Prm5	smoke	

Output 15.5 contains the initial parameter estimates. To generate a starting solution, the GENMOD procedure first treats all of the measurements as independent and fits a generalized linear model. These parameter estimates are then used as the starting values for the GEE solution.

15.4. Passive Smoking Example

Output 15.5 Initial Parameter Estimates

Analysis Of Initial Parameter Estimates								
Parameter		DF	Estimate	Standard Error	Wald 95% Confidence Limits		Wald Chi-Square	Pr > ChiSq
Intercept		1	2.4161	1.8673	-1.2438	6.0760	1.67	0.1957
city	greenhil	1	0.0017	0.4350	-0.8508	0.8543	0.00	0.9968
city	steelcit	0	0.0000	0.0000	0.0000	0.0000	.	.
age		1	-0.3283	0.1914	-0.7035	0.0468	2.94	0.0863
smoke		1	0.5598	0.2952	-0.0188	1.1385	3.60	0.0579
Scale		0	1.0000	0.0000	1.0000	1.0000		

Note: The scale parameter was held fixed.

The beginning of the output produced by the GEE analysis is the general model information that is displayed in Output 15.6. Since there are 25 subjects with repeated measures, there are 25 clusters. Each subject has 4 measures, and the data are complete. Thus, the minimum and maximum cluster size is 4.

Output 15.6 General GEE Model Information

GEE Model Information	
Correlation Structure	Exchangeable
Subject Effect	id (25 levels)
Number of Clusters	25
Correlation Matrix Dimension	4
Maximum Cluster Size	4
Minimum Cluster Size	4

Output 15.7 contains the Type 3 analysis results for the model effects.

The results indicate that city is not a factor in wheezing status. However, smoking exposure has a nearly significant association ($p = 0.0583$). Age is marginally influential ($p = 0.0981$).

Output 15.7 Type 3 Analysis

Score Statistics For Type 3 GEE Analysis			
Source	DF	Chi-Square	Pr > ChiSq
city	1	0.01	0.9388
age	1	2.74	0.0981
smoke	1	3.59	0.0583

Output 15.8 contains the parameter estimates produced by the GEE analysis. The table also supplies standard errors, confidence intervals, Z statistics, and p-values. The empirical standard errors are produced by default. Since the effects reported in the Type 3 analysis are single degree of freedom effects, the score statistics in that table assess the same hypotheses as the Z statistics in this table. Note that the p-value for the Z for smoking is 0.0211, compared to the 0.0583 reported for the score statistic in the Type 3 table. In a strict testing situation, you would assess the null hypothesis with the score statistic. When the number of clusters is only moderate, the Z and Wald statistic generally produce overly small p-values compared to the p-values produced by the corresponding score statistic. Particularly for small sample sizes, you would want to report the more accurate score statistic.

Output 15.8 GEE Parameter Estimates

Analysis Of GEE Parameter Estimates							
Empirical Standard Error Estimates							
Parameter		Estimate	Standard Error	95% Confidence Limits		Z	Pr > |Z|
Intercept		2.2615	2.0243	-1.7060	6.2290	1.12	0.2639
city	greenhil	0.0418	0.5435	-1.0234	1.1070	0.08	0.9387
city	steelcit	0.0000	0.0000	0.0000	0.0000	.	.
age		-0.3201	0.1884	-0.6894	0.0492	-1.70	0.0893
smoke		0.6506	0.2821	0.0978	1.2035	2.31	0.0211

Since the Type 3 score statistics reported correspond to effects with one df, you can compare the p-values in Output 15.7 with those for the Z statistics for the parameter estimates in Output 15.8. If they differ to some degree, especially when $|Z| > 1$, then you might want to consider a rescaling of the standard errors reported in the parameter estimates table by the absolute value of Z divided by the square root of the score statistic. Apply this quantity

$$\frac{|Z|}{\sqrt{Q_S}}$$

to the standard error, and then use that rescaled standard error to recompute the 95% confidence limits.

For example, consider the effect for SMOKE. Compute

$$\frac{2.31}{\sqrt{3.59}} = 1.22$$

and rewrite the rescaled 95% confidence limits as $0.6506 \pm 1.96(0.2821)(1.22)$, which produces the limits $(-0.0024, 1.3252)$. These confidence limits are thus in harmony with the score test results for SMOKE ($p = 0.0583$). Note that it is also feasible to apply the same adjustments to odds ratios and their confidence intervals in principle; however, that is beyond the scope of this discussion.

The GENMOD procedure prints both the empirical and model-based covariance matrix of the parameter estimates; these are displayed in Output 15.9. Note that their values are often similar, especially for large samples. If these matrices are very similar, you may have some confidence that you have correctly specified the correlation structure and the estimates are relatively efficient. However, recall that, even if you have misspecified the correlation structure, both the parameter estimates and their empirical standard errors are consistent, provided that the specification is correct for the explanatory variables. Another way to compare working correlation models is with QIC statistics, which is illustrated in Section 15.7.

Output 15.9 Covariance Matrix Estimates

Covariance Matrix (Model-Based)				
	Prm1	Prm2	Prm4	Prm5
Prm1	3.26069	-0.16313	-0.32274	-0.12257
Prm2	-0.16313	0.24015	0.002520	0.03422
Prm4	-0.32274	0.002520	0.03379	0.004471
Prm5	-0.12257	0.03422	0.004471	0.09533

Covariance Matrix (Empirical)				
	Prm1	Prm2	Prm4	Prm5
Prm1	4.09770	-0.55261	-0.37280	-0.29397
Prm2	-0.55261	0.29538	0.03719	0.09143
Prm4	-0.37280	0.03719	0.03550	0.02064
Prm5	-0.29397	0.09143	0.02064	0.07957

Finally, the exchangeable working correlation matrix is also produced, and it is displayed in Output 15.10. The estimated correlation is fairly low at 0.0883.

Output 15.10 Working Correlation Matrix

Working Correlation Matrix				
	Col1	Col2	Col3	Col4
Row1	1.0000	0.0883	0.0883	0.0883
Row2	0.0883	1.0000	0.0883	0.0883
Row3	0.0883	0.0883	1.0000	0.0883
Row4	0.0883	0.0883	0.0883	1.0000

Since this is a logistic regression based on reference cell coding, you can exponentiate the parameter estimates to obtain estimates of odds ratios for various explanatory factors. Since the parameter estimate for age is -0.3201, the odds of symptoms for those at a higher age group are $e^{-0.3201} = 0.7261$ times the odds of symptoms for those children in the lower age group, or have

27.4% less odds of having symptoms.

The GENMOD procedure can produce the odds ratio estimate with its ESTIMATE statement, along with 95% confidence limits. You can produce estimates for any estimable linear combination of the parameters from the GEE analysis.

Since the age effect is represented by a single parameter, you place the coefficient 1 after listing a label and the AGE variable. The EXP option requests that the estimate be exponentiated, which, in the case of reference parameterization, produces the odds ratio estimate.

```
ods select Estimates;
proc genmod data=children descending;
   class id city;
   model symptom = city age smoke /
                  link=logit dist=bin type3;
   repeated subject=id / type=exch covb corrw;
   estimate 'age' age 1 / exp;
run;
```

Output 15.11 displays the results from the ESTIMATE statement. The point estimate for the odds ratio is 0.7261 (from the "L'Beta" column) with 95% confidence limits of (0.5019, 1.0504) for the extent of increased odds of symptoms per category of increase in age level. The confidence limits are based on the Wald statistic. The confidence interval does contain the value 1, which is consistent with the $p = 0.0893$ reported for the Wald test in the table of parameter estimates.

Output 15.11 ESTIMATE Results

	Contrast Estimate Results							
		Mean					L'Beta	
Label	Mean Estimate	Confidence Limits		L'Beta Estimate	Standard Error	Alpha	Confidence Limits	
age	0.4206	0.3342	0.5123	-0.3201	0.1884	0.05	-0.6894	0.0492
Exp(age)				0.7261	0.1368	0.05	0.5019	1.0504

Contrast Estimate Results		
Label	Chi-Square	Pr > ChiSq
age	2.89	0.0893
Exp(age)		

Specifying the coefficients in the ESTIMATE statement for the default parameterization can be more involved with CLASS variables, due to the less than full rank parameterization used by PROC GENMOD. If you were to compute the odds ratio for the variable CITY, with two levels, you would need to write an ESTIMATE statement like the following:

```
estimate 'city' city 1 -1 /exp;
```

Refer to the *SAS/STAT User's Guide* for more information regarding the ESTIMATE statement and parameterization.

15.5 Using a Modified Wald Statistic to Assess Model Effects

Section 15.3 discusses the possibility of using an adjusted Wald statistic to evaluate model effects in the GEE approach. Shah, Holt, and Folsom (1977) describe a modification of the Wald statistic based on a Hotelling T^2 type of transformation of Q_C:

$$\frac{(n-c)Q_C}{c(n-1)} \quad \text{is distributed as} \quad F_{c,n-c}$$

The quantity c is equal to the number of rows of the contrast. Thus, for tests concerning effects of explanatory factors, it is equal to the corresponding number of df. This test is more conservative than the Wald test; LaVange, Koch, and Schwartz (2000) suggest that you can use the Wald and F-transform statistic p-values as the lower and upper bounds for judging the robustness of the actual p-value. As the number of clusters becomes very large, these statistics produce very similar conclusions.

Appendix B contains a macro that produces these F-transform statistics and appends a table containing them to the end of the PROC GENMOD output. The input are ODS output data sets from PROC GENMOD that contains the number of clusters, the appropriate Wald statistics, and the df for the Type 3 analysis.

The following statements call this macro to produce these statistics for the previous analysis. The statements assume that the macro GEEF has been included in a file named MACROS.SAS that is stored in the directory in which you are running your SAS program. The ODS OUTPUT statement puts the GEE model information, including the number of clusters, into a SAS data set named CLUSTOUT, and it puts the Type 3 analysis results, which include the Wald chi-square values and their df, into a SAS data set named SCOREOUT (the names CLUSTOUT and SCOREOUT must be used).

The PROC GENMOD invocation is exactly the same as before, except that the WALD option is specified in the MODEL statement to produce Wald statistics in the Type 3 analysis, and the ODS OUTPUT statement is added.

```
proc genmod data=children descending;
   class id city;
   model symptom = city age smoke  /
              link=logit dist=bin type3 wald;
   repeated subject=id / type=exch covb corrw;
   ods output GEEModInfo=clustout Type3=scoreout;
run;
```

The statements produce the usual PROC GENMOD output, including the model information, the initial parameter estimates, the GEE information, the GEE parameter estimates, and the GEE Type 3 analysis. Output 15.12 contains the resulting tables.

Output 15.12 Type 3 Analyses

Wald Statistics For Type 3 GEE Analysis			
Source	DF	Chi-Square	Pr > ChiSq
city	1	0.01	0.9387
age	1	2.89	0.0893
smoke	1	5.32	0.0211

F-Statistics for Type 3 GEE Analysis			
Source	DF	F Value	Pr > F
city	1	0.01	0.9393
age	1	2.89	0.1023
smoke	1	5.32	0.0300

Note that the values of the F statistics are the same as the values for the Wald statistics for the single degree of freedom tests. However, all of the p-values are more conservative. You might choose to generate the transformed F statistics when you are dealing with a small number of clusters, especially when you have marginal significance.

15.6 Crossover Example

Crossover designs provide another form of repeated measurements. In a crossover design, subjects serve as their own controls and receive two or more treatments or conditions in two or more consecutive periods. You can use the GEE method to analyze such data, managing the subjects as clusters and managing the treatment as a time-varying covariate.

The data in Table 15.2 are from a two-period crossover study investigating three treatments. These data were analyzed with conditional logistic regression in Chapter 10.

15.6. Crossover Example

Table 15.2 Crossover Design Data

Age	Sequence	Response Profiles				Total
		FF	FU	UF	UU	
older	A:B	12	12	6	20	50
older	B:P	8	5	6	31	50
older	P:A	5	3	22	20	50
younger	B:A	19	3	25	3	50
younger	A:P	25	6	6	13	50
younger	P:B	13	5	21	11	50

As described in Chapter 10, this study design is a two-period crossover design where patients have been stratified to two age groups and, within age group, have been assigned to one of three treatment sequences. These data can be modeled with parameters for period effect, effects for Drug A and Drug B relative to the placebo (P), carryover effects for Drug A and Drug B, and interactions of period with age and drug with age.

The following DATA step enters the data into SAS data set CROSS. The variable AGE contains information on whether the subject is older or younger, and the variable SEQUENCE contains two letters describing the sequence of treatments for that group. For example, the value AB means that treatment A was received in the first period and treatment B was received in the second, and the value BP means that treatment B was received in the first period and the placebo was received in the second. The variables TIME1 and TIME2 have the values F and U depending on whether the treatment produced a favorable or unfavorable response. The data are frequency counts, and the variable COUNT contains the frequency for each response profile for each sequence and age combination. The DATA step creates an observation for each subject.

```
data cross (drop=count);
   input age $ sequence $ time1 $ time2 $ count;
   do i=1 to count;
      output;
   end;
   datalines;
older AB F F 12
older AB F U 12
older AB U F 6
older AB U U 20
older BP F F 8
older BP F U 5
older BP U F 6
older BP U U 31
older PA F F 5
older PA F U 3
older PA U F 22
older PA U U 20
younger BA F F 19
younger BA F U 3
younger BA U F 25
younger BA U U 3
younger AP F F 25
younger AP F U 6
```

```
younger AP U F  6
younger AP U U 13
younger PB F F 13
younger PB F U  5
younger PB U F 21
younger PB U U 11
;
```

The next DATA step creates an observation for each response in each period so that the data are in the correct data structure for the GEE analysis. The variable PERIOD is an indicator variable for whether the observation is from the first period. The RESPONSE variable contains the value 1 if the response was favorable and 0 if it was not.

```
data cross2;
   set cross;
   subject=_n_;
      period=1;
         drug = substr(sequence, 1, 1);
         carry='N';
         response = time1;
         output;
      period=0;
         drug  = substr(sequence, 2, 1);
         carry = substr(sequence, 1, 1);
         if carry='P' then carry='N';
         response = time2;
         output;
;

proc print data=cross2(obs=15);
run;
```

The variable CARRY takes the value N (no) if the observation is from the first period; it takes the value A or B if it comes from the second period and the treatment in the first period is A or B, respectively. If the subject received the placebo in the first period, the value of CARRY is also set to N for the observations in the second period.

Output 15.13 displays the first 15 observations of SAS data set CROSS2.

Output 15.13 First 15 Observations of Data Set CROSS2

Obs	age	sequence	time1	time2	i	subject	period	drug	carry	response
1	older	AB	F	F	1	1	1	A	N	F
2	older	AB	F	F	1	1	0	B	A	F
3	older	AB	F	F	2	2	1	A	N	F
4	older	AB	F	F	2	2	0	B	A	F
5	older	AB	F	F	3	3	1	A	N	F
6	older	AB	F	F	3	3	0	B	A	F
7	older	AB	F	F	4	4	1	A	N	F
8	older	AB	F	F	4	4	0	B	A	F
9	older	AB	F	F	5	5	1	A	N	F
10	older	AB	F	F	5	5	0	B	A	F
11	older	AB	F	F	6	6	1	A	N	F
12	older	AB	F	F	6	6	0	B	A	F
13	older	AB	F	F	7	7	1	A	N	F
14	older	AB	F	F	7	7	0	B	A	F
15	older	AB	F	F	8	8	1	A	N	F

The following PROC GENMOD statements fit the GEE model. Since there are 300 subjects in the crossover study, there are 300 clusters or experimental units in the GEE analysis, an entirely adequate sample size. With responses for both periods, the cluster size is two. There are no missing values, so both the minimum and maximum cluster size is two.

A logistic regression analysis is appropriate for these data, so DIST=BIN is specified in the MODEL statement. The logit link is used by default. SUBJECT, AGE, DRUG, and CARRY are specified in the CLASS statement. The model includes main effects for period, age, drug, and carryover effects and interactions for period and age and for drug and age. The option TYPE=UNSTR specifies the unstructured correlation structure. Since there are only two measurements per subject, this is the same as the exchangeable structure.

```
proc genmod data=cross2;
   class subject age drug carry;
   model response = period age drug
                   period*age carry
                   drug*age / dist=bin type3;
   repeated subject=subject/type=unstr;
run;
```

The "Class Level Information" table in Output 15.14 lists the variables treated as classification variables and their values.

Output 15.14 Class Level Information

Class Level Information		
Class	Levels	Values
subject	300	1 2 3 4 5 6 7 8 9 10 11 12 13 14 15 16 17 18 19 20 21 22 23 24 25 26 27 28 29 30 31 32 33 34 35 36 37 38 39 40 41 42 43 44 45 46 47 48 49 50 51 52 53 54 55 56 57 58 59 60 61 62 63 64 65 66 67 68 69 70 71 72 73 74 75 76 77 78 79 80 81 82 83 84 85 86 87 ...
age	2	older younger
drug	3	A B P
carry	3	A B N

The "Parameter Information" table in Output 15.15 lists the 18 parameters.

Output 15.15 Parameter Information

Parameter Information				
Parameter	Effect	age	drug	carry
Prm1	Intercept			
Prm2	period			
Prm3	age	older		
Prm4	age	younger		
Prm5	drug		A	
Prm6	drug		B	
Prm7	drug		P	
Prm8	period*age	older		
Prm9	period*age	younger		
Prm10	carry			A
Prm11	carry			B
Prm12	carry			N
Prm13	age*drug	older	A	
Prm14	age*drug	older	B	
Prm15	age*drug	older	P	
Prm16	age*drug	younger	A	
Prm17	age*drug	younger	B	
Prm18	age*drug	younger	P	

After the initial estimates are printed, the "GEE Model Information" table is displayed and confirms that you have 300 clusters, each containing two responses.

Output 15.16 GEE Model Information

GEE Model Information	
Correlation Structure	Unstructured
Subject Effect	subject (300 levels)
Number of Clusters	300
Correlation Matrix Dimension	2
Maximum Cluster Size	2
Minimum Cluster Size	2

Output 15.17 contains the QIC statistics; QIC takes the value 749.8178.

Output 15.17 QIC Fit Statistics

GEE Fit Criteria	
QIC	749.8178
QICu	750.3226

Output 15.18 contains the results of the Type 3 effect tests for all of the terms specified in the MODEL statement. When you are conducting an analysis of crossover data, you hope that there are no carryover effects. Having such effects greatly complicates the model and interpretation. In this analysis, the carryover effect is not significant. The score statistic for the two-level CARRY variable is 1.15 with p-value equal to 0.5626. In addition, the age × drug interaction appears to be unimportant, with a score chi-square statistic of 0.72 for 2 df ($p = 0.6981$).

Output 15.18 Type 3 Table

Score Statistics For Type 3 GEE Analysis			
Source	DF	Chi-Square	Pr > ChiSq
period	1	4.61	0.0318
age	1	36.03	<.0001
drug	2	27.66	<.0001
period*age	1	4.69	0.0303
carry	2	1.15	0.5626
age*drug	2	0.72	0.6981

You can use the CONTRAST statement to obtain the joint test for CARRY and the AGE*DRUG interaction. You submit the following statements. The ODS SELECT statement restricts the output to the test results. The contrast labeled 'joint' is the joint test for both the CARRY and AGE*DRUG effects. The contrasts labeled 'carry' and 'inter' are the separate effects tests and should match the results displayed in the Type 3 analysis for those effects.

```
ods select Contrasts;
proc genmod data=cross2;
   class subject age drug carry;
   model response = period age drug
                    period*age carry
                    drug*age / dist=bin type3;
   repeated subject=subject/type=unstr;
   contrast 'carry' carry 1 0 -1,
                    carry 0 1 -1;
   contrast 'inter' age*drug 1 0 -1 -1 0  1 ,
                    age*drug 0 1 -1  0 -1 1 ;
   contrast 'joint' carry 1 0 -1,
                    carry 0 1 -1,
                    age*drug 1 0 -1 -1 0  1 ,
                    age*drug 0 1 -1  0 -1 1 ;
run;
```

Output 15.19 contains the results of these tests. The joint test is definitely nonsignificant, with a chi-square value of 1.31 for 4 df and a *p*-value of 0.8595.

Output 15.19 Contrast Results

Contrast Results for GEE Analysis				
Contrast	DF	Chi-Square	Pr > ChiSq	Type
carry	2	1.15	0.5626	Score
inter	2	0.72	0.6981	Score
joint	4	1.31	0.8595	Score

A reduced model was then specified, with main effects for period, age, and drug, as well as the period × age interaction. The terms of the reduced model are listed in the MODEL statement, and the CORRW option requests that the estimate of the working correlation matrix be printed.

```
proc genmod data=cross2;
   class subject age drug;
   model response = period age drug
                    period*age
      / dist=bin type3;
   repeated subject=subject/type=unstr corrw;
run;
```

Output 15.20 displays the QIC statistics. The QIC value of 743.1093 is lower than that for the previous model, indicating a better fit with the reduced model.

Output 15.20 QIC Fit Statistics

GEE Fit Criteria	
QIC	743.1093
QICu	743.6522

Output 15.21 contains the Type 3 results. These tests indicate that period, age, and drug are highly significant. And with a p-value of 0.0240, the period × age interaction cannot be dismissed.

Output 15.21 Type 3 Table

Score Statistics For Type 3 GEE Analysis			
Source	DF	Chi-Square	Pr > ChiSq
period	1	24.98	<.0001
age	1	35.53	<.0001
drug	2	39.31	<.0001
period*age	1	5.10	0.0240

The parameter estimates are displayed in Output 15.22.

Output 15.22 Parameter Estimates

Analysis Of GEE Parameter Estimates							
Empirical Standard Error Estimates							
Parameter		Estimate	Standard Error	95% Confidence Limits		Z	Pr > |Z|
Intercept		0.5127	0.2063	0.1084	0.9170	2.49	0.0129
period		-1.1553	0.2304	-1.6069	-0.7037	-5.01	<.0001
age	older	-1.4994	0.2583	-2.0056	-0.9931	-5.80	<.0001
age	younger	0.0000	0.0000	0.0000	0.0000	.	.
drug	A	1.2542	0.2010	0.8602	1.6483	6.24	<.0001
drug	B	0.3404	0.2016	-0.0546	0.7355	1.69	0.0912
drug	P	0.0000	0.0000	0.0000	0.0000	.	.
period*age	older	0.7088	0.3131	0.0951	1.3224	2.26	0.0236
period*age	younger	0.0000	0.0000	0.0000	0.0000	.	.

Finally, Output 15.23 displays the working correlation matrix. As discussed, the unstructured correlation structure is the same as the exchangeable correlation structure when you have two responses per cluster. The correlation is estimated to be 0.2274.

Output 15.23 Working Correlation Matrix

Working Correlation Matrix		
	Col1	Col2
Row1	1.0000	0.2274
Row2	0.2274	1.0000

These results are similar to those presented for the conditional logistic analysis in Chapter 10. Most of the time, the general conclusions for a GEE analysis and the corresponding conditional logistic regression are the same; the *p*-values are similar, but estimates may be somewhat different. The conditional logistic model is a subject-specific model, producing odds ratio estimates for the individual, while the GEE model is a marginal model, producing odds ratios "on the average."

You may be interested in comparing the two drugs, A and B. This is done with a CONTRAST statement. You test to see if the difference of the two parameters for drugs A and B is equal to zero. The following statements request a contrast test for drug A versus drug B. The ODS SELECT statement restricts the output to just the CONTRAST statement results.

```
ods select Contrasts;
proc genmod data=cross2;
   class subject age drug;
   model response = period age drug
                    period*age
                 / dist=bin type3;
   repeated subject=subject/type=unstr;
   contrast 'A versus B' drug 1 -1 0;
run;
```

Output 15.24 displays these results.

Output 15.24 Contrast Results

Contrast Results for GEE Analysis				
Contrast	DF	Chi-Square	Pr > ChiSq	Type
A versus B	1	19.15	<.0001	Score

This is a single degree of freedom test, and the chi-square value of 19.15 for the score test is highly significant. If you want the Wald statistic instead of the score statistic, you specify the WALD option in the CONTRAST statement.

15.7 Respiratory Data

A clinical trial compared two treatments for a respiratory illness (Koch et al. 1990). In each of two centers, eligible patients were randomly assigned to active treatment or placebo. During treatment, respiratory status was determined at baseline and four visits and recorded on a five-point scale of 0 for terrible to 4 for excellent. Potential explanatory variables were center, sex, and baseline respiratory status, as well as age (in years) at the time of study entry. There were 111 patients (54 active, 57 placebo), with no missing data for responses or covariates. One direction of analysis was to focus on the dichotomous response of good outcome (response is 3 or 4) versus poor outcome (response is less than 3) with baseline dichotomized as well. Table 15.3 displays partial data from Center 1, with baseline and visit outcomes dichotomized.

Table 15.3 Partial Respiratory Disorder Data from Center 1

				Respiratory Status (0=poor, 1=good)				
Patient	Treatment	Sex	Age	Baseline	Visit 1	Visit 2	Visit 3	Visit 4
1	P	M	46	0	0	0	0	0
2	P	M	28	0	0	0	0	0
3	A	M	23	1	1	1	1	1
4	P	M	44	1	1	1	1	0
5	P	F	13	1	1	1	1	1
6	A	M	34	0	0	0	0	0
7	P	M	43	0	1	0	1	1
8	A	M	28	0	0	0	0	0
9	A	M	31	1	1	1	1	1
10	P	M	37	1	0	1	1	0
11	A	M	30	1	1	1	1	1
12	A	M	14	0	1	1	1	0
13	P	M	23	1	1	0	0	0
14	P	M	30	0	0	0	0	0
15	P	M	20	1	1	1	1	1
16	A	M	22	0	0	0	0	1
17	P	M	25	0	0	0	0	0
18	A	F	47	0	0	1	1	1
19	P	F	31	0	0	0	0	0
20	A	M	20	1	1	0	1	0
21	A	M	26	0	1	0	1	0
22	A	M	46	1	1	1	1	1
23	A	M	32	1	1	1	1	1
24	A	M	48	0	1	0	0	0
.	.	.	.	.	.	.	.	.
.	.	.	.	.	.	.	.	.

The following SAS DATA step inputs the respiratory data and creates an observation for each response. The baseline and follow-up responses are actually measured on a five-point scale, from terrible to excellent, and this ordinal response is analyzed later in the chapter. For this analysis, the dichotomous outcome of whether the patient experienced good or excellent response is analyzed with a logistic regression. The second DATA step creates the SAS data set RESP2 and computes response variable DICHOT and dichotomous baseline variable DI_BASE. Note that the baseline variable, which was recorded on a five-point scale, could be managed as either ordinal or dichotomous.

```
data resp;
   input center id treatment $ sex $ age baseline
   visit1-visit4 @@;
   visit=1; outcome=visit1; output;
   visit=2; outcome=visit2; output;
   visit=3; outcome=visit3; output;
```

516 Chapter 15: Generalized Estimating Equations

```
         visit=4;  outcome=visit4;  output;
      datalines;
1   53 A F  32 1  2 2 4 2  2  30 A F  37 1  3 4 4 4
1   18 A F  47 2  2 3 4 4  2  52 A F  39 2  3 4 4 4
1   54 A M  11 4  4 4 4 2  2  23 A F  60 4  4 3 3 4
1   12 A M  14 2  3 3 3 2  2  54 A F  63 4  4 4 4 4
1   51 A M  15 0  2 3 3 3  2  12 A M  13 4  4 4 4 4
1   20 A M  20 3  3 2 3 1  2  10 A M  14 1  4 4 4 4
1   16 A M  22 1  2 2 2 3  2  27 A M  19 3  3 2 3 3
1   50 A M  22 2  1 3 4 4  2  16 A M  20 2  4 4 4 3
1    3 A M  23 3  3 4 4 3  2  47 A M  20 2  1 1 0 0
1   32 A M  23 2  3 4 4 4  2  29 A M  21 3  3 4 4 4

      ... more lines ...

1   46 P M  49 2  2 2 2 2  2  42 P M  66 3  3 3 4 4
1   47 P M  63 2  2 2 2 2
;
data resp2; set resp;
   dichot=(outcome=3 or outcome=4);
   di_base = (baseline=3 or baseline=4);
run;
```

While there are a number of explanatory variables in the model, including age, a continuous variable, the GEE analysis can handle them reasonably well with 111 clusters. The preliminary analysis includes all of the main effects plus the visit × treatment interaction and the treatment × center interaction. The exchangeable working correlation structure is thought to be a reasonable choice, so it is specified with the TYPE=EXCH option in the REPEATED statement.

```
proc genmod descending;
   class id center sex treatment visit;
   model dichot = treatment sex age center di_base
                  visit visit*treatment treatment*center/
                  link=logit dist=bin type3;
   repeated subject=id*center / type=exch;
run;
```

Output 15.25 displays the general model information. Because the DESCENDING option is specified, the probability that the response DICHOT is 1 is modeled.

Output 15.25 Model Information

Model Information	
Data Set	WORK.RESP2
Distribution	Binomial
Link Function	Logit
Dependent Variable	dichot

PROC GENMOD is modeling the probability that dichot='1'.

In the "GEE Model Information" table in Output 15.26, you can see that there are 111 clusters in the analysis, with all clusters having responses for each of the four visits.

Output 15.26 GEE Model Information

GEE Model Information	
Correlation Structure	Exchangeable
Subject Effect	id*center (111 levels)
Number of Clusters	111
Correlation Matrix Dimension	4
Maximum Cluster Size	4
Minimum Cluster Size	4

The Type 3 analysis displayed in Output 15.27 indicates that the two interaction terms are nonsignificant. The TREAT*VISIT interaction has a score test statistic of 3.10 and a p-value of 0.3760 with 3 df. The CENTER*TREATMENT interaction has a score test statistic value of 2.46 with a p-value of 0.1169 and 1 df.

Output 15.27 Type 3 Tests for Model with Interactions

Score Statistics For Type 3 GEE Analysis			
Source	DF	Chi-Square	Pr > ChiSq
treatment	1	12.85	0.0003
sex	1	0.24	0.6247
age	1	2.23	0.1351
center	1	3.32	0.0683
di_base	1	23.06	<.0001
visit	3	3.33	0.3429
treatment*visit	3	3.10	0.3760
center*treatment	1	2.46	0.1169

Output 15.28 displays a QIC value of 515.5623.

Output 15.28 QIC Fit Statistics

GEE Fit Criteria	
QIC	515.5623
QICu	504.5310

A reduced model is fit that eliminates the interactions.

```
proc genmod descending;
   class id center sex treatment visit;
   model dichot = center sex treatment age di_base
              visit / link=logit dist=bin type3;
   repeated subject=id*center / type=exch;
run;
```

Output 15.29 displays the resulting Type 3 analysis. Visit does not appear to be influential ($p = 0.3251$), and neither does sex ($p = 0.7565$) or age ($p = 0.1345$).

Output 15.29 Type 3 Tests for Reduced Model

Score Statistics For Type 3 GEE Analysis			
Source	DF	Chi-Square	Pr > ChiSq
center	1	3.24	0.0720
sex	1	0.10	0.7565
treatment	1	12.11	0.0005
age	1	2.24	0.1345
di_base	1	22.53	<.0001
visit	3	3.47	0.3251

The next model includes all of the main effects except for visit. Since sex and age were identified as covariates for this analysis ahead of time, they remain in the analysis. The following statements produce the desired GEE analysis:

```
proc genmod descending;
   class id center sex treatment;
   model dichot = center sex treatment age di_base
              / link=logit dist=bin type3;
   repeated subject=id*center / type=exch corrw;
run;
```

Output 15.30 displays the Type 3 tests for the final model.

Output 15.30 Type 3 Tests for Final Model

Score Statistics For Type 3 GEE Analysis			
Source	DF	Chi-Square	Pr > ChiSq
center	1	3.11	0.0780
sex	1	0.10	0.7562
treatment	1	12.52	0.0004
age	1	2.28	0.1312
di_base	1	22.97	<.0001

There is a very significant treatment effect. As seen in the parameter estimates table in Output 15.32, active treatment increases the odds of a good or excellent response. Baseline is also very influential, with a *p*-value of less than 0.0001. Sex and age remain nonsignificant, and center is marginally influential with a *p*-value of 0.0780.

The QIC value of 512.5723 displayed in Output 15.31 supports this choice of model.

Output 15.31 QIC Fit Statistics

GEE Fit Criteria	
QIC	512.5723
QICu	499.4873

Using the parameter estimates displayed in Output 15.32, you see that patients on active treatment have, on the average, $e^{1.2654} = 3.5$ times greater odds of a good or excellent response than those patients on placebo, adjusted for the other effects in the model.

Output 15.32 Parameter Estimates

| Analysis Of GEE Parameter Estimates ||||||||
| Empirical Standard Error Estimates ||||||||
Parameter		Estimate	Standard Error	95% Confidence Limits || Z	Pr > |Z|
Intercept		-0.2066	0.5776	-1.3388	0.9255	-0.36	0.7206
center	1	-0.6495	0.3532	-1.3418	0.0428	-1.84	0.0660
center	2	0.0000	0.0000	0.0000	0.0000	.	.
sex	F	0.1368	0.4402	-0.7261	0.9996	0.31	0.7560
sex	M	0.0000	0.0000	0.0000	0.0000	.	.
treatment	A	1.2654	0.3467	0.5859	1.9448	3.65	0.0003
treatment	P	0.0000	0.0000	0.0000	0.0000	.	.
age		-0.0188	0.0130	-0.0442	0.0067	-1.45	0.1480
di_base		1.8457	0.3460	1.1676	2.5238	5.33	<.0001

Note that visit is usually considered part of the design configuration and generally would be kept in the model, particularly in a clinical trials type of analysis. The design is balanced, and you would not gain that much precision by deleting effects such as visit, age, and sex. However, in the case of an observational study, in which the design was not planned, you will probably encounter collinearity in the predictors and may need to simplify your model to some extent in order to reduce the "noise" and make very real gains in precision. However, such simplification should not be excessive in order to avoid potential bias from overfitting.

The estimated exchangeable working correlation matrix is displayed in Output 15.33.

Output 15.33 Working Correlation Matrix

Working Correlation Matrix				
	Col1	Col2	Col3	Col4
Row1	1.0000	0.3270	0.3270	0.3270
Row2	0.3270	1.0000	0.3270	0.3270
Row3	0.3270	0.3270	1.0000	0.3270
Row4	0.3270	0.3270	0.3270	1.0000

There may be some interest in considering the unstructured working correlation matrix, since there are four visits per subject. This requires the estimation of more parameters, but that might be appropriate given that the model only contains five terms. The following PROC GENMOD invocation fits the same model but specifies the unstructured working correlation matrix. Note that using the unstructured correlation here, for four responses, requires you to ensure that your responses are in a consistent order, that is, the first observation in a cluster contains the first response, followed by the observation containing the second response, and so on. The DATA step used to create data set RESP on page 515 creates the proper ordering. However, if your data are not ordered correctly, then you need to create a variable that can be used by the GENMOD procedure to identify the correct sequence of responses. You use the WITHINSUBJECT option to specify that variable in the REPEATED statement.

```
proc genmod descending;
   class id center sex treatment;
   model dichot = center sex treatment age di_base
                / link=logit dist=bin type3;
   repeated subject=id*center / type=unstr corrw;
run;
```

Output 15.34 displays the estimated correlation matrix. You can see that there is reasonable homogeneity in the various visit-wise correlations.

Output 15.34 Unstructured Working Correlation Matrix

Working Correlation Matrix				
	Col1	Col2	Col3	Col4
Row1	1.0000	0.3351	0.2140	0.2953
Row2	0.3351	1.0000	0.4429	0.3581
Row3	0.2140	0.4429	1.0000	0.3964
Row4	0.2953	0.3581	0.3964	1.0000

Output 15.35 displays the parameter estimates that result from this model.

15.7. Respiratory Data

Output 15.35 Parameter Estimates for Unstructured Working Correlation Structure

Analysis Of GEE Parameter Estimates								
Empirical Standard Error Estimates								
Parameter			Estimate	Standard Error	95% Confidence Limits		Z	Pr > |Z|
Intercept			-0.2324	0.5763	-1.3620	0.8972	-0.40	0.6868
center	1		-0.6558	0.3512	-1.3442	0.0326	-1.87	0.0619
center	2		0.0000	0.0000	0.0000	0.0000	.	.
sex	F		0.1128	0.4408	-0.7512	0.9768	0.26	0.7981
sex	M		0.0000	0.0000	0.0000	0.0000	.	.
treatment	A		1.2442	0.3455	0.5669	1.9214	3.60	0.0003
treatment	P		0.0000	0.0000	0.0000	0.0000	.	.
age			-0.0175	0.0129	-0.0427	0.0077	-1.36	0.1728
di_base			1.8981	0.3441	1.2237	2.5725	5.52	<.0001

Compare these estimates to those in Output 15.32. The parameter estimates themselves are quite similar, and, while most of the standard errors are a little smaller for the unstructured correlation model, there really is very little gain in efficiency.

The similarity in results is further confirmed with the QIC statistic displayed in Output 15.36. The value 512.3416 is not very different from the value of 512.5723 displayed in Output 15.31.

Output 15.36 QIC Fit Statistics

GEE Fit Criteria	
QIC	512.3416
QICu	499.6081

Thus, for this example, your choice of working correlation structures depends on what you believe is most realistic.

If you have no idea of what to specify for your correlation structure, you might want to consider the independent working correlation matrix for these data. Many analysts regularly use the independent working structure with GEE analysis and don't attempt to postulate a correlation structure. They rely on the GEE properties that both the parameter estimates and their standard errors are consistent even if the correlation structure has not been correctly specified. They are not that concerned about the potential loss of efficiency. If you have a smaller number of clusters, you might consider simpler structures for the working correlation matrix since it would mean fewer parameters to estimate. If you have more missing data, you might want to consider the unstructured working correlation matrix since that would make a stronger argument for the MAR (missing at random) assumption of GEE analysis.

Output 15.37 displays the parameter estimates and standard errors that result when you repeat this analysis with the independent working correlation matrix. Again, the parameter estimates are very similar to those obtained by specifying the exchangeable and unstructured correlation structures, respectively.

Output 15.37 Parameter Estimates for Independent Working Correlation Structure

Analysis Of GEE Parameter Estimates							
Empirical Standard Error Estimates							
Parameter		Estimate	Standard Error	95% Confidence Limits		Z	Pr > |Z|
Intercept		-0.2066	0.5776	-1.3388	0.9255	-0.36	0.7206
center	1	-0.6495	0.3532	-1.3418	0.0428	-1.84	0.0660
center	2	0.0000	0.0000	0.0000	0.0000	.	.
sex	F	0.1368	0.4402	-0.7261	0.9996	0.31	0.7560
sex	M	0.0000	0.0000	0.0000	0.0000	.	.
treatment	A	1.2654	0.3467	0.5859	1.9448	3.65	0.0003
treatment	P	0.0000	0.0000	0.0000	0.0000	.	.
age		-0.0188	0.0130	-0.0442	0.0067	-1.45	0.1480
di_base		1.8457	0.3460	1.1676	2.5238	5.33	<.0001

Output 15.38 contains the QIC results, which are identical to those for the exchangeable working correlation structure.

Output 15.38 QIC Fit Statistics

GEE Fit Criteria	
QIC	512.5723
QICu	499.4873

For this example, which has complete data and a smaller number of clusters, as well as relatively small correlations, the use of the independent working correlation matrix is indicated. This use is also supported by the QIC fit statistic.

15.8 Diagnostic Data

The diagnostic data analyzed in Chapter 10 are now analyzed with the GEE method. Recall that subjects had assessments for test and standard procedures at two times, and researchers recorded response as positive or negative. Besides analyzing these assessments with conditional logistic regression and repeated measures WLS, you can also analyze these data with the GEE method. There are 793 clusters (corresponding to the number of subjects) and four measurements per subject (corresponding to the two types of tests at two times).

The following DATA steps input the diagnosis data and create an observation for each measurement so that the GEE facilities in the GENMOD procedure can be used. In addition, an indicator variable is created for time. The variable PROCEDURE takes the values of the standard or test procedures.

```
data diagnos;
    input std1 $ test1 $ std2 $ test2 $ count;
    do i=1 to count;
     output;
    end;
    datalines;
Neg Neg Neg Neg 509
Neg Neg Neg Pos   4
Neg Neg Pos Neg  17
Neg Neg Pos Pos   3
Neg Pos Neg Neg  13
Neg Pos Neg Pos   8
Neg Pos Pos Neg   0
Neg Pos Pos Pos   8
Pos Neg Neg Neg  14
Pos Neg Neg Pos   1
Pos Neg Pos Neg  17
Pos Neg Pos Pos   9
Pos Pos Neg Neg   7
Pos Pos Neg Pos   4
Pos Pos Pos Neg   9
Pos Pos Pos Pos 170
;
data diagnos2;
    set diagnos;
     drop std1 test1 std2 test2;
    subject=_n_;
    time=1; procedure='standard';
    response=std1; output;
     time=1; procedure='test';
    response=test1; output;
    time=2; procedure='standard';
     response=std2; output;
     time=2; procedure='test';
    response=test2; output;
run;
```

The model consists of time and procedure main effects as well as their interaction. The exchangeable working correlation structure is specified with the TYPE=EXCH option. Logistic regression is requested with the LINK=LOGIT and DIST=BIN options in the MODEL statement. The model is based on the probability of the positive response since the DESCENDING option is used in the PROC statement.

```
proc genmod descending;
   class subject time procedure;
   model response = time procedure time*procedure /
                    link=logit dist=bin type3;
   repeated subject=subject /type=exch;
run;
```

Output 15.39 displays the model information. Note that 'Pos' is the first ordered response value.

Output 15.39 Model Information

Model Information	
Data Set	WORK.DIAGNOS2
Distribution	Binomial
Link Function	Logit
Dependent Variable	response

PROC GENMOD is modeling the probability that response='Pos'.

Output 15.40 defines the parameters.

Output 15.40 Parameter Information

Parameter Information			
Parameter	Effect	time	procedure
Prm1	Intercept		
Prm2	time	1	
Prm3	time	2	
Prm4	procedure		standard
Prm5	procedure		test
Prm6	time*procedure	1	standard
Prm7	time*procedure	1	test
Prm8	time*procedure	2	standard
Prm9	time*procedure	2	test

The model information in Output 15.41 indicates that 793 clusters are analyzed with a cluster size of four and no missing data; the exchangeable working correlation structure is requested.

Output 15.41 GEE Model Information

GEE Model Information	
Correlation Structure	Exchangeable
Subject Effect	subject (793 levels)
Number of Clusters	793
Correlation Matrix Dimension	4
Maximum Cluster Size	4
Minimum Cluster Size	4

The score statistics in the Type 3 analysis in Output 15.42 indicate that the time × procedure interaction is not significant, using an $\alpha = 0.05$ criterion.

Output 15.42 Type 3 Test Results

Score Statistics For Type 3 GEE Analysis			
Source	DF	Chi-Square	Pr > ChiSq
time	1	0.91	0.3390
procedure	1	8.17	0.0043
time*procedure	1	2.49	0.1142

The reduced model fit consists of the main effects only.

```
proc genmod descending;
   class subject time procedure;
   model response = time procedure /
                 link=logit dist=bin type3;
   repeated subject=subject / type=exch corrw;
run;
```

The Type 3 analysis for the reduced model in Output 15.43 finds procedure significant with a chi-square value of 8.11 and 1 df ($p = 0.0044$).

Output 15.43 Reduced Model

Score Statistics For Type 3 GEE Analysis			
Source	DF	Chi-Square	Pr > ChiSq
time	1	0.85	0.3573
procedure	1	8.11	0.0044

The QIC statistics are displayed in Output 15.44.

Output 15.44 QIC Fit Statistics

GEE Fit Criteria	
QIC	3770.4988
QICu	3768.7541

Output 15.45 displays the parameter estimates and their test statistics. The first procedure, the standard, is associated with higher odds of getting the positive response as compared to the test treatment. The odds of the positive response with the standard procedure are $e^{0.1188}$ or 1.13 times higher than the odds for the test procedure.

Output 15.45 Parameter Estimates (Exchangeable R)

Analysis Of GEE Parameter Estimates							
Empirical Standard Error Estimates							
Parameter		Estimate	Standard Error	95% Confidence Limits		Z	Pr > |Z|
Intercept		-1.0173	0.0792	-1.1726	-0.8621	-12.84	<.0001
time	1	0.0313	0.0340	-0.0353	0.0978	0.92	0.3573
time	2	0.0000	0.0000	0.0000	0.0000	.	.
procedure	standard	0.1188	0.0415	0.0373	0.2002	2.86	0.0042
procedure	test	0.0000	0.0000	0.0000	0.0000	.	.

Output 15.46 contains the estimated exchangeable working correlation matrix. Note that with over 700 clusters and four measurements, you may want to specify the unstructured correlation matrix for a possible gain in precision of the estimates.

Output 15.46 Estimated Exchangeable Correlation

Working Correlation Matrix				
	Col1	Col2	Col3	Col4
Row1	1.0000	0.8041	0.8041	0.8041
Row2	0.8041	1.0000	0.8041	0.8041
Row3	0.8041	0.8041	1.0000	0.8041
Row4	0.8041	0.8041	0.8041	1.0000

When you resubmit the preceding PROC GENMOD statements with the following REPEATED statement, you obtain the parameter estimates with the unstructured correlation matrix.

```
proc genmod descending;
   class subject time procedure;
   model response = time procedure /
                    link=logit dist=bin type3;
   repeated subject=subject /type=unstr corrw;
run;
```

Output 15.47 displays these parameter estimates.

Output 15.47 Parameter Estimates (Unstructured R)

Analysis Of GEE Parameter Estimates							
Empirical Standard Error Estimates							
Parameter		Estimate	Standard Error	95% Confidence Limits		Z	Pr > |Z|
Intercept		-1.0208	0.0793	-1.1762	-0.8654	-12.88	<.0001
time	1	0.0344	0.0339	-0.0321	0.1009	1.01	0.3103
time	2	0.0000	0.0000	0.0000	0.0000	.	.
procedure	standard	0.1240	0.0414	0.0429	0.2052	3.00	0.0027
procedure	test	0.0000	0.0000	0.0000	0.0000	.	.

These have minimally smaller standard errors than those for the exchangeable structure. When you compare the estimated unstructured correlation matrix in Output 15.48 with the estimated exchangeable correlation matrix in Output 15.46, you can see that they are fairly similar.

Output 15.48 Estimated Unstructured Correlation

Working Correlation Matrix				
	Col1	Col2	Col3	Col4
Row1	1.0000	0.7855	0.8369	0.7763
Row2	0.7855	1.0000	0.7691	0.8560
Row3	0.8369	0.7691	1.0000	0.8163
Row4	0.7763	0.8560	0.8163	1.0000

Output 15.49 confirms that these models are very similar.

Output 15.49 QIC Fit Statistics

GEE Fit Criteria	
QIC	3770.5001
QICu	3768.7607

The results of the analyses for these data—a GEE analysis, the WLS repeated analysis, and the conditional logistic regression analysis—all provided similar conclusions. The WLS analysis focused on the marginal proportion of the negative response, the GEE analysis was a marginal analysis of the logit function, and the conditional logistic analysis was a subject-specific analysis of the logit function. The conditional analysis, through its subject-specific focus, found the standard procedure effect to be stronger, with the odds of positive response for the standard procedure being nearly twice the odds of positive response for the test procedure. However, note that the conditional analysis produces a subject-specific odds ratio, whereas the GEE odds ratio is a population-averaged odds ratio. Your choice of strategy depends on the overall objectives of the study analysis.

15.9 Using GEE for Count Data

Sometimes, categorical data come in the form of count data. For example, you may record the number of acute pain episodes in a time interval in a clinical trial evaluating treatments. Other examples might be the number of insurance claims registered during the year or the number of unscheduled medical visits made during a study of a new protocol for asthma medication. Often, Poisson regression is the appropriate strategy for analyzing such data. See Chapter 12, "Poisson Regression and Related Loglinear Models," for further discussion of Poisson regression.

Since Poisson regression is an application of the generalized linear model with the Poisson distribution and the log link function, you can fit models for clustered or repeated data with GEE methods. In this example, researchers evaluated a new drug to treat osteoporosis in women past menopause. In a double-blind study, a group of women were assigned the treatment and a group of women were assigned the placebo. Both groups of women were provided with calcium supplements, given nutritional counseling, and encouraged to be physically active through the exercise programs made available to them.

The study ran for three years, and the number of fractures occurring in each of those years was recorded. The length of each of the years, the corresponding risk periods, is 12 months. However, there were a few drop-outs in the third year, and those risk periods were set at 6 months. The offset variable is log of months at risk, as contained in the variable LMONTHS.

The following DATA step inputs the fracture data.

```
data fracture;
    input ID age center $ treatment $ year1 year2 year3 @@;
    total=year1+year2+year3;
    lmonths=log(12);
    datalines;
 1 56 A p 0 0 0    2 71 A p 1 0 0    3 60 A p 0 0 1    4 71 A p 0 1 0
 5 78 A p 0 0 0    6 67 A p 0 0 0    7 49 A p 0 0 0
 9 75 A p 1 0 0    8 68 A p 0 0 0   11 82 A p 0 0 0
13 56 A p 0 0 0   12 71 A p 0 0 0   15 66 A p 1 0 0
17 78 A p 0 0 0   16 63 A p 0 2 0   19 61 A p 0 0 0
21 75 A p 1 0 0   20 68 A p 0 0 0   23 63 A p 1 1 1
25 54 A p 0 0 0   24 65 A p 0 0 0   27 71 A p 0 0 0
29 56 A p 0 0 0   28 64 A p 0 0 0   31 78 A p 0 0 2
```

```
 33 76 A p 0 0 0    32 61 A p 0 0 0    35 76 A p 0 0 0
 37 74 A p 0 0 0    36 56 A p 0 0 0    39 62 A p 0 0 0
 41 56 A p 0 0 0    40 72 A p 0 0 1    43 76 A p 0 0 0
 45 75 A p 0 0 0    44 77 A p 2 2 0    47 78 A p 0 0 0
 49 71 A p 0 0 0    48 68 A p 0 0 0    51 74 A p 0 0 0
 53 69 A p 0 0 0    52 78 A p 1 0 0    55 81 A p 2 0 1
 57 68 A p 0 0 0    56 77 A p 0 0 0    59 77 A p 0 0 0
 61 75 A p 0 0 0    60 83 A p 0 0 0    63 72 A p 0 0 0    64 88 A p 0 0 0
 65 69 A p 0 0 0    66 55 A p 0 0 0    67 76 A p 0 0 0    68 55 A p 0 0 0
 69 63 A t 0 0 2    70 52 A t 0 0 0    71 56 A t 0 0 0    72 52 A t 0 0 0
 73 74 A t 0 0 0    74 61 A t 0 0 0    75 69 A t 0 0 0    76 61 A t 0 0 0
 77 84 A t 0 0 0    78 76 A t 0 1 0    79 59 A t 0 0 1    80 76 A t 0 0 0
 81 66 A t 0 0 1    82 78 A t 0 0 1    83 77 A t 0 0 0    84 75 A t 1 0 0
 85 75 A t 0 0 0    86 62 A t 0 0 0    87 67 A t 0 0 0    88 62 A t 0 0 0
 89 71 A t 0 0 0    90 63 A t 0 0 0                       92 68 A t 0 0 0
 93 69 A t 0 0 0    94 61 A t 0 0 0                       96 61 A t 0 0 0
 97 67 A t 0 0 0    98 77 A t 0 0 0  91 70 A t 0 0 1  102 81 A t 0 0 0
 95 49 A t 0 0 0   106 55 A t 0 0 0
 99 63 A t 2 1 0   100 52 A t 0 0 0  101 48 A t 0 0 0
103 71 A t 0 0 0   104 61 A t 0 0 0  105 74 A t 0 0 0
107 67 A t 0 0 0   108 56 A t 0 0 0  109 54 A t 0 0 0
111 56 A t 0 0 0   112 77 A t 1 0 0  113 65 A t 0 0 0
115 66 A t 0 0 0   116 71 A t 0 0 0  117 71 A t 0 0 0  128 71 A t 0 0 0
119 86 A t 1 0 0   120 81 A t 0 0 0  121 64 A t 0 0 0  132 76 A t 0 0 0
123 71 A t 0 0 0   124 76 A t 0 0 0  125 66 A t 0 0 0  136 76 A t 0 0 0
  1 68 B p 0 0 0     2 63 B p 0 0 0    3 66 B p 0 0 0    4 63 B p 0 0 0
  5 70 B p 0 1 0     6 62 B p 0 0 0    7 54 B p 1 0 0    8 66 B p 0 0 0
  9 71 B p 0 0 0    10 76 B p 0 0 0   11 72 B p 0 0 1   12 65 B p 0 1 0
 13 55 B p 0 1 0    14 59 B p 0 0 2   15 61 B p 1 0 0   16 56 B p 0 1 0
 17 54 B p 0 0 0    18 68 B p 0 0 0   19 68 B p 0 0 0   20 81 B p 0 0 0
 21 81 B p 1 0 0    22 61 B p 2 0 1   23 72 B p 1 0 0   24 67 B p 0 0 0
 25 56 B p 0 0 0    26 66 B p 0 0 0   27 71 B p 0 1 0   28 75 B p 0 1 0
 29 76 B p 0 0 0    30 73 B p 2 0 0   31 56 B p 0 0 0   32 89 B p 0 0 0
 33 56 B p 0 0 0    34 78 B p 0 0 0   35 55 B p 0 0 0   36 73 B p 0 0 1
 37 71 B p 0 0 0    38 56 B p 0 0 0   39 69 B p 0 0 0   40 77 B p 0 0 0
 41 89 B p 0 0 0    42 63 B p 0 0 0   43 67 B p 0 0 0   44 73 B p 0 0 0
 45 60 B p 0 0 0    46 67 B p 0 0 0   47 56 B p 0 0 0   48 78 B p 0 0 0
 49 73 B t 1 0 0    50 76 B t 0 0 0   51 61 B t 0 0 0   52 81 B t 0 0 0
 53 55 B t 0 0 0    54 82 B t 0 0 0   55 78 B t 0 0 0   56 60 B t 0 0 0
 57 56 B t 0 0 0    58 83 B t 0 0 0   59 55 B t 0 0 0   60 60 B t 0 0 0
 61 80 B t 0 0 0    62 78 B t 0 0 0   63 67 B t 0 0 0   64 67 B t 0 0 0
 65 56 B t 0 0 0    66 72 B t 0 0 0   67 71 B t 0 0 0   68 83 B t 0 0 0
 69 66 B t 0 0 0    70 71 B t 0 0 1   71 78 B t 1 0 2   72 61 B t 0 0 0
 73 56 B t 0 0 0    74 61 B t 0 0 0   75 55 B t 0 0 0   76 69 B t 1 1 0
 77 71 B t 0 0 0    78 76 B t 0 0 0   79 56 B t 0 0 0   80 75 B t 0 0 0
 81 89 B t 0 0 0    82 77 B t 0 0 0   83 77 B t 1 0 0   84 73 B t 0 0 0
 85 60 B t 0 0 0    86 61 B t 0 0 0   87 79 B t 0 0 0   88 71 B t 0 0 0
 89 61 B t 0 0 0    90 79 B t 0 0 0   91 87 B t 1 0 0   92 55 B t 0 0 0
 93 55 B t 0 0 0    94 79 B t 0 0 0   95 66 B t 0 0 0   96 49 B t 0 0 0
 97 56 B t 0 0 0    98 64 B t 0 0 0   99 88 B t 0 0 0  100 62 B t 1 0 0
101 80 B t 0 0 1   102 65 B t 0 0 0  103 57 B t 0 0 1  104 85 B t 0 0 0
;
```

The next DATA step creates one observation per year. It also sets the variable LMONTHS to a value of log(6) for the known drop-outs in the third year. Counts were recorded for that year for each subject, for the time they were still in the study.

```
data fracture2;
   set fracture;
   drop year1-year3;
   year=1; fractures=year1; output;
   year=2; fractures=year2; output;
   do; if center = A then do;
      if (ID=85 or ID=66 or ID=124 or ID=51) then lmonths=log(6); end;
      if center = B then do;
      if (ID=29 or ID=45 or ID=55) then lmonths=log(6); end;
   end;
   year=3; fractures=year3; output;
run;
```

The specification for the Poisson GEE analysis is straightforward. You specify the link function, LINK=LOG, and also specify the distribution with the DIST=POISSON option. The response variable is FRACTURES and the variables in the model include CENTER, TREATMENT, AGE, YEAR, the TREATMENT*CENTER interaction, and the TREATMENT*YEAR interaction. In addition, in Poisson regression you usually specify an offset variable. In this situation, the offset is the variable LMONTHS, which is the log length of time at risk in each year. The offset is specified with the OFFSET=LMONTHS option in the MODEL statement. Since there is not a unique subject identifier, you use the crossing of ID and CENTER with the SUBJECT= option to create unique values that determine the experimental units.

```
proc genmod;
   class id treatment center year;
   model fractures = center treatment age year treatment*center
                    treatment*year/
                    dist=poisson type3 offset=lmonths;
   repeated subject=id*center / type=exch corrw;
run;
```

Output 15.50 displays the general information about the model being fit: the Poisson distribution is requested, the offset variable is LMONTHS, and the response variable is FRACTURES.

Output 15.50 Model Information

Model Information	
Data Set	WORK.FRACTURE2
Distribution	Poisson
Link Function	Log
Dependent Variable	fractures
Offset Variable	lmonths

In Output 15.51, you can see that there are 214 clusters in the analysis, with all clusters having responses for each of the three years (even though some of the responses for the third year were for a reduced risk period).

15.9. Using GEE for Count Data

Output 15.51 GEE Model Information

GEE Model Information	
Correlation Structure	Exchangeable
Subject Effect	ID*center (214 levels)
Number of Clusters	214
Correlation Matrix Dimension	3
Maximum Cluster Size	3
Minimum Cluster Size	3

The Type 3 analysis displayed in Output 15.52 finds the two interaction terms to be nonsignificant. The treatment × center interaction has a score test statistic of 0.04 and a *p*-value of 0.8364 with 1 df. The treatment × year interaction has a score test statistic value of 3.15 with a *p*-value of 0.2074 and 2 df.

Output 15.52 Type 3 Tests for Model with Interactions

Score Statistics For Type 3 GEE Analysis			
Source	DF	Chi-Square	Pr > ChiSq
center	1	0.02	0.8750
treatment	1	4.69	0.0303
age	1	2.44	0.1180
year	2	7.64	0.0220
treatment*center	1	0.04	0.8364
treatment*year	2	3.15	0.2074

The same analysis was repeated with just the main effects. The nesting of ID and CENTER in the SUBJECT= option is just another way to specify a unique set of values with which to identify the individual experimental units. The CORRW option requests that the working correlation matrix be printed.

```
proc genmod;
   class id treatment center year;
   model fractures = center treatment age year /
                     dist=poisson type3 offset=lmonths;
   repeated subject=id(center) / type=exch corrw;
run;
```

Output 15.53 displays the resulting Type 3 analysis.

Output 15.53 Type 3 Tests for Reduced Model

Score Statistics For Type 3 GEE Analysis			
Source	DF	Chi-Square	Pr > ChiSq
center	1	0.02	0.8930
treatment	1	3.41	0.0647
age	1	2.22	0.1359
year	2	4.71	0.0948

Treatment is nearly significant here, with a *p*-value of 0.0647. Year also has some modest influence, with a *p*-value of 0.0948.

Output 15.54 contains the parameter estimates.

Output 15.54 Parameter Estimates

Analysis Of GEE Parameter Estimates							
Empirical Standard Error Estimates							
Parameter		Estimate	Standard Error	95% Confidence Limits || Z	Pr > |Z|
Intercept		-6.6379	1.1201	-8.8333	-4.4424	-5.93	<.0001
center	A	0.0400	0.2968	-0.5416	0.6216	0.13	0.8928
center	B	0.0000	0.0000	0.0000	0.0000	.	.
treatment	p	0.5715	0.3042	-0.0248	1.1678	1.88	0.0603
treatment	t	0.0000	0.0000	0.0000	0.0000	.	.
age		0.0223	0.0147	-0.0065	0.0512	1.52	0.1294
year	1	0.2763	0.2940	-0.2999	0.8524	0.94	0.3473
year	2	-0.3830	0.3747	-1.1173	0.3513	-1.02	0.3067
year	3	0.0000	0.0000	0.0000	0.0000	.	.

The placebo increases the log fracture rate by 0.5715; the test treatment lowers the log fracture rate by −0.5715.

Output 15.55 contains the estimate of the working correlation matrix. It indicates small, but not ignorable, correlations among the respective years.

Output 15.55 Working Correlation Matrix

	Working Correlation Matrix		
	Col1	Col2	Col3
Row1	1.0000	0.1049	0.1049
Row2	0.1049	1.0000	0.1049
Row3	0.1049	0.1049	1.0000

15.10 Fitting the Proportional Odds Model

Recall that the respiratory data analyzed in Section 15.7 contained an ordinal response that ranged from 0 for poor to 4 for excellent. (The responses were dichotomized in the previous analyses.) The proportional odds model provides a strategy that takes into account the ordinality of the data. See Chapter 9, "Logistic Regression II: Polytomous Response," for a discussion of the proportional odds model in the univariate case, and refer to Lipsitz, Kim, and Zhao (1994) and Miller, Davis, and Landis (1993) for discussions on fitting the proportional odds model with GEE.

The following statements request a proportional odds model to be fit with the GEE method. The SAS data set RESP is the same as created in Section 15.7.

```
proc genmod data=resp descending;
   class id center sex treatment visit;
   model outcome = treatment sex center age baseline
                   visit visit*treatment /
                   link=clogit dist=mult type3;
   repeated subject=id*center / type=ind;
run;
```

The variable OUTCOME has five levels, ranging from 0 to 4, for poor to excellent. Since interest lies in assessing how much better the subjects receiving the active treatment were, you form the cumulative logits that focus on the comparison of better to poorer outcomes. By default, the GENMOD procedure forms the cumulative logits based on the ratio of the probability of the lower ordered response values to the probability of the higher ordered response values. In this case, this would be poorer outcomes compared to better outcomes. To reverse this ordering, you simply specify the DESCENDING option in the PROC statement.

You specify the LINK=CLOGIT option to request the cumulative logit link and the DIST=MULT option to request the multinomial distribution. Together, these options specify the proportional odds model. The preliminary model includes TREATMENT, SEX, CENTER, AGE, BASELINE, VISIT, and the VISIT*TREATMENT interaction as the explanatory variables. Note that the BASELINE variable also lies on a 0–4 scale.

Since a unique patient identification requires the ID value and the CENTER value, you specify the SUBJECT=ID*CENTER option in the REPEATED statement. The TYPE=IND option specifies the independent working correlation matrix, which is currently the only correlation structure available with the ordinal response model. Output 15.56 displays the class level and response profile information. Since the ordered values are listed in descending order, the cumulative logits are modeling the better outcomes compared to the poorer outcomes.

Output 15.56 Class Level and Response Information

Class Level Information		
Class	Levels	Values
id	56	1 2 3 4 5 6 7 8 9 10 11 12 13 14 15 16 17 18 19 20 21 22 23 24 25 26 27 28 29 30 31 32 33 34 35 36 37 38 39 40 41 42 43 44 45 46 47 48 49 50 51 52 53 54 55 56
center	2	1 2
sex	2	F M
treatment	2	A P
visit	4	1 2 3 4

Response Profile		
Ordered Value	outcome	Total Frequency
1	4	152
2	3	96
3	2	116
4	1	40
5	0	40

The parameter information is displayed in Output 15.57.

15.10. Fitting the Proportional Odds Model

Output 15.57 Parameter Information

Parameter	Effect	center	sex	treatment	visit
Prm1	treatment			A	
Prm2	treatment			P	
Prm3	sex		F		
Prm4	sex		M		
Prm5	center	1			
Prm6	center	2			
Prm7	age				
Prm8	baseline				
Prm9	visit				1
Prm10	visit				2
Prm11	visit				3
Prm12	visit				4
Prm13	treatment*visit			A	1
Prm14	treatment*visit			A	2
Prm15	treatment*visit			A	3
Prm16	treatment*visit			A	4
Prm17	treatment*visit			P	1
Prm18	treatment*visit			P	2
Prm19	treatment*visit			P	3
Prm20	treatment*visit			P	4

From the GEE Model Information table in Output 15.58, you can see that there are 111 clusters, with four visit outcomes in each of the clusters.

Output 15.58 GEE Model Information

GEE Model Information	
Correlation Structure	Independent
Subject Effect	id*center (111 levels)
Number of Clusters	111
Correlation Matrix Dimension	4
Maximum Cluster Size	4
Minimum Cluster Size	4

The Type 3 analysis displayed in Output 15.59 indicates that the treatment × visit interaction is significant, at least at the $\alpha = 0.05$ level. Gender doesn't appear to be an important factor, and neither does age. Center also appears to be non-influential, but as a pre-stated design covariate, it stays in the model regardless.

Baseline and treatment (for visit 4) have strongly significant effects.

Output 15.59 Type 3 Test Results

Score Statistics For Type 3 GEE Analysis			
Source	DF	Chi-Square	Pr > ChiSq
treatment	1	15.33	<.0001
sex	1	0.53	0.4664
center	1	1.33	0.2481
age	1	2.68	0.1016
baseline	1	21.60	<.0001
visit	3	0.66	0.8837
treatment*visit	3	10.47	0.0150

The next PROC GENMOD invocation simplifies the model by excluding the AGE and SEX terms. Since CENTER is part of the study design, it remains in the model. The VISIT*TREATMENT term also stays.

```
proc genmod data=resp descending;
   class id center sex treatment visit;
   model outcome = treatment center baseline
                   visit visit*treatment /
                   link=clogit dist=mult type3;
   repeated subject=id*center / type=ind;
run;
```

The treatment × visit interaction remains important in this simplified model, as indicated in Output 15.60. Note that the analysis of the dichotomous outcome did not find the treatment × visit interaction to be noteworthy.

Output 15.60 Reduced Model

Score Statistics For Type 3 GEE Analysis			
Source	DF	Chi-Square	Pr > ChiSq
treatment	1	16.40	<.0001
center	1	1.25	0.2636
baseline	1	21.27	<.0001
visit	3	0.54	0.9106
treatment*visit	3	10.50	0.0148

Baseline and treatment (for visit 4) remain extremely significant in this reduced model, with p-values less than 0.001.

Output 15.61 contains the parameter estimates.

Output 15.61 Parameter Estimates

Analysis Of GEE Parameter Estimates								
Empirical Standard Error Estimates								
Parameter			Estimate	Standard Error	95% Confidence Limits		Z	Pr > |Z|
Intercept1			-3.3645	0.5766	-4.4945	-2.2345	-5.84	<.0001
Intercept2			-2.2049	0.5412	-3.2657	-1.1441	-4.07	<.0001
Intercept3			-0.6060	0.5193	-1.6239	0.4119	-1.17	0.2433
Intercept4			0.2929	0.5643	-0.8131	1.3988	0.52	0.6037
treatment	A		0.9995	0.3625	0.2891	1.7100	2.76	0.0058
treatment	P		0.0000	0.0000	0.0000	0.0000	.	.
center	1		-0.3491	0.3023	-0.9415	0.2434	-1.15	0.2482
center	2		0.0000	0.0000	0.0000	0.0000	.	.
baseline			0.8993	0.1670	0.5719	1.2266	5.38	<.0001
visit	1		0.2581	0.2501	-0.2321	0.7484	1.03	0.3021
visit	2		-0.2505	0.2303	-0.7019	0.2010	-1.09	0.2768
visit	3		-0.0360	0.1615	-0.3525	0.2806	-0.22	0.8238
visit	4		0.0000	0.0000	0.0000	0.0000	.	.
treatment*visit	A	1	-0.3049	0.3927	-1.0746	0.4648	-0.78	0.4375
treatment*visit	A	2	0.7247	0.3547	0.0296	1.4198	2.04	0.0410
treatment*visit	A	3	0.2990	0.3321	-0.3519	0.9500	0.90	0.3679
treatment*visit	A	4	0.0000	0.0000	0.0000	0.0000	.	.
treatment*visit	P	1	0.0000	0.0000	0.0000	0.0000	.	.
treatment*visit	P	2	0.0000	0.0000	0.0000	0.0000	.	.
treatment*visit	P	3	0.0000	0.0000	0.0000	0.0000	.	.
treatment*visit	P	4	0.0000	0.0000	0.0000	0.0000	.	.

15.11 GEE Analyses for Data with Missing Values

One of the main advantages of the GEE method is that it addresses the possibility of missing values. The number of responses per subject, or cluster, can vary; recall that you can have t_i

538 Chapter 15: Generalized Estimating Equations

responses per subject, where t_i depends on the ith subject. While the data sets analyzed in previous sections were complete, or balanced, you are faced with missing data in many situations, especially for observational data that are longitudinal. Loss to follow-up is a common problem for planned studies that involve repeated visits. The GEE method works nicely for many of these data situations. However, the GEE method does assume that the missing values are missing completely at random, or MCAR.

15.11.1 Crossover Study with Missing Data

Consider a two-period crossover study on treatments for a skin disorder where patients were given sequences of the standard drug A, a new drug B, and a placebo. Investigators introduced a skin irritant and then applied topical treatments. Subjects were stratified by gender. The first session included 300 patients, but 50 patients failed to attend the second session one week later. Investigators determined that none of the losses to follow-up were actually due to the failure of the treatments, but they were due to the usual attrition plus a breakdown in the communication to emphasize the importance of the return visit. For this reason, although much more missing data occurred than was expected, the analysis proceeded.

You can analyze these data in a similar manner to the way in which the crossover data were analyzed in Section 15.6. The design is exactly the same; the only difference is that the 50 subjects with only one measurement have a cluster size of 1. Note the number of missing values for the second period. These data are input into SAS data set SKINCROSS.

```
data skincross;
    input subject gender $ sequence $ Time1 $ Time2 $ @@;
    datalines;
 1  m AB  Y  Y    101 m PA  Y  Y    201 f AP  Y  Y
 2  m AB  Y  .    102 m PA  Y  Y    202 f AP  Y  Y
 3  m AB  Y  Y    103 m PA  Y  Y    203 f AP  Y  Y
 4  m AB  Y  .    104 m PA  Y  Y    204 f AP  Y  Y
 5  m AB  Y  Y    105 m PA  Y  Y    205 f AP  Y  Y
 6  m AB  Y  .    106 m PA  Y  N    206 f AP  Y  Y
 7  m AB  Y  .    107 m PA  Y  .    207 f AP  Y  Y
 8  m AB  Y  Y    108 m PA  Y  N    208 f AP  Y  Y
 9  m AB  Y  Y    109 m PA  N  .    209 f AP  Y  Y
10  m AB  Y  Y    110 m PA  N  Y    210 f AP  Y  Y
11  m AB  Y  .    111 m PA  N  Y    211 f AP  Y  Y
12  m AB  Y  Y    112 m PA  N  Y    212 f AP  Y  Y
13  m AB  Y  N    113 m PA  N  .    213 f AP  Y  Y
14  m AB  Y  N    114 m PA  N  .    214 f AP  Y  .
15  m AB  Y  N    115 m PA  N  Y    215 f AP  Y  .
16  m AB  Y  N    116 m PA  N  Y    216 f AP  Y  .
17  m AB  Y  N    117 m PA  N  Y    217 f AP  Y  Y
18  m AB  Y  N    118 m PA  N  Y    218 f AP  Y  Y
19  m AB  Y  .    119 m PA  N  Y    219 f AP  Y  Y
20  m AB  Y  N    120 m PA  N  Y    220 f AP  Y  Y
21  m AB  Y  N    121 m PA  N  Y    221 f AP  Y  .
22  m AB  Y  N    122 m PA  N  Y    222 f AP  Y  Y
23  m AB  Y  .    123 m PA  N  Y    223 f AP  Y  Y
24  m AB  Y  N    124 m PA  N  Y    224 f AP  Y  Y
```

```
   ... more lines ...
99  m BP    N   N   199 f   BA   N   N   299 f   PB   N   N
100 m BP    N   .   200 f   BA   N   N   300 f   PB   N   N
;
```

The next step manipulates the data the same way as in Section 15.6. The DATA step creates observations for each period and creates indicator variables for the carryover effects.

```
data skincross2;
   set skincross;
   period=1;
   treatment=substr(sequence, 1, 1);
   carry='N';
   response=Time1;
   output;
   period=2;
   Treatment=substr(sequence, 2, 1);
   carry = substr(sequence, 1, 1);
   if carry='P' then carry='N';
   response=Time2;
   output;
run;
```

The following PROC GENMOD invocation requests a GEE analysis for a model including effects for treatment, period, gender, carryover, and the period × gender interaction. The DESCENDING option specifies that the probability of a 'yes' response is to be modeled. Logistic regression is used along with the exchangeable working correlation structure.

```
proc genmod data=skincross2 descending;
   class subject treatment period gender carry;
   model response = treatment period gender carry
                   gender*period /type3
                   dist=bin link=logit;
   repeated subject=subject / type=exch;
run;
```

The "GEE Model Information" table in Output 15.62 shows that there are 300 clusters total and 50 clusters with missing data; 250 clusters have two measurements and 50 clusters have only one measurement corresponding to the first period.

Output 15.62 GEE Model Information

GEE Model Information	
Correlation Structure	Exchangeable
Subject Effect	subject (300 levels)
Number of Clusters	300
Clusters With Missing Values	50
Correlation Matrix Dimension	2
Maximum Cluster Size	2
Minimum Cluster Size	1

The Type 3 analysis displayed in Output 15.63 suggests that the carryover effect is not influential.

Output 15.63 Type 3 Analysis

Score Statistics For Type 3 GEE Analysis			
Source	DF	Chi-Square	Pr > ChiSq
treatment	2	29.38	<.0001
period	1	7.11	0.0077
gender	1	29.94	<.0001
carry	2	1.08	0.5841
period*gender	1	4.21	0.0401

The next model fit includes the treatment, period, and gender main effects as well as the gender × period interaction. The ESTIMATE statement specifies that odds ratio estimates be computed to compare the effect of drug A to the placebo effect, the effect of drug B to the placebo effect, and the effect of drug A to the effect of drug B.

```
proc genmod data=skincross2 descending;
    class subject treatment period gender;
    model response = treatment period gender gender*period
                /type3
                 dist=bin link=logit;
    repeated subject=subject / type=exch;
    estimate 'OR:A-B' treatment 1 -1 0 /exp;
    estimate 'OR:A-P' treatment 1 0 -1 / exp;
    estimate 'OR:B-P' treatment 0 1 -1 / exp;
run;
```

The main effects and gender × period interaction remain important, as indicated in Output 15.64.

Output 15.64 Type 3 Analysis

Score Statistics For Type 3 GEE Analysis			
Source	DF	Chi-Square	Pr > ChiSq
treatment	2	40.02	<.0001
period	1	20.97	<.0001
gender	1	28.89	<.0001
period*gender	1	3.93	0.0474

The parameter estimates are displayed in Output 15.65.

Output 15.65 Parameter Estimates

Analysis Of GEE Parameter Estimates								
Empirical Standard Error Estimates								
Parameter			Estimate	Standard Error	95% Confidence Limits		Z	Pr > |Z|
Intercept			-0.9287	0.2249	-1.3696	-0.4879	-4.13	<.0001
treatment	A		1.2622	0.2079	0.8548	1.6696	6.07	<.0001
treatment	B		0.1722	0.2141	-0.2473	0.5918	0.80	0.4210
treatment	P		0.0000	0.0000	0.0000	0.0000	.	.
period	1		-0.4520	0.2257	-0.8944	-0.0095	-2.00	0.0453
period	2		0.0000	0.0000	0.0000	0.0000	.	.
gender	f		1.4443	0.2816	0.8925	1.9961	5.13	<.0001
gender	m		0.0000	0.0000	0.0000	0.0000	.	.
period*gender	1	f	-0.6505	0.3289	-1.2951	-0.0059	-1.98	0.0480
period*gender	1	m	0.0000	0.0000	0.0000	0.0000	.	.
period*gender	2	f	0.0000	0.0000	0.0000	0.0000	.	.
period*gender	2	m	0.0000	0.0000	0.0000	0.0000	.	.

The odds ratio estimates displayed in Output 15.66 show that subjects receiving drug A had almost 3.5 higher odds of improvement as subjects on the placebo. Subjects receiving drug A have almost three times higher odds of improvement as subjects receiving drug B. The odds ratio estimate comparing the odds of drug B and placebo is 1.1880 and its confidence limits contain the value 1. Subjects on drug B did no better than subjects on the placebo.

Output 15.66 Odds Ratio Estimates

		Contrast Estimate Results						
		Mean					L'Beta	
Label	Mean Estimate	Confidence Limits		L'Beta Estimate	Standard Error	Alpha	Confidence Limits	
OR:A-B	0.7484	0.6593	0.8205	1.0899	0.2193	0.05	0.6601	1.5198
Exp(OR:A-B)				2.9741	0.6523	0.05	1.9350	4.5713
OR:A-P	0.7794	0.7016	0.8415	1.2622	0.2079	0.05	0.8548	1.6696
Exp(OR:A-P)				3.5331	0.7344	0.05	2.3508	5.3099
OR:B-P	0.5430	0.4385	0.6438	0.1722	0.2141	0.05	-0.2473	0.5918
Exp(OR:B-P)				1.1880	0.2543	0.05	0.7809	1.8072

Contrast Estimate Results		
Label	Chi-Square	Pr > ChiSq
OR:A-B	24.70	<.0001
Exp(OR:A-B)		
OR:A-P	36.87	<.0001
Exp(OR:A-P)		
OR:B-P	0.65	0.4210
Exp(OR:B-P)		

15.12 Alternating Logistic Regression

There are some limitations of the correlation approach to fit models to binary data. The data influence the range of the correlation since the estimates of r_{jk} are constrained by the means, $\mu_{ij} = \Pr\{y_{ij} = 1\}$. Consider:

$$\text{Corr}(Y_{ij}, Y_{ik}) = r_{jk} = \frac{Pr(Y_{ij} = 1, Y_{ik} = 1) - \mu_{ij}\mu_{ik}}{\sqrt{\mu_{ij}(1 - \mu_{ij})\mu_{ik}(1 - \mu_{ik})}}$$

The odds ratio appears to be a more natural choice for modeling the association in binary data as they are not constrained by the means.

$$\text{OR}(Y_{ij}, Y_{ik}) = \frac{Pr(Y_{ij} = 1, Y_{ik} = 1)Pr(Y_{ij} = 0, Y_{ik} = 0)}{Pr(Y_{ij} = 1, Y_{ik} = 0)Pr(Y_{ij} = 0, Y_{ik} = 1)}$$

15.12. Alternating Logistic Regression

In GEE, the correlations are treated as nuisance parameters, and the use of correlations versus odds ratios usually has little influence on inference on β, the regression parameters for the marginal mean model.

In some applications, you may want your analysis to focus on both regressing the outcome on the explanatory variables and describing the association between the outcomes. The generalized estimating equations discussed in this chapter are known as the first-order estimating equations, and they are efficient for the estimation of β but not necessarily efficient in the estimation of the association parameters, which are the correlations estimated in PROC GENMOD with the method of moments. Prentice (1988) describes second-order estimating equations and the simultaneous modeling of the responses and all pairwise products as a method of producing more efficient estimation of the association parameters. However, as the number of clusters grows large, this method can become computationally infeasible.

Carey, Zeger, and Diggle (1993) describe the alternating logistic regression (ALR) algorithm, which provides a means of both fitting the first-order GEE model and simultaneously modeling the association in a manner that produces relatively efficient estimators. With this method, you obtain $\hat{\beta}$ and also obtain estimates of the association parameters that relate to log odds ratios, as well as their standard errors and confidence intervals.

The ALR algorithm models the log of the odds ratio as

$$\psi_{ijk} = z'\alpha$$

where $\psi_{ijk} = \log(OR(y_{ij}, y_{ik}))$, α is a $q \times 1$ vector of regression parameters, and z_{ijk} is a fixed vector of coefficients. The method switches between the first-order GEE estimation of the β and a modified (with offset) logistic regression estimate of the α until convergence, updating the GEE with product-moments from the newly estimated OR, and then updating the offsets in the association model with the new $\hat{\beta}$s. Thus, you are applying alternating logistic regressions, one for α and one for β.

There are numerous choices for modeling the log odds ratio: you can choose to specify the log odds ratio as a constant across clusters; for pairs (j, k), you can specify that the log odds ratio is a constant within different levels of a blocking factor such as clinics; and you can specify fully parameterized clusters in which each cluster is parameterized the same way. There are numerous other possibilities for model structures for the log odds ratio. For more information on the motivation and the details of the ALR approach, refer to Carey, Zeger, and Diggle (1993), Lipsitz, Laird, and Harrington (1991), and Firth (1992).

The GENMOD procedure produces the ALR algorithm for binary data. The following log odds ratio structures are available:

- exchangeable (constant over all clusters)
- covariate (block effect)
- fully parameterized within cluster (parameter for each pair)
- nested (one parameter for pairs within same subcluster, one for between subclusters)
- user-specified Z-matrix

The ALR algorithm provides a reasonable approach when the focus of your analysis is estimating association as much as modeling the response; this method provides estimates of association with more efficiency than the usual GEE method. The resulting parameter estimates β are consistent (Pickles 1998) and so is the estimated covariance matrix of β. That is, you retain the robustness properties of the first-order GEE even if the association structure is misspecified. A possible limitation of ALR is that no covariance estimator has been suggested that is analogous to the model-based estimator of GEE.

15.12.1 Respiratory Data

You may recall the respiratory data analyzed in previous sections of this book. In this example, the ALR algorithm is applied in modeling the dichotomous outcome of whether respiratory symptoms were good or excellent at the four visits. The first analysis models the log odds ratios as exchangeable: in this case, α is the common log odds ratio.

$$\log(\text{OR}(Y_{ij}, Y_{ik})) = \alpha \quad \text{for all } i, j \neq k$$

The following SAS statements produce this analysis. The DATA step creating RESP2 is listed on page 515. To specify the ALR algorithm, you include the LOGOR option in the REPEATED statement. Here, the exchangeable structure for the log odds ratio is requested.

```
proc genmod data=resp2 descending;
   class id treatment sex center visit;
   model dichot = center sex treatment age di_base visit
                 / dist=bin type3 link=logit;
   repeated subject=id*center / logor=exch;
run;
```

Output 15.67 contains information about the GEE modeling; it tells you that the log OR structure is exchangeable.

Output 15.67 GEE Model Information

GEE Model Information	
Log Odds Ratio Structure	Exchangeable
Subject Effect	id*center (111 levels)
Number of Clusters	111
Correlation Matrix Dimension	4
Maximum Cluster Size	4
Minimum Cluster Size	4

Output 15.68 contains Type 3 tests; the results are very similar to those obtained in the first-order GEE based on the exchangeable structure defined with the Pearson correlations (Output 15.29).

Output 15.68 Type 3 Analysis

Score Statistics For Type 3 GEE Analysis			
Source	DF	Chi-Square	Pr > ChiSq
center	1	3.13	0.0767
sex	1	0.09	0.7642
treatment	1	12.56	0.0004
age	1	2.03	0.1542
di_base	1	22.48	<.0001
visit	3	2.99	0.3932

Output 15.68 displays the parameter estimates for the elements of β. The estimates are also similar to those obtained in the standard GEE model; the parameter labeled "Alpha 1" is the estimate of the common log odds ratio and has the value 1.7524. Note that you need to interpret the "Alpha" estimates somewhat cautiously since they assume their model specification is correct (as compared to the estimates of β, which are robust).

Output 15.69 Parameter Estimates

Analysis Of GEE Parameter Estimates								
Empirical Standard Error Estimates								
Parameter			Estimate	Standard Error	95% Confidence Limits		Z	Pr > |Z|
Intercept			-0.4137	0.5760	-1.5428	0.7153	-0.72	0.4726
center	1		-0.6590	0.3517	-1.3483	0.0303	-1.87	0.0610
center	2		0.0000	0.0000	0.0000	0.0000	.	.
sex	F		0.1329	0.4365	-0.7226	0.9884	0.30	0.7608
sex	M		0.0000	0.0000	0.0000	0.0000	.	.
treatment	A		1.2696	0.3432	0.5969	1.9423	3.70	0.0002
treatment	P		0.0000	0.0000	0.0000	0.0000	.	.
age			-0.0180	0.0127	-0.0429	0.0068	-1.42	0.1552
di_base			1.8381	0.3439	1.1642	2.5121	5.35	<.0001
visit	1		0.3138	0.2494	-0.1751	0.8027	1.26	0.2084
visit	2		0.1065	0.2409	-0.3657	0.5786	0.44	0.6585
visit	3		0.3269	0.2314	-0.1266	0.7804	1.41	0.1577
visit	4		0.0000	0.0000	0.0000	0.0000	.	.
Alpha1			1.7524	0.2767	1.2102	2.2947	6.33	<.0001

Another approach with the ALR strategy is to estimate a separate log odds ratio for each center. The following PROC GENMOD statements produce that analysis. The LOGOR=LOGORVAR(CENTER) specifies that each center has its own log odds ratio.

```
proc genmod data=resp2 descending;
    class id treatment sex center visit;
    model dichot = center sex treatment age di_base visit
                 / dist=bin type3 link=logit;
    repeated subject=id*center / logor=logorvar(center) corrw;
run;
```

Output 15.70 indicates that, this time, the exchangeable structure for the log OR is based on the CENTER variable in this specification.

Output 15.70 GEE Model Information

GEE Model Information	
Log Odds Ratio Covariate	center
Subject Effect	id*center (111 levels)
Number of Clusters	111
Correlation Matrix Dimension	4
Maximum Cluster Size	4
Minimum Cluster Size	4

Output 15.71 is produced when the log OR structure has more than one α parameter; this table lists the group(center) levels associated with the log OR parameters. In this case, there is a common log OR for all clusters in Center 1 and a common log OR for all clusters in Center 2.

Output 15.71 Log Odds Ratio Parameter Information

Log Odds Ratio Parameter Information	
Parameter	Group
Alpha1	1
Alpha2	2

Output 15.72 contains Type 3 tests; the results are still similar to those obtained in the usual GEE model.

Output 15.72 Type 3 Analysis

Score Statistics For Type 3 GEE Analysis			
Source	DF	Chi-Square	Pr > ChiSq
center	1	2.96	0.0852
sex	1	0.12	0.7325
treatment	1	12.31	0.0004
age	1	2.32	0.1276
di_base	1	22.53	<.0001
visit	3	2.77	0.4292

Output 15.73 displays the parameter estimates.

Output 15.73 Parameter Estimates

Analysis Of GEE Parameter Estimates							
Empirical Standard Error Estimates							
Parameter		Estimate	Standard Error	95% Confidence Limits		Z	Pr > |Z|
Intercept		-0.3729	0.5693	-1.4887	0.7430	-0.65	0.5125
center	1	-0.6423	0.3527	-1.3335	0.0489	-1.82	0.0686
center	2	0.0000	0.0000	0.0000	0.0000	.	.
sex	F	0.1506	0.4318	-0.6958	0.9969	0.35	0.7273
sex	M	0.0000	0.0000	0.0000	0.0000	.	.
treatment	A	1.2392	0.3396	0.5735	1.9048	3.65	0.0003
treatment	P	0.0000	0.0000	0.0000	0.0000	.	.
age		-0.0194	0.0126	-0.0440	0.0052	-1.54	0.1226
di_base		1.9210	0.3396	1.2555	2.5866	5.66	<.0001
visit	1	0.3017	0.2511	-0.1904	0.7937	1.20	0.2295
visit	2	0.0640	0.2398	-0.4060	0.5341	0.27	0.7895
visit	3	0.2782	0.2310	-0.1745	0.7310	1.20	0.2284
visit	4	0.0000	0.0000	0.0000	0.0000	.	.
Alpha1		1.3677	0.3191	0.7423	1.9930	4.29	<.0001
Alpha2		2.0886	0.4349	1.2362	2.9410	4.80	<.0001

The parameter labeled "Alpha 1" is the common log odds ratio for the Center 1 subjects; the parameter labeled "Alpha 2" is the common log odds ratio for the Center 2 subjects. In Center

1, the common log odds ratio is 1.3677 with a 95% confidence interval of (0.7423, 1.9930). For Center 2, the common log odds ratio is 2.0886 with a 95% confidence interval of (1.2362, 2.9410). The association would appear to be slightly stronger in Center 2.

Note that inserting the following statement in the previous PROC GENMOD invocation requests a fully parameterized cluster model for the log odds ratio parameters:

```
proc genmod data=resp2 descending;
    class id treatment sex center visit;
    model dichot = center sex treatment age di_base visit
                 / dist=bin type3 link=logit;
    repeated subject=id*center / logor=fullclust;
run;
```

Information about what the parameters mean is presented in Output 15.74.

Output 15.74 Log OR Parameter Information

Log Odds Ratio Parameter Information	
Parameter	Group
Alpha1	(1, 2)
Alpha2	(1, 3)
Alpha3	(1, 4)
Alpha4	(2, 3)
Alpha5	(2, 4)
Alpha6	(3, 4)

The estimated parameters are presented in Output 15.75.

Output 15.75 Parameter Estimates

Analysis Of GEE Parameter Estimates

Empirical Standard Error Estimates

Parameter		Estimate	Standard Error	95% Confidence Limits		Z	Pr > |Z|
Intercept		-0.4890	0.5745	-1.6151	0.6371	-0.85	0.3947
center	1	-0.6439	0.3484	-1.3268	0.0390	-1.85	0.0646
center	2	0.0000	0.0000	0.0000	0.0000	.	.
sex	F	0.0982	0.4344	-0.7533	0.9497	0.23	0.8212
sex	M	0.0000	0.0000	0.0000	0.0000	.	.
treatment	A	1.2498	0.3407	0.5821	1.9175	3.67	0.0002
treatment	P	0.0000	0.0000	0.0000	0.0000	.	.
age		-0.0161	0.0126	-0.0407	0.0085	-1.28	0.2006
di_base		1.8905	0.3406	1.2230	2.5579	5.55	<.0001
visit	1	0.3123	0.2505	-0.1786	0.8033	1.25	0.2125
visit	2	0.1072	0.2418	-0.3668	0.5812	0.44	0.6576
visit	3	0.3160	0.2320	-0.1387	0.7708	1.36	0.1732
visit	4	0.0000	0.0000	0.0000	0.0000	.	.
Alpha1		1.6230	0.4961	0.6507	2.5953	3.27	0.0011
Alpha2		1.0681	0.4831	0.1213	2.0149	2.21	0.0270
Alpha3		1.6292	0.4858	0.6771	2.5814	3.35	0.0008
Alpha4		2.1529	0.5106	1.1522	3.1535	4.22	<.0001
Alpha5		1.8752	0.4676	0.9588	2.7916	4.01	<.0001
Alpha6		2.1839	0.5101	1.1842	3.1837	4.28	<.0001

The relative magnitudes of these six estimated log odds ratios have a pattern similar to that of the unstructured working correlation estimates presented on page 520, and this is what you expect. This pattern seems consistent with exchangeable structure.

15.13 Using GEE to Account for Overdispersion: Univariate Outcome

Section 8.2.7 mentions overdispersion in the case of logistic regression. Overdispersion occurs when the observed variance is larger than the nominal variance for a particular distribution. It occurs with some regularity in the analysis of proportions and discrete counts. This is not surprising for the assumed distributions (binomial and Poisson, respectively) because their respective variances are fixed by a single parameter, the mean. Overdispersion can have a major impact on inference so it needs to be taken into account. Underdispersion also occurs. See McCullagh and Nelder (1989) and Dean (1998) for more detail on overdispersion.

One way to manage overdispersion is to assume a more flexible distribution, such as the negative binomial in the case of overdispersed Poisson data, as illustrated in Chapter 12. You can also adjust the covariance matrix of a Poisson-based analysis with a scaling factor, which is the method PROC GENMOD uses with the SCALE= option in the MODEL statement. Then the covariance matrix is pre-multiplied by the scaling factor ϕ, and the scaled deviance and the log likelihood ratio tests are divided by ϕ.

Another way of managing the overdispersion is to take the generalized estimating approach. Recall that the robust, or empirical, covariance matrix estimated by the GEE method is robust to the misspecification of the covariance structure, and misspecification is occurring in the case of overdispersion. The variance is not "acting" as it should; it does not take the form for data from a Poisson distribution. With GEE estimation, you are using a subject-to-subject measure for variance estimation instead of a model-based one. The robustness comes from the fact that the variance estimation process involves aggregates at the cluster level. While GEE was devised for the analysis of correlated data with more than one response per subject, you can also use it for the analysis of single outcomes and derive the benefits of the robust standard errors. This section describes the use of GEE for adjusting for overdispersion in the univariate case.

Researchers studying the incidence of lower respiratory illness in infants took repeated observations of infants over one year. They studied 284 children and examined them every two weeks. Explanatory variables evaluated included passive smoking (one or more smokers in the household), socioeconomic status, and crowding. See LaVange et al. (1994) for more information on the study and a discussion of the analysis of incidence densities. One outcome of interest was the total number of times, or counts, of lower respiratory infection recorded for the year. The strategy was to model these counts with Poisson regression. However, it is reasonable to expect some overdispersion since the children that have an infection are more likely to have other infections.

The following DATA step inputs the data into a SAS data set named LRI. The variable COUNT is the total number of infections that year, and the variable RISK is the number of weeks during that year for which the child is considered at risk (when a lower respiratory infection is ongoing, the child is not considered to be at risk for a new one). The variable CROWDING is an indicator variable for whether crowded conditions occur in the household, and SES is an indicator variable for whether the family's socioeconomic status was considered low (0), medium (1), or high (2). The variable RACE is an indicator variable for whether the child was white (1) or not (0), and the variable PASSIVE is an indicator variable for whether the child was exposed to cigarette smoking. Finally, the AGEGROUP variable takes the values 1, 2, and 3 for under four, four to six, or more than six months.

```
data lri;
    input id count risk passive crowding ses agegroup race @@;
    logrisk =log(risk/52);
    datalines;
1 0 42 1 0 2 2 0   96 1 41 1 0 1 2 0    191 0 44 1 0 0 2 0
2 0 43 1 0 0 2 0   97 1 26 1 1 2 2 0    192 0 45 0 0 0 2 1
3 0 41 1 0 1 2 0   98 0 36 0 0 0 2 0    193 0 42 0 0 0 2 0
```

15.13. Using GEE to Account for Overdispersion: Univariate Outcome

```
 4 1 36 0 1 0 2 0    99 0 34 0 0 0 2 0    194 1 31 0 0 0 2 1
 5 1 31 0 0 0 2 0   100 1  3 1 1 2 3 1    195 0 35 0 0 0 2 0
 6 0 43 1 0 0 2 0   101 0 45 1 0 0 2 0    196 1 35 1 0 0 2 0
 7 0 45 0 0 0 2 0   102 0 38 0 0 1 2 0    197 1 27 1 0 1 2 0
 8 0 42 0 0 0 2 1   103 0 41 1 1 1 2 1    198 1 33 0 0 0 2 0
 9 0 45 0 0 0 2 1   104 1 37 0 1 0 2 0    199 0 39 1 0 1 2 0
10 0 35 1 1 0 2 0   105 0 40 0 0 0 2 0    200 3 40 0 1 2 2 0
11 0 43 0 0 0 2 0   106 1 35 1 0 0 2 0    201 4 26 1 0 1 2 0
12 2 38 0 0 0 2 0   107 0 28 0 1 2 2 0    202 0 14 1 1 1 1 1
13 0 41 0 0 0 2 0   108 3 33 0 1 2 2 0    203 0 39 0 1 1 2 0
14 0 12 1 1 0 1 0   109 0 38 0 0 0 2 0    204 0  4 1 1 1 3 0
15 0  6 0 0 0 3 0   110 0 42 1 1 2 2 1    205 1 27 1 1 1 2 1
16 0 43 0 0 0 2 0   111 0 40 1 1 2 2 0    206 0 36 1 0 0 2 1
17 2 39 1 0 1 2 0   112 0 38 0 0 0 2 0    207 0 30 1 0 2 2 1
18 0 43 0 1 0 2 0   113 2 37 0 1 1 2 0    208 0 34 0 1 0 2 0
19 2 37 0 0 0 2 1   114 1 42 0 1 0 2 0    209 1 40 1 1 1 2 0
20 0 31 1 1 1 2 0   115 5 37 1 1 1 2 1    210 0  6 1 0 1 1 1
21 0 45 0 1 0 2 0   116 0 38 0 0 0 2 0    211 1 40 1 1 1 2 0
22 1 29 1 1 1 2 1   117 0  4 0 0 0 3 0    212 2 43 0 1 0 2 0
23 1 35 1 1 1 2 0   118 2 37 1 1 1 2 0    213 0 36 1 1 1 2 0

   ... more lines ...

94 0 35 1 0 0 2 0   189 0 36 1 0 0 2 0    284 0 35 0 0 0 2 1
95 3 37 1 0 0 2 0   190 0 39 0 1 0 2 0
;
```

The following SAS statements request the analysis. To produce the Poisson regression, options LINK=LOG and DIST=POISSON are specified. The variable LOGRISK is the offset, and the main effects model is requested.

```
proc genmod data=lri;
   class ses id race agegroup;
   model count = passive crowding ses race agegroup/
              dist=poisson link=log offset=logrisk type3;
run;
```

Output 15.76 contains the general model information.

Output 15.76 Model Information

Model Information	
Data Set	WORK.LRI
Distribution	Poisson
Link Function	Log
Dependent Variable	count
Offset Variable	logrisk

Output 15.77 contains the goodness-of-fit statistics, along with the ratios of their values to their degrees of freedom. With values of 1.4788 for the deviance/df and 1.7951 for Pearson/df, there is

evidence of overdispersion. The model-based estimates of standard errors may not be appropriate and therefore any inference is questionable. (When this ratio is close to 1, you conclude that little evidence of over- or underdispersion exists.) The next step is to account for this overdispersion with the GEE-generated robust covariances.

Output 15.77 Goodness-of-Fit Statistics

Criteria For Assessing Goodness Of Fit			
Criterion	DF	Value	Value/DF
Deviance	276	408.1549	1.4788
Scaled Deviance	276	408.1549	1.4788
Pearson Chi-Square	276	495.4493	1.7951
Scaled Pearson X2	276	495.4493	1.7951
Log Likelihood		-260.4117	
Full Log Likelihood		-337.5776	
AIC (smaller is better)		691.1551	
AICC (smaller is better)		691.6788	
BIC (smaller is better)		720.3469	

The following statements produce the desired GEE analysis. In this case, the subject is the cluster and there is only one measurement per cluster. The working independent correlation structure is specified with the TYPE=IND option although, with a cluster size of 1, the estimates will be the same if you specify exchangeable or unstructured. Otherwise, the model specification is the same as in the previous analysis.

```
proc genmod data=lri;
   class id ses race agegroup;
   model count = passive crowding ses race agegroup /
                 dist=poisson link=log offset=logrisk type3;
   repeated subject=id / type=ind;
run;
```

Output 15.78 reports that the GEE analysis involves one measurement per subject, and that there are 284 subjects, or clusters, in the analysis.

15.13. Using GEE to Account for Overdispersion: Univariate Outcome

Output 15.78 GEE Model Information

GEE Model Information	
Correlation Structure	Independent
Subject Effect	id (284 levels)
Number of Clusters	284
Correlation Matrix Dimension	1
Maximum Cluster Size	1
Minimum Cluster Size	1

Output 15.79 contains the parameter estimates. They are the same as displayed for the unadjusted GLM analysis in Chapter 12, as you would expect, but the standard errors are different. They are larger than the corresponding standard errors in the GLM analysis; this is also what you would expect because overdispersion means that the data are exhibiting additional variance.

Output 15.79 GEE Parameter Estimates

Analysis Of GEE Parameter Estimates							
Empirical Standard Error Estimates							
Parameter		Estimate	Standard Error	95% Confidence Limits		Z	Pr > |Z|
Intercept		0.6047	0.5564	-0.4858	1.6952	1.09	0.2771
passive		0.4310	0.2105	0.0184	0.8436	2.05	0.0406
crowding		0.5199	0.2367	0.0559	0.9839	2.20	0.0281
ses	0	-0.3970	0.2977	-0.9805	0.1865	-1.33	0.1824
ses	1	-0.0681	0.2520	-0.5619	0.4258	-0.27	0.7871
ses	2	0.0000	0.0000	0.0000	0.0000	.	.
race	0	0.1402	0.2211	-0.2931	0.5736	0.63	0.5259
race	1	0.0000	0.0000	0.0000	0.0000	.	.
agegroup	1	-0.4792	0.6033	-1.6617	0.7033	-0.79	0.4270
agegroup	2	-0.9919	0.4675	-1.9082	-0.0756	-2.12	0.0339
agegroup	3	0.0000	0.0000	0.0000	0.0000	.	.

Output 15.80 contains the Type 3 analysis. SES, race, and age group are non-influential. Crowding and smoking exposure are significant at the $\alpha = 0.05$ level.

Output 15.80 Type 3 Analysis

Score Statistics For Type 3 GEE Analysis			
Source	DF	Chi-Square	Pr > ChiSq
passive	1	3.90	0.0484
crowding	1	4.72	0.0298
ses	2	2.11	0.3478
race	1	0.42	0.5176
agegroup	2	2.79	0.2484

Thus, this section provides an alternative strategy for adjusting for overdispersion to the scaling factor adjustment discussed in Chapter 12. WIth the GEE method, you are using a measure of variability based on the data to do the adjustment, rather than a single parameter (scaling factor) applied to the covariance matrix. With the GEE method, using the robust variances, you are providing a measure of variability for each parameter you estimate together with the corresponding covariances, all based on your data. In many situations, this strategy may be a practical approach to handling overdispersion.

Appendix A: Steps to Find the GEE Solution

Finding the GEE solution requires a number of steps, including specifying the marginal model for the first moment, specifying the variance function for the relationship between the first and second moments, choosing a working correlation matrix, computing an initial estimate of β, and then using this estimate in an iterative estimation process. In detail:

The **first step** of the GEE method is to relate the marginal response $\mu_{ij} = E(y_{ij})$ to a linear combination of the covariates: $g(\mu_{ij}) = x'_{ij}\beta$, where $\beta = (\beta_1, \ldots, \beta_p)'$ is a $p \times 1$ vector of unknown parameters and g is a known link function. Common link functions are the logit function $g(x) = \log(x/(1-x))$ for binary responses and the log function $g(x) = \log(x)$ for Poisson counts. The $p \times 1$ parameter vector β characterizes how the cross-sectional response distribution depends on the explanatory variables.

The **second step** is to describe the variance of y_{ij} as a function of the mean: $\text{Var}(y_{ij}) = v(\mu_{ij})\phi$, where v is a known variance function and ϕ is a possibly unknown scale parameter. For binary responses, $v(\mu_{ij}) = \mu_{ij}(1 - \mu_{ij})$; and for Poisson responses, $v(\mu_{ij}) = \mu_{ij}$. For these two types of response variables, $\phi = 1$. Overdispersion ($\phi > 1$) may exist for binomial-like or count data, but use of the empirical covariance matrix for the GEE procedure is robust to this overdispersion.

The **third step** is to choose the form of a $t \times t$ working correlation matrix $\mathbf{R}_i(\alpha)$ for each $y_i = (y_{i1}, \ldots, y_{it_i})'$, with $t_i \le t$. The (j, j') element of $\mathbf{R}_i(\alpha)$ is the known, hypothesized, or estimated correlation between y_{ij} and $y_{ij'}$. This working correlation matrix may depend on a vector of unknown parameters α, which is the same for all subjects. You assume that $\mathbf{R}_i(\alpha)$ is known except for a fixed number of parameters α that must be estimated from the data. Although this correlation matrix can differ from subject to subject, you commonly use a working correlation matrix $\mathbf{R}(\alpha)$ that approximates the average dependence among repeated observations over subjects.

The GEE method yields consistent estimates of the regression coefficients and their variances, even with misspecification of the structure of the covariance matrix. In addition, the loss of efficiency from an incorrect choice of $\mathbf{R}$ is usually not consequential when the number of subjects is large.

The **fourth step** of the GEE method is to estimate the parameter vector β and its covariance matrix. First, let $\mathbf{A}_i$ be the $t_i \times t_i$ diagonal matrix with $v(\mu_{ij})$ as the jth diagonal element. The working covariance matrix for y_i is $\mathbf{V}_i(\alpha) = \phi \mathbf{A}_i^{1/2} \mathbf{R}_i(\alpha) \mathbf{A}_i^{1/2}$. The GEE estimate of β is the solution of the estimating equation

$$U(\beta) = \sum_{i=1}^{n} \left(\frac{\partial \mu_i}{\partial \beta}\right)' \left[\mathbf{V}_i(\hat{\alpha})\right]^{-1} (y_i - \mu_i) = \mathbf{0}_p$$

where $\mu_i = (\mu_{i1}, \ldots, \mu_{it_i})'$, $\mathbf{0}_p$ is the $p \times 1$ vector $(0, \ldots, 0)'$, and $\hat{\alpha}$ is a consistent estimate of α.

The estimating equation is solved by iterating between quasi-likelihood methods for estimating β and method of moments estimation of α as a function of β, as follows:

1. Compute an initial estimate of β, using a GLM model or some other method.

2. Compute the standardized Pearson residuals

$$r_{ij} = \frac{y_{ij} - \hat{\mu}_{ij}}{\sqrt{v(\hat{\mu}_{ij})}}$$

and obtain the estimates for the nuisance parameters ϕ and α using moment estimation.

3. Update $\hat{\beta}$ with

$$\hat{\beta} - \left[\sum_{i=1}^{n} \frac{\partial \mu_i}{\partial \beta}' V_i^{-1} \frac{\partial \mu_i}{\partial \beta}\right]^{-1} \left[\sum_{i=1}^{n} \frac{\partial \mu_i}{\partial \beta}' V_i^{-1} (Y_i - \mu_i)\right]$$

4. Iterate until convergence.

15.14 Appendix B: Macro for Adjusted Wald Statistic

The following macro is used in Section 15.5.

```
%macro geef;
data temp1;
   set clustout;
   drop Label1 cvalue1;
   if Label1='Number of Clusters';
```

```
      run;
      data temp2;
         set scoreout;
         drop ProbChiSq;
      run;
      data temp3;
         merge temp1 temp2;
      run;
      data temp4; set temp3;
         retain nclusters; drop nvalue1;
         if _n_=1 then nclusters=nvalue1;
      run;
      data temp5;
         set temp4;
         drop ChiSq nclusters d;
         d=nclusters-1;
         NewF= ((d-df+1)*ChiSq)/(d*df);
         ProbF=1-cdf('F', NewF,df,d-df+1);
      run;

/* Set the ODS path to include your store first (this
   sets the search path order so that ODS looks in your
   store first, followed by the default store */

ods path sasuser.templat (update)
         sashelp.tmplmst (read);

/* Print the path to the log to make sure you will get
   what you expect */

*ods path show;

/* Define your table, and store it */
   proc template;
      define table GEEType3F;
      parent=Stat.Genmod.Type3GEESc;
      header "#F-Statistics for Type 3 GEE Analysis##";
      column Source DF i NewF ProbF;
      define NewF;
      parent = Common.ANOVA.FValue;
    end;
   end;
   run;
   title1;
   data _null_;
      set temp5;
      file print ods=(template='GEEType3F');
      put _ods_;
   run;
;
%mend geef;
```

References

Agresti, A. (1988). Logit models for repeated ordered categorical response data. *Proceedings of the Thirteenth Annual SAS Users Group International Conference*, 997–1005. Cary, NC: SAS Institute Inc.

Agresti, A. (1989). A survey of models for repeated ordered categorical response data. *Statistics in Medicine* 8: 1209–1224.

Agresti, A. (1992). A survey of exact inference for contingency tables. *Statistical Science* 7 (1): 131–177.

Agresti, A. (1996). *An Introduction to Categorical Data Analysis*. New York: John Wiley & Sons.

Agresti, A. (2007). *An Introduction to Categorical Data Analysis*. 2nd ed. New York: John Wiley & Sons.

Agresti, A. (2002). *Categorical Data Analysis*. 2nd ed. New York: John Wiley & Sons. (3rd ed. available Fall 2012)

Agresti, A. (2010). *Analysis of Ordinal Categorical Analysis*. 2nd ed. New York: John Wiley & Sons.

Agresti, A. (2011). Score and pseudo-score confidence intervals for categorical data analysis. *Statistics in Biopharmaceutical Research* 3 (2): 163–172.

Agresti, A. and Caffo, B. (2000). Simple and effective confidence intervals for proportions and differences of proportions from adding two successes and two failures. *American Statistician* 54: 280–288.

Agresti, A. and Coull, B. A. (1998). Approximate is better than "exact" for interval estimation of binomial proportions. *American Statistician* 52: 119–126.

Agresti, A. and Min, Y. (2001). On small-sample confidence intervals for parameters in discrete distributions. *Biometrics* 57: 963–971.

Agresti, A., Wackerly, D., and Boyett, J. M. (1979). Exact conditional tests for cross-classifications: Approximation of attained significance levels. *Psychometrika* 44: 75–83.

Ahn, C., Koch, G. G., Paynter, L., Preisser, J. S., and Seiller-Moiseiwitsch, F. (2007) *Journal of Statistical Planning and Inference*, 137: 3227–3239.

Akaike, H. (1974). A new look at statistical model identification. *IEEE Transactions on Automatic Control* 19: 716–723.

Albert, A. and Anderson, J. A. (1984). On the existence of maximum likelihood estimates in logistic regression models. *Biometrika* 71: 1–10.

Albert, P. and McShane, L. (1995). A generalized estimating equations approach for spatially correlated binary data: Applications to the analysis of neuroimaging data. *Biometrics* 51: 627–638.

Andersen, E. B. (1991). *The Statistical Analysis of Categorical Data*. 2nd ed. Berlin: Springer-Verlag.

Armitage, P. (1955). Tests for linear trends in proportions and frequencies. *Biometrics* 11: 375–386.

Baglivo, J., Olivier, D., and Pagano, M. (1988). Methods for the analysis of contingency tables with large and small cell counts. *Journal of the American Statistical Association* 83: 1006–1013.

Bangdiwala, K. and Bryan, H. (1987). Using SAS software graphical procedures for the observer agreement chart. *Proceedings of the Twelfth SAS Users Group International Conference*, 1083–1088.

Barnhart, H. and Williamson, J. (1998). Goodness-of-fit tests for GEE modeling with binary responses. *Biometrics* 54: 720–729.

Bauman, K. E., Koch, G. G., and Lentz, M. (1989). Parent characteristics, perceived health risk, and smokeless tobacco use among white adolescent males. *NI Monographs* 8: 43–48.

Benard, A. and van Elteren, P. (1953). A generalization of the method of m rankings. *Proceedings Koninklijke Nederlands Akademie van Wetenschappen (A)* 56: 358–369.

Berkson, J. (1951). Why I prefer logits to probits. *Biometrics* 7: 327–339.

Berry, D. A. (1987). Logarithmic transformations in ANOVA. *Biometrics* 43: 439–456.

Birch, M. W. (1963). Maximum likelihood in three-way contingency tables. *Journal of the Royal Statistical Society, Series B* 25: 220–233.

Bishop, Y. M. M., Fienberg, S. E., and Holland, P. W. (1975). *Discrete Multivariate Analysis*. Cambridge, MA: MIT Press.

Bock, R. D. (1975). *Multivariate Statistical Methods in Behavioral Research*. New York: McGraw-Hill.

Bock, R. D. and Jones, L. V. (1968). *The Measurement and Prediction of Judgment and Choice*. San Francisco: Holden-Day.

Boos, D. (1992). On generalized score tests. *American Statistician* 46: 327–333.

Bowdre, J. H., Hull, J. H., and Coccetto, D. M. (1983). Antibiotic efficacy against Vibrio vulnificus in the mouse: superiority of tetracycline. *Journal of Pharmacology and Experimental Therapeutics*. 225: 595–598.

Bowker, A. H. (1948). Bowker's test for symmetry. *Journal of the American Statistical Association* 43: 572–574.

Breslow, N. E. and Day, N. E. (1980). *Statistical Methods in Cancer Research*, vol. 1: *The Analysis of Case-Control Studies*. Lyon: International Agency for Research on Cancer.

Bruce, R. A., Kusumi, F., and Hosmer, D. (1973). Maximal oxygen intake and nomographic assessment of functional aerobic impairment in cardiovascular disease. *American Heart Journal* 65: 546–562.

Carey, V., Zeger, S. L., and Diggle, P. J. (1993). Modelling multivariate binary data with alternating logistic regressions. *Biometrika* 80: 517–526.

Carr, G. J., Hafner, K. B., and Koch, G. G. (1989). Analysis of rank measures of association for ordinal data from longitudinal studies. *Journal of the American Statistical Association* 84: 797–804.

Cartwright, H. V., Lindahl, R. L., and Bawden, J. W. (1968). Clinical findings on the effectiveness of stannous fluoride and acid phosphate fluoride as caries reducing agents in children. *Journal of Dentistry for Children* 35: 36–40.

Cawson M. J., Anderson, A. B. M., Turnbull, A. C., and Lampe, L. (1974). Cortisol, cortisone, and 11-deoxycortisol levels in human umbilical and maternal plasma in relation to the onset of labour. *Journal of Obstetrics and Gynaecology of the British Commonwealth* 81: 737–745.

Chan, I. S. F. and Zhang, Z. (1999). Test-based exact confidence intervals for the difference of two binomial proportions. *Biometrics* 55: 1202–1209.

Charnes, A., Frome, E. L., and Yu, P. L. (1976). The equivalence of generalized least squares and maximum likelihood estimates in the exponential family. *Journal of the American Statistical Association* 71: 169–172.

Clopper, C. J. and Pearson, E. S. (1934). The use of confidence or fiducial limits illustrated in the case of the binomial. *Biometrika* 26: 404–413.

Cochran, W. G. (1940). The analysis of variance when experimental errors follow the Poisson or binomial laws. *Annals of Mathematical Statistics* 11: 335–347.

Cochran, W. G. (1950). The comparison of percentages in matched samples. *Biometrika* 37: 256–266.

Cochran, W. G. (1954). Some methods of strengthening the common χ^2 tests. *Biometrics* 10: 417–451.

Cohen, J. (1960). A coefficient of agreement for nominal data. *Educational and Psychological Measurement* 20: 37–46.

Collett, D. (2003). *Modelling Binary Data*. 2nd ed. London: Chapman & Hall.

Cook, R. D. and Weisberg, S. (1982). *Residuals and Influence in Regression*. London: Chapman & Hall.

Cornoni-Huntley, J., Brock, D. B., Ostfeld, A., Taylor, J. O., and Wallace, R. B. (1986). *Established Populations for Epidemiologic Studies of the Elderly: Resource Data Book*. Bethesda, MD: National Institutes of Health (NIH Pub. No. 86-2443).

Cox, D. R. (1970). *Analysis of Binary Data*. London: Chapman & Hall.

Cox, D. R. (1972). Regression models and life tables. *Journal of the Royal Statistical Society, Series B* 34: 187–220.

Cox, M. A. A. and Plackett, R. L. (1980). Small samples in contingency tables. *Biometrika* 67: 1–13.

Davis, C. S. (1992). Analysis of incomplete categorical repeated measures. *Proceedings of the Seventeenth Annual SAS Users Group International Conference*, 1374–1379. Cary, NC: SAS Institute Inc.

Dean, C. B. (1998). Overdispersion. In *Encyclopedia of Biostatistics*, vol. 4, ed. P. Armitage and T. Colton. New York: John Wiley & Sons.

Deddens, J. A. and Koch, G. G. (1988). Survival analysis, grouped data. In *Encyclopedia of Statistical Sciences*, vol. 9, ed. S. Kotz and N. L. Johnson. New York: John Wiley & Sons.

Deming, W. E. and Stephan, F. F. (1940). On a least squares adjustment of a sample frequency table when the expected marginal totals are known. *Annals of Mathematical Statistics* 11: 427–444.

Derr, Robert E. (2009). Performing exact logistic regression with the SAS System (revised). *Proceedings of the Twenty-Fifth Annual SAS Users Group International Conference*. http://support.sas.com/rnd/app/papers/multipleimputation.pdf.

Diggle, P. J. (1992). Discussion of paper by K. Y. Liang, S. L. Zeger, and B. Qaqish. *Journal of the Royal Statistical Society, Series B* 45: 28–29.

Diggle, P. J., Liang, K. Y., and Zeger, S. L. (1994). *Analysis of Longitudinal Data*. Oxford: Clarendon Press.

Dobson, A. J. (1990). *An Introduction to Generalized Linear Models*. London: Chapman & Hall.

Draper, N. R. and Smith, H. (1998). *Applied Regression Analysis*. 3rd ed. New York: John Wiley & Sons.

Durbin, J. (1951). Incomplete blocks in ranking experiments. *British Journal of Mathematical and Statistical Psychology* 4: 85–90.

Dyke, G. V. and Patterson, H. D. (1952). Analysis of factorial arrangements when the data are proportions. *Biometrics* 8: 1–12.

Elashoff, J. D. and Koch, G. G. (1991). Statistical methods in trials of anti-ulcer drugs. In *Ulcer Disease: Investigation and Basis for Therapy*, ed. E. A. Swabb and S. Szabo, 375–406. New York: Marcel Dekker.

Farrington, C. P. and Manning, G. (1990). Test statistics and sample size formulae for comparative binomial trials with null hypothesis of non-zero risk difference or non-unity relative risk. *Statistics in Medicine* 9: 1447–1454.

Fienberg, S. E. (2011). The analysis of contingency tables: from chi-squared tests and log-linear models to models of mixed membership. *Statistics in Biopharmaceutical Research* 3 (2): 173–184.

Finney, D. J. (1978). *Statistical Methods in Biological Assay.* 3rd ed. New York: Macmillan.

Firth, D. (1992). Discussion of paper by K. Y. Liang, S. L. Zeger, and B. Qaqish. *Journal of the Royal Statistical Society, Series B* 45: 24–26.

Firth, D. (1993). Bias reduction of maximum likelihood estimates. *Biometrika* 80: 27–38.

Fisher, L. D. and van Belle, G. (1993). *Biostatistics: A Methodology for the Health Sciences.* New York: John Wiley & Sons.

Fleiss, J. L. (1975). Measuring agreement between two judges on the presence or absence of a trait. *Biometrics* 31: 651–659.

Fleiss, J. L. (1986). *The Design and Analysis of Clinical Experiments.* New York: John Wiley & Sons.

Fleiss, J. L. and Cohen, J. (1973). The equivalence of weighted kappa and the intraclass correlation coefficient as measures of reliability. *Educational and Psychological Measurement* 33: 613–619.

Fleiss, J. L., Levin, B., and Paik, M. C. (2003). *Statistical Methods for Rates and Proportions.* New York: John Wiley & Sons.

Friedman, M. (1937). The use of ranks to avoid the assumption of normality implicit in the analysis of variance. *Journal of the American Statistical Association* 32: 675–701.

Frome, E. L. (1981). Poisson regression analysis. *American Statistician* 35: 262–263.

Frome, E. L., Kutner, M. H., and Beauchamp, J. J. (1973). Regression analysis of Poisson-distributed data. *Journal of the American Statistical Association* 68: 935–940.

Gail, M. (1978). The analysis of heterogeneity for indirect standardized mortality ratios. *Journal of the Royal Statistical Society, Series A* 141, Part 2: 224–234.

Gansky, S. A., Koch, G. G., and Wilson, J. (1994). Statistical evaluation of relationships between analgesic dose and ordered ratings of pain relief over an eight-hour period. *Journal of Biopharmaceutical Statistics* 4: 233–265.

Gart, J. J. (1971). The comparison of proportions: A review of significance tests, confidence intervals and adjustments for stratification. *Review of the International Statistical Institute* 39 (2): 148–169.

Goodman, L. A. (1968). The analysis of cross-classified data: Independence, quasi-independence, and interactions in contingency tables with or without missing entries. *Journal of the American Statistical Association* 63: 1091–1131.

Goodman, L. A. (1970). The multivariate analysis of qualitative data: Interaction among multiple classifications. *Journal of the American Statistical Association* 65: 226–256.

Govindarajulu, Z. (1988). *Statistical Methods in Bioassay.* Basel: Karger.

Graubard, B. I. and Korn, E. L. (1987). Choice of column scores for testing independence in ordered $2 \times k$ contingency tables. *Biometrics* 43: 471–476.

Grizzle, J. E., Starmer, C. F., and Koch, G. G. (1969). Analysis of categorical data by linear models. *Biometrics* 25: 489–504.

Harrell, F. E. (1989). Analysis of repeated measurements. Unpublished course notes.

Heinze, G. and Schemper, M. (2002). A solution to the problem of separation in logistic regression. *Statistics in Medicine* 21: 2409–2419.

Hendricks, S. A., Wassell, J. T., Collins, J. W., and Sedlak, S. L. (1996). Power determination for geographically clustered data using generalized estimating equations. *Statistics in Medicine* 15: 1951–1960.

Herman-Giddens, M. E., Slora, E. J., Wasserman, R. C., Bourdony, C. J., Bhapkar, M. V., Koch, G. G., and Hasemeier, C. M. (1997). Secondary sexual characteristics and menses in young girls seen in office practice: A study from the pediatric research in office settings network. *Pediatrics* 99: 505–512.

Higgins, J. E. and Koch, G. G. (1977). Variable selection and generalized chi-square analysis of categorical data applied to a large cross-sectional occupational health survey. *International Statistical Review* 45: 51–62.

Hirji, K. F., Mehta, C. R., and Patel, N. R. (1987). Computing distributions for exact logistic regression. *Journal of the American Statistical Association* 82: 1110–1117.

Hirji, K. F., Vollset, S. E., Reis, I. M., and Afifi, A. A. (1996). Exact Tests for Interaction in Several Tables. *Journal of Computational and Graphical Statistics* 5: 209–224.

Hochberg, Y., Stutts, J. C., and Reinfurt, D. W. (1977). Observed shoulder belt usage of drivers in North Carolina: A follow-up. *University of North Carolina Highway Safety Research Center Report*, May.

Hodges, J. L. and Lehmann, E. L. (1962). Rank methods for combination of independent experiments in analysis of variance. *Annals of Mathematical Statistics* 33: 482–497.

Hollander, M. and Wolfe, D. A. (1973). *Nonparametric Statistical Methods*. New York: John Wiley & Sons.

Hosmer, D. W. and Lemeshow, S. (2000). *Applied Logistic Regression*. 2nd ed. New York: John Wiley & Sons.

Imrey, P. B. (2011). Koch's contribution to statistical practice, 1985–1985. *Statistics in Biopharmaceutical Research* 3 (2): 139–162.

Imrey, P. B., Koch, G. G., and Stokes, M. E. (1982). Categorical data analysis: Some reflections on the log linear model and logistic regression, Part II. *International Statistical Review* 50: 35–64.

Johnson, W. D. and Koch, G. G. (1978). Linear models analysis of competing risks for grouped survival times. *International Statistical Review* 46: 21–51.

Karim, M. R. and Zeger, S. L. (1988). GEE: A SAS macro for longitudinal data analysis. Technical Report No. 674, Department of Biostatistics, Johns Hopkins University.

Kawaguchi, A., Koch, G. G., and Wang, X. (2011). Stratified multivariate Mann-Whitney estimators for the comparison of two treatments with randomization based covariance adjustment. *Statistics in Biopharmaceutical Research* 3 (2): 217–231.

Kendall, M. G. and Stuart, A. (1961). *The Advanced Theory of Statistics*, vol. 2: *Inference and Relationship*. London: Charles Griffin and Company.

Keene, O. N., Jones, R. K. and Lane, P. W. (2007). Analysis of exacerbation rates in asthma and chronic obstructive pulmonary disease: example from the TRISTAN study. *Pharmaceutical Sciences*, 6: 89–97.

Koch, G. G., Amara, I. A., Davis, G. W., and Gillings, D. B. (1982). A review of some statistical methods for covariance analysis of categorical data. *Biometrics* 38: 563–595.

Koch, G. G., Amara, I. A., and Singer, J. M. (1985). A two-stage procedure for the analysis of ordinal categorical data. In *Statistics in Biomedical, Public Health and Environmental Sciences*, ed. P. K. Sen, 357–387. Amsterdam: North-Holland.

Koch, G. G., Atkinson, S. S., and Stokes, M. E. (1986). Poisson regression. In *Encyclopedia of Statistical Sciences*, vol. 7, ed. S. Kotz and N. L. Johnson. New York: John Wiley & Sons.

Koch, G. G., Carr, G. J., Amara, I. A., Stokes, M. E., and Uryniak, T. J. (1990). Categorical data analysis. In *Statistical Methodology in the Pharmaceutical Sciences*, ed. D. A. Berry, 391–475. New York: Marcel Dekker.

Koch, G. G. and Edwards, S. (1985). Logistic regression. In *Encyclopedia of Statistics*, vol. 5. New York: John Wiley & Sons.

Koch, G. G. and Edwards, S. (1988). Clinical efficacy trials with categorical data. In *Biopharmaceutical Statistics for Drug Development*, ed. K. E. Peace, 403–451. New York: Marcel Dekker.

Koch, G. G., Gillings, D. B., and Stokes, M. E. (1980). Biostatistical implications of design, sampling, and measurement to health science data. *Annual Review of Public Health* 1: 163–225.

Koch, G. G., Imrey, P. B., Singer, J. M., Atkinson, S. S., and Stokes, M. E. (1985). *Analysis of Categorical Data*. Montreal: Les Presses de L'Université de Montreal.

Koch, G. G., Landis, J. R., Freeman, J. L., Freeman, D. H., and Lehnen, R. G. (1977). A general methodology for the analysis of experiments with repeated measurement of categorical data. *Biometrics* 33: 133–158.

Koch, G. G., McCanless, I. and Ward, J. F. (1984). Interpretation of statistical methodology associated with maintenance trials. *American Journal of Medicine* 77: 43–50.

Koch, G. G. and Reinfurt, D. W. (1971). The analysis of categorical data from mixed models. *Biometrics* 27: 157–173.

Koch, G. G. and Sen, P. K. (1968). Some aspects of the statistical analysis of the mixed model. *Biometrics* 24: 27–48.

Koch, G. G., Sen, P. K., and Amara, I. A. (1985). Log-rank scores, statistics, and tests. In *Encyclopedia of Statistics*, vol. 5. New York: John Wiley & Sons.

Koch, G. G., Singer, J. M., Stokes, M. E., Carr, G. J., Cohen, S. B., and Forthofer, R. N. (1989). Some aspects of weighted least squares analysis for longitudinal categorical data. In *Statistical Models for Longitudinal Studies of Health*, ed. J. H. Dywer, 215–258. Oxford: Oxford University Press.

Koch, G. G. and Stokes, M. E. (1979). Annotated computer applications of weighted least squares methods for illustrative analyses of examples involving health survey data. Technical report prepared for the U.S. National Center for Health Statistics.

Koch, G. G., Tangen, C. M., Jung, J.-W., and Amara, I. A. (1998). Issues for covariance analysis of dichotomous and ordered categorical data from randomized clinical trials and non-parametric strategies for addressing them. *Statistics in Medicine* 17: 1863–1892.

Kruskal, W. H. and Wallis, W. A. (1952). Use of ranks in one-criterion variance analysis. *Journal of the American Statistical Association* 47: 583–621.

Kuritz, S. J., Landis, J. R., and Koch, G. G. (1988). A general overview of Mantel-Haenszel methods: Applications and recent developments. *Annual Review of Public Health*: 123–160.

Lafata, J. E., Koch, G. G., and Weissert, W. G. (1994). Estimating activity limitation in the noninstitutionalized population: A method for small areas. *American Journal of Public Health* 84: 1813–1817.

Landis, J. R., Heyman, E. R., and Koch, G. G. (1978). Average partial association in three-way contingency tables: A review and discussion of alternative tests. *International Statistical Review* 46: 237–254.

Landis, J.R., King, T. S., Choi, J. W., Chinchilli, V. M., and Koch, G. G. (2011). Measures of agreement and concordance with clinical research applications. *Statistics in Biopharmaceutical Research* 3 (2): 185–209.

Landis, J. R. and Koch, G. G. (1977). The measurement of observer agreement for categorical data. *Biometrics* 33: 159–174.

Landis, J. R. and Koch, G. G. (1979). The analysis of categorical data on longitudinal studies of behavioral development. In *Longitudinal Research in the Study of Behavior and Development*, ed. J. R. Nesselroade and P. B. Bates, 231–261. New York: Academic Press.

Landis, J. R., Miller, M. E., Davis, C. S., and Koch, G. G. (1988). Some general methods for the analysis of categorical data in longitudinal studies. *Statistics in Medicine* 7: 233–261.

Landis, J. R., Sharp, T. J., Kuritz, S. J., and Koch, G. G. (1998). Mantel-Haenszel methods. In *Encyclopedia of Biostatistics*, vol. 3, ed. P. Armitage and T. Colton. New York: John Wiley & Sons.

Landis, J. R., Stanish, W. M., and Koch, G. G. (1976). A computer program for the generalized chi-square analysis of categorical data using weighted least squares (GENCAT). *Computer Programs in Biomedicine* 6: 196–231.

LaVange, L. M., Keyes, L. L., Koch, G. G., and Margolis, P. E. (1994). Application of sample survey methods for modelling ratios to incidence densities. *Statistics in Medicine* 13: 343–355.

LaVange, L. M. and Koch, G. G. (2008). Randomization-based nonparametric (ANCOVA). *Encyclopedia of Clinical Trials*.

LaVange, L. M., Koch, G. G., and Schwartz, T. A. (2001). Applying sample survey methods to clinical trials data. *Statistics in Medicine* 20: 2609–2623.

Lehmann, E. L. (1975). *Nonparametrics: Statistical Methods Based on Ranks*. San Francisco: Holden-Day.

Lemeshow, S. and Hosmer, D. (1989). *Applied Logistic Regression Analysis*. New York: John Wiley & Sons.

Liang, K. Y. and Zeger, S. L. (1986). Longitudinal data analysis using generalized linear models. *Biometrika* 73: 13–22.

Liang, K. Y., Zeger, S. L., and Qaqish, B. (1992). Multivariate regression analyses for categorical data (with discussion). *Journal of the Royal Statistical Society, Series B* 54: 3–40.

Lipsitz, S. R., Fitzmaurice, G. M., Orav, E. J., and Laird, N. M. (1994). Performance of generalized estimating equations in practical situations. *Biometrics* 50: 270–278.

Lipsitz, S. R., Kim, K., and Zhao, L. (1994). Analysis of repeated categorical data using generalized estimating equations. *Statistics in Medicine* 13: 1149–1163.

Lipsitz, S. R., Laird, N. M., and Harrington, D. P. (1991). Generalized estimating equations for correlated binary data: Using the odds ratio as a measure of association. *Biometrika* 78: 153–160.

Loomis, D., Dufort, V., Kleckner, R. C., and Savitz, D. A. (1999). Fatal occupational injuries among electric power company workers. *American Journal of Industrial Medicine* 35: 302–309.

Lund, A. K., Williams, A. F., and Zador, P. (1986). High school driver education: Further evaluation of the De Kalb County study. *Accident Analysis and Prevention* 18: 349–357.

Luta, G., Koch, G. G., Cascio, W. E., and Smith, W. T. (1998). An application of methods for clustered binary responses to a cardiovascular study with small sample size. *Journal of Biopharmaceutical Statistics* 8: 87–102.

Mack, T. M., Pike, M. C., Henderson, B. E., Pfeffer, R. I., Gerkins, V. R., Arthur, M., and Brown, S. E. (1976). Estrogens and endometrial cancer in a retirement community. *New England Journal of Medicine* 294 (23): 1262–1267.

Macknin, M. L., Mathew, S., and Medendorp, S. V. (1990). Effect of inhaling heated vapor on symptoms of the common cold. *Journal of the American Medical Association* 264: 989–991.

MacMillan, J., Becker, C., Koch, G. G., Stokes, M. E., and Vandivire, H. M. (1981). An application of weighted least squares methods to the analysis of measurement process components of variability in an observational study. *American Statistical Association Proceedings of Survey Research Methods*: 680–685.

Madansky, A. (1963). Test of homogeneity for correlated samples. *Journal of the American Statistical Association* 58: 97–119.

Mann, H. B. and Whitney, D. R. (1947). On a test of whether one of two random variables is stochastically larger than the other. *Annals of Mathematical Statistics* 18: 50–60.

Mantel, N. (1963). Chi-square tests with one degree of freedom: Extensions of the Mantel-Haenszel procedure. *Journal of the American Statistical Association* 58: 690–700.

Mantel, N. (1966). Evaluation of survival data and two new rank order statistics arising in its consideration. *Cancer Chemotherapy Report* 50: 163–170.

Mantel, N. and Fleiss, J. (1980). Minimum expected cell size requirements for the Mantel-Haenszel one-degree of freedom chi-square test and a related rapid procedure. *American Journal of Epidemiology* 112: 129–143.

Mantel, N. and Haenszel, W. (1959). Statistical aspects of the analysis of data from retrospective studies of disease. *Journal of the National Cancer Institute* 22: 719–748.

Margolin, B. H., Kaplan, N., and Zeiger, E. (1981). *Proceedings of the National Academy of Sciences, USA* 78: 3779–3783.

McCullagh, P. (1980). Regression models for ordinal data (with discussion). *Journal of the Royal Statistical Society, Series B* 42: 109–142.

McCullagh, P. and Nelder, J. A. (1989). *Generalized Linear Models.* 2nd ed. London: Chapman & Hall.

McNemar, Q. (1947). Note on the sampling error of the difference between correlated proportions or percentages. *Psychometrika* 12: 153–157.

Mee, R. W. (1984). Confidence bounds for the difference between two probabilities. *Biometrics* 40: 1175–1176.

Mehta, C. R. and Patel, N. R. (1983). A network algorithm for performing Fisher's exact test in r by c contingency tables. *Journal of the American Statistical Association* 78: 427–434.

Mehta, C. R. and Patel, N. R. (1995). Exact logistic regression: Theory and examples. *Statistics in Medicine* 13: 2143–2160.

Mehta, C. R., Patel, N. R., and Tsiatis, A. A. (1984). Exact significance testing to establish treatment equivalence with ordered categorical data. *Biometrics* 40: 427–434.

Miettinen, O. S. and Nurminen, M. (1985). Comparative analysis of two rates. *Statistics in Medicine* 4: 213–226.

Miller, M. E., Davis, C. S., and Landis, J. R. (1993). The analysis of longitudinal polytomous data: Generalized estimating equations and connections with weighted least squares. *Biometrics* 49: 1033–1044.

Moradi, T., Nyrén, O., Bergström, R., Gridley, G., Linet, M., Wolk, A., Dosemeci, M., and Adami,

H. (1998). Risk for endometrial cancer in relation to occupational physical activity: A nationwide cohort study in Sweden. *International Journal of Cancer* 76: 665–670.

Moulton, L. H. and Zeger, S. L. (1989). Analyzing repeated measures on generalized linear models via the bootstrap. *Biometrics* 45: 381–394.

Murphy, L., Schwartz, T. A., Melmick, C. G., Renner, J. N., Tudor, G., Koch, G., Dragomir, A., Kalsbeek, W. D., Luta, G., and Jordan, J. M. (2008). Lifetime risk of symptomatic knee osteoarthritis. *Arthritis and Rheumatism (Arthritis Care and Research)* 59 (9): 1207–1213.

Nelder, J. A. and Wedderburn, R. W. M. (1972). Generalized linear models. *Journal of the Royal Statistical Society, Series A* 135: 370–384.

Nemeroff, C. B., Bissette, G., Prange, A. J., Loosen, P. Y., Barlow, F. S., and Lipton, M. A. (1977). Neurotensin: Central nervous system effects of a hypothalamic peptide. *Brain Research*, 128, 485–496.

Newcombe, R. G. (1998). Interval estimation for the difference between independent proportions: Comparison of eleven methods. *Statistics in Medicine* 17: 873–890.

Newcombe, R. G. and Nurminen, M. N. (2011). In defence of score intervals for proportions and their differences. *Communications in Statistics—Theory and Methods* 40: 1271–1282.

Neyman, J. (1949). Contributions to the theory of the χ^2 test. In *Proceedings of the Berkeley Symposium of Mathematical Statistics and Probability*, 239–273. Berkeley: University of California Press.

Odeh, R. E., Owen, D. B., Birnbaum, Z. W., and Fisher, L. D. (1977). *Pocket Book of Statistical Tables*. New York: Marcel Dekker.

Ogilvie, J. C. (1965). Paired comparison models with tests for interaction. *Biometrics* 21: 651–654.

Pagano, M. and Halvorsen, K. T. (1981). An algorithm for finding the exact significance levels of r × c contingency tables. *Journal of the American Statistical Association* 76: 931–934.

Pan, W. (2001). Akaike's information criterion in generalized estimating equations. *Biometrics* 57: 120–125.

Peterson, B. and Harrell, F. E. (1990). Partial proportional odds models for ordinal response variables. *Applied Statistics* 39: 205–217.

Pickles, A. (1998). Generalized estimating equations. In *Encyclopedia of Biostatistics*, vol. 2, ed. P. Armitage and T. Colton. New York: John Wiley & Sons.

Pirie, W. (1983). Jonckheere tests for ordered alternatives. In *Encyclopedia of Statistical Sciences*, vol. 4, ed. S. Kotz and N. L. Johnson. New York: John Wiley & Sons.

Pitman, E. J. G. (1948). Lecture notes on nonparametric statistics. Mimeograph. New York: Columbia University.

Pregibon, D. (1981). Logistic regression diagnostics. *Annals of Statistics* 9: 705–724.

Preisser, J. S. and Koch, G. G. (1997). Categorical data analysis in public health. *Annual Review of Public Health* 18: 51–82.

Preisser, J. S. and Qaqish, B. F. (1996). Deletion diagnostics for generalised estimating equations. *Biometrika* 83 (3): 551–562.

Prentice, R. L. (1988). Correlated binary regression with covariate specific to each binary observation. *Biometrics* 44: 1033–1048.

Quade, D. (1967). Rank analysis of covariance. *Journal of the American Statistical Association* 62: 1187–1200.

Quade, D. (1982). Nonparametric analysis of covariance by matching. *Biometrics* 38: 597–611.

Rao, C. R. (1961). Asymptotic efficiency and limiting information. *Proceedings of the Fourth Berkeley Symposium on Mathematical Statistics and Probability* 1: 531–545.

Rao, C. R. (1962). Efficient estimates and optimum inference procedures in large samples (with discussion). *Journal of the Royal Statistical Society, Series B* 24: 46–72.

Read, C. B. (1983). Fieller's theorem. In *Encyclopedia of Statistical Sciences*, vol. 3, ed. S. Kotz and N. L. Johnson. New York: John Wiley & Sons.

Roberts, G., Martyn, A. L., Dobson, A. J., and McCarthy, W. H. (1981). Tumour thickness and histological type in malignant melanoma in New South Wales, Australia. *Pathology* 13: 763–770.

Robins, J., Breslow, N., and Greenland, S. (1986). Estimators of the Mantel-Haenszel variance consistent in both sparse data and large-strata limiting models. *Biometrics* 42: 311–323.

Rotnitzky, A. and Jewell, N. P. (1990). Hypothesis testing of regression parameters in semiparametric generalized linear models for cluster correlated data. *Biometrika* 77: 485–497.

Roy, S. N. and Kastenbaum, M. A. (1956). On the hypothesis of no "interaction" in a multiway contingency table. *Annals of Mathematical Statistics* 27: 749–757.

Roy, S. N. and Mitra, S. K. (1956). An introduction to some nonparametric generalizations of analysis of variance and multivariate analysis. *Biometrika* 43: 361–376.

Royall, R. M. (1986). Model robust confidence intervals using maximum likelihood estimators. *International Statistical Review* 54: 221–226.

SAS Institute Inc. (1999). *SAS/STAT User's Guide, Version 8*. Cary, NC: SAS Institute Inc.

Saville, B. R., LaVange, L. M., and Koch, G. G. (2011). Estimating covariate-adjusted incidence density ratios for multiple time intervals in clinical trials using nonparametric randomization-based ANCOVA. *Statistics in Biopharmaceutical Research* 3 (2): 242–252.

Schechter, P. J., Horwitz, D., and Henkin, R. I. (1973). Sodium chloride preference in essential hypertension. *Journal of the American Medical Association* 225: 1311–1315.

Semenya, K. A. and Koch, G. G. (1979). Linear models analysis for rank functions of ordinal categorical data. In *Proceedings of the Statistical Computing Section of the American Statistical*

Association, 271–276.

Semenya, K. A. and Koch, G. G. (1980). Compound function and linear model methods for the multivariate analysis of ordinal categorical data. *Institute of Statistics Mimeo Series No. 1323*. Chapel Hill: University of North Carolina.

Shah, B. V., Holt, M. M., and Folsom, R. E. (1977). Inference about regression models from sample survey data. *Bulletin of the International Statistical Institute* 47: 43–57.

Shahpar, C. and Guohua, L. (1999). Homicide mortality in the United States, 1935–1994: Age, period, and cohort effects. *American Journal of Epidemiology* 150: 1213–1222.

Silvapulle, M. J. (1981). On the existence of maximum likelihood estimators for the binomial response models. *Journal of the Royal Statistical Society, Series B* 43: 310–313.

Simpson, E. H. (1951). The interpretation of interaction in contingency tables. *Journal of the Royal Statistical Society, Series B* 13: 238–241.

Stanish, W. M. (1986). Categorical data analysis strategies using SAS software. In *Computer Science and Statistics: Proceedings of the Seventeenth Symposium on the Interface*, ed. D. M. Allen. New York: Elsevier Science.

Stanish, W. M., Gillings, D. B., and Koch, G. G. (1978). An application of multivariate ratio methods for the analysis of a longitudinal clinical trial with missing data. *Biometrics* 34: 305–317.

Stanish, W. M. and Koch, G. G. (1984). The use of CATMOD for repeated measurement analysis of categorical data. *Proceedings of the Ninth Annual SAS Users Group International Conference*, 761–770. Cary, NC: SAS Institute Inc.

Stewart, J. R. (1975). An analysis of automobile accidents to determine which variables are most strongly associated with driver injury: Relationships between driver injury and vehicle model year. *University of North Carolina Highway Safety Research Center Technical Report*.

Stock, J. R., Weaver, J. K., Ray, H. W., Brink, J. R., and Sadof, M. G. (1983). *Evaluation of Safe Performance Secondary School Driver Education Curriculum Demonstration Project*. Washington, DC: U.S. Department of Transportation, National Highway Traffic Safety Administration.

Stokes, M. E. (1986). An application of categorical data analysis to a large environmental data set with repeated measurements and missing values. *Institute of Statistics Mimeo Series No. 1807T*. Chapel Hill: University of North Carolina.

Stram, D. O., Wei, L. J., and Ware, J. H. (1988). Analysis of repeated ordered categorical outcomes with possibly missing observations and time-dependent covariates. *Journal of the American Statistical Association* 83: 631–637.

Tardif, S. (1980). On the asymptotic distribution of a class of aligned rank order test statistics in randomized block designs. *Canadian Journal of Statistics* 8: 7–25.

Tardif, S. (1981). On the almost sure convergence of the permutation distribution for aligned rank test statistics in randomized block designs. *Annals of Statistics* 9: 190–193.

Tardif, S. (1985). On the asymptotic efficiency of aligned-rank tests in randomized block designs. *Canadian Journal of Statistics* 13: 217–232.

Thomas, D. G. (1971). Algorithm AS-36: Exact confidence limits for the odds ratio in a 2×2 table. *Applied Statistics* 20: 105–110.

Tritchler, D. (1984). An algorithm for exact logistic regression. *Journal of the American Statistical Association* 79: 709–711.

Tsutakawa, R. K. (1982). Statistical methods in bioassay. In *Encyclopedia of Statistical Sciences*, vol. 1, ed. S. Kotz and N. L. Johnson. New York: John Wiley & Sons.

Tudor, G., Koch, G. G., and Catellier, D. (2000). Statistical methods for crossover designs in bioenvironmental and public health studies. In *Handbook of Statistics*, vol. 18: *Bioenvironmental and Public Health Statistics*, ed. P. K. Sen and C. R. Rao. Amsterdam: Elsevier Science.

van Elteren, P. H. (1960). On the combination of independent two-sample tests of Wilcoxon. *Bulletin of the International Statistical Institute* 37: 351–361.

Vesikari, T., Itzler, R., Matson, D. O., Santosham, M., Christie, C. D. C., Coia, M., Cook, J. R., Koch, G., and Heaton, P. (2007). Efficacy of a pentavalent rotavirus vaccine in reducing rotavirus-assocated health care utilization across three regions (11 countries). *International Journal of Infectious Diseases* 11: 528–534.

Vine, M. F., Schoenbach, V., Hulka, B. S., Koch, G. G., and Samsa, G. (1990). Atypical metaplasia as a risk factor for bronchogenic carcinoma. *American Journal of Epidemiology* 131: 781–793.

Wald, A. (1943). Tests of statistical hypotheses concerning general parameters when the number of observations is large. *Transactions of the American Mathematical Society* 54: 426–482.

Ware, J. H., Lipsitz, S., and Speizer, F. E. (1988). Issues in the analysis of repeated categorical outcomes. *Statistics in Medicine* 7: 95–107.

Wei, L. J. and Stram, D. O. (1988). Analyzing repeated measurements with possibly missing observations by modelling marginal distributions. *Statistics in Medicine* 7: 139–148.

Wilcoxon, F. (1945). Individual comparison by ranking methods. *Biometrics* 1: 80–83.

Wilson, E. B. (1927). Probable inference, the law of succession, and statistical inference. *Journal of the American Statistical Association* 22: 209–212.

Yates, F. (1934). Contingency table involving small numbers and the χ^2 test. *Supplement to the Journal of the Royal Statistical Society* 1 (2): 217–235.

Yule, G. U. (1903). Notes on the theory of association of attributes in statistics. *Biometrika* 2: 121–134.

Zeger, S. L. (1988). Commentary. *Statistics in Medicine* 7: 161–168.

Zeger, S. L. and Liang, K. Y. (1986). Longitudinal data analysis for discrete and continuous outcomes. *Biometrics* 42: 121–130.

Zeger, S. L., Liang, K. Y., and Albert, P. S. (1988). Models for longitudinal data: A generalized estimating equation approach. *Biometrics* 44: 1049–1060.

Zelen, M. (1971). The Analysis of 2 × 2 Contingency Tables. *Biometrika* 58: 129–137.

Zerbe, G. O. (1978). On Fieller's theorem and the general linear model. *American Statistician* 32 (3): 103–105.

Zhao, L. P. and Prentice, R. L. (1990). Correlated binary regression using a quadratic exponential model. *Biometrika* 77: 642–648.

Index

AGGREGATE option
    MODEL statement (LOGISTIC), 195
AGGREGATE= option
    MODEL statement (LOGISTIC), 207
AGREE option
    TABLES statement (FREQ), 42
agreement plot, 132
aligned ranks test, 181
ALL option
    TABLES statement (FREQ), 36
ALPHA= option
    TABLES statement (FREQ), 35
alternating logistic regression (ALR) algorithm, 543
ANOVA statistic, 78
association
    2×2 table, 15, 16
    general, 107
    ordered rows and columns, 115
    ordinal columns, 113
association statistics
    exact tests, 119
average partial association, 49, 62

Bayes' Theorem, 339
bioassay, 345
Breslow-Day statistic, 63

categorized survival data, 409
CATMOD procedure
    CONTRAST statement, 444, 447, 471
    direct input of response functions, 450
    FACTOR statement, 451
    marginal proportions response function, 466
    mean response function, 433
    POPULATION statement, 436
    REPEATED statement, 468
    RESPONSE statement, 433
    WEIGHT statement, 433
changeover study, 308
chi-square statistic
    2×2 table, 17
    continuity-adjusted, 23
CHISQ keyword
    EXACT statement, 24
CHISQ option
    TABLES statement (FREQ), 18
CLASS statement (LOGISTIC)
    PARAM=REF option, 245

CMH option
    TABLES statement (FREQ), 50
CMH1 option
    TABLES statement (FREQ), 170
Cochran-Armitage test, 90, 91
cohort study, 36
collinearity, 206
common odds ratio, 62
    logit estimator, 62
    Mantel-Haenszel estimator, 62
conditional likelihood, 298, 299, 327, 338, 340
conditional likelihood for matched pairs, 339
conditional logistic regression, 297, 298
    highly stratified cohort study, 298
    LOGISTIC procedure, 327
    $1:m$ matching analysis, 331
    $1:1$ matching analysis, 327
    retrospective matched study, 326
conditional probability, 299
confidence interval
    common odds ratio, 63
    log LD50, 348, 351
    odds ratio, 32
    odds ratio, logistic regression, 199
    proportions, difference in, 25
confounding, 65
continuity-adjusted chi-square, 23
CONTRAST statement
    GEE analysis, 514
CONTRAST statement (GENMOD)
    WALD option, 256
contrast tests
    CATMOD procedure, 446
    GENMOD procedure, 255
    LOGISTIC procedure, 218, 221
CORRECT option
    TABLES statement (FREQ), 30
correlated data
    generalized estimating equations, 488–490
    generalized linear models, 489
correlation coefficients
    exact tests, 127
correlation statistic, 90
correlation test
    $s \times r$ table, 115
CORRW option
    REPEATED statement (GENMOD), 499, 512

COV option
    MODEL statement (CATMOD), 468
COVB option
    MODEL statement (LOGISTIC), 359
    REPEATED statement (GENMOD), 499
COVOUT option
    MODEL statement (LOGISTIC), 359
crossover design study, 308, 506
cumulative logits, 260
    GENMOD procedure, 533
cumulative probabilities, 260

DETAILS option
    MODEL statement (LOGISTIC), 230
deviance, 194
diagnostics, LOGISTIC procedure
    input requirements, 236
diagnostics, logistic regression, 235
dichotomous response, 2
direct input of response functions
    CATMOD procedure, 450
direction of effect, 52
discrete counts, 3
DIST=BIN option
    MODEL statement (GENMOD), 509, 523
DIST=BINOMIAL option
    MODEL statement (GENMOD), 253
DIST=MULT option
    MODEL statement (GENMOD), 533
DIST=NB option
    MODEL statement (GENMOD), 388
DIST=POISSON option
    MODEL statement (GENMOD), 380, 530
dummy coding, 195
Durbin's test, 185

ED50, 347
effect plot, 216
ESTIMATE statement
    GEE analysis, 504
ESTIMATE=BOTH option
    EXACT statement (LOGISTIC), 241, 246
EVENT= option
    PROC LOGISTIC statement, 195
events/trials syntax
    MODEL statement (GENMOD), 253
    MODEL statement (LOGISTIC), 236
exact p-values, trend test, 93
exact p-values
    Jonckheere-Terpstra test, 139
    Q for general association, 120
exact computations, 108, 121
exact confidence interval
    incidence density ratio, 44

exact confidence limits
    odds ratio, 38
exact logistic regression, 241, 292
EXACT option
    TABLES statement (FREQ), 119
exact p-values
    likelihood ratio test, 23
    Pearson chi-square, 23
EXACT statement
    CHISQ keyword, 24
EXACT statement (FREQ)
    JT keyword, 139
    MAXTIME option, 121
    MCNEM keyword, 43
    OR keyword, 38
EXACT statement (LOGISTIC)
    ESTIMATE=BOTH option, 241, 246
exact tests
    association statistics, 119
    correlation coefficients, 127
    Fisher's exact test, 20
    kappa statistics, 134
    Monte Carlo estimation, 121
    $s \times r$ table, 119
EXACTONLY option
    PROC LOGISTIC statement, 241
    PROC statement (LOGISTIC), 335
EXPECTED option
    TABLES statement (FREQ), 18
explanatory variables, continuous
    logistic regression, 229
explanatory variables, ordinal
    logistic regression, 229
extended Mantel-Haenszel general association
    statistic, 144
extended Mantel-Haenszel mean score statistic, 78
extended Mantel-Haenszel statistics, 144
    summary table, 145

FACTOR statement (CATMOD)
    PROFILE= option, 451
    READ option, 451
Fieller's theorem, 360, 371
FIRTH option
    MODEL statement (LOGISTIC), 243
Firth penalized likelihood method, 240
Firth's penalized likelihood, 240
Fisher's exact test, 20
    one-sided, 21
    two-sided, 21
FREQ option
    MODEL statement (CATMOD), 433
FREQ procedure, 18

EXACT statement, 23
TABLES statement, 18, 94
WEIGHT statement, 18
Friedman's chi-square test, 178
Friedman's chi-square test, generalization, 180

GEE analysis
　CONTRAST statement, 514
　ESTIMATE statement, 504
　testing contrasts, 514
GEE methodology
　example, 497, 506, 528
general association
　$s \times r$ table, 108
generalized estimating equations
　data structure, 491
　GENMOD procedure, 498, 506
　marginal model, 496
　methodology, 487, 488, 490, 491
　missing values, 495
　proportional odds model, 533
　working correlation matrix, 491–495
generalized estimating equations (GEE)
　working correlation matrix, 521
generalized linear models, 489
generalized logit, 280
GENMOD procedure, 253
　alternating logistic regression (ALR)
　　algorithm, 543
　CLASS statement, 499
　CONTRAST statement, 255, 514
　cumulative logits, 533
　ESTIMATE statement, 504
　generalized estimating equations, 488, 498, 506
　MODEL statement, 253, 499
　REPEATED statement, 499
goodness of fit
　CATMOD procedure, 435
　expanded model, 200
　Hosmer and Lemeshow statistic, 230
　logistic regression, 194, 229
　PROC LOGISTIC output, 196
　Wald tests, 431
graphs
　agreement plot, 132
　effect plot, 216
　mosaic plot, 110
　odds ratio plot, 60
　predicted probabilities plot, 368
　proportion difference plot, 60
　survival plot, 413
grouped survival data, 409
grouped survival times, 4

highly stratified cohort study, 298
homogeneity of odds ratios, 63
Hosmer and Lemeshow statistic
　PROC LOGISTIC output, 233
hypergeometric distribution, 20
hypothesis testing
　GENMOD procedure, 255
　LOGISTIC procedure, 218
　logistic regression, 218

incidence densities, 43
INCLUDE= option
　MODEL statement (LOGISTIC), 231
incremental effects, 193
index plot, 236
indicator coding, 195
infinite parameter estimates, 238
INFLUENCE option
　MODEL statement (LOGISTIC), 236
inputting model matrix
　MODEL statement (CATMOD), 455
integer scores, 79
interactions
　logistic regression, 221
interchangeability, 157
INVERSECL option
　MODEL statement (PROBIT), 351

Jonckheere-Terpstra test, 136
　exact p-values, 139
JT keyword
　EXACT statement (FREQ), 139
JT option
　TABLES statement (FREQ), 138

kappa coefficient, 132
kappa statistics
　exact tests, 134
Kruskal-Wallis test
　MH mean score equivalent, 176

LACKFIT option
　MODEL statement (PROBIT), 368
LD50, 347
life table method, 410
LIFETEST procedure
　STRATA statement, 413
　TIME statement, 413
likelihood ratio (Q_L), 194
likelihood ratio statistic
　PROC FREQ output, 20
　PROC GENMOD output, 254
likelihood ratio test
　exact p-values, 23
LINK=CLOGIT

MODEL statement (LOGISTIC), 276
LINK=CLOGIT option
    MODEL statement (GENMOD), 533
LINK=GLOGIT
    MODEL statement (LOGISTIC), 282
LINK=LOG option
    MODEL statement (GENMOD), 530
LINK=LOGIT option
    MODEL statement (GENMOD), 253, 523
location shifts, 75
log likelihood
    PROC LOGISTIC output, 196
logistic model, 191
    partial proportional odds model, 276
LOGISTIC procedure, 194
    CLASS statement, 202, 245
    DESCENDING option, 199
    deviation from the mean parameterization, 207
    EXACT statement, 241
    FREQ statement, 195
    MODEL statement, 195
    nominal effects, 212
    ordering of response value, 195
    proportional odds model, 264
    STRATA statement, 335
    UNITS statement, 233
logistic regression
    diagnostics, 235
    exact conditional, stratified, 334
    Firth penalized likelihood method, 240
    GENMOD procedure, 253
    interpretation of parameters, 193
    loglinear model correspondence, 405
    methodology, 257
    model fitting, 192
    model interpretation, 192
    model matrix, 193
    nominal response, 280
    nominal variables, 210
    ordinal response, 260
    parameter estimation, 192
    qualitative variables, 210
logit, 32, 192
loglinear model
    logistic model correspondence, 405
    odds ratio, 399
    three-way contingency table, 394
LOGOR= option
    REPEATED statement (GENMOD), 546
LOGOR=EXCH option
    REPEATED statement (GENMOD), 544
LOGOR=FULLCLUST option
    REPEATED statement (GENMOD), 548

logrank scores, 79
longitudinal studies, 463

Mann-Whitney rank measure, 82, 457
Mantel-Cox test, 416
Mantel-Fleiss criterion, 49
Mantel-Haenszel methodology, 142
Mantel-Haenszel statistics
    assumptions, 142
    overview, 142
    PROC FREQ output summary, 101
    relationships, 101
Mantel-Haenszel strategy
    repeated measurements, continuous, 182
    repeated measurements, missing data, 168
    repeated measurements, ordinal, 164
    sets of 2×2 tables, 49
    sets of $2 \times r$ tables, 77
    sets of $s \times 2$ tables, 93
marginal homogeneity, 157, 467, 473
marginal proportions response function
    CATMOD procedure, 466
MARGINALS keyword
    RESPONSE statement (CATMOD), 467
matched pairs, 41, 158, 326
matched studies, 298
maximum likelihood estimation
    problems in logistic regression, 238
MAXTIME option
    EXACT statement (FREQ), 121
MCNEM keyword
    EXACT statement (FREQ), 43
McNemar's Test, 41, 157, 300, 301
mean response function
    CATMOD procedure, 433
mean score statistic, 74, 113, 144
    $s \times r$ table, 112
MEANS keyword
    RESPONSE statement (CATMOD), 433
measures of association
    nominal, 129
    ordinal, 124
    standard error, 125
MEASURES option
    TABLES statement (FREQ), 33
METHOD=LT option
    PROC statement (LIFETEST), 413
Miettinen-Nurminen interval, 29
$m:n$ matching, 326
MN option
    TABLES statement (FREQ), 29
model assessment
    generalized estimating equations, 495
model fitting, logistic regression, 192, 202

choosing parameterization, 211
    keeping marginal effects, 206
model interpretation
    logistic regression, 192
MODEL statement
    effects specification, 433
MODEL statement (CATMOD)
    COV option, 468
    FREQ option, 433
    nested effects, 445
    ONEWAY option, 468
    PROB option, 433
    repeated measurements, 467
    RESPONSE keyword, 468
MODEL statement (GENMOD)
    DIST=BIN option, 509, 523
    DIST=BINOMIAL option, 253
    DIST=MULT option, 533
    DIST=NB option, 388
    DIST=POISSON option, 380, 530
    events/trials syntax, 253
    LINK=CLOGIT option, 533
    LINK=LOG option, 530
    LINK=LOGIT option, 253, 523
    OFFSET= option, 380
MODEL statement (LOGISTIC)
    AGGREGATE option, 195
    AGGREGATE= option, 207
    COVB option, 359
    COVOUT option, 359
    DETAILS option, 230
    events/trials syntax, 236
    FIRTH option, 243
    INCLUDE= option, 231
    INFLUENCE option, 236
    LINK=CLOGIT, 276
    LINK=GLOGIT, 282
    NOINT option, 328
    SCALE= option, 202
    SCALE=NONE option, 195
    SELECTION=FORWARD option, 230
    UNEQUALSLOPES option, 276
MODEL statement (PROBIT)
    INVERSECL option, 351
    LACKFIT option, 368
modified ridit scores, 79
Monte Carlo estimation
    exact tests, 121
mosaic plot, 110
multivariate hypergeometric distribution, 109

nested effects
    MODEL statement (CATMOD), 445
Newcombe hybrid score interval, 29

NEWCOMBE option
    TABLES statement (FREQ), 30
NOCOL option
    TABLES statement (FREQ), 21
NOINT option
    MODEL statement (LOGISTIC), 328
nominal effects
    LOGISTIC procedure, 212
nominal variables, logistic regression, 210
nonparametric methods, 175
Nonzero Correlation statistic, 95
NOPCT option
    TABLES statement (FREQ), 33
nuisance parameters, 298, 496, 555

observer agreement, 107, 131
observer agreement studies, 107
odds ratio, 31
    common, 62
    computing using effect parameterization, 208
    exact confidence limits, 38
    generalized logit, 287
    homogeneity, 63
    interpretation, 198
    logistic regression parameters, 193
    loglinear model, 399
    PROC FREQ output, 35
    profile likelihood confidence intervals, 216
    proportional odds model, 262
    units of change, 234
    Wald confidence interval, 199
odds ratio estimation
    GEE analysis, 504
odds ratio plot, 60
odds ratios
    interactions, 221
ODS SELECT statement, 27
offset
    Poisson regression, 375
OFFSET= option
    MODEL statement (GENMOD), 380
ONEWAY option
    MODEL statement (CATMOD), 468
OR keyword
    EXACT statement (FREQ), 38
ORDER=DATA option
    PROC FREQ statement, 11, 33
ordered differences, 136
ordering of response value
    LOGISTIC procedure, 195
ordinal response, 2
    choosing scores for MH strategy, 78
    Mantel-Haenszel tests, 77

OUTEST= option
    PROC LOGISTIC statement, 359
overdispersion, 202
    GEE adjustment, 549
    scaling adjustment, 384

paired data
    crossover design study, 308
    highly stratified cohort study, 298
    retrospective matched study, 326
parallel lines assay, 354
PARAM=REF option
    CLASS statement (LOGISTIC), 245
parameter interpretation
    GENMOD procedure, 253
parameterization
    CATMOD procedure, 432
    deviation from the mean, 207, 432
    incremental effects, 193
partial proportional odds model, 276
Pearson chi-square
    exact p-values, 23
Pearson chi-square (Q_P), 194
Pearson chi-square statistic
    2×2 table, 18
Pearson correlation coefficient, 26
Pearson residuals, 235
piecewise exponential model, 419
    example, 421
PLOTS= option
    PROC statement (LIFETEST), 413
PLOTS=MOSAICPLOT option
    TABLES statement (FREQ), 110
Poisson regression, 373, 374, 376
    example, 379
    offset, 375
population profiles
    CATMOD procedure, 433
population-averaged models, 322
PROB option
    MODEL statement (CATMOD), 433
probit, 346
probit analysis, 368
PROBIT procedure, 368
PROC FREQ statement
    ORDER=DATA option, 11, 33
PROC LOGISTIC statement
    DESCENDING option, 199
    EXACTONLY option, 241
    OUTEST= option, 359
PROC statement (LOGISTIC)
    EXACTONLY option, 335
profile likelihood confidence intervals
    odds ratio, 216

regression parameters, 216
PROFILE= option
    FACTOR statement (CATMOD), 451
proportion difference plot, 60
proportional odds assumption, 261, 265
proportional odds model, 261, 533
    generalized estimating equations, 533
proportions, difference in, 25
    confidence interval, 25
$\hat{\psi}_L$, 62

Q
    PROC FREQ output, 24
Q
    PROC FREQ output, 112
    quadratic form, 110
Q for general association
    exact p-values, 120
Q_C, 432
QIC
    GENMOD procedure, 496
Q_L
    PROC FREQ output, 24
Q_L, 194
Q_{MH}, 49
Q_P
    PROC FREQ output, 24
Q_P, 109, 194
Q_{RS}, 230
    PROC LOGISTIC output, 231
Q_S, 74
    PROC FREQ output, 76, 115
    $s \times r$ table, 113
qualitative variables, logistic regression, 210
Q_W, 431

random samples, 15
    2×2 table, 15
randomization
    assignment to treatments, 15
randomization Q
    PROC FREQ output, 19
randomization Q
    2×2 table, 17
    $s \times r$ table, 109
randomized complete blocks, 178, 181
rank analysis of covariance, 186
rank measures of association, 457
rank scores, 79
rare outcome assumption, 32
READ option
    FACTOR statement (CATMOD), 451
relative potency, 355
relative risk, 31

PROC FREQ output, 35
repeated measurements
    GEE methodology, 487, 488, 491, 492, 494
    Mantel-Haenszel strategy, 155
    marginal homogeneity, 467
    marginal logit function, 478
    mean response, 476
    MODEL statement (CATMOD), 467
    single population, dichotomous response, 466
    two populations, polytomous response, 471
    WLS methodology, 465
repeated measurements studies, 463
REPEATED statement (CATMOD)
    RESPONSE keyword, 468
REPEATED statement (GENMOD)
    CORRW option, 499, 512
    COVB option, 499
    LOGOR= option, 546
    LOGOR=EXCH option, 544
    LOGOR=FULLCLUST option, 548
    SUBJECT= option, 499, 530
    TYPE=EXCH option, 499, 516, 523
    TYPE=UNSTR option, 509
residual score statistic, 230
residual variation
    weighted least squares, 438
residuals
    deviance, 235
    Pearson, 235
RESPONSE=keyword
    FACTOR statement (CATMOD), 451
RESPONSE keyword
    MODEL statement (CATMOD), 468
    REPEATED statement (CATMOD), 468
response profiles
    CATMOD procedure, 433
RESPONSE statement (CATMOD)
    MARGINALS keyword, 467
    MEANS keyword, 433
    specifying scores, 442
response variable
    character-valued (LOGISTIC), 202
RISKDIFF option
    TABLES statement (FREQ), 27
Row Mean Scores Differ statistic, 81

$s \times r$ table, 108
$s \times r$ tables, sets of, 142
$s \times 2$ tables, sets of, 93
sample size
    goodness of fit, Q_L, 194
    goodness of fit, Q_P, 194
    logistic regression, 194

proportional odds model, 265
Q_P, 18
Q_P, 109
Q_{RS}, 230
Q_S, 75, 115
    small, 20, 119
    weighted least squares, 431
sampling framework, 4
    experimental data, 5
    historical data, 4
    survey samples, 5
saturated model
    weighted least squares, 432
scale of measurement, 2
SCALE= option
    MODEL statement (LOGISTIC), 202
SCALE=NONE option
    MODEL statement (LOGISTIC), 195
SCALE=PEARSON option
    MODEL statement (GENMOD), 385
scores
    comparison, 167
scores for MH statistics
    integer, 79
    logrank, 79
    modified ridit, 79
    rank, 79
    specifying in PROC FREQ, 79
    standardized midranks, 79
SCORES=MODRIDIT option
    TABLES statement (FREQ), 81, 146
SCORES=RANK option
    TABLES statement (FREQ), 165
SELECTION=FORWARD option
    MODEL statement (LOGISTIC), 230
sensitivity, 39
Simpson's Paradox, 54
small sample size
    applicability of exact tests, 20
specificity, 39
specifying scores
    RESPONSE statement (CATMOD), 442
standardized midranks, 79
strata, specifying in PROC FREQ, 50
stratified analysis, 48
    overview, 141
stratified random sample, 15
subject-specific models, 322
SUBJECT= option
    REPEATED statement (GENMOD), 499, 530
summary statistics
    sets of 2×2 tables, 50
survey data analysis, 449

survival rate, 410
TABLES statement (FREQ)
    AGREE option, 42
    ALL option, 36
    ALPHA= option, 35
    CHISQ option, 18
    CMH option, 50
    CORRECT option, 30
    EXACT option, 119
    EXPECTED option, 18
    JT option, 138
    MEASURES option, 33
    MN option, 29
    NEWCOMBE option, 30
    NOCOL option, 21
    NOPCT option, 33
    RISKDIFF option, 27
    SCORES=MODRIDIT option, 81, 146
    TREND option, 90
target population, 190
test for general association, 144
test for linear association, 145
testing contrasts
    GEE analysis, 514
time-to-event data, 409
tolerance distribution, 346
TREND option
    TABLES statement (FREQ), 90
trend test, 90, 91
$2 \times r$ tables, sets of, 77
2×2 tables, sets of, 47
TYPE=EXCH option
    REPEATED statement (GENMOD), 499, 516, 523
TYPE=UNSTR option
    REPEATED statement (GENMOD), 509

UNEQUALSLOPES option
    MODEL statement (LOGISTIC), 276

Wald confidence interval
    odds ratio, 199
WALD option
    CONTRAST statement (GENMOD), 256
Wald test
    goodness of fit, 431
    hypothesis testing, 218, 431
weighted kappa coefficient, 132
weighted least squares, 429, 430
    methodology, 482
    residual variation, 438
withdrawal, 409
WITHINSUBJECT= option
    REPEATED statement (GENMOD), 520

working correlation matrix, 490–495, 521, 554